Nissan Pick-ups, Xterra & Pathfinder Automotive Repair Manual

by Jeff Kibler
and John H Haynes
Member of the Guild of Motoring Writers

Models covered:
Frontier pick-ups - 1998 through 2001
Xterra - 2000 and 2001
Pathfinder - 1996 through 2001

Two- and four-wheel drive versions

Does not include information specific to supercharged engine models

(9E2 - 72031)

ABCDE
FGHIJ
KLMNO
PQ

Haynes Publishing Group
Sparkford Nr Yeovil
Somerset BA22 7JJ England

Haynes North America, Inc
861 Lawrence Drive
Newbury Park
California 91320 USA

Acknowledgements

Technical writers who contributed to this project include Eric Godfrey and Bob Henderson. Technical consultants include Jaime Sarté Jr. and John Wegmann. Wiring diagrams originated exclusively for Haynes North America, Inc. by Valley Forge Technical Information Services.

© **Haynes North America, Inc. 2001**
With permission from J.H. Haynes & Co. Ltd.

A book in the Haynes Automotive Repair Manual Series

Printed in the U.S.A.

All rights reserved. No part of this book may be reproduced or transmitted in any form or by any means, electronic or mechanical, including photocopying, recording or by any information storage or retrieval system, without permission in writing from the copyright holder.

ISBN 1 56392 396 3

Library of Congress Control Number 00-110806

While every attempt is made to ensure that the information in this manual is correct, no liability can be accepted by the authors or publishers for loss, damage or injury caused by any errors in, or omissions from, the information given.

Contents

Introductory pages

About this manual	0-5
Introduction to the Nissan Frontier, Xterra and Pathfinder	0-5
Vehicle identification numbers	0-6
Buying parts	0-7
Maintenance techniques, tools and working facilities	0-7
Booster battery (jump) starting	0-13
Jacking and towing	0-14
Automotive chemicals and lubricants	0-15
Conversion factors	0-16
Fraction/decimal/millimeter equivalents	0-17
Safety first!	0-18
Troubleshooting	0-19

Chapter 1
Tune-up and routine maintenance — 1-1

Chapter 2   Part A
2.4L four-cylinder engine — 2A-1

Chapter 2   Part B
3.3L V6 engine — 2B-1

Chapter 2   Part C
3.5L V6 engine — 2C-1

Chapter 2   Part D
General engine overhaul procedures — 2D-1

Chapter 3
Cooling, heating and air conditioning systems — 3-1

Chapter 4
Fuel and exhaust systems — 4-1

Chapter 5
Engine electrical systems — 5-1

Chapter 6
Emissions and engine control systems — 6-1

Chapter 7   Part A
Manual transmission — 7A-1

Chapter 7   Part B
Automatic transmission — 7B-1

Chapter 7   Part C
Transfer case — 7C-1

Chapter 8
Clutch and driveline — 8-1

Chapter 9
Brakes — 9-1

Chapter 10
Suspension and steering systems — 10-1

Chapter 11
Body — 11-1

Chapter 12
Chassis electrical system — 12-1

Wiring diagrams — 12-22

Index — IND-1

Haynes photographer, mechanic and author with 2001 Nissan Pathfinder

About this manual

Its purpose

The purpose of this manual is to help you get the best value from your vehicle. It can do so in several ways. It can help you decide what work must be done, even if you choose to have it done by a dealer service department or a repair shop; it provides information and procedures for routine maintenance and servicing; and it offers diagnostic and repair procedures to follow when trouble occurs.

We hope you use the manual to tackle the work yourself. For many simpler jobs, doing it yourself may be quicker than arranging an appointment to get the vehicle into a shop and making the trips to leave it and pick it up. More importantly, a lot of money can be saved by avoiding the expense the shop must pass on to you to cover its labor and overhead costs. An added benefit is the sense of satisfaction and accomplishment that you feel after doing the job yourself.

Using the manual

The manual is divided into Chapters. Each Chapter is divided into numbered Sections, which are headed in bold type between horizontal lines. Each Section consists of consecutively numbered paragraphs.

At the beginning of each numbered Section you will be referred to any illustrations which apply to the procedures in that Section. The reference numbers used in illustration captions pinpoint the pertinent Section and the Step within that Section. That is, illustration 3.2 means the illustration refers to Section 3 and Step (or paragraph) 2 within that Section.

Procedures, once described in the text, are not normally repeated. When it's necessary to refer to another Chapter, the reference will be given as Chapter and Section number. Cross references given without use of the word "Chapter" apply to Sections and/or paragraphs in the same Chapter. For example, "see Section 8" means in the same Chapter.

References to the left or right side of the vehicle assume you are sitting in the driver's seat, facing forward.

Even though we have prepared this manual with extreme care, neither the publisher nor the author can accept responsibility for any errors in, or omissions from, the information given.

NOTE

A **Note** provides information necessary to properly complete a procedure or information which will make the procedure easier to understand.

CAUTION

A **Caution** provides a special procedure or special steps which must be taken while completing the procedure where the Caution is found. Not heeding a Caution can result in damage to the assembly being worked on.

WARNING

A **Warning** provides a special procedure or special steps which must be taken while completing the procedure where the Warning is found. Not heeding a Warning can result in personal injury.

Introduction to the Nissan Frontier, Xterra and Pathfinder

Nissan Frontier pick-ups are available in regular cab, king cab and crew cab (four-door) body styles. Xterra and Pathfinder models are available in four-door body styles only. Frontiers and Xterras are available with an inline four-cylinder engine or a 3.3L V6 engine. Pathfinders are available with V6 engines only. All engines used in these vehicles are equipped with electronic fuel injection.

The chassis layout is conventional with the engine mounted at the front and the power being transmitted through either a four-speed automatic or five-speed manual transmission and driveshaft to the rear axle. On 4WD models a transfer case transmits the power through a driveshaft to the front differential, which powers the front wheels through independent driveaxles.

All models have independent front suspension - Frontier and Xterra models use torsion bars and shock absorbers, while Pathfinders have MacPherson struts. All models have a solid rear axle - Frontier and Xterra models are suspended by leaf springs, while Pathfinders have coil spring rear suspension.

Frontier and Xterra models all have conventional recirculating-ball type steering gear. Pathfinders all have rack-and-pinion steering. Most models have power steering, the exception being base-model 2WD pick-ups.

The brakes are disc at the front and drums at the rear, with power assist standard. 2WD Frontiers are equipped with a rear Anti-lock Brake System (ABS). Xterras, Pathfinders and 4WD Frontiers all have four-wheel ABS.

Vehicle identification numbers

Modifications are a continuing and unpublicized process in vehicle manufacturing. Since spare parts manuals and lists are compiled on a numerical basis, the individual vehicle numbers are essential to correctly identify the component required.

Vehicle Identification Number (VIN)

The Vehicle Identification Number (VIN), which appears on the Vehicle Certificate of Title and Registration, is also embossed on a plate located in the left (driver's side) corner of the dashboard, near the windshield **(see illustration)**. The VIN tells you when and where a vehicle was manufactured, its country of origin, make, type, passenger safety system, line, series, body style, engine and assembly plant.

VIN engine and model year codes

Two particularly important pieces of information found in the VIN are the engine code and the model year code. Counting from the left, the engine code letter designation is the 4th character and the model year code is the 10th character.

On the models covered by this manual the engine codes are:
Frontier and Xterra
    D........ 2.4L four-cylinder (KA24DE)
    E........ 3.3L V6 (VG33E)
Pathfinder
    A........ 3.3L V6 (VG33E)
    D........ 3.5L V6 (VQ35DE)

On the models covered by this manual the model year codes are:
T.. 1996
V.. 1997
W... 1998
X.. 1999
Y.. 2000
1.. 2001

Vehicle Safety Certification label

The Vehicle Safety Certification label is attached to the rear edge of the driver's door or on the door post **(see illustration)**. The label contains the name of the manufacturer, the month and year of production, the Gross Vehicle Weight Rating (GVWR), the Gross Axle Weight Rating (GAWR) and the certification statement. On most models, the label also includes the OEM tire sizes and pressures.

Engine Identification Number (EIN)

The Engine Identification Number (EIN) on V6 engines is stamped into the engine block on a machined surface just behind the right (passenger's side) cylinder head. On four-cylinder engines the EIN is located on the left rear side of the engine block, just below the cylinder head and rearward of the exhaust manifold.

Transmission Identification Number (TIN)

The manual Transmission Identification Number (TIN) is stamped into the top of the transmission bellhousing. On automatic transmissions, the number is stamped into a tag and fastened to the transmission with a bolt **(see illustrations)**.

Transfer case identification label

The transfer case identification information is stamped into the top of the case.

The VIN plate is visible from the outside of the vehicle, through the driver's side of the windshield

The Vehicle Safety Certification label is affixed to the driver's side door end or post

Automatic transmission identification tag - 2WD four-cylinder Frontier

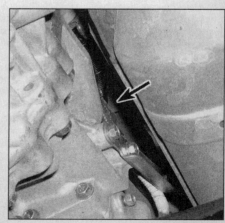

Automatic transmission identification tag - 4WD Pathfinder

Buying parts

Replacement parts are available from many sources, which generally fall into one of two categories - authorized dealer parts departments and independent retail auto parts stores. Our advice concerning these parts is as follows:

Retail auto parts stores: Good auto parts stores will stock frequently needed components which wear out relatively fast, such as clutch components, exhaust systems, brake parts, tune-up parts, etc. These stores often supply new or reconditioned parts on an exchange basis, which can save a considerable amount of money. Discount auto parts stores are often very good places to buy materials and parts needed for general vehicle maintenance such as oil, grease, filters, spark plugs, belts, touch-up paint, bulbs, etc. They also usually sell tools and general accessories, have convenient hours, charge lower prices and can often be found not far from home.

Authorized dealer parts department: This is the best source for parts which are unique to the vehicle and not generally available elsewhere (such as major engine parts, transmission parts, trim pieces, etc.).

Warranty information: If the vehicle is still covered under warranty, be sure that any replacement parts purchased - regardless of the source - do not invalidate the warranty!

To be sure of obtaining the correct parts, have engine and chassis numbers available and, if possible, take the old parts along for positive identification.

Maintenance techniques, tools and working facilities

Maintenance techniques

There are a number of techniques involved in maintenance and repair that will be referred to throughout this manual. Application of these techniques will enable the home mechanic to be more efficient, better organized and capable of performing the various tasks properly, which will ensure that the repair job is thorough and complete.

Fasteners

Fasteners are nuts, bolts, studs and screws used to hold two or more parts together. There are a few things to keep in mind when working with fasteners. Almost all of them use a locking device of some type, either a lockwasher, locknut, locking tab or thread adhesive. All threaded fasteners should be clean and straight, with undamaged threads and undamaged corners on the hex head where the wrench fits. Develop the habit of replacing all damaged nuts and bolts with new ones. Special locknuts with nylon or fiber inserts can only be used once. If they are removed, they lose their locking ability and must be replaced with new ones.

Rusted nuts and bolts should be treated with a penetrating fluid to ease removal and prevent breakage. Some mechanics use turpentine in a spout-type oil can, which works quite well. After applying the rust penetrant, let it work for a few minutes before trying to loosen the nut or bolt. Badly rusted fasteners may have to be chiseled or sawed off or removed with a special nut breaker, available at tool stores.

If a bolt or stud breaks off in an assembly, it can be drilled and removed with a special tool commonly available for this purpose. Most automotive machine shops can perform this task, as well as other repair procedures, such as the repair of threaded holes that have been stripped out.

Flat washers and lockwashers, when removed from an assembly, should always be replaced exactly as removed. Replace any damaged washers with new ones. Never use a lockwasher on any soft metal surface (such as aluminum), thin sheet metal or plastic.

Bolt strength marking (standard/SAE/USS; bottom - metric)

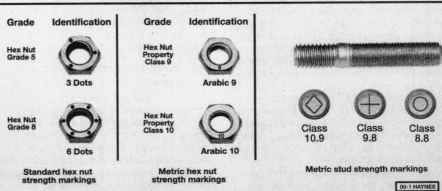

Fastener sizes

For a number of reasons, automobile manufacturers are making wider and wider use of metric fasteners. Therefore, it is important to be able to tell the difference between standard (sometimes called U.S. or SAE) and metric hardware, since they cannot be interchanged.

All bolts, whether standard or metric, are sized according to diameter, thread pitch and length. For example, a standard 1/2 - 13 x 1 bolt is 1/2 inch in diameter, has 13 threads per inch and is 1 inch long. An M12 - 1.75 x 25 metric bolt is 12 mm in diameter, has a thread pitch of 1.75 mm (the distance between threads) and is 25 mm long. The two bolts are nearly identical, and easily confused, but they are not interchangeable.

In addition to the differences in diameter, thread pitch and length, metric and standard bolts can also be distinguished by examining the bolt heads. To begin with, the distance across the flats on a standard bolt head is measured in inches, while the same dimension on a metric bolt is sized in millimeters (the same is true for nuts). As a result, a standard wrench should not be used on a metric bolt and a metric wrench should not be used on a standard bolt. Also, most standard bolts have slashes radiating out from the center of the head to denote the grade or strength of the bolt, which is an indication of the amount of torque that can be applied to it. The greater the number of slashes, the greater the strength of the bolt. Grades 0 through 5 are commonly used on automobiles. Metric bolts have a property class (grade) number, rather than a slash, molded into their heads to indicate bolt strength. In this case, the higher the number, the stronger the bolt. Property class numbers 8.8, 9.8 and 10.9 are commonly used on automobiles.

Strength markings can also be used to distinguish standard hex nuts from metric hex nuts. Many standard nuts have dots stamped into one side, while metric nuts are marked with a number. The greater the number of dots, or the higher the number, the greater the strength of the nut.

Metric studs are also marked on their ends according to property class (grade). Larger studs are numbered (the same as metric bolts), while smaller studs carry a geometric code to denote grade.

It should be noted that many fasteners, especially Grades 0 through 2, have no distinguishing marks on them. When such is the case, the only way to determine whether it is standard or metric is to measure the thread pitch or compare it to a known fastener of the same size.

Standard fasteners are often referred to as SAE, as opposed to metric. However, it should be noted that SAE technically refers to a non-metric fine thread fastener only. Coarse thread non-metric fasteners are referred to as USS sizes.

Since fasteners of the same size (both standard and metric) may have different strength ratings, be sure to reinstall any bolts, studs or nuts removed from your vehicle in their original locations. Also, when replacing a fastener with a new one, make sure that the new one has a strength rating equal to or greater than the original.

	Ft-lbs	Nm
Metric thread sizes		
M-6	6 to 9	9 to 12
M-8	14 to 21	19 to 28
M-10	28 to 40	38 to 54
M-12	50 to 71	68 to 96
M-14	80 to 140	109 to 154
Pipe thread sizes		
1/8	5 to 8	7 to 10
1/4	12 to 18	17 to 24
3/8	22 to 33	30 to 44
1/2	25 to 35	34 to 47
U.S. thread sizes		
1/4 - 20	6 to 9	9 to 12
5/16 - 18	12 to 18	17 to 24
5/16 - 24	14 to 20	19 to 27
3/8 - 16	22 to 32	30 to 43
3/8 - 24	27 to 38	37 to 51
7/16 - 14	40 to 55	55 to 74
7/16 - 20	40 to 60	55 to 81
1/2 - 13	55 to 80	75 to 108

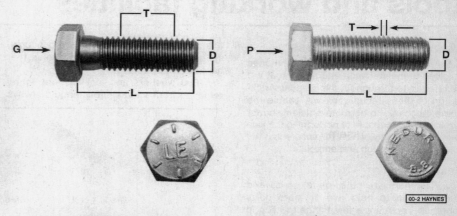

Standard (SAE and USS) bolt dimensions/grade marks
- G Grade marks (bolt strength)
- L Length (in inches)
- T Thread pitch (number of threads per inch)
- D Nominal diameter (in inches)

Metric bolt dimensions/grade marks
- P Property class (bolt strength)
- L Length (in millimeters)
- T Thread pitch (distance between threads in millimeters)
- D Diameter

Tightening sequences and procedures

Most threaded fasteners should be tightened to a specific torque value (torque is the twisting force applied to a threaded component such as a nut or bolt). Overtightening the fastener can weaken it and cause it to break, while undertightening can cause it to eventually come loose. Bolts, screws and studs, depending on the material they are made of and their thread diameters, have specific torque values, many of which are noted in the Specifications at the beginning of each Chapter. Be sure to follow the torque recommendations closely. For fasteners not assigned a specific torque, a general torque value chart is presented here as a guide. These torque values are for dry (unlubricated) fasteners threaded into steel or cast iron (not aluminum). As was previously mentioned, the size and grade of a fastener determine the amount of torque that can safely be applied to it. The figures listed here are approximate for Grade 2 and Grade 3 fasteners. Higher grades can tolerate higher torque values.

Fasteners laid out in a pattern, such as cylinder head bolts, oil pan bolts, differential cover bolts, etc., must be loosened or tightened in sequence to avoid warping the com-

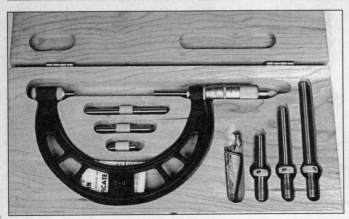
Micrometer set

Dial indicator set

ponent. This sequence will normally be shown in the appropriate Chapter. If a specific pattern is not given, the following procedures can be used to prevent warping.

Initially, the bolts or nuts should be assembled finger-tight only. Next, they should be tightened one full turn each, in a criss-cross or diagonal pattern. After each one has been tightened one full turn, return to the first one and tighten them all one-half turn, following the same pattern. Finally, tighten each of them one-quarter turn at a time until each fastener has been tightened to the proper torque. To loosen and remove the fasteners, the procedure would be reversed.

Component disassembly

Component disassembly should be done with care and purpose to help ensure that the parts go back together properly. Always keep track of the sequence in which parts are removed. Make note of special characteristics or marks on parts that can be installed more than one way, such as a grooved thrust washer on a shaft. It is a good idea to lay the disassembled parts out on a clean surface in the order that they were removed. It may also be helpful to make sketches or take instant photos of components before removal.

When removing fasteners from a component, keep track of their locations. Sometimes threading a bolt back in a part, or putting the washers and nut back on a stud, can prevent mix-ups later. If nuts and bolts cannot be returned to their original locations, they should be kept in a compartmented box or a series of small boxes. A cupcake or muffin tin is ideal for this purpose, since each cavity can hold the bolts and nuts from a particular area (i.e. oil pan bolts, valve cover bolts, engine mount bolts, etc.). A pan of this type is especially helpful when working on assemblies with very small parts, such as the carburetor, alternator, valve train or interior dash and trim pieces. The cavities can be marked with paint or tape to identify the contents.

Whenever wiring looms, harnesses or connectors are separated, it is a good idea to identify the two halves with numbered pieces of masking tape so they can be easily reconnected.

Gasket sealing surfaces

Throughout any vehicle, gaskets are used to seal the mating surfaces between two parts and keep lubricants, fluids, vacuum or pressure contained in an assembly.

Many times these gaskets are coated with a liquid or paste-type gasket sealing compound before assembly. Age, heat and pressure can sometimes cause the two parts to stick together so tightly that they are very difficult to separate. Often, the assembly can be loosened by striking it with a soft-face hammer near the mating surfaces. A regular hammer can be used if a block of wood is placed between the hammer and the part. Do not hammer on cast parts or parts that could be easily damaged. With any particularly stubborn part, always recheck to make sure that every fastener has been removed.

Avoid using a screwdriver or bar to pry apart an assembly, as they can easily mar the gasket sealing surfaces of the parts, which must remain smooth. If prying is absolutely necessary, use an old broom handle, but keep in mind that extra clean up will be necessary if the wood splinters.

After the parts are separated, the old gasket must be carefully scraped off and the gasket surfaces cleaned. Stubborn gasket material can be soaked with rust penetrant or treated with a special chemical to soften it so it can be easily scraped off. A scraper can be fashioned from a piece of copper tubing by flattening and sharpening one end. Copper is recommended because it is usually softer than the surfaces to be scraped, which reduces the chance of gouging the part. Some gaskets can be removed with a wire brush, but regardless of the method used, the mating surfaces must be left clean and smooth. If for some reason the gasket surface is gouged, then a gasket sealer thick enough to fill scratches will have to be used during reassembly of the components. For most applications, a non-drying (or semi-drying) gasket sealer should be used.

Hose removal tips

Warning: *If the vehicle is equipped with air conditioning, do not disconnect any of the A/C hoses without first having the system depressurized by a dealer service department or a service station.*

Hose removal precautions closely parallel gasket removal precautions. Avoid scratching or gouging the surface that the hose mates against or the connection may leak. This is especially true for radiator hoses. Because of various chemical reactions, the rubber in hoses can bond itself to the metal spigot that the hose fits over. To remove a hose, first loosen the hose clamps that secure it to the spigot. Then, with slip-joint pliers, grab the hose at the clamp and rotate it around the spigot. Work it back and forth until it is completely free, then pull it off. Silicone or other lubricants will ease removal if they can be applied between the hose and the outside of the spigot. Apply the same lubricant to the inside of the hose and the outside of the spigot to simplify installation.

As a last resort (and if the hose is to be replaced with a new one anyway), the rubber can be slit with a knife and the hose peeled from the spigot. If this must be done, be careful that the metal connection is not damaged.

If a hose clamp is broken or damaged, do not reuse it. Wire-type clamps usually weaken with age, so it is a good idea to replace them with screw-type clamps whenever a hose is removed.

Tools

A selection of good tools is a basic requirement for anyone who plans to maintain and repair his or her own vehicle. For the owner who has few tools, the initial investment might seem high, but when compared to the spiraling costs of professional auto maintenance and repair, it is a wise one.

To help the owner decide which tools are needed to perform the tasks detailed in this manual, the following tool lists are offered: *Maintenance and minor repair, Repair/overhaul* and *Special*.

The newcomer to practical mechanics

0-10 Maintenance techniques, tools and working facilities

Dial caliper

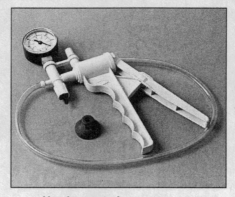
Hand-operated vacuum pump

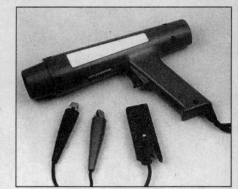
Timing light

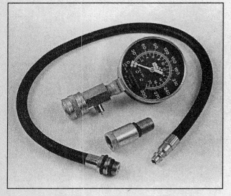
Compression gauge with spark plug hole adapter

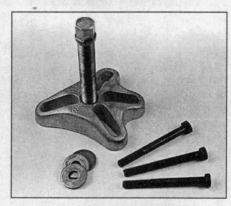
Damper/steering wheel puller

General purpose puller

Hydraulic lifter removal tool

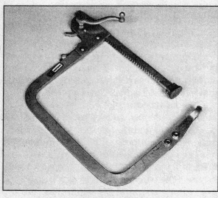
Valve spring compressor

Valve spring compressor

Ridge reamer

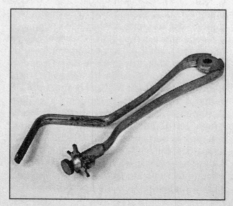
Piston ring groove cleaning tool

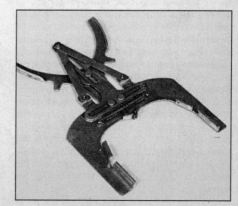
Ring removal/installation tool

Maintenance techniques, tools and working facilities

Ring compressor

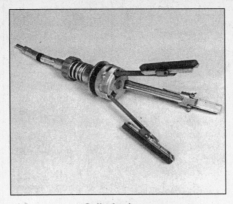
Cylinder hone

Brake hold-down spring tool

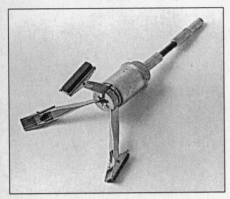
Brake cylinder hone

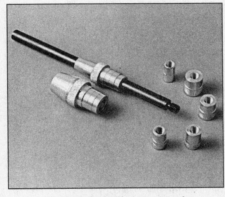
Clutch plate alignment tool

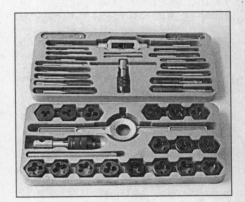
Tap and die set

should start off with the *maintenance and minor repair* tool kit, which is adequate for the simpler jobs performed on a vehicle. Then, as confidence and experience grow, the owner can tackle more difficult tasks, buying additional tools as they are needed. Eventually the basic kit will be expanded into the *repair and overhaul* tool set. Over a period of time, the experienced do-it-yourselfer will assemble a tool set complete enough for most repair and overhaul procedures and will add tools from the special category when it is felt that the expense is justified by the frequency of use.

Maintenance and minor repair tool kit

The tools in this list should be considered the minimum required for performance of routine maintenance, servicing and minor repair work. We recommend the purchase of combination wrenches (box-end and open-end combined in one wrench). While more expensive than open end wrenches, they offer the advantages of both types of wrench.

Combination wrench set (1/4-inch to 1 inch or 6 mm to 19 mm)
Adjustable wrench, 8 inch
Spark plug wrench with rubber insert
Spark plug gap adjusting tool
Feeler gauge set
Brake bleeder wrench
Standard screwdriver (5/16-inch x 6 inch)
Phillips screwdriver (No. 2 x 6 inch)
Combination pliers - 6 inch
Hacksaw and assortment of blades
Tire pressure gauge
Grease gun
Oil can
Fine emery cloth
Wire brush
Battery post and cable cleaning tool
Oil filter wrench
Funnel (medium size)
Safety goggles
Jackstands (2)
Drain pan

Note: *If basic tune-ups are going to be part of routine maintenance, it will be necessary to purchase a good quality stroboscopic timing light and combination tachometer/dwell meter. Although they are included in the list of special tools, it is mentioned here because they are absolutely necessary for tuning most vehicles properly.*

Repair and overhaul tool set

These tools are essential for anyone who plans to perform major repairs and are in addition to those in the maintenance and minor repair tool kit. Included is a comprehensive set of sockets which, though expensive, are invaluable because of their versatility, especially when various extensions and drives are available. We recommend the 1/2-inch drive over the 3/8-inch drive. Although the larger drive is bulky and more expensive, it has the capacity of accepting a very wide range of large sockets. Ideally, however, the mechanic should have a 3/8-inch drive set and a 1/2-inch drive set.

Socket set(s)
Reversible ratchet
Extension - 10 inch
Universal joint
Torque wrench (same size drive as sockets)
Ball peen hammer - 8 ounce
Soft-face hammer (plastic/rubber)
Standard screwdriver (1/4-inch x 6 inch)
Standard screwdriver (stubby - 5/16-inch)
Phillips screwdriver (No. 3 x 8 inch)
Phillips screwdriver (stubby - No. 2)
Pliers - vise grip
Pliers - lineman's
Pliers - needle nose
Pliers - snap-ring (internal and external)
Cold chisel - 1/2-inch
Scribe
Scraper (made from flattened copper tubing)
Centerpunch
Pin punches (1/16, 1/8, 3/16-inch)
Steel rule/straightedge - 12 inch
Allen wrench set (1/8 to 3/8-inch or 4 mm to 10 mm)
A selection of files
Wire brush (large)
Jackstands (second set)
Jack (scissor or hydraulic type)

Maintenance techniques, tools and working facilities

Note: *Another tool which is often useful is an electric drill with a chuck capacity of 3/8-inch and a set of good quality drill bits.*

Special tools

The tools in this list include those which are not used regularly, are expensive to buy, or which need to be used in accordance with their manufacturer's instructions. Unless these tools will be used frequently, it is not very economical to purchase many of them. A consideration would be to split the cost and use between yourself and a friend or friends. In addition, most of these tools can be obtained from a tool rental shop on a temporary basis.

This list primarily contains only those tools and instruments widely available to the public, and not those special tools produced by the vehicle manufacturer for distribution to dealer service departments. Occasionally, references to the manufacturer's special tools are included in the text of this manual. Generally, an alternative method of doing the job without the special tool is offered. However, sometimes there is no alternative to their use. Where this is the case, and the tool cannot be purchased or borrowed, the work should be turned over to the dealer service department or an automotive repair shop.

- *Valve spring compressor*
- *Piston ring groove cleaning tool*
- *Piston ring compressor*
- *Piston ring installation tool*
- *Cylinder compression gauge*
- *Cylinder ridge reamer*
- *Cylinder surfacing hone*
- *Cylinder bore gauge*
- *Micrometers and/or dial calipers*
- *Hydraulic lifter removal tool*
- *Balljoint separator*
- *Universal-type puller*
- *Impact screwdriver*
- *Dial indicator set*
- *Stroboscopic timing light (inductive pick-up)*
- *Hand operated vacuum/pressure pump*
- *Tachometer/dwell meter*
- *Universal electrical multimeter*
- *Cable hoist*
- *Brake spring removal and installation tools*
- *Floor jack*

Buying tools

For the do-it-yourselfer who is just starting to get involved in vehicle maintenance and repair, there are a number of options available when purchasing tools. If maintenance and minor repair is the extent of the work to be done, the purchase of individual tools is satisfactory. If, on the other hand, extensive work is planned, it would be a good idea to purchase a modest tool set from one of the large retail chain stores. A set can usually be bought at a substantial savings over the individual tool prices, and they often come with a tool box. As additional tools are needed, add-on sets, individual tools and a larger tool box can be purchased to expand the tool selection. Building a tool set gradually allows the cost of the tools to be spread over a longer period of time and gives the mechanic the freedom to choose only those tools that will actually be used.

Tool stores will often be the only source of some of the special tools that are needed, but regardless of where tools are bought, try to avoid cheap ones, especially when buying screwdrivers and sockets, because they won't last very long. The expense involved in replacing cheap tools will eventually be greater than the initial cost of quality tools.

Care and maintenance of tools

Good tools are expensive, so it makes sense to treat them with respect. Keep them clean and in usable condition and store them properly when not in use. Always wipe off any dirt, grease or metal chips before putting them away. Never leave tools lying around in the work area. Upon completion of a job, always check closely under the hood for tools that may have been left there so they won't get lost during a test drive.

Some tools, such as screwdrivers, pliers, wrenches and sockets, can be hung on a panel mounted on the garage or workshop wall, while others should be kept in a tool box or tray. Measuring instruments, gauges, meters, etc. must be carefully stored where they cannot be damaged by weather or impact from other tools.

When tools are used with care and stored properly, they will last a very long time. Even with the best of care, though, tools will wear out if used frequently. When a tool is damaged or worn out, replace it. Subsequent jobs will be safer and more enjoyable if you do.

How to repair damaged threads

Sometimes, the internal threads of a nut or bolt hole can become stripped, usually from overtightening. Stripping threads is an all-too-common occurrence, especially when working with aluminum parts, because aluminum is so soft that it easily strips out.

Usually, external or internal threads are only partially stripped. After they've been cleaned up with a tap or die, they'll still work. Sometimes, however, threads are badly damaged. When this happens, you've got three choices:

1) Drill and tap the hole to the next suitable oversize and install a larger diameter bolt, screw or stud.
2) Drill and tap the hole to accept a threaded plug, then drill and tap the plug to the original screw size. You can also buy a plug already threaded to the original size. Then you simply drill a hole to the specified size, then run the threaded plug into the hole with a bolt and jam nut. Once the plug is fully seated, remove the jam nut and bolt.
3) The third method uses a patented thread repair kit like Heli-Coil or Slimsert. These easy-to-use kits are designed to repair damaged threads in straight-through holes and blind holes. Both are available as kits which can handle a variety of sizes and thread patterns. Drill the hole, then tap it with the special included tap. Install the Heli-Coil and the hole is back to its original diameter and thread pitch.

Regardless of which method you use, be sure to proceed calmly and carefully. A little impatience or carelessness during one of these relatively simple procedures can ruin your whole day's work and cost you a bundle if you wreck an expensive part.

Working facilities

Not to be overlooked when discussing tools is the workshop. If anything more than routine maintenance is to be carried out, some sort of suitable work area is essential.

It is understood, and appreciated, that many home mechanics do not have a good workshop or garage available, and end up removing an engine or doing major repairs outside. It is recommended, however, that the overhaul or repair be completed under the cover of a roof.

A clean, flat workbench or table of comfortable working height is an absolute necessity. The workbench should be equipped with a vise that has a jaw opening of at least four inches.

As mentioned previously, some clean, dry storage space is also required for tools, as well as the lubricants, fluids, cleaning solvents, etc. which soon become necessary.

Sometimes waste oil and fluids, drained from the engine or cooling system during normal maintenance or repairs, present a disposal problem. To avoid pouring them on the ground or into a sewage system, pour the used fluids into large containers, seal them with caps and take them to an authorized disposal site or recycling center. Plastic jugs, such as old antifreeze containers, are ideal for this purpose.

Always keep a supply of old newspapers and clean rags available. Old towels are excellent for mopping up spills. Many mechanics use rolls of paper towels for most work because they are readily available and disposable. To help keep the area under the vehicle clean, a large cardboard box can be cut open and flattened to protect the garage or shop floor.

Whenever working over a painted surface, such as when leaning over a fender to service something under the hood, always cover it with an old blanket or bedspread to protect the finish. Vinyl covered pads, made especially for this purpose, are available at auto parts stores.

Booster battery (jump) starting

Observe these precautions when using a booster battery to start a vehicle:

a) Before connecting the booster battery, make sure the ignition switch is in the Off position.
b) Turn off the lights, heater and other electrical loads.
c) Your eyes should be shielded. Safety goggles are a good idea.
d) Make sure the booster battery is the same voltage as the dead one in the vehicle.
e) The two vehicles MUST NOT TOUCH each other!
f) Make sure the transaxle is in Neutral (manual) or Park (automatic).
g) If the booster battery is not a maintenance-free type, remove the vent caps and lay a cloth over the vent holes.

Connect the red-colored jumper cable to the positive (+) terminal of the booster battery and the other end to the positive (+) terminal of the dead battery. Then connect one end of the black jumper cable to the negative (-) terminal of the booster battery, and the other end of the cable to a good ground, such as a bolt or bracket.

Start the engine using the booster battery, then run the booster vehicle at a fast idle for a few minutes to instill some charge in the dead battery. Let the engine idle, then disconnect the jumper cables in the reverse order of connection. The vehicle with the dead battery may have to be driven for 20 minutes or more to sufficiently recharge the battery for independent starting.

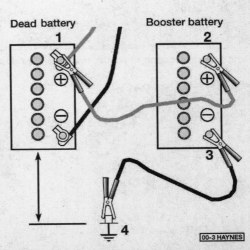

Make the booster battery cable connections in the numerical order shown (note that the negative cable of the booster battery is NOT attached to the negative terminal of the dead battery)

Jacking and towing

Jacking

The jack supplied with the vehicle should only be used for raising the vehicle when changing a tire or placing jackstands under the frame. NEVER work under the vehicle or start the engine when the vehicle supported only by a jack.

The vehicle should be parked on level ground with the wheels blocked, the parking brake applied and the transmission in Park (automatic) or Reverse (manual). If the vehicle is parked alongside the roadway, or in any other hazardous situation, turn on the emergency hazard flashers. If a tire is to be changed, loosen the lug nuts one-half turn before raising off the ground.

Place the jack under the vehicle in the indicated positions (see illustrations). Operate the jack with a slow, smooth motion until the wheel is raised off the ground. Remove the lug nuts, pull off the wheel, install the spare and thread the lug nuts back on with the beveled side facing in. Tighten the lug nuts snugly, lower the vehicle until some weight is on the wheel, then tighten them completely in a criss-cross pattern and remove the jack.

Towing

Equipment specifically designed for towing should be used and attached to the main structural members of the vehicle. Optional tow hooks may be attached to the frame at both ends of the vehicle; they are intended for emergency use only, for rescuing a stranded vehicle. Do not use the tow hooks for highway towing. Stand clear when using tow straps or chains; they may break, causing serious injury.

The manufacturer recommends that these vehicles be towed only by wheel-lift equipment or a flatbed car-carrier.

Safety is a major consideration when towing and all applicable state and local laws must be obeyed. In addition to a tow bar, a safety chain must be used for all towing.

Two-wheel drive vehicles with automatic transmission may be towed with the rear wheels on a towing dolly with no mileage restriction (at posted highway speeds). If the vehicle is being towed with four wheels on the ground, speed should be no more than 35 mph (56 km/h) for no more than 50 miles (80 km), or damage may be done to the automatic transmission.

Two-wheel drive vehicles with a manual transmission can be towed up to 500 miles with the ignition lock in the Off position and the transmission in Neutral.

Vehicles with part-time four-wheel drive should be towed with the transfer case in Neutral. Towing is restricted to a maximum of 500 miles (800 km) distance.

Pathfinders with all-mode 4WD should only be towed with all wheels off the ground (using a towing dolly and tow truck or a flatbed carrier).

If any vehicle is to be towed with the front wheels on the ground and the rear wheels raised, the ignition key must be turned to the OFF position to unlock the steering column and a steering wheel clamping device designed for towing must be used or damage to the steering column lock may occur.

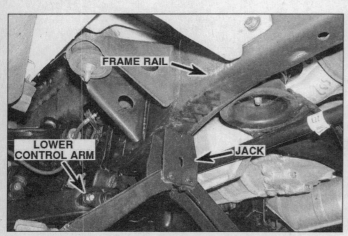

Front jacking location - Frontier and Xterra models

Rear jacking location - Frontier and Xterra models

Front jacking location - Pathfinder models

Rear jacking location - Pathfinder models

Automotive chemicals and lubricants

A number of automotive chemicals and lubricants are available for use during vehicle maintenance and repair. They include a wide variety of products ranging from cleaning solvents and degreasers to lubricants and protective sprays for rubber, plastic and vinyl.

Cleaners

Carburetor cleaner and choke cleaner is a strong solvent for gum, varnish and carbon. Most carburetor cleaners leave a dry-type lubricant film which will not harden or gum up. Because of this film it is not recommended for use on electrical components.

Brake system cleaner is used to remove grease and brake fluid from the brake system, where clean surfaces are absolutely necessary. It leaves no residue and often eliminates brake squeal caused by contaminants.

Electrical cleaner removes oxidation, corrosion and carbon deposits from electrical contacts, restoring full current flow. It can also be used to clean spark plugs, carburetor jets, voltage regulators and other parts where an oil-free surface is desired.

Demoisturants remove water and moisture from electrical components such as alternators, voltage regulators, electrical connectors and fuse blocks. They are non-conductive, non-corrosive and non-flammable.

Degreasers are heavy-duty solvents used to remove grease from the outside of the engine and from chassis components. They can be sprayed or brushed on and, depending on the type, are rinsed off either with water or solvent.

Lubricants

Motor oil is the lubricant formulated for use in engines. It normally contains a wide variety of additives to prevent corrosion and reduce foaming and wear. Motor oil comes in various weights (viscosity ratings) from 0 to 50. The recommended weight of the oil depends on the season, temperature and the demands on the engine. Light oil is used in cold climates and under light load conditions. Heavy oil is used in hot climates and where high loads are encountered. Multi-viscosity oils are designed to have characteristics of both light and heavy oils and are available in a number of weights from 5W-20 to 20W-50.

Gear oil is designed to be used in differentials, manual transmissions and other areas where high-temperature lubrication is required.

Chassis and wheel bearing grease is a heavy grease used where increased loads and friction are encountered, such as for wheel bearings, balljoints, tie-rod ends and universal joints.

High-temperature wheel bearing grease is designed to withstand the extreme temperatures encountered by wheel bearings in disc brake equipped vehicles. It usually contains molybdenum disulfide (moly), which is a dry-type lubricant.

White grease is a heavy grease for metal-to-metal applications where water is a problem. White grease stays soft under both low and high temperatures (usually from -100 to +190-degrees F), and will not wash off or dilute in the presence of water.

Assembly lube is a special extreme pressure lubricant, usually containing moly, used to lubricate high-load parts (such as main and rod bearings and cam lobes) for initial start-up of a new engine. The assembly lube lubricates the parts without being squeezed out or washed away until the engine oiling system begins to function.

Silicone lubricants are used to protect rubber, plastic, vinyl and nylon parts.

Graphite lubricants are used where oils cannot be used due to contamination problems, such as in locks. The dry graphite will lubricate metal parts while remaining uncontaminated by dirt, water, oil or acids. It is electrically conductive and will not foul electrical contacts in locks such as the ignition switch.

Moly penetrants loosen and lubricate frozen, rusted and corroded fasteners and prevent future rusting or freezing.

Heat-sink grease is a special electrically non-conductive grease that is used for mounting electronic ignition modules where it is essential that heat is transferred away from the module.

Sealants

RTV sealant is one of the most widely used gasket compounds. Made from silicone, RTV is air curing, it seals, bonds, waterproofs, fills surface irregularities, remains flexible, doesn't shrink, is relatively easy to remove, and is used as a supplementary sealer with almost all low and medium temperature gaskets.

Anaerobic sealant is much like RTV in that it can be used either to seal gaskets or to form gaskets by itself. It remains flexible, is solvent resistant and fills surface imperfections. The difference between an anaerobic sealant and an RTV-type sealant is in the curing. RTV cures when exposed to air, while an anaerobic sealant cures only in the absence of air. This means that an anaerobic sealant cures only after the assembly of parts, sealing them together.

Thread and pipe sealant is used for sealing hydraulic and pneumatic fittings and vacuum lines. It is usually made from a Teflon compound, and comes in a spray, a paint-on liquid and as a wrap-around tape.

Chemicals

Anti-seize compound prevents seizing, galling, cold welding, rust and corrosion in fasteners. High-temperature anti-seize, usually made with copper and graphite lubricants, is used for exhaust system and exhaust manifold bolts.

Anaerobic locking compounds are used to keep fasteners from vibrating or working loose and cure only after installation, in the absence of air. Medium strength locking compound is used for small nuts, bolts and screws that may be removed later. High-strength locking compound is for large nuts, bolts and studs which aren't removed on a regular basis.

Oil additives range from viscosity index improvers to chemical treatments that claim to reduce internal engine friction. It should be noted that most oil manufacturers caution against using additives with their oils.

Gas additives perform several functions, depending on their chemical makeup. They usually contain solvents that help dissolve gum and varnish that build up on carburetor, fuel injection and intake parts. They also serve to break down carbon deposits that form on the inside surfaces of the combustion chambers. Some additives contain upper cylinder lubricants for valves and piston rings, and others contain chemicals to remove condensation from the gas tank.

Miscellaneous

Brake fluid is specially formulated hydraulic fluid that can withstand the heat and pressure encountered in brake systems. Care must be taken so this fluid does not come in contact with painted surfaces or plastics. An opened container should always be resealed to prevent contamination by water or dirt.

Weatherstrip adhesive is used to bond weatherstripping around doors, windows and trunk lids. It is sometimes used to attach trim pieces.

Undercoating is a petroleum-based, tar-like substance that is designed to protect metal surfaces on the underside of the vehicle from corrosion. It also acts as a sound-deadening agent by insulating the bottom of the vehicle.

Waxes and polishes are used to help protect painted and plated surfaces from the weather. Different types of paint may require the use of different types of wax and polish. Some polishes utilize a chemical or abrasive cleaner to help remove the top layer of oxidized (dull) paint on older vehicles. In recent years many non-wax polishes that contain a wide variety of chemicals such as polymers and silicones have been introduced. These non-wax polishes are usually easier to apply and last longer than conventional waxes and polishes.

Conversion factors

Length (distance)
Inches (in)	X	25.4	= Millimetres (mm)	X 0.0394	= Inches (in)
Feet (ft)	X	0.305	= Metres (m)	X 3.281	= Feet (ft)
Miles	X	1.609	= Kilometres (km)	X 0.621	= Miles

Volume (capacity)
Cubic inches (cu in; in^3)	X	16.387	= Cubic centimetres (cc; cm^3)	X 0.061	= Cubic inches (cu in; in^3)
Imperial pints (Imp pt)	X	0.568	= Litres (l)	X 1.76	= Imperial pints (Imp pt)
Imperial quarts (Imp qt)	X	1.137	= Litres (l)	X 0.88	= Imperial quarts (Imp qt)
Imperial quarts (Imp qt)	X	1.201	= US quarts (US qt)	X 0.833	= Imperial quarts (Imp qt)
US quarts (US qt)	X	0.946	= Litres (l)	X 1.057	= US quarts (US qt)
Imperial gallons (Imp gal)	X	4.546	= Litres (l)	X 0.22	= Imperial gallons (Imp gal)
Imperial gallons (Imp gal)	X	1.201	= US gallons (US gal)	X 0.833	= Imperial gallons (Imp gal)
US gallons (US gal)	X	3.785	= Litres (l)	X 0.264	= US gallons (US gal)

Mass (weight)
Ounces (oz)	X	28.35	= Grams (g)	X 0.035	= Ounces (oz)
Pounds (lb)	X	0.454	= Kilograms (kg)	X 2.205	= Pounds (lb)

Force
Ounces-force (ozf; oz)	X	0.278	= Newtons (N)	X 3.6	= Ounces-force (ozf; oz)
Pounds-force (lbf; lb)	X	4.448	= Newtons (N)	X 0.225	= Pounds-force (lbf; lb)
Newtons (N)	X	0.1	= Kilograms-force (kgf; kg)	X 9.81	= Newtons (N)

Pressure
Pounds-force per square inch (psi; lbf/in^2; lb/in^2)	X	0.070	= Kilograms-force per square centimetre (kgf/cm^2; kg/cm^2)	X 14.223	= Pounds-force per square inch (psi; lbf/in^2; lb/in^2)
Pounds-force per square inch (psi; lbf/in^2; lb/in^2)	X	0.068	= Atmospheres (atm)	X 14.696	= Pounds-force per square inch (psi; lbf/in^2; lb/in^2)
Pounds-force per square inch (psi; lbf/in^2; lb/in^2)	X	0.069	= Bars	X 14.5	= Pounds-force per square inch (psi; lbf/in^2; lb/in^2)
Pounds-force per square inch (psi; lbf/in^2; lb/in^2)	X	6.895	= Kilopascals (kPa)	X 0.145	= Pounds-force per square inch (psi; lbf/in^2; lb/in^2)
Kilopascals (kPa)	X	0.01	= Kilograms-force per square centimetre (kgf/cm^2; kg/cm^2)	X 98.1	= Kilopascals (kPa)

Torque (moment of force)
Pounds-force inches (lbf in; lb in)	X	1.152	= Kilograms-force centimetre (kgf cm; kg cm)	X 0.868	= Pounds-force inches (lbf in; lb in)
Pounds-force inches (lbf in; lb in)	X	0.113	= Newton metres (Nm)	X 8.85	= Pounds-force inches (lbf in; lb in)
Pounds-force inches (lbf in; lb in)	X	0.083	= Pounds-force feet (lbf ft; lb ft)	X 12	= Pounds-force inches (lbf in; lb in)
Pounds-force feet (lbf ft; lb ft)	X	0.138	= Kilograms-force metres (kgf m; kg m)	X 7.233	= Pounds-force feet (lbf ft; lb ft)
Pounds-force feet (lbf ft; lb ft)	X	1.356	= Newton metres (Nm)	X 0.738	= Pounds-force feet (lbf ft; lb ft)
Newton metres (Nm)	X	0.102	= Kilograms-force metres (kgf m; kg m)	X 9.804	= Newton metres (Nm)

Vacuum
Inches mercury (in. Hg)	X	3.377	= Kilopascals (kPa)	X 0.2961	= Inches mercury
Inches mercury (in. Hg)	X	25.4	= Millimeters mercury (mm Hg)	X 0.0394	= Inches mercury

Power
Horsepower (hp)	X	745.7	= Watts (W)	X 0.0013	= Horsepower (hp)

Velocity (speed)
Miles per hour (miles/hr; mph)	X	1.609	= Kilometres per hour (km/hr; kph)	X 0.621	= Miles per hour (miles/hr; mph)

Fuel consumption*
Miles per gallon, Imperial (mpg)	X	0.354	= Kilometres per litre (km/l)	X 2.825	= Miles per gallon, Imperial (mpg)
Miles per gallon, US (mpg)	X	0.425	= Kilometres per litre (km/l)	X 2.352	= Miles per gallon, US (mpg)

Temperature
Degrees Fahrenheit = (°C x 1.8) + 32

Degrees Celsius (Degrees Centigrade; °C) = (°F - 32) x 0.56

*It is common practice to convert from miles per gallon (mpg) to litres/100 kilometres (l/100km), where mpg (Imperial) x l/100 km = 282 and mpg (US) x l/100 km = 235

Fraction/Decimal/Millimeter Equivalents

DECIMALS TO MILLIMETERS

Decimal	mm	Decimal	mm
0.001	0.0254	0.500	12.7000
0.002	0.0508	0.510	12.9540
0.003	0.0762	0.520	13.2080
0.004	0.1016	0.530	13.4620
0.005	0.1270	0.540	13.7160
0.006	0.1524	0.550	13.9700
0.007	0.1778	0.560	14.2240
0.008	0.2032	0.570	14.4780
0.009	0.2286	0.580	14.7320
		0.590	14.9860
0.010	0.2540		
0.020	0.5080		
0.030	0.7620		
0.040	1.0160	0.600	15.2400
0.050	1.2700	0.610	15.4940
0.060	1.5240	0.620	15.7480
0.070	1.7780	0.630	16.0020
0.080	2.0320	0.640	16.2560
0.090	2.2860	0.650	16.5100
		0.660	16.7640
0.100	2.5400	0.670	17.0180
0.110	2.7940	0.680	17.2720
0.120	3.0480	0.690	17.5260
0.130	3.3020		
0.140	3.5560		
0.150	3.8100	0.700	17.7800
0.160	4.0640	0.710	18.0340
0.170	4.3180	0.720	18.2880
0.180	4.5720	0.730	18.5420
0.190	4.8260	0.740	18.7960
		0.750	19.0500
0.200	5.0800	0.760	19.3040
0.210	5.3340	0.770	19.5580
0.220	5.5880	0.780	19.8120
0.230	5.8420	0.790	20.0660
0.240	6.0960		
0.250	6.3500		
0.260	6.6040	0.800	20.3200
0.270	6.8580	0.810	20.5740
0.280	7.1120	0.820	21.8280
0.290	7.3660	0.830	21.0820
		0.840	21.3360
0.300	7.6200	0.850	21.5900
0.310	7.8740	0.860	21.8440
0.320	8.1280	0.870	22.0980
0.330	8.3820	0.880	22.3520
0.340	8.6360	0.890	22.6060
0.350	8.8900		
0.360	9.1440		
0.370	9.3980		
0.380	9.6520		
0.390	9.9060	0.900	22.8600
		0.910	23.1140
0.400	10.1600	0.920	23.3680
0.410	10.4140	0.930	23.6220
0.420	10.6680	0.940	23.8760
0.430	10.9220	0.950	24.1300
0.440	11.1760	0.960	24.3840
0.450	11.4300	0.970	24.6380
0.460	11.6840	0.980	24.8920
0.470	11.9380	0.990	25.1460
0.480	12.1920	1.000	25.4000
0.490	12.4460		

FRACTIONS TO DECIMALS TO MILLIMETERS

Fraction	Decimal	mm	Fraction	Decimal	mm
1/64	0.0156	0.3969	33/64	0.5156	13.0969
1/32	0.0312	0.7938	17/32	0.5312	13.4938
3/64	0.0469	1.1906	35/64	0.5469	13.8906
1/16	0.0625	1.5875	9/16	0.5625	14.2875
5/64	0.0781	1.9844	37/64	0.5781	14.6844
3/32	0.0938	2.3812	19/32	0.5938	15.0812
7/64	0.1094	2.7781	39/64	0.6094	15.4781
1/8	0.1250	3.1750	5/8	0.6250	15.8750
9/64	0.1406	3.5719	41/64	0.6406	16.2719
5/32	0.1562	3.9688	21/32	0.6562	16.6688
11/64	0.1719	4.3656	43/64	0.6719	17.0656
3/16	0.1875	4.7625	11/16	0.6875	17.4625
13/64	0.2031	5.1594	45/64	0.7031	17.8594
7/32	0.2188	5.5562	23/32	0.7188	18.2562
15/64	0.2344	5.9531	47/64	0.7344	18.6531
1/4	0.2500	6.3500	3/4	0.7500	19.0500
17/64	0.2656	6.7469	49/64	0.7656	19.4469
9/32	0.2812	7.1438	25/32	0.7812	19.8438
19/64	0.2969	7.5406	51/64	0.7969	20.2406
5/16	0.3125	7.9375	13/16	0.8125	20.6375
21/64	0.3281	8.3344	53/64	0.8281	21.0344
11/32	0.3438	8.7312	27/32	0.8438	21.4312
23/64	0.3594	9.1281	55/64	0.8594	21.8281
3/8	0.3750	9.5250	7/8	0.8750	22.2250
25/64	0.3906	9.9219	57/64	0.8906	22.6219
13/32	0.4062	10.3188	29/32	0.9062	23.0188
27/64	0.4219	10.7156	59/64	0.9219	23.4156
7/16	0.4375	11.1125	15/16	0.9375	23.8125
29/64	0.4531	11.5094	61/64	0.9531	24.2094
15/32	0.4688	11.9062	31/32	0.9688	24.6062
31/64	0.4844	12.3031	63/64	0.9844	25.0031
1/2	0.5000	12.7000	1	1.0000	25.4000

Safety first!

Regardless of how enthusiastic you may be about getting on with the job at hand, take the time to ensure that your safety is not jeopardized. A moment's lack of attention can result in an accident, as can failure to observe certain simple safety precautions. The possibility of an accident will always exist, and the following points should not be considered a comprehensive list of all dangers. Rather, they are intended to make you aware of the risks and to encourage a safety conscious approach to all work you carry out on your vehicle.

Essential DOs and DON'Ts

DON'T rely on a jack when working under the vehicle. Always use approved jackstands to support the weight of the vehicle and place them under the recommended lift or support points.

DON'T attempt to loosen extremely tight fasteners (i.e. wheel lug nuts) while the vehicle is on a jack - it may fall.

DON'T start the engine without first making sure that the transmission is in Neutral (or Park where applicable) and the parking brake is set.

DON'T remove the radiator cap from a hot cooling system - let it cool or cover it with a cloth and release the pressure gradually.

DON'T attempt to drain the engine oil until you are sure it has cooled to the point that it will not burn you.

DON'T touch any part of the engine or exhaust system until it has cooled sufficiently to avoid burns.

DON'T siphon toxic liquids such as gasoline, antifreeze and brake fluid by mouth, or allow them to remain on your skin.

DON'T inhale brake lining dust - it is potentially hazardous (see *Asbestos* below).

DON'T allow spilled oil or grease to remain on the floor - wipe it up before someone slips on it.

DON'T use loose fitting wrenches or other tools which may slip and cause injury.

DON'T push on wrenches when loosening or tightening nuts or bolts. Always try to pull the wrench toward you. If the situation calls for pushing the wrench away, push with an open hand to avoid scraped knuckles if the wrench should slip.

DON'T attempt to lift a heavy component alone - get someone to help you.

DON'T rush or take unsafe shortcuts to finish a job.

DON'T allow children or animals in or around the vehicle while you are working on it.

DO wear eye protection when using power tools such as a drill, sander, bench grinder, etc. and when working under a vehicle.

DO keep loose clothing and long hair well out of the way of moving parts.

DO make sure that any hoist used has a safe working load rating adequate for the job.

DO get someone to check on you periodically when working alone on a vehicle.

DO carry out work in a logical sequence and make sure that everything is correctly assembled and tightened.

DO keep chemicals and fluids tightly capped and out of the reach of children and pets.

DO remember that your vehicle's safety affects that of yourself and others. If in doubt on any point, get professional advice.

Asbestos

Certain friction, insulating, sealing, and other products - such as brake linings, brake bands, clutch linings, torque converters, gaskets, etc. - may contain asbestos. Extreme care must be taken to avoid inhalation of dust from such products, since it is hazardous to health. If in doubt, assume that they do contain asbestos.

Fire

Remember at all times that gasoline is highly flammable. Never smoke or have any kind of open flame around when working on a vehicle. But the risk does not end there. A spark caused by an electrical short circuit, by two metal surfaces contacting each other, or even by static electricity built up in your body under certain conditions, can ignite gasoline vapors, which in a confined space are highly explosive. Do not, under any circumstances, use gasoline for cleaning parts. Use an approved safety solvent.

Always disconnect the battery ground (-) cable at the battery before working on any part of the fuel system or electrical system. Never risk spilling fuel on a hot engine or exhaust component. It is strongly recommended that a fire extinguisher suitable for use on fuel and electrical fires be kept handy in the garage or workshop at all times. Never try to extinguish a fuel or electrical fire with water.

Fumes

Certain fumes are highly toxic and can quickly cause unconsciousness and even death if inhaled to any extent. Gasoline vapor falls into this category, as do the vapors from some cleaning solvents. Any draining or pouring of such volatile fluids should be done in a well ventilated area.

When using cleaning fluids and solvents, read the instructions on the container carefully. Never use materials from unmarked containers.

Never run the engine in an enclosed space, such as a garage. Exhaust fumes contain carbon monoxide, which is extremely poisonous. If you need to run the engine, always do so in the open air, or at least have the rear of the vehicle outside the work area.

If you are fortunate enough to have the use of an inspection pit, never drain or pour gasoline and never run the engine while the vehicle is over the pit. The fumes, being heavier than air, will concentrate in the pit with possibly lethal results.

The battery

Never create a spark or allow a bare light bulb near a battery. They normally give off a certain amount of hydrogen gas, which is highly explosive.

Always disconnect the battery ground (-) cable at the battery before working on the fuel or electrical systems.

If possible, loosen the filler caps or cover when charging the battery from an external source (this does not apply to sealed or maintenance-free batteries). Do not charge at an excessive rate or the battery may burst.

Take care when adding water to a non maintenance-free battery and when carrying a battery. The electrolyte, even when diluted, is very corrosive and should not be allowed to contact clothing or skin.

Always wear eye protection when cleaning the battery to prevent the caustic deposits from entering your eyes.

Household current

When using an electric power tool, inspection light, etc., which operates on household current, always make sure that the tool is correctly connected to its plug and that, where necessary, it is properly grounded. Do not use such items in damp conditions and, again, do not create a spark or apply excessive heat in the vicinity of fuel or fuel vapor.

Secondary ignition system voltage

A severe electric shock can result from touching certain parts of the ignition system (such as the spark plug wires) when the engine is running or being cranked, particularly if components are damp or the insulation is defective. In the case of an electronic ignition system, the secondary system voltage is much higher and could prove fatal.

Troubleshooting

Contents

Symptom	Section
Engine	
Engine backfires	13
Engine diesels (continues to run) after switching off	15
Engine hard to start when cold	4
Engine hard to start when hot	5
Engine lacks power	12
Engine lopes while idling or idles erratically	8
Engine misses at idle speed	9
Engine misses throughout driving speed range	10
Engine rotates but will not start	2
Engine stalls	11
Engine starts but stops immediately	7
Engine will not rotate when attempting to start	1
Pinging or knocking engine sounds during acceleration or uphill	14
Starter motor noisy or excessively rough in engagement	6
Starter motor operates without rotating engine	3
Engine electrical system	
Battery will not hold a charge	16
Alternator light fails to come on when key is turned on	18
Alternator light fails to go out	17
Fuel system	
Excessive fuel consumption	19
Fuel leakage and/or fuel odor	20
Cooling system	
Coolant loss	25
External coolant leakage	23
Internal coolant leakage	24
Overcooling	22
Overheating	21
Poor coolant circulation	26
Clutch	
Clutch pedal stays on floor when disengaged	32
Clutch slips (engine speed increases with no increase in vehicle speed)	28
Fails to release (pedal pressed to the floor - shift lever does not move freely in and out of Reverse)	27
Grabbing (chattering) as clutch is engaged	29
Squeal or rumble with clutch fully disengaged (pedal depressed)	31
Squeal or rumble with clutch fully engaged (pedal released)	30
Manual transmission	
Difficulty in engaging gears	37
Noisy in all gears	34
Noisy in Neutral with engine running	33
Noisy in one particular gear	35
Oil leakage	38
Slips out of high gear	36
Automatic transmission	
Fluid leakage	42
General shift mechanism problems	39
Transmission slips, shifts rough, is noisy or has no drive in forward or reverse gears	41
Transmission will not downshift with accelerator pedal pressed to the floor	40
Transfer case	
Lubricant leaks from the vent or output shaft seals	46
Noisy or jumps out of four-wheel drive Low range	45
Transfer case is difficult to shift into the desired range	43
Transfer case noisy in all gears	44
Driveshaft	
Knock or clunk when the transmission is under initial load (just after transmission is put into gear)	48
Metallic grinding sound consistent with vehicle speed	49
Oil leak at front of driveshaft	47
Vibration	50
Axles	
Noise	51
Oil leakage	53
Vibration	52
Driveaxles (4WD)	
Clicking	54
Shudder on acceleration	55
Vibration on highway	56
Brakes	
Brake pedal feels spongy when depressed	60
Brake pedal pulsates during brake application	63
Excessive brake pedal travel	59
Excessive effort required to stop vehicle	61
Noise (high-pitched squeal with the brakes applied)	58
Pedal travels to the floor with little resistance	62
Vehicle pulls to one side during braking	64
Suspension and steering systems	
Excessive pitching and/or rolling around corners or during braking	66
Excessive play in steering	68
Excessive tire wear (not specific to one area)	70
Excessive tire wear on inside edge	72
Excessive tire wear on outside edge	71
Excessively stiff steering	67
Lack of power assistance	69
Shimmy, shake or vibration	65
Tire tread worn in one place	73
Vehicle pulls to one side	64

Troubleshooting

This section provides an easy reference guide to the more common problems that may occur during the operation of your vehicle. These problems and possible causes are grouped under various components or systems; i.e. Engine, Cooling System, etc., and also refer to the Chapter and/or Section that deals with the problem.

Remember that successful troubleshooting is not a mysterious black art practiced only by professional mechanics. It's simply the result of a bit of knowledge combined with an intelligent, systematic approach to the problem. Always work by a process of elimination, starting with the simplest solution and working through to the most complex - and never overlook the obvious. Anyone can forget to fill the gas tank or leave the lights on overnight, so don't assume that you are above such oversights.

Finally, always get clear in your mind why a problem has occurred and take steps to ensure that it doesn't happen again. If the electrical system fails because of a poor connection, check all other connections in the system to make sure that they don't fail as well. If a particular fuse continues to blow, find out why - don't just go on replacing fuses. Remember, failure of a small component can often be indicative of potential failure or incorrect functioning of a more important component or system.

Engine

1 Engine will not rotate when attempting to start

1   Battery terminal connections loose or corroded. Check the cable terminals at the battery. Tighten the cable or remove corrosion as necessary.
2   Battery discharged or faulty. If the cable connections are clean and tight on the battery posts, turn the key to the On position and switch on the headlights and/or windshield wipers. If they fail to function, the battery is discharged.
3   Automatic transmission not completely engaged in Park or Neutral or clutch pedal not completely depressed.
4   Broken, loose or disconnected wiring in the starting circuit. Inspect all wiring and connectors at the battery, starter solenoid and ignition switch.
5   Starter motor pinion jammed in flywheel ring gear. If manual transmission, place transmission in gear and rock the vehicle to manually turn the engine. Remove starter and inspect pinion and flywheel at earliest convenience (Chapter 5).
6   Starter solenoid faulty (Chapter 5).
7   Starter motor faulty (Chapter 5).
8   Ignition switch faulty (Chapter 12).

2 Engine rotates but will not start

1   Fuel tank empty, fuel filter plugged or fuel line restricted.
2   Fault in the fuel injection system (Chapter 4).
3   Battery discharged (engine rotates slowly). Check the operation of electrical components as described in the previous Section.
4   Battery terminal connections loose or corroded (see previous Section).
5   Fuel pump faulty (Chapter 4).
6   Excessive moisture on, or damage to, ignition components (see Chapter 5).
7   Worn, faulty or incorrectly gapped spark plugs (Chapter 1).
8   Broken, loose or disconnected wiring in the starting circuit (see previous Section).
9   Broken, loose or disconnected wires at the ignition coils (Chapter 5).

3 Starter motor operates without rotating engine

1   Starter pinion sticking. Remove the starter (Chapter 5) and inspect.
2   Starter pinion or flywheel teeth worn or broken. Remove the flywheel/driveplate access cover and inspect.

4 Engine hard to start when cold

1   Discharged or low battery. Check as described in Section 1.
2   Fault in the fuel or ignition systems (Chapters 4 and 5).
3   Injector(s) leaking (Chapter 4).
4   Distributor rotor carbon-tracked (Chapter 1).

5 Engine hard to start when hot

1   Air filter clogged (Chapter 1).
2   Fault in the fuel or ignition systems (Chapters 4 and 5).
3   Fuel not reaching the injectors (see Chapter 4).
4   Low cylinder compression (Chapter 2D).
5   Malfunctioning EVAP system (Chapter 6).

6 Starter motor noisy or excessively rough in engagement

1   Pinion or flywheel gear teeth worn or broken. Remove the cover at the rear of the engine (if equipped) and inspect.
2   Starter motor mounting bolts loose or missing.

7 Engine starts but stops immediately

1   Loose or faulty electrical connections at the distributor (V6), coils or alternator.
2   Fault in the fuel or ignition systems (Chapters 4 and 5).
3   Vacuum leak at the gasket surfaces of the intake manifold. Make sure all mounting bolts/nuts are tightened securely and all vacuum hoses connected to the manifold are positioned properly and in good condition.
4   Restricted intake or exhaust systems (Chapter 4).

8 Engine lopes while idling or idles erratically

1   Vacuum leakage. Check the mounting bolts/nuts at the throttle body and intake manifold for tightness. Make sure all vacuum hoses are connected and in good condition. Use a stethoscope or a length of fuel hose held against your ear to listen for vacuum leaks while the engine is running. A hissing sound will be heard. A soapy water solution will also detect leaks.
2   Fault in the fuel or ignition systems (Chapters 4 and 5).
3   Plugged PCV valve or hose (see Chapters 1 and 6).
4   Air filter clogged (Chapter 1).
5   Fuel pump not delivering sufficient fuel to the fuel injectors (see Chapter 4).
6   Leaking head gasket. Perform a compression check (Chapter 2).
7   Camshaft lobes worn (Chapter 2).

9 Engine misses at idle speed

1   Spark plugs worn, fouled or not gapped properly (Chapter 1).
2   Fault in the fuel or ignition systems (Chapters 4 and 5).
3   Faulty spark plug wires (Chapter 1).
4   Vacuum leaks at intake or hose connections. Check as described in Section 8.
5   Uneven or low cylinder compression. Check compression as described in Chapter 2D.

10 Engine misses throughout driving speed range

1   Fuel filter clogged and/or impurities in the fuel system (Chapter 1).
2   Faulty or incorrectly gapped spark plugs (Chapter 1).
3   Fault in the fuel or ignition systems (Chapters 4 and 5).
4   Defective spark plug wires (Chapter 1).
5   Faulty emissions system components (Chapter 6).
6   Low or uneven cylinder compression pressures. Check compression as described

Troubleshooting 0-21

in Chapter 2D
8 Vacuum leaks at the throttle body, intake manifold or vacuum hoses (see Section 8).

11 Engine stalls

1 Fuel filter clogged and/or water and impurities in the fuel system (Chapter 1).
2 Fault in the fuel system or sensors (Chapters 4 and 6).
3 Faulty emissions system components (Chapter 6).
4 Faulty or incorrectly gapped spark plugs (Chapter 1). Also check the spark plug wires (Chapter 1).
5 Vacuum leak at the throttle body, intake manifold or vacuum hoses. Check as described in Section 8.

12 Engine lacks power

1 Fault in the fuel or ignition systems (Chapters 4 and 5).
2 Faulty or incorrectly gapped spark plugs (Chapter 1).
3 Faulty coils (Chapter 5).
4 Brakes binding (Chapter 1).
5 Automatic transmission fluid level incorrect (Chapter 1).
6 Clutch slipping (Chapter 8).
7 Fuel filter clogged and/or impurities in the fuel system (Chapter 1).
8 Emissions control system not functioning properly (Chapter 6).
9 Use of substandard fuel. Fill the tank with the proper fuel.
10 Low or uneven cylinder compression pressures (Chapter 2D).
11 Restriction in the intake or exhaust system (Chapter 4).

13 Engine backfires

1 Emissions system not functioning properly (Chapter 6).
2 Fault in the fuel or ignition systems (Chapters 4 and 5).
3 Faulty secondary ignition system (cracked spark plug insulator or faulty plug wires) (Chapters 1 and 5).
4 Fuel injection system not functioning properly (Chapter 4).
5 Vacuum leak at the throttle body, intake manifold or vacuum hoses. Check as described in Section 8.
6 Valves sticking (Chapter 2).
7 Crossed plug wires (Chapter 1).

14 Pinging or knocking engine sounds during acceleration or uphill

1 Incorrect grade of fuel. Fill the tank with fuel of the proper octane rating.
2 Fault in the fuel or ignition systems (Chapters 4 and 5).
3 Improper spark plugs. Check the plug type against the VECI label located in the engine compartment. Also check the plugs and wires for damage (Chapter 1).
4 Faulty emissions system or knock sensor (Chapter 6).
5 Vacuum leak. Check as described in Section 9.

15 Engine diesels (continues to run) after switching off

1 Idle speed too high (Chapter 4).
2 Fault in the fuel or ignition systems (Chapters 4 and 5).
3 Excessive engine operating temperature. Probable causes of this are a low coolant level (see Chapter 1), malfunctioning thermostat, clogged radiator or faulty water pump (see Chapter 3).

Engine electrical system

16 Battery will not hold a charge

1 Alternator drivebelt defective or not adjusted properly (Chapter 1).
2 Electrolyte level low or battery discharged (Chapter 1).
3 Battery terminals loose or corroded (Chapter 1).
4 Alternator not charging properly (Chapter 5).
5 Loose, broken or faulty wiring in the charging circuit (Chapter 5).
6 Short in the vehicle wiring causing a continuous drain on the battery (refer to Chapter 12 and the Wiring Diagrams).
7 Battery defective internally.

17 Alternator light fails to go out

1 Fault in the alternator or charging circuit (Chapter 5).
2 Alternator drivebelt defective or not properly adjusted (Chapter 1).

18 Alternator light fails to come on when key is turned on

1 Instrument cluster warning light bulb defective (Chapter 12).
2 Alternator faulty (Chapter 5).
3 Fault in the instrument cluster printed circuit, dashboard wiring or bulb holder (Chapter 12).

Fuel system

19 Excessive fuel consumption

1 Dirty or clogged air filter element (Chapter 1).
2 Emissions system not functioning properly (Chapter 6).
3 Fault in the fuel or ignition systems (Chapters 4 and 5).
4 Low tire pressure or incorrect tire size (Chapter 1).
5 Restricted exhaust system (Chapter 4).

20 Fuel leakage and/or fuel odor

1 Leak in a fuel feed or vent line (Chapter 4).
2 Tank overfilled. Fill only to automatic shut-off.
3 Evaporative emissions system canister clogged (Chapter 6).
4 Vapor leaks from system lines or injectors (Chapter 4).

Cooling system

21 Overheating

1 Insufficient coolant in the system (Chapter 1).
2 Water pump drivebelt defective or not adjusted properly (Chapter 1).
3 Radiator core blocked or radiator grille dirty and restricted (see Chapter 3).
4 Thermostat faulty (Chapter 3).
5 Fan blades broken or cracked (Chapter 3).
6 Radiator cap not maintaining proper pressure. Have the cap pressure tested by a gas station or repair shop.
7 Defective fan clutch (see Chapter 3).

22 Overcooling

1 Thermostat faulty (Chapter 3).
2 Inaccurate temperature gauge (Chapter 12).
3 Defective fan clutch (see Chapter 3).

23 External coolant leakage

1 Deteriorated or damaged hoses or loose clamps. Replace hoses and/or tighten the clamps at the hose connections (Chapter 1).
2 Water pump seals defective (Chapter 3).
3 Leakage from the radiator core or tank(s). This will require the radiator to be professionally repaired (see Chapter 3 for removal procedures).

Troubleshooting

4 Engine drain plug(s) leaking (Chapter 1) or water jacket core plugs leaking (see Chapter 2).
5 Leakage at the heater core. Signs of leakage should show up on interior carpeting (Chapter 3).

24 Internal coolant leakage

Note: *Internal coolant leaks can usually be detected by examining the oil. Check the dipstick and inside of the valve cover for water deposits and an oil consistency like that of a milkshake.*

1 Leaking cylinder head gasket. Have the cooling system pressure tested.
2 Cracked cylinder bore or cylinder head. Remove the head(s) and inspect (Chapter 2).
3 Leaking intake manifold gasket (Chapter 2).

25 Coolant loss

1 Too much coolant in the system (Chapter 1).
2 Coolant boiling away due to overheating (see Section 15).
3 External or internal leakage (see Sections 23 and 24).
4 Faulty radiator cap. Have the cap pressure tested.

26 Poor coolant circulation

1 Inoperative water pump. A quick test is to pinch the top radiator hose closed with your hand while the engine is idling, then let it loose. You should feel the surge of coolant if the pump is working properly (see Chapter 1).
2 Restriction in the cooling system. Drain, flush and refill the system (Chapter 1). If necessary, remove the radiator (Chapter 3) and have it reverse flushed.
3 Water pump drivebelt defective or not adjusted properly (Chapter 1).
4 Thermostat sticking (Chapter 3).

Clutch

27 Fails to release (pedal pressed to the floor - shift lever does not move freely in and out of Reverse)

1 Leak in the clutch hydraulic system. Check the master cylinder, slave cylinder and lines (Chapters 1 and 8).
2 Clutch plate warped or damaged (Chapter 8).
3 Broken release bearing (Chapter 8).

28 Clutch slips (engine speed increases with no increase in vehicle speed)

1 Clutch plate oil-soaked or lining worn. Remove clutch (Chapter 8) and inspect.
2 Clutch plate not seated. It may take 30 or 40 normal starts for a new one to seat.
3 Pressure plate worn (Chapter 8).

29 Grabbing (chattering) as clutch is engaged

1 Oil on clutch plate lining. Remove (Chapter 8) and inspect. Correct any leakage source.
2 Worn or loose engine or transmission mounts. Inspect the mounts and bolts (Chapter 2).
3 Worn splines on clutch plate hub. Remove the clutch components (Chapter 8) and inspect.
4 Warped pressure plate or flywheel. Remove the clutch components and inspect.

30 Squeal or rumble with clutch fully engaged (pedal released)

Release bearing binding on transmission bearing retainer. Remove clutch components (Chapter 8) and check slave cylinder and release bearing assembly. Remove any burrs or nicks; clean and relubricate before installing.

31 Squeal or rumble with clutch fully disengaged (pedal depressed)

1 Worn, defective or broken release bearing (Chapter 8).
2 Worn or broken pressure plate springs (or diaphragm fingers) (Chapter 8).

32 Clutch pedal stays on floor when disengaged

1 Release bearing binding, or a fault in the hydraulic system (Chapter 8).
2 Clutch master cylinder faulty (Chapter 8).

Manual transmission

Note: *All the following references are in Chapter 7A, unless noted.*

33 Noisy in Neutral with engine running

1 Input shaft bearing worn.
2 Damaged main drive gear bearing.
3 Worn countershaft bearings.
4 Worn or damaged countershaft endplay shims.

34 Noisy in all gears

1 Any of the above causes, and/or:
2 Insufficient lubricant (see the checking procedures in Chapter 1).

35 Noisy in one particular gear

1 Worn, damaged or chipped gear teeth for that particular gear.
2 Worn or damaged synchronizer for that particular gear.

36 Slips out of high gear

1 Transmission loose on clutch housing.
2 Dirt between the transmission case and engine or misalignment of the transmission.

37 Difficulty in engaging gears

1 Clutch not releasing completely (see clutch adjustment in Chapter 1).
2 Loose or damaged shifter. Make a thorough inspection, replacing parts as necessary.

38 Oil leakage

1 Excessive amount of lubricant in the transmission (see Chapter 1 for correct checking procedures). Drain lubricant as required.
2 Transmission oil seal in need of replacement.

Automatic transmission

Note: *Due to the complexity of the automatic transmission, it's difficult for the home mechanic to properly diagnose and service this component. For problems other than the following, the vehicle should be taken to a dealer service department or a transmission shop.*

39 General shift mechanism problems

1 Chapter 7B deals with checking and adjusting the shift cable on automatic transmissions. Common problems that may be attributed to poorly adjusted cable are:

a) *Engine starting in gears other than Park or Neutral.*

Troubleshooting

b) Indicator on shifter pointing to a gear other than the one actually being selected.
c) Vehicle moves when in Park.

2   Refer to Chapter 7B to adjust the cable.

40 Transmission will not downshift with accelerator pedal pressed to the floor

1   Misadjusted or broken Throttle Valve cable (four-cylinder Frontier models) (Chapter 7B).
2   Transmission pressure control solenoid valve faulty. Check for Diagnostic Trouble Codes (Chapters 6 and 7B).

41 Transmission slips, shifts rough, is noisy or has no drive in forward or reverse gears

1   Of the many probable causes for the above problems, the home mechanic should be concerned with only one possibility - fluid level.
2   Before taking the vehicle to a repair shop, check the level and condition of the fluid as described in Chapter 1. Correct fluid level as necessary or change the fluid and filter if needed. If the problem persists, have a professional diagnose the probable cause.
3   If the transmission shifts late and the shifts are harsh, suspect a faulty transmission pressure control solenoid valve. Check for Diagnostic Trouble Codes (Chapters 6 and 7B).

42 Fluid leakage

1   Automatic transmission fluid is a deep red color. Fluid leaks should not be confused with engine oil, which can easily be blown by airflow to the transmission.
2   To pinpoint a leak, first remove all built-up dirt and grime from around the transmission. Degreasing agents and/or steam cleaning will achieve this. With the underside clean, drive the vehicle at low speeds so airflow will not blow the leak far from its source. Raise the vehicle and determine where the leak is coming from. Common areas of leakage are:

a) *Pan:* Tighten the mounting bolts and/or replace the pan gasket as necessary (see Chapter 1).
b) *Filler pipe:* Replace the rubber seal where the pipe enters the transmission case.
c) *Transmission oil lines:* Tighten the connectors where the lines enter the transmission case and/or replace the lines.
d) *Vent pipe:* Transmission overfilled and/or water in fluid (see checking procedures, Chapter 1).

e) *Speed sensor connector:* Replace the O-ring where the vehicle speed sensor enters the transmission case (Chapter 6).

Transfer case

43 Transfer case is difficult to shift into the desired range

1   Speed may be too great to permit engagement. Stop the vehicle and shift into the desired range.
2   If the vehicle has been driven on a paved surface for some time, the driveline torque can make shifting difficult. Stop and shift into two-wheel drive on paved or hard surfaces.
3   Insufficient or incorrect grade of lubricant. Drain and refill the transfer case with the specified lubricant. (Chapter 1).
4   Worn or damaged internal components. Disassembly and overhaul of the transfer case, by a qualified shop, may be necessary.
5   Fault in the electrical system of the all-mode transfer case (on Pathfinders so equipped).

44 Transfer case noisy in all gears

Insufficient or incorrect grade of lubricant. Drain and refill (Chapter 1).

45 Noisy or jumps out of four-wheel drive Low range

1   Transfer case not fully engaged. Stop the vehicle, shift into Neutral and then engage 4L.
2   Shift linkage loose, worn or binding. Tighten, repair or lubricate linkage as necessary.
3   Shift fork cracked, inserts worn or fork binding on the rail. Disassemble and repair as necessary (Chapter 7C).
4   Fault in the electrical system of the all-mode transfer case (on Pathfinders so equipped).

46 Lubricant leaks from the vent or output shaft seals

1   Transfer case is overfilled. Drain to the proper level (Chapter 1).
2   Vent is clogged or jammed closed. Clear or replace the vent.
3   Output shaft seal incorrectly installed or damaged. Replace the seal and check contact surfaces for nicks and scoring.

Driveshaft

47 Oil leak at seal end of driveshaft

Defective transmission or transfer case oil seal. See Chapter 7 for replacement procedures. While this is done, check the splined yoke for burrs or a rough condition that may be damaging the seal. Burrs can be removed with crocus cloth or a fine whetstone.

48 Knock or clunk when the transmission is under initial load (just after transmission is put into gear)

1   Loose or disconnected rear suspension components. Check all mounting bolts, nuts and bushings (see Chapter 10).
2   Loose driveshaft bolts. Inspect all bolts and nuts and tighten them to the specified torque.
3   Worn or damaged universal joint bearings. Check for wear (see Chapter 8).

49 Metallic grinding sound consistent with vehicle speed

Pronounced wear in the universal joint bearings. Check as described in Chapter 8.

50 Vibration

Note: *Before assuming that the driveshaft is at fault, make sure the tires are perfectly balanced and perform the following test.*

1   Install a tachometer inside the vehicle to monitor engine speed as the vehicle is driven. Drive the vehicle and note the engine speed at which the vibration (roughness) is most pronounced. Now shift the transmission to a different gear and bring the engine speed to the same point.
2   If the vibration occurs at the same engine speed (rpm) regardless of which gear the transmission is in, the driveshaft is NOT at fault since the driveshaft speed varies.
3   If the vibration decreases or is eliminated when the transmission is in a different gear at the same engine speed, refer to the following probable causes.
4   Bent or dented driveshaft. Inspect and replace as necessary (see Chapter 8).
5   Undercoating or built-up dirt, etc. on the driveshaft. Clean the shaft thoroughly and recheck.
6   Worn universal joint bearings. Remove and inspect (see Chapter 8).
7   Driveshaft and/or companion flange out of balance. Check for missing weights on the shaft. Remove the driveshaft (see Chapter 8) and reinstall 180-degrees from original position, then retest. Have the driveshaft profes-

sionally balanced if the problem persists.
8   Center support bearing worn out (two-piece driveshafts) (Chapter 8).

Axles

51 Noise

1   Road noise. No corrective procedures available.
2   Tire noise. Inspect tires and check tire pressures (Chapter 1).
3   Rear axle bearings worn or damaged (Chapter 8).

52 Vibration

See probable causes under *Driveshaft*. Proceed under the guidelines listed for the driveshaft. If the problem persists, check the rear wheel bearings by raising the rear of the vehicle and spinning the rear wheels by hand. Listen for evidence of rough (noisy) bearings. Remove and inspect (see Chapter 8).

53 Oil leakage

1   Pinion seal damaged (see Chapter 8).
2   Axleshaft oil seals damaged (see Chapter 8).
3   Differential inspection cover leaking. Tighten the bolts or replace the gasket as required (see Chapter 8).

Driveaxles (4WD)

54 Clicking noise on turns

Worn or damaged outboard CV joints (Chapter 8).

55 Shudder or vibration during acceleration

1   Excessive toe-in. Have alignment checked.
2   Incorrect spring heights (Chapter 10).
3   Worn or damaged inboard or outboard CV joints (Chapter 8).
4   Sticking inboard CV joint assembly (Chapter 8).

56 Vibration at highway speeds

1   Out-of-balance front wheels and/or tires (Chapters 1 and 10).
2   Out-of-round front tires (Chapters 1 and 10).
3   Worn CV joints (Chapter 8).

Brakes

Note: *Before assuming that a brake problem exists, make sure that the tires are in good condition and inflated properly* (see Chapter 1), *that the front-end alignment is correct and that the vehicle is not loaded with weight in an unequal manner.*

57 Vehicle pulls to one side during braking

1   Defective, damaged or oil contaminated disc brake pads or shoes on one side. Inspect as described in Chapter 9.
2   Excessive wear of brake pad material or drum/disc on one side. Inspect and correct as necessary.
3   Loose or disconnected front suspension components. Inspect and tighten all bolts to the specified torque (Chapter 10).
4   Defective brake caliper assembly. Remove the caliper and inspect for a stuck piston or other damage (Chapter 9).
5   Inadequate lubrication of front brake caliper slide pins. Remove caliper and lubricate slide pins (Chapter 9).

58 Noise (high-pitched squeal with the brakes applied)

1   Disc brake pads worn out. The noise comes from the wear sensor rubbing against the disc (does not apply to all vehicles) or the actual pad backing plate itself if the material is completely worn away. Replace the pads with new ones immediately (Chapter 9). If the pad material has worn completely away, the brake discs should be inspected for damage as described in Chapter 9.
2   Missing or damaged brake pad insulators. Replace pad insulators (see Chapter 9).
3   Linings contaminated with dirt or grease. Replace pads.
4   Incorrect linings. Replace with correct linings.

59 Excessive brake pedal travel

1   Partial brake system failure. Inspect the entire system (Chapter 9) and correct as required.
2   Insufficient fluid in the master cylinder. Check (Chapter 1), add fluid and bleed the system if necessary (Chapter 9).

60 Brake pedal feels spongy when depressed

1   Air in the hydraulic lines. Bleed the brake system (Chapter 9).
2   Faulty flexible hoses. Inspect all system hoses and lines. Replace parts as necessary.
3   Master cylinder mounting bolts/nuts loose.
4   Master cylinder defective (Chapter 9).

61 Excessive effort required to stop vehicle

1   Power brake booster not operating properly (see check in Chapter 1, repairs in Chapter 9).
2   Excessively worn pads or shoes. Inspect and replace if necessary (Chapter 9).
3   One or more caliper pistons seized or sticking. Inspect and replace as required (Chapter 9).
4   One or more wheel cylinder pistons seized or sticking. Inspect and replace as required (Chapter 9).
5   Brake pads or shoes contaminated with oil or grease. Inspect and replace as required (Chapter 9).
6   New pads or shoes installed and not yet seated. It will take a while for the new material to seat against the disc or drum.

62 Pedal travels to the floor with little resistance

1   Little or no fluid in the master cylinder reservoir caused by leaking caliper piston(s), loose, damaged or disconnected brake lines. Inspect the entire system and correct as necessary.
2   Worn master cylinder seals (Chapter 9).

63 Brake pedal pulsates during brake application

1   Caliper improperly installed. Remove and inspect (Chapter 9).
2   Disc(s) defective. Remove (Chapter 9) and check for excessive lateral runout and parallelism. Have the disc(s) resurfaced or replace it with a new one.

Suspension and steering systems

64 Vehicle pulls to one side

1   Tire pressures uneven or tires mismatched (Chapter 1).
2   Defective tire (Chapter 1).
3   Excessive wear in suspension or steering components (Chapter 10).
4   Front end in need of alignment.
5   Front brakes dragging. Inspect the brakes as described in Chapter 9.

65 Shimmy, shake or vibration

1   Tire or wheel out-of-balance or out-of-round. Have professionally balanced.
2   Loose, worn or out-of-adjustment front wheel bearings (Chapter 1).
3   Shock absorbers and/or suspension components worn or damaged (Chapter 10).

Troubleshooting

66 Excessive pitching and/or rolling around corners or during braking

1 Defective shock absorbers. Replace as a set (Chapter 10).
2 Broken or weak springs and/or suspension components. Inspect as described in Chapters 1 and 10.

67 Excessively stiff steering

1 Lack of fluid in power steering fluid reservoir (Chapter 1).
2 Incorrect tire pressures (Chapter 1).
3 Lack of lubrication at steering joints (see Chapter 1).
4 Front end out of alignment.
5 Lack of power assistance (see Section 69).

68 Excessive play in steering

1 Loose front wheel bearings (Chapters 1 and 10).
2 Excessive wear in suspension or steering components (Chapter 10).
3 Steering gear damaged or out of adjustment (Chapter 10).

69 Lack of power assistance

1 Steering pump drivebelt faulty or not adjusted properly (Chapter 1).
2 Fluid level low (Chapter 1).
3 Hoses or lines restricted. Inspect and replace parts as necessary.
4 Air in power steering system. Bleed the system (Chapter 10).

70 Excessive tire wear (not specific to one area)

1 Incorrect tire pressures (Chapter 1).
2 Tires out-of-balance. Have professionally balanced.
3 Wheels damaged. Inspect and replace as necessary.
4 Suspension or steering components excessively worn (Chapter 10).

71 Excessive tire wear on outside edge

1 Inflation pressures incorrect (Chapter 1).
2 Excessive speed in turns.
3 Front-end alignment incorrect. Have the front end professionally aligned.
4 Suspension arm bent or twisted (Chapter 10).

72 Excessive tire wear on inside edge

1 Inflation pressures incorrect (Chapter 1).
2 Front-end alignment incorrect. Have the front end professionally aligned.
3 Loose or damaged steering components (Chapter 10).

73 Tire tread worn in one place

1 Tires out-of-balance.
2 Damaged or buckled wheel. Inspect and replace if necessary.
3 Defective tire (Chapter 1).

Notes

Chapter 1
Tune-up and routine maintenance

Contents

Section		Section
Air filter replacement	25	Fuel system check ... 21
Automatic transmission fluid and filter change	30	Introduction ... 2
Automatic transmission fluid level check	7	Maintenance schedule ... 1
Battery check, maintenance and charging	11	Manual transmission lubricant change ... 35
Brake fluid change	24	Manual transmission lubricant level check ... 18
Brake system check	23	Positive Crankcase Ventilation (PCV) valve and hose check
Chassis lubrication	20	and replacement ... 32
CHECK ENGINE light	See Chapter 6	Power steering fluid level check ... 6
Cooling system check	14	Seat belt check ... 9
Cooling system servicing (draining, flushing and refilling)	29	Spark plug replacement ... 28
Differential lubricant change	37	Spark plug wire, distributor cap and rotor check
Differential lubricant level check	17	and replacement ... 33
Drivebelt check, adjustment and replacement	12	Suspension, steering and driveaxle boot check ... 22
Engine oil and filter change	8	Tire and tire pressure checks ... 5
Evaporative emissions control system check	27	Tire rotation ... 15
Exhaust Gas Recirculation (EGR) system check	34	Transfer case lubricant change (4WD models) ... 36
Exhaust system check	16	Transfer case lubricant level check (4WD models) ... 19
Fluid level checks	4	Tune-up general information ... 3
Front hub and wheel bearing check, repack and adjustment	31	Underhood hose check and replacement ... 13
Fuel filter replacement	26	Wiper blade inspection and replacement ... 10

Specifications

Recommended lubricants and fluids

Note: *Listed here are manufacturer recommendations at the time this manual was written. Manufacturers occasionally upgrade their fluid and lubricant specifications, so check with your local auto parts store for current recommendations.*

Engine oil .... API "certified for gasoline engines"
Viscosity .... See accompanying chart

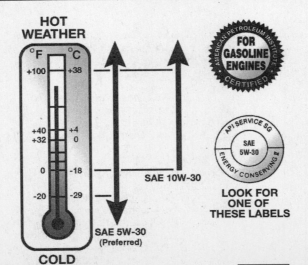

Engine oil viscosity chart

Chapter 1 Tune-up and routine maintenance

Recommended lubricants and fluids

Fuel	Unleaded gasoline, 87 octane minimum
Manual transmission lubricant	75W-90 GL-4 gear oil
Automatic transmission fluid	DEXRON III automatic transmission fluid
Transfer case	DEXRON III automatic transmission fluid
Differential	
Standard	80W90 GL-5 gear oil
Limited slip	80W90 GL-5 LSD gear oil approved for Nissan limited slip differentials
Power steering fluid	DEXRON III automatic transmission fluid
Manual steering gear lubricant	80W-90 GL-4 gear oil
Brake fluid	DOT 3 brake fluid
Engine coolant	50/50 mixture of ethylene glycol-based antifreeze and distilled or demineralized water
Wheel bearing grease	NLGI No. 2 disc brake wheel bearing grease
Free-running hub grease (4WD models)	NLGI No. 2 disc brake wheel bearing grease
Chassis lubrication grease	NLGI No. 2 moly-based chassis grease
Key lock cylinder lubricant	Graphite spray

Capacities*

Engine oil (including filter)	
Four-cylinder engine	
2WD models	3-3/4 quarts (3.5 liters)
4WD models	4-1/8 quarts (3.9 liters)
3.3L V6 engine	
Frontier and Xterra models	3-1/2 quarts (3.3 liters)
Pathfinder models	3-7/8 quarts (3.7 liters)
3.5L V6 engine	5-1/4 quarts (5.0 liters)
Manual transmission	
FS5W71C (four-cylinder models)	
2WD models	4-1/4 pints (2.0 liters)
4WD models	10-3/8 pints (4.9 liters)
FS5R30A (V6 models)	
2WD models	
All except 2001 Pathfinder	5-1/8 pints (2.4 liters)
2001 Pathfinder	5-7/8 pints (2.8 liters)
4WD models	10-3/4 pints (5.1 liters)
Automatic transmission	
Fluid and filter change	The best way to determine the amount of fluid to add during a routine fluid change is to measure the amount drained.
From dry, including torque converter	
2WD models	8-3/4 quarts (8.3 liters)
4WD models	9 quarts (8.5 liters)
Transfer case	
TX10A (part-time 4WD) case	2-3/8 quarts (2.2 liters)
ATX14A (All-Mode 4WD) case	3-1/8 quarts (3.0 liters)
Cooling system	
Frontier	
1998	
2WD	8-5/8 quarts (8.1 liters)
4WD	9-1/2 quarts (9.0 liters)
1999 and later	
Four-cylinder engine	
2WD models	
Manual transmission	9-5/8 quarts (9.15 liters)
Automatic transmission	9-1/2 quarts (8.95 liters)
4WD models	9-3/4 quarts (9.25 liters)
V6 engine	11-5/8 quarts (10.95 liters)
Xterra	
2000	
V6 engine	11-5/8 quarts (10.95 liters)
Four-cylinder engine	
2000	9-3/4 quarts (9.25 liters)
2001	7-3/4 quarts (7.3 liters)
V6 engine	11-5/8 quarts (10.95 liters)
Pathfinder	
1996 through 1999	11-1/4 quarts (10.6 liters)
2000	10-3/4 quarts (10.2 liters)
2001	9-3/4 quarts (9.2 liters)

*All capacities approximate. Add as necessary to bring to appropriate level.

Chapter 1 Tune-up and routine maintenance

Brakes

Disc brake pad lining thickness (minimum)	3/32 inch (2.4 mm)
Drum brake shoe lining thickness (minimum)	1/16 inch (1.6 mm)
Brake pedal adjustments	
Frontier	
1998	
Free height	
Manual transmission	8-15/64 to 8-5/8 inches (209 to 219 mm)
Automatic transmission	8-23/64 to 8-3/4 inches (212 to 222 mm)
Reserve distance	4-23/32 inches (120 mm)
Freeplay	5/32 to 15/32 inch (4 to 12 mm)
1999 and later	
Free height	
Manual transmission	7-33/64 to 7-29/32 inches (191 to 201 mm)
Automatic transmission	7-29/32 to 8-5/16 inches (201 to 211 mm)
Reserve distance	
Four-cylinder engine	
Manual transmission	4-1/8 inch (105 mm)
Automatic transmission	4-17/32 inch (115 mm)
V6 engine	
Manual transmission	3-15/16 inch (100 mm)
Automatic transmission	4-21/64 inch (110 mm)
Freeplay	3/64 to 1/8 inch (1 to 3 mm)
Xterra	
Free height	
Manual transmission	7-33/64 to 7-29/32 inches (191 to 201 mm)
Automatic transmission	7-29/32 to 8-5/16 inches (201 to 211 mm)
Reserve distance	
Four-cylinder engine	
Manual transmission	4-1/8 inch (105 mm)
Automatic transmission	4-17/32 inch (115 mm)
V6 engine	
Manual transmission	3-15/16 inch (100 mm)
Automatic transmission	4-21/64 inch (110 mm)
Freeplay	3/64 to 1/8 inch (1 to 3 mm)
Pathfinder	
Free height	
Manual transmission	6-1/2 to 6-7/8 inches (165 to 175 mm)
Automatic transmission	6-7/8 to 7-9/32 inches (175 to 185 mm)
Reserve distance	
Manual transmission	2-9/16 inches (65 mm)
Automatic transmission	2-3/4 inches (70 mm)
Freeplay	
1996 through 1999	5/32 to 15/32 inch (4 to 12 mm)
2000 and later	3/64 to 1/8 inch (1 to 3 mm)

Ignition system

Ignition timing	See Chapter 5
Spark plug type	
Frontier and Xterra	
1998	
Standard	NGK BKR5E-11 or equivalent
Cold*	NGK BKR6E-11 or equivalent
1999 through 2000	
Standard	NGK FR5AP-10 or equivalent
Cold*	NGK FR6AP-10 or equivalent
2001	
Standard	NGK PFR5G-11 or equivalent
Cold*	NGK PFR6G-11 or equivalent
Pathfinder	
1996 and 1997	
Standard	NGK BKR5ES-11 or equivalent
Cold*	NGK BKR6ES-11 or equivalent
1998	
Standard	NGK BKR5E-11 or equivalent
Cold*	NGK BKR6E-11 or equivalent
1999 and 2000	
Standard	NGK FR5AP-10 or equivalent
Cold*	NGK FR6AP-10 or equivalent

Ignition system (continued)

Spark plug type (continued)
    Pathfinder (continued)
        2001
            Standard.. NGK PLFR5A-11 or equivalent
            Cold*... NGK PLFR6A-11 or equivalent
Spark plug gap (all)... 0.039 to 0.043 inch (1 to 1.1 mm)
Firing order
    Four-cylinder engine 1-3-4-2
    V6 engine... 1-2-3-4-5-6

If spark knock occurs during extended high-speed driving

Wheel bearing preload

2WD Pathfinders and all 4WD models .... 1.59 to 4.72 lbs (7.06 to 20.99 N)

Torque specifications

	Ft-lbs (unless otherwise indicated)	Nm
Engine oil drain plug	22 to 29	29 to 39
Automatic transmission		
Pan bolts	61 to 78 in-lbs	7 to 9
Drain plug	22 to 29	29 to 39
Manual transmission drain/fill plugs	18 to 25	25 to 34
Transfer case drain/fill plugs		
TX10A (part time 4WD) case	18 to 25	25 to 34
ATX14A (All-mode 4WD) case	87 to 174 in-lbs	10 to 20
Differential drain/fill plugs	43	59
Spark plugs	14 to 22	20 to 29
Wheel lug nuts	87 to 108	118 to 147

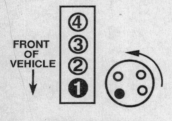

Cylinder numbering and distributor rotation diagram - four-cylinder engine

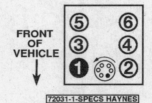

Cylinder numbering and distributor rotation diagram - 3.3L V6 engine

Cylinder numbering diagram - 3.5L V6 engine

The blackened terminal shown on the distributor cap indicates the Number One spark plug wire position

Frontier/Xterra/Pathfinder Maintenance schedule

The following maintenance intervals are based on the assumption that the vehicle owner will be doing the maintenance or service work, as opposed to having a dealer service department do the work. These are the minimum maintenance intervals recommended by the factory for vehicles that are driven daily. If you wish to keep your vehicle in peak condition at all times, you may wish to perform some of these procedures even more often. Because frequent maintenance enhances the efficiency, performance and resale value of your car, we encourage you to do so. If you drive in dusty areas, tow a trailer, idle or drive at low speeds for extended periods or drive for short distances (less than four miles) in below freezing temperatures, shorter intervals are also recommended.

When the vehicle is new, follow the maintenance schedule to the letter, record the maintenance performed in your owner's manual and keep all receipts to protect the new vehicle warranty. In many cases the initial maintenance check is done at no cost to the owner (check with your dealer service department for more information).

Every 250 miles or weekly, whichever comes first

    Check the engine oil level (Section 4)
    Check the coolant level (Section 4)
    Check the windshield washer fluid level (Section 4)
    Check the brake and clutch fluid levels (Section 4)
    Check the tires and tire pressures (Section 5)

Chapter 1 Tune-up and routine maintenance

Every 3000 miles or 3 months, whichever comes first

All items listed above, plus . . .
    Check the power steering fluid level (Section 6)
    Check the automatic transmission fluid level (Section 7)
    Change the engine oil and filter (Section 8)
    Repack the front hubs and wheel bearings (only if driven frequently through deep water; otherwise, service at 30,000 miles/24 months) (Section 31)

Every 6000 miles or 6 months, whichever comes first

All items listed above, plus . . .
    Check the seat belts (Section 9)
    Inspect the windshield wiper blades (Section 10)
    Check and service the battery (Section 11)
    Check the engine drivebelt (Section 12)
    Inspect underhood hoses (Section 13)
    Check the cooling system (Section 14)
    Rotate the tires (Section 15)
    Check the exhaust system (Section 16)

Every 15,000 miles or 12 months, whichever comes first

All items listed above, plus . . .
    Check the lubricant level in the front (4x4) and rear axles (Section 17)
    Check the manual transmission lubricant level (Section 18)
    Check the transfer case lubricant level - 4WD (Section 19)
    Lubricate the chassis (Section 20)
    Check the fuel system (Section 21)
    Check the suspension, steering, and driveaxle boots (Section 22)
    Check the brake system (Section 23)*

Every 30,000 miles or 24 months, whichever comes first

All items listed above, plus . . .
    Change the brake fluid (Section 24)
    Replace the air filter (Section 25)*
    Replace the fuel filter (Section 26)
    Check the evaporative emissions control system (Section 27)
    Replace the spark plugs (conventional, non-platinum) (Section 28)
    Service the cooling system (drain, flush and refill) (green-colored ethylene glycol anti-freeze only) (Section 29)
    Change the automatic transmission fluid and filter (Section 30)**
    Inspect and repack the front wheel bearings (and free-running hubs, on 4WD models) (Section 31)
    Replace the Positive Crankcase Ventilation (PCV) valve (Section 32)
    If noisy, check and, if necessary, adjust the valve clearances (four-cylinder and 3.5L V6 engines) (Chapter 2A or 2C).

Every 60,000 miles or 48 months, whichever comes first

    Inspect/replace the spark plug wires/distributor cap/rotor (Section 33)
    Inspect the Exhaust Gas Recirculation (EGR) valve (Section 34)
    Change the manual transmission lubricant (Section 35)
    Change the transfer case lubricant (Section 36)
    Change the differential lubricant (Section 37)**

Every 100,000 miles or 60 months, whichever comes first

    Replace the spark plugs (platinum type) (Section 28)
    Replace the timing belt (3.3L V6 engine) (Chapter 2, Part B)

* This item is affected by "severe" operating conditions, as described below. If the vehicle is operated under severe conditions, perform all maintenance indicated with an asterisk (*) at half the indicated intervals. Severe conditions exist if you mainly operate the vehicle . . .

    in dusty areas
    towing a trailer
    idling for extended periods
    driving at low speeds when outside temperatures remain below freezing and most trips are less than four miles long

** Perform this procedure at half the recommended interval if operated under one or more of the following conditions:

    in heavy city traffic where the outside temperature regularly reaches 90-degrees F or higher in hilly or mountainous terrain
    frequent trailer towing
    if the vehicle has been driven through deep water

Chapter 1 Tune-up and routine maintenance

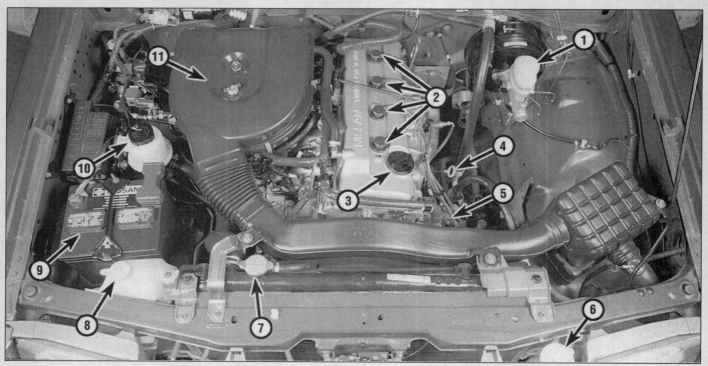

Engine compartment components - Frontier with a 2.4L four-cylinder engine

1. Brake fluid reservoir
2. Spark plugs
3. Engine oil filler cap
4. Engine oil dipstick
5. Distributor cap
6. Windshield washer fluid reservoir
7. Radiator cap
8. Coolant reservoir
9. Battery
10. Power steering fluid reservoir
11. Air filter housing

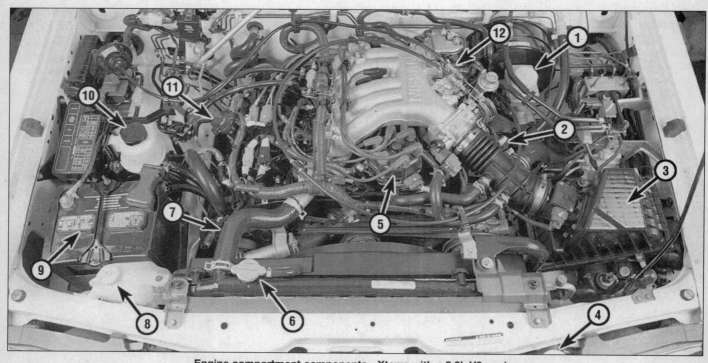

Engine compartment components - Xterra with a 3.3L V6 engine

1. Brake fluid reservoir
2. Engine oil dipstick
3. Air filter housing
4. Windshield/rear window washer fluid reservoir
5. Distributor cap
6. Radiator cap
7. Radiator hose
8. Coolant reservoir
9. Battery
10. Power steering fluid reservoir
11. Engine oil filler cap
12. PCV valve

Chapter 1 Tune-up and routine maintenance   1-7

Engine compartment components - 2001 Pathfinder with a 3.5L V6 engine

1. Engine oil dipstick
2. Brake fluid reservoir
3. Engine oil filler cap
4. Air filter housing
5. Coolant reservoir
6. Radiator cap
7. Radiator hose
8. Windshield/rear window washer fluid reservoir
9. Battery
10. Power steering fluid reservoir
11. Automatic transmission fluid dipstick

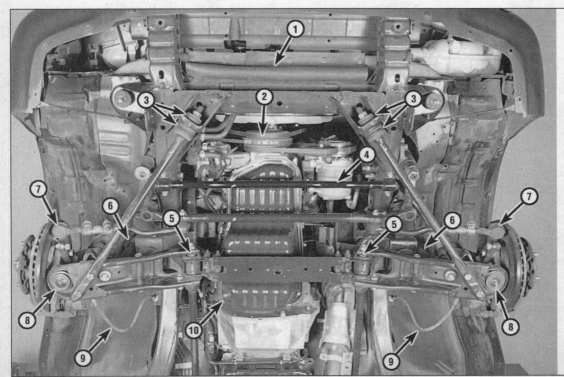

Engine compartment underside components - 2WD Frontier

1. Radiator drain plug (behind cover)
2. Drivebelts
3. Tension rod bushings
4. Stabilizer bar
5. Lower control arm bushings
6. Shock absorber
7. Tie-rod end
8. Lower control arm balljoint
9. Brake hose
10. Engine oil drain plug

Chapter 1 Tune-up and routine maintenance

Engine compartment underside components - 2WD Xterra (4WD Frontier and Xterra similar)

1. Radiator drain plug (behind cover)
2. Drivebelts
3. Engine oil filter
4. Stabilizer bar
5. Tie-rod end
6. Lower control arm bushings
7. Lower control arm balljoint
8. Engine oil drain plug
9. Manual transmission check/fill plug
10. Manual transmission drain plug

Engine compartment underside components - 2WD Pathfinder

1. Engine oil filter
2. Stabilizer bar
3. Tie-rod end
4. Control arm bushings
5. Balljoint
6. Engine oil drain plug
7. Automatic transmission fluid drain plug

Chapter 1 Tune-up and routine maintenance   1-9

Rear underside components - Frontier (Xterra similar)

1. Spare tire
2. Brake hose
3. Shock absorber
4. Differential lubricant check/fill plug
5. Differential lubricant drain plug
6. Muffler
7. Universal joint
8. Fuel tank

Rear underside components - Pathfinder

1. Charcoal canister
2. Post muffler
3. Shock absorber
4. Differential lubricant check/fill plug
5. Differential lubricant drain plug
6. Main muffler
7. Driveshaft
8. Fuel tank

Chapter 1 Tune-up and routine maintenance

4.2a Location of the engine oil dipstick - four-cylinder engine

4.2b Location of the engine oil dipstick - 3.3L V6 engine

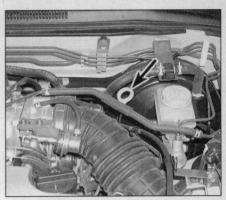

4.2c Location of the engine oil dipstick - 3.5L V6 engine

2 Introduction

This Chapter is designed to help the home mechanic maintain the Nissan Frontier/Xterra/Pathfinder vehicle with the goals of maximum performance, economy, safety and reliability in mind.

Included is a master maintenance schedule, followed by procedures dealing specifically with each item on the schedule. Visual checks, adjustments, component replacement and other helpful items are included. Refer to the accompanying illustrations of the engine compartment and the underside of the vehicle for the locations of various components.

Servicing your vehicle in accordance with the mileage/time maintenance schedule and the step-by-step procedures will result in a planned maintenance program that should produce a long and reliable service life. Keep in mind that it's a comprehensive plan, so maintaining some items but not others at the specified intervals will not produce the same results.

As you service your vehicle, you will discover that many of the procedures can - and should - be grouped together because of the nature of the particular procedure you're performing or because of the close proximity of two otherwise unrelated components to one another.

For example, if the vehicle is raised for chassis lubrication, you should inspect the exhaust, suspension, steering and fuel systems while you're under the vehicle. When you're rotating the tires, it makes good sense to check the brakes since the wheels are already removed. Finally, let's suppose you have to borrow or rent a torque wrench. Even if you only need it to tighten the spark plugs, you might as well check the torque of as many critical fasteners as time allows.

The first step in this maintenance program is to prepare yourself before the actual work begins. Read through all the procedures you're planning to do, then gather up all the parts and tools needed. If it looks like you might run into problems during a particular job, seek advice from a mechanic or an experienced do-it-yourselfer.

3 Tune-up general information

The term tune-up is used in this manual to represent a combination of individual operations rather than one specific procedure that will maintain a gasoline engine in proper tune.

If, from the time the vehicle is new, the routine maintenance schedule is followed closely and frequent checks are made of fluid levels and high wear items, as suggested throughout this manual, the engine will be kept in relatively good running condition and the need for additional work will be minimized.

More likely than not, however, there may be times when the engine is running poorly due to lack of regular maintenance. This is even more likely if a used vehicle, which has not received regular and frequent maintenance checks, is purchased. In such cases, an engine tune-up will be needed outside of the regular routine maintenance intervals.

The first step in any tune-up or diagnostic procedure to help correct a poor running engine is a cylinder compression check. A compression check (see Chapter 2D) will help determine the condition of internal engine components and should be used as a guide for tune-up and repair procedures. If, for instance, the compression check indicates serious internal engine wear, a conventional tune-up won't improve the performance of the engine and would be a waste of time and money. Because of its importance, the compression check should be done by someone with the right equipment and the knowledge to use it properly.

The following procedures are those most often needed to bring a generally poor running engine back into a proper state of tune.

Minor tune-up

Check all engine related fluids (Section 4)
Clean, inspect and test the battery (Section 11)
Check and adjust the drivebelts (Section 12)
Check all underhood hoses (Section 13)
Check the cooling system (Section 14)
Check the air filter (Section 25)
Replace the spark plugs (Section 28)
Inspect the spark plug wires (Section 33)

Major tune-up

All items listed under Minor tune-up, plus . . .
Replace the air filter (Section 25)
Replace the spark plug wires (and distributor cap and rotor, if so equipped (Section 33)
Check the ignition system (Chapter 5)
Check the charging system (Chapter 5)

4 Fluid level checks (every 250 miles or weekly)

Note: *The following are fluid level checks to be done on a 250 mile or weekly basis. Additional fluid level checks can be found in specific maintenance procedures that follow. Regardless of intervals, be alert to fluid leaks under the vehicle, which would indicate a fault to be corrected immediately.*

1 Fluids are an essential part of the lubrication, cooling, brake, clutch and windshield washer systems. Because the fluids gradually become depleted and/or contaminated during normal operation of the vehicle, they must be periodically replenished. See *Recommended lubricants and fluids* at the beginning of this Chapter before adding fluid to any of the following components. **Note:** *The vehicle must be on level ground when fluid levels are checked.*

Engine oil

Refer to illustrations 4.2a, 4.2b, 4.2c, 4.4 and 4.6

2 The engine oil level is checked with a dipstick that extends through a tube and into the oil pan at the bottom of the engine **(see illustrations).**

3 The oil level should be checked before the vehicle has been driven, or about 5 minutes after the engine has been shut off. If the oil is checked immediately after driving the vehicle, some of the oil will remain in the upper engine components, resulting in an inaccurate reading on the dipstick.

4 Pull the dipstick out of the tube and wipe all the oil from the end with a clean rag or paper towel. Insert the clean dipstick all the way back into the tube, then pull it out again. Note the oil at the end of the dipstick. Add oil

Chapter 1 Tune-up and routine maintenance

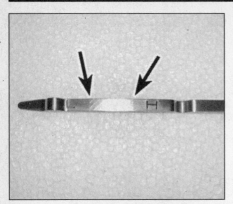

4.4 The oil level must be maintained between the marks at all times - it takes one quart of oil to raise the level from the L to the H mark

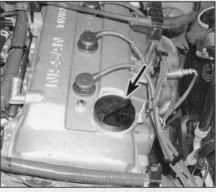

4.6 Oil is added to the engine after unscrewing the oil filler cap (arrow) from the valve cover - always make sure the area around the opening is clean before removing the cap to prevent dirt from contaminating the engine (four-cylinder engine shown)

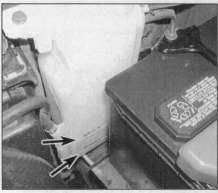

4.8a Keep the coolant level near the MAX mark or MIN mark on the side of the reservoir, depending on engine temperature - this reservoir, located near the battery, is on a four-cylinder model (3.3L V6 engines similar) . . .

4.8b . . . on Pathfinders with a 3.5L V6 engine the coolant reservoir marks are seen between the grille and the radiator, on the driver's side of the vehicle

4.14 On Frontiers and Xterras the windshield washer fluid reservoir is located on the left side of the engine compartment on Pathfinders, it's on the right side)

as necessary to keep the level between the L and H marks or within the cross-hatched zone on the dipstick **(see illustration)**.

5 Do not overfill the engine by adding too much oil since this may result in oil-fouled spark plugs, oil leaks or oil seal failures.

6 Oil is added to the engine after unscrewing a cap from the valve cover **(see illustration)**. A funnel may help to reduce spills.

7 Checking the oil level is an important preventive maintenance step. A consistently low oil level indicates oil leakage through damaged seals, defective gaskets or past worn rings or valve guides. If the oil looks milky or has water droplets in it, the cylinder head gasket(s) may be blown or the head(s) or block may be cracked. The engine should be checked immediately. The condition of the oil should also be checked. Whenever you check the oil level, slide your thumb and index finger up the dipstick before wiping off the oil. If you see small dirt or metal particles clinging to the dipstick, the oil should be changed (see Section 8).

Engine coolant

Refer to illustrations 4.8a and 4.8b

Warning: *Do not allow antifreeze to come in contact with your skin or painted surfaces of the vehicle. Rinse off spills immediately with plenty of water. Antifreeze is highly toxic if ingested. Never leave antifreeze lying around in an open container or in puddles on the floor; children and pets are attracted by its sweet smell and may drink it. Check with local authorities on disposing of used anti-freeze. Many communities have collection centers that will see that antifreeze is disposed of safely.*

Note: *Non-toxic antifreeze is now manufactured and available at local auto parts stores, but even this type should be disposed of properly.*

8 All vehicles covered by this manual are equipped with a pressurized coolant recovery system. A white plastic coolant reservoir located in the engine compartment is connected by a hose to the radiator filler neck **(see illustrations)**.

9 The coolant level in the reservoir should be checked regularly. **Warning:** *Do not remove the pressure cap to check the coolant level when the engine is warm.* The level of coolant in the reservoir varies with the temperature of the engine. When the engine is cold, the coolant level should be at or slightly above the MIN mark on the reservoir. Once the engine has warmed up, the level should be at or near the MAX mark. If it isn't, add coolant to the reservoir. To add coolant simply twist open the cap and add a 50/50 mixture of ethylene glycol based antifreeze and water.

10 Drive the vehicle and recheck the coolant level. If only a small amount of coolant is required to bring the system up to the proper level, water can be used. However, repeated additions of water will dilute the antifreeze and water solution. In order to maintain the proper ratio of antifreeze and water, always top up the coolant level with the correct mixture. An empty plastic milk jug or bleach bottle makes an excellent container for mixing coolant. Do not use rust inhibitors or additives.

11 If the coolant level drops consistently, there may be a leak in the system. Inspect the radiator, hoses, filler cap, drain plugs and water pump (see Section 14). If no leaks are noted, have the radiator cap tested by a service station.

12 If you have to remove the radiator cap, wait until the engine has cooled completely, then wrap a thick cloth around the cap and turn it to the first stop. If coolant or steam escapes, let the engine cool down longer, then remove the cap.

13 Check the condition of the coolant as well. It should be relatively clear. If it is brown or rust colored, the system should be drained, flushed and refilled. Even if the coolant appears to be normal, the corrosion inhibitors wear out, so it must be replaced at the specified intervals.

Windshield washer fluid

Refer to illustration 4.14

14 Fluid for the windshield washer system is located in a plastic reservoir near the left or right headlight housing, depending on model **(see illustration)**.

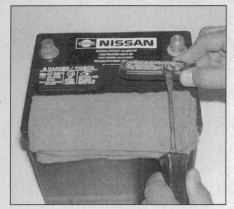

4.16 The electrolyte level can be checked on original equipment batteries; some have individual cell plugs that can be unscrewed, while others have caps which must be pried off

4.18a Never let the brake fluid level drop below the MIN mark

4.18b Clutch fluid reservoir level marks

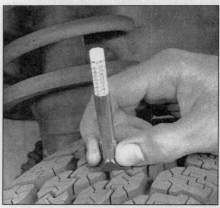

5.2 Use a tire tread depth indicator to monitor tire wear - they are available at auto parts stores and service stations and cost very little

15 In milder climates, plain water can be used in the reservoir, but it should be kept no more than 2/3 full to allow for expansion if the water freezes. In colder climates, use windshield washer system antifreeze, available at any auto parts store, to lower the freezing point of the fluid. Mix the antifreeze with water in accordance with the manufacturer's directions on the container. **Caution:** *Don't use cooling system antifreeze - it will damage the vehicle's paint.* **Note:** *To help prevent icing in cold weather, warm the windshield with the defroster before using the washer.*

Battery electrolyte

Refer to illustration 4.16

16 On models not equipped with a sealed battery, check the electrolyte level of all six battery cells. On models with a translucent battery case, minimum and maximum level marks are present on the side of the case; keep the electrolyte level at the MAX mark. On models with an opaque case, carefully remove the cell caps off to check the level or add water. Some batteries have six individual cell plugs that can be unscrewed, but others have two cell caps that must be carefully pried off **(see illustration)**.

17 If the level is low, add *distilled* water until the level is up to the MAX mark (translucent batteries) or up to the bottom of the split-ring indicators (opaque batteries).

Brake and clutch fluid

Refer to illustrations 4.18a and 4.18b

18 The brake master cylinder is mounted on the front of the power brake booster, on the left (driver's) side of the engine compartment firewall **(see illustration)**. The clutch master cylinder used on manual transmission models is mounted next to the brake master cylinder **(see illustration)**.

19 The translucent plastic reservoir allows the fluid inside to be checked without removing the cap. Be sure to wipe the area around either reservoir cap with a clean rag to prevent contamination of the brake and/or clutch system before removing the cap. Keep the fluid level at or near the MAX mark.

20 When adding fluid, pour it carefully into the reservoir to avoid spilling it on surrounding painted surfaces. Be sure the specified fluid is used, since mixing different types of brake fluid can cause damage to the system. See *Recommended lubricants and fluids* at the front of this Chapter or your owner's manual. **Warning:** *Brake fluid can harm your eyes and damage painted surfaces, so use extreme caution when handling or pouring it. Do not use brake fluid that has been standing open or is more than one year old. Brake fluid absorbs moisture from the air. Moisture in the system can cause a dangerous loss of brake performance.*

21 At this time, the fluid and master cylinder can be inspected for contamination. The system should be drained and refilled if deposits, dirt particles or water droplets are seen in the fluid.

22 After filling the reservoir to the proper level, make sure the cap is on tight to prevent fluid leakage.

23 The brake fluid level in the master cylinder will drop slightly as the pads at the front wheels wear down during normal operation. If the master cylinder requires repeated additions to keep it at the proper level, it's an indication of leakage in the brake system, which should be corrected immediately. Check all brake lines and connections (see Section 23 for more information).

24 If, upon checking the master cylinder fluid level, you discover the reservoir empty or nearly empty, the brake system should be bled and thoroughly inspected (see Chapter 9).

5 Tire and tire pressure checks (every 250 miles or weekly)

Refer to illustrations 5.2, 5.3, 5.4a, 5.4b and 5.8

1 Periodic inspection of the tires may spare you the inconvenience of being stranded with a flat tire. It can also provide you with vital information regarding possible problems in the steering and suspension systems before major damage occurs.

2 The original tires on this vehicle are equipped with 1/2-inch wide wear bands that will appear when tread depth reaches 1/16-inch, at which point the tires can be considered worn out. Tread wear can be monitored with a simple, inexpensive device known as a tread depth indicator **(see illustration)**.

3 Note any abnormal tread wear **(see illustration)**. Tread pattern irregularities such as cupping, flat spots and more wear on one side than the other are indications of front end alignment and/or balance problems. If any of these conditions are noted, take the vehicle to a tire shop or service station to correct the problem.

4 Look closely for cuts, punctures and embedded nails or tacks. Sometimes a tire will hold air pressure for a short time or leak down very slowly after a nail has embedded itself in the tread. If a slow leak persists, check the valve stem core to make sure it's tight **(see illustration)**. Examine the tread for an object that may have embedded itself in the tire or for a "plug" that may have begun to leak (radial tire punctures are repaired with a plug that's installed in a puncture). If a puncture is suspected, it can be easily verified by spraying a solution of soapy water onto the puncture area **(see illustration)**. The soapy

UNDERINFLATION

CUPPING

Cupping may be caused by:
- Underinflation and/or mechanical irregularities such as out-of-balance condition of wheel and/or tire, and bent or damaged wheel.
- Loose or worn steering tie-rod or steering idler arm.
- Loose, damaged or worn front suspension parts.

OVERINFLATION

INCORRECT TOE-IN OR EXTREME CAMBER

FEATHERING DUE TO MISALIGNMENT

5.3 This chart will help you determine the condition of the tires and the probable cause(s) of abnormal wear

solution will bubble if there's a leak. Unless the puncture is unusually large, a tire shop or service station can usually repair the tire.

5 Carefully inspect the inner sidewall of each tire for evidence of brake fluid leakage. If you see any, inspect the brakes immediately.

6 Correct air pressure adds miles to the lifespan of the tires, improves mileage and enhances overall ride quality. Tire pressure cannot be accurately estimated by looking at a tire, especially if it's a radial. A tire pressure gauge is essential. Keep an accurate gauge in the vehicle. The pressure gauges attached to the nozzles of air hoses at gas stations are often inaccurate.

7 Always check tire pressure when the tires are cold. Cold, in this case, means the vehicle has not been driven over a mile in the three hours preceding a tire pressure check. A pressure rise of four to eight pounds is not uncommon once the tires are warm.

8 Unscrew the valve cap protruding from the wheel or hubcap and push the gauge firmly onto the valve stem **(see illustration)**. Note the reading on the gauge and compare the figure to the recommended tire pressure shown on the placard on the driver's side door pillar. Be sure to reinstall the valve cap to keep dirt and moisture out of the valve stem mechanism. Check all four tires and, if necessary, add enough air to bring them up to the recommended pressure.

9 Don't forget to keep the spare tire inflated to the specified pressure (refer to your owner's manual or the tire sidewall).

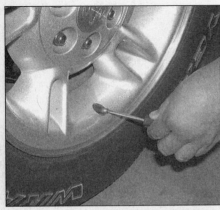
5.4a If a tire loses air on a steady basis, check the valve stem core first to make sure it's snug (special inexpensive wrenches are commonly available at auto parts stores)

5.4b If the valve stem core is tight, raise the corner of the vehicle with the low tire and spray a soapy water solution onto the tread as the tire is turned slowly - leaks will cause small bubbles to appear

5.8 To extend the life of the tires, check the air pressure at least once a week with an accurate gauge (don't forget the spare!)

1-14 Chapter 1 Tune-up and routine maintenance

6.5 On models with a translucent reservoir, the power steering fluid level can be checked without removing the cap

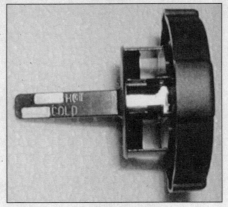

6.6 The power steering fluid dipstick, on models without a translucent reservoir, has marks on it so the fluid can be checked hot or cold

6 Power steering fluid level check (every 3000 miles or 3 months)

Refer to illustrations 6.5 and 6.6

1 Unlike manual steering, the power steering system relies on fluid which may, over a period of time, require replenishing.
2 All models have remote power steering fluid reservoirs mounted on the right (passenger's) side of the engine compartment. Some models have translucent reservoirs which allow the fluid to be checked without removing the cap; models without translucent reservoirs incorporate a dipstick on the underside of the cap.
3 For the check, the front wheels should be pointed straight ahead and the engine should be off.
4 Use a clean rag to wipe off the reservoir cap and the area around the cap. This will help prevent any foreign matter from entering the reservoir during the check.
5 Feel the reservoir to check the temperature of the fluid. On models with a translucent reservoir, check the level of the fluid on the side of the reservoir (see illustration). The level should be at the HOT or MAX mark if the reservoir was hot to the touch. If the reservoir felt cool, the level should be at the COLD mark (but not below the MIN level).
6 On models with a dipstick, unscrew the cap, wipe off the fluid with a clean rag, reinsert the dipstick, then withdraw it and read the fluid level. The fluid should be at the proper level, depending on whether it was checked hot or cold (see illustration). Never allow the fluid level to drop below the lower mark on the dipstick.
7 If additional fluid is required, pour the specified type directly into the reservoir, using a funnel to prevent spills.
8 If the reservoir requires frequent fluid additions, all power steering hoses, hose connections, steering gear and the power steering pump should be carefully checked for leaks.

7 Automatic transmission fluid level check (every 3000 miles or 3 months)

Refer to illustrations 7.3a, 7.3b and 7.6

1 The automatic transmission fluid level should be carefully maintained. Low fluid level can lead to slipping or loss of drive, while overfilling can cause foaming and loss of fluid.
2 With the parking brake set, start the engine, then move the shift lever through all the gear ranges, ending in Park. The fluid level must be checked with the vehicle level and the engine running at idle. Note: *Incorrect fluid level readings will result if the vehicle has just been driven at high speeds for an extended period, in hot weather in city traffic, or if it has been pulling a trailer. If any of these conditions apply, wait until the fluid has cooled (about 30 minutes).*
3 With the transmission at normal operating temperature, remove the dipstick from the filler tube. The dipstick is located at the rear of the engine compartment on the passenger's side (see illustrations).
4 Wipe the fluid from the dipstick with a clean rag and push it back into the filler tube until the cap seats.
5 Pull the dipstick out again and note the fluid level.
6 If the fluid is cool to warm, the level should be in the crosshatched area on the COLD side of the dipstick (see illustration). If it's hot, the level should be in the crosshatched area on the HOT side of the dipstick. If additional fluid is required, add it directly into the tube using a funnel. It takes about one pint to raise the level from the bottom of the crosshatched area to the top with a hot transmission, so add the fluid a little at a time and keep checking the level until it's correct.
7 The condition of the fluid should also be checked along with the level. If the fluid at the end of the dipstick is a dark reddish-brown color, or if it smells burned, it should be changed. If you are in doubt about the condition of the fluid, purchase some new fluid and compare the two for color and smell.

8 Engine oil and filter change (every 3000 miles or 3 months)

Refer to illustrations 8.3, 8.9, 8.14a, 8.14b, 8.14c and 8.18

1 Frequent oil changes are the most

7.3a Location of the automatic transmission fluid dipstick (arrow) on a four-cylinder Frontier model - the air filter housing has been removed for clarity

7.3b Dipstick location - Pathfinder with a 3.5L V6 engine

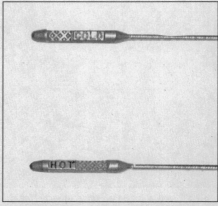

7.6 The automatic transmission fluid must be kept in the cross-hatched area, depending on the fluid temperature

Chapter 1 Tune-up and routine maintenance

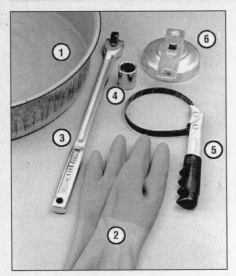

8.3 These tools are required when changing the engine oil and filter

1. **Drain pan** - It should be fairly shallow in depth, but wide to prevent spills
2. **Rubber gloves** - When removing the drain plug and filter, you will get oil on your hands (the gloves will prevent burns)
3. **Breaker bar** - Sometimes the oil drain plug is tight, and a long breaker bar is needed to loosen it
4. **Socket** - To be used with the breaker bar or a ratchet (must be the correct size to fit the drain plug - six-point preferred)
5. **Filter wrench** - This is a metal band-type wrench, which requires clearance around the filter to be effective
6. **Filter wrench** - This type fits on the bottom of the filter and can be turned with a ratchet or breaker bar (different-size wrenches are available for different types of filters)

important preventive maintenance procedures that can be done by the home mechanic. As engine oil ages, it becomes diluted and contaminated, which leads to premature engine wear.

2 Although some sources recommend oil filter changes every other oil change, we feel that the minimal cost of an oil filter and the relative ease with which it is installed dictate that a new filter be installed every time the oil is changed.

3 Gather together all necessary tools and materials before beginning this procedure **(see illustration)**.

4 You should have plenty of clean rags and newspapers handy to mop up any spills. Access to the under side of the vehicle may be improved if the vehicle can be lifted on a hoist, driven onto ramps or supported by jackstands. **Warning:** *Do not work under a vehicle which is supported only by a jack.*

5 If this is your first oil change, familiarize yourself with the locations of the oil drain plug and the oil filter.

6 Warm the engine to normal operating temperature. If the new oil or any tools are needed, use this warm-up time to gather everything necessary for the job. The correct type of oil for your application can be found in *Recommended lubricants and fluids* at the beginning of this Chapter.

7 With the engine oil warm (warm engine oil will drain better and more built-up sludge will be removed with it), raise and support the vehicle. Make sure it's safely supported!

8 Move all necessary tools, rags and newspapers under the vehicle. Set the drain pan under the drain plug. Keep in mind that the oil will initially flow from the pan with some force; position the pan accordingly.

9 Being careful not to touch any of the hot exhaust components, use a box-end wrench or a socket to remove the drain plug near the bottom of the oil pan **(see illustration)**. Depending on how hot the oil is, you may want to wear gloves while unscrewing the plug the final few turns.

10 Allow the oil to drain into the pan. It may be necessary to move the pan as the oil flow slows to a trickle.

11 After all the oil has drained, wipe off the drain plug with a clean rag. Small metal particles may cling to the plug and would immediately contaminate the new oil.

12 Clean the area around the drain plug opening and reinstall the plug. Tighten the plug securely with the wrench. If a torque wrench is available, use it to tighten the plug to the torque listed in this Chapter's Specifications.

13 Move the drain pan into position under the oil filter. If you're working on a model with a four-cylinder engine, remove the right (passenger's) side inner fender panel. It isn't necessary to remove the wheel, but doing so makes removing the oil filter easier.

14 Use the oil filter wrench to loosen the oil filter **(see illustrations)**.

15 Completely unscrew the old filter. Be careful: it's full of oil. Empty the oil inside the filter into the drain pan, then lower the filter.

16 Compare the old filter with the new one to make sure they're the same type.

17 Use a clean rag to remove all oil, dirt and sludge from the area where the oil filter mounts to the engine. Check the old filter to make sure the rubber gasket isn't stuck to the engine. If the gasket is stuck to the engine (use a flashlight if necessary), remove it.

18 Apply a light coat of clean oil to the rubber gasket on the new oil filter **(see illustration)**.

8.9 Use a proper size box-end wrench or socket to remove the oil drain plug and avoid rounding it off

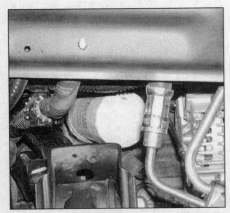

8.14a The oil filter on four-cylinder models is accessed through the right-side wheel well

8.14b Oil filter location - 3.3L V6 engine

8.14c On 3.5L V6 engines the oil filter is directly below the crankshaft pulley

8.18 Lubricate the oil filter gasket with clean engine oil before installing the filter on the engine

19 Attach the new filter to the engine, following the tightening directions printed on the filter canister or packing box. Most filter manufacturers recommend against using a filter wrench due to the possibility of overtightening and damage to the seal.
20 If you're working on a four-cylinder model, install the right side inner fender panel. Remove all tools, rags, etc. from under the vehicle, being careful not to spill the oil in the drain pan, then lower the vehicle.
21 Move to the engine compartment and locate the oil filler cap.
22 Pour the fresh oil through the filler opening. A funnel may be helpful.
23 Refer to the engine oil capacity in this Chapter's Specifications and add the proper amount of fresh oil into the engine. Wait a few minutes to allow the oil to drain into the pan, then check the level on the oil dipstick (see Section 4 if necessary). If the oil level is above the hatched area, start the engine and allow the new oil to circulate.
24 Run the engine for only about a minute and then shut it off. Immediately look under the vehicle and check for leaks at the oil pan drain plug and around the oil filter.
25 With the new oil circulated and the filter now completely full, recheck the level on the dipstick and add more oil as necessary.
26 During the first few trips after an oil change, make it a point to check frequently for leaks and proper oil level.
27 The old oil drained from the engine cannot be reused in its present state and should be disposed of. Check with your local auto parts store, disposal facility or environmental agency to see if they will accept the oil for recycling. After the oil has cooled it can be drained into a container (capped plastic jugs, topped bottles, milk cartons, etc.) for transport to one of these disposal sites. Don't dispose of the oil by pouring it on the ground or down a drain!

9 Seat belt check (every 6000 miles or 6 months)

1 Check the seat belts, buckles, latch plates and guide loops for obvious damage and signs of wear.
2 Where the seat belt receptacle bolts to the floor of the vehicle, check that the bolts are secure.
2 See if the seat belt reminder light comes on when the key is turned to the Run or Start position. A chime should also sound.

10 Wiper blade inspection and replacement (every 6000 miles or 6 months)

Refer to illustrations 10.3 and 10.4

1 The windshield wiper blade elements should be checked periodically for cracks and deterioration.
2 Lift the wiper blade assembly away from the glass.
3 Press the release lever and slide the blade assembly out of the hook in the end of the wiper arm **(see illustration)**.
4 Squeeze the two rubber prongs at the end of the blade element, then slide the element out of the frame **(see illustration)**.
5 Compare the new element with the old for length, design, etc. Some replacement elements come in a three-piece design (two metal strips, one on either side of the rubber) that is held together by several small plastic sleeves. Keep the sleeves in place on this design until you start sliding the element into the frame. Remove each of the plastic sleeves as needed when they reach the frame.
6 Slide the new element into the frame, notched end last and secure the clips into the notches of the frame.
7 Reinstall the blade assembly on the arm, wet the windshield and test for proper operation.

11 Battery check, maintenance and charging (every 6000 miles or 6 months)

Refer to illustrations 11.1, 11.5, 11.6a, 11.6b, 11.7a and 11.7b
Warning: *Certain precautions must be followed when checking and servicing the battery. Hydrogen gas, which is highly flammable, is always present in the battery cells, so keep lighted tobacco and all other open flames and sparks away from the battery. The electrolyte inside the battery is actually dilute sulfuric acid, which will cause injury if splashed on your skin or in your eyes. It will also ruin clothes and painted surfaces. When removing the battery cables, always detach the negative cable first and hook it up last!*
1 A routine preventive maintenance program for the battery in your vehicle is the only way to ensure quick and reliable starts. But before performing any battery maintenance, make sure that you have the proper equipment necessary to work safely around the battery **(see illustration)**.
2 There are also several precautions that should be taken whenever battery maintenance is performed. Before servicing the battery, always turn the engine and all accessories off and disconnect the cable from the negative terminal of the battery.
3 The battery produces hydrogen gas, which is both flammable and explosive. Never create a spark, smoke or light a match around the battery. Always charge the battery in a ventilated area.
4 Electrolyte contains poisonous and corrosive sulfuric acid. Do not allow it to get in your eyes, on your skin or on your clothes. Never ingest it. Wear protective safety glasses when working near the battery. Keep children away from the battery.
5 Note the external condition of the battery. If the positive terminal and cable clamp on your vehicle's battery is equipped with a rubber protector, make sure that it's not torn or damaged. It should completely cover the terminal. Look for any corroded or loose connections, cracks in the case or cover or loose hold-down clamps. Also check the entire length of each cable for cracks and frayed conductors **(see illustration)**.

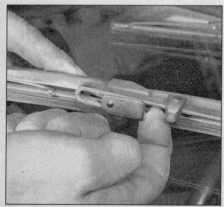

10.3 Depress the release lever (finger is on it here) and slide the wiper assembly down the wiper arm and out of the hook in the end of the arm

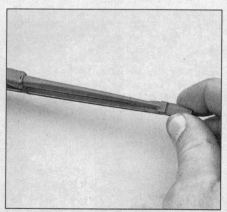

10.4 Squeeze the metal prongs on the blade element to allow it to slide out of the assembly - you may need pliers

Chapter 1 Tune-up and routine maintenance

11.1 Tools and materials required for battery maintenance

1. **Face shield/safety goggles** - When removing corrosion with a brush, the acidic particles can easily fly up into your eyes
2. **Baking soda** - A solution of baking soda and water can be used to neutralize corrosion
3. **Petroleum jelly** - A layer of this on the battery posts will help prevent corrosion
4. **Battery post/cable cleaner** - This wire brush cleaning tool will remove all traces of corrosion from the battery posts and cable clamps
5. **Treated felt washers** - Placing one of these on each post, directly under the cable clamps, will help prevent corrosion
6. **Puller** - Sometimes the cable clamps are very difficult to pull off the posts, even after the nut/bolt has been completely loosened. This tool pulls the clamp straight up and off the post without damage
7. **Battery post/cable cleaner** - Here is another cleaning tool that is a slightly different version of Number 4 above, but it does the same thing
8. **Rubber gloves** - Another safety item to consider when servicing the battery; remember that's acid inside the battery!

6 If corrosion, which looks like white, fluffy deposits is evident, particularly around the terminals, the battery should be removed for cleaning **(see illustration)**. Loosen the cable nuts with a wrench or battery pliers, being careful to remove the ground cable first, and slide them off the terminals **(see illustration)**. Then disconnect the hold-down clamp bolt and nut, remove the clamp and lift the battery from the engine compartment.

7 Clean the cable ends thoroughly with a battery brush or a terminal cleaner and a solution of warm water and baking soda. Wash the terminals and the battery case with the same solution but make sure that the solution doesn't get into the battery. When

Terminal end corrosion or damage.

Insulation cracks.

Chafed insulation or exposed wires.

Burned or melted insulation.

11.5 Typical battery cable problems

11.7a When cleaning the cable clamps, all corrosion must be removed (the inside of the clamp is tapered to match the taper on the post, so don't remove too much material)

cleaning the cables, terminals and battery case, wear safety goggles and rubber gloves to prevent any solution from coming in contact with your eyes or hands. Wear old clothes too - even diluted, sulfuric acid splashed onto clothes will burn holes in them. If the terminals have been corroded, clean them up with a terminal cleaner **(see illustrations)**. Thoroughly wash all cleaned areas with plain water.

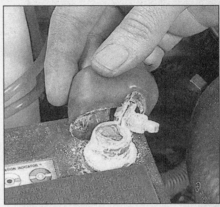

11.6a Battery terminal corrosion usually appears as light, fluffy powder

11.6b If the battery cable terminal is in good shape and not corroded, it can usually be loosened with a wrench - sometimes special battery pliers are required if corrosion has caused deterioration of the nut hex (always remove the ground cable first and hook it up last!)

11.7b Regardless of the type of tool used on the battery posts, a clean, shiny surface should be the result

8 Make sure that the battery tray is in good condition and the hold-down clamp bolts are tight. If the battery is removed from the tray, make sure no parts remain in the bottom of the tray when the battery is rein-

1-18   Chapter 1 Tune-up and routine maintenance

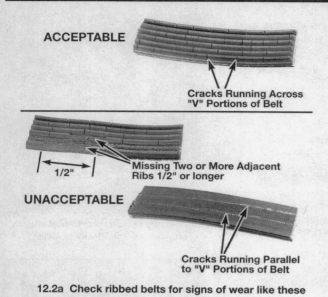

12.2a Check ribbed belts for signs of wear like these

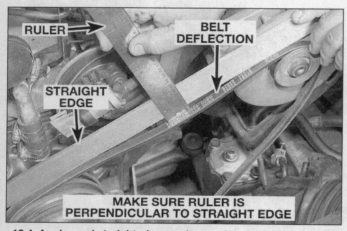

12.4 A ruler and straightedge can be used to determine the belt deflection (tension) between two pulleys

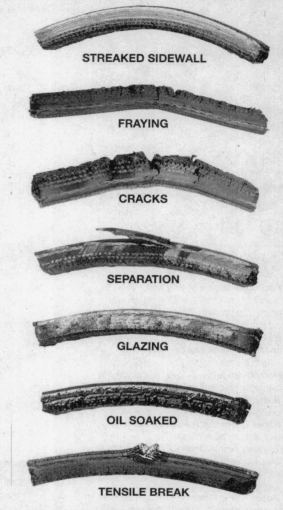

12.2b Look for these signs of wear or damage on V-belt drivebelts

stalled. When reinstalling the hold-down clamp bolts, do not overtighten them.

9 Any metal parts of the vehicle damaged by corrosion should be covered with a zinc-based primer, then painted.

10 Information on removing and installing the battery can be found in Chapter 5. Information on jump starting can be found at the front of this manual. For more detailed battery checking procedures, refer to the *Haynes Automotive Electrical Manual*.

Charging

Warning: *When batteries are being charged, hydrogen gas, which is very explosive and flammable, is produced. Do not smoke or allow open flames near a charging or a recently charged battery. Wear eye protection when near the battery during charging. Also, make sure the charger is unplugged before connecting or disconnecting the battery from the charger.*

Note: *It is recommended that the battery be removed from the vehicle for charging because the gas that escapes during this procedure can damage the paint. Fast charging with the battery cables connected can result in damage to the electrical system.*

11 Slow-rate charging is the best way to restore a battery that's discharged to the point where it will not start the engine. It's also a good way to maintain the battery charge in a vehicle that's only driven a few miles between starts. Maintaining the battery charge is particularly important in the winter when the battery must work harder to start the engine and electrical accessories that drain the battery are in greater use.

12 It's best to use a one or two-amp battery charger (sometimes called a "trickle" charger). They are the safest and put the least strain on the battery. They are also the least expensive. For a faster charge, you can use a higher amperage charger, but don't use one rated more than 1/10th the amp/hour rating of the battery. Rapid boost charges that claim to restore the power of the battery in one to two hours are hardest on the battery and can damage batteries not in good condition. This type of charging should only be used in emergency situations.

13 The average time necessary to charge a battery should be listed in the instructions that come with the charger. As a general rule, a trickle charger will charge a battery in 12 to 16 hours.

14 Remove all the cell caps (if equipped - see Section 4) and cover the holes with a clean cloth to prevent spattering electrolyte. Disconnect the negative battery cable and hook the battery charger cable clamps up to the battery posts (positive-to-positive, negative-to-negative), then plug in the charger. Make sure it is set at 12-volts if it has a selector switch.

15 If you're using a charger with a rate higher than two amps, check the battery regularly during charging to make sure it doesn't overheat. If you're using a trickle charger, you can safely let the battery charge overnight after you've checked it regularly for the first couple of hours.

16 If the battery has removable cell caps, measure the specific gravity with a hydrometer every hour during the last few hours of the

Chapter 1 Tune-up and routine maintenance

1-19

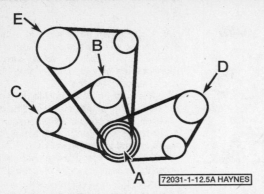

12.5a Drivebelt routing diagram - four-cylinder engine

A Crankshaft pulley
B Water pump
C Alternator
D Air conditioning compressor
E Power steering pump

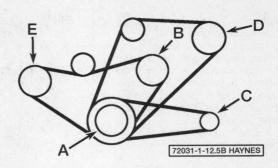

12.5b Drivebelt routing diagram - 3.3L V6 engine (Frontier/Xterra)

A Crankshaft pulley
B Water pump
C Alternator
D Air conditioning compressor
E Power steering pump

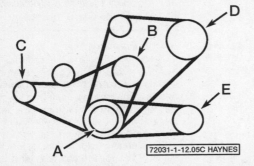

12.5c Drivebelt routing diagram - 3.3L V6 engine (Pathfinder)

A Crankshaft pulley
B Water pump
C Alternator
D Air conditioning compressor
E Power steering pump

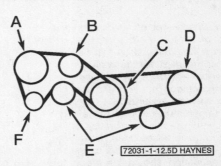

12.5d Drivebelt routing diagram - 3.5L V6 engine (Pathfinder)

A Power steering pump
B Fan pulley
C Crankshaft pulley
D Air conditioning compressor
E Idler pulley
F Alternator

charging cycle. Hydrometers are available inexpensively from auto parts stores - follow the instructions that come with the hydrometer. Consider the battery charged when there's no change in the specific gravity reading for two hours and the electrolyte in the cells is gassing (bubbling) freely. The specific gravity reading from each cell should be very close to the others. If not, the battery probably has a bad cell(s).

17 Some batteries with sealed tops have built-in hydrometers on the top that indicate the state of charge by the color displayed in the hydrometer window. Normally, a bright-colored hydrometer indicates a full charge and a dark hydrometer indicates the battery still needs charging.

18 If the battery has a sealed top and no built-in hydrometer, you can hook up a digital voltmeter across the battery terminals to check the charge. A fully charged battery should read 12.5 volts or higher.

19 Further information on the battery and jump-starting can be found in Chapter 5 and at the front of this manual.

12 Drivebelt check, adjustment and replacement (every 6000 miles or 6 months)

Refer to illustrations 12.2a, 12.2b, 12.4, 12.5a, 12.5b, 12.5c, 12.5d, 12.5e and 12.5f

1 Drivebelts are located at the front of the engine and play an important role in the overall operation of the engine and its components. Due to their function and material make up, the belts are prone to wear and should be periodically inspected. Most models have three belts, while some models only have two.

2 With the engine off, open the hood and use your fingers (and a flashlight, if necessary), to move along the belt checking for cracks and separation of the belt plies. Also check for fraying and glazing, which gives the belt a shiny appearance **(see illustrations)**. Both sides of the belt should be inspected, which means you will have to twist the belt to check the underside.

3 Check the ribs on the underside of multi-ribbed belts. They should all be the same depth, with none of the surface uneven.

4 Belt tension can be checked manually, by pushing on the belt at a distance halfway between two pulleys. Push firmly with your thumb and see how much the belt moves (deflects) **(see illustration)**. As rule of thumb, if the distance from pulley center-to-pulley center is between 7 and 11 inches, the belt should deflect 1/4-inch. If the belt travels between pulleys spaced 12 to 16 inches apart, the belt should deflect 1/4 to 1/2-inch.

5 Refer to the accompanying illustrations for the belt routing diagram for your vehicle **(see illustrations)**. Belts that are routed over an idler pulley are adjusted by loosening the nut in the center of the idler pulley and turning the adjusting bolt **(see illustration)**. Belts

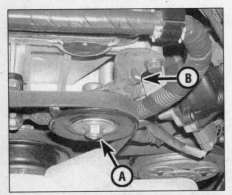

12.5e Loosen the nut (A) in the center of the idler pulley, then turn the adjustment bolt (B) to tension (or loosen) the belt

that do not have an idler pulley are adjusted by moving the driven component (typically the alternator or power steering pump) in its bracket **(see illustration)**.

6 To replace the belts, loosen the idler or component until the belt can be removed from the various pulleys. On multiple-belt applications, outer belts will have to be removed to access inner belts, but as a general rule, all belts should be replaced at the same time anyway.

7 Route the new belt over the various pulleys, then adjust the tension.

13 Underhood hose check and replacement (every 6000 miles or 6 months)

General

Caution: *Replacement of air conditioning hoses must be left to a dealer service department or air conditioning shop that has the equipment to depressurize the system safely and recover the refrigerant. Never remove air conditioning components or hoses until the system has been depressurized.*

1 High temperatures in the engine compartment can cause the deterioration of the rubber and plastic hoses used for engine, accessory and emission systems operation. Periodic inspection should be made for cracks, loose clamps, material hardening and leaks. Information specific to the cooling system hoses can be found in Section 14.

2 Some, but not all, hoses are secured to their fittings with clamps. Where clamps are used, check to be sure they haven't lost their tension, allowing the hose to leak. If clamps aren't used, make sure the hose has not expanded and/or hardened where it slips over the fitting, allowing it to leak.

Vacuum hoses

3 It's quite common for vacuum hoses, especially those in the emissions system, to be color-coded or identified by colored stripes molded into them. Various systems require hoses with different wall thickness, collapse resistance and temperature resistance. When replacing hoses, be sure the new ones are made of the same material.

4 Often the only effective way to check a hose is to remove it completely from the vehicle. If more than one hose is removed, be sure to label the hoses and fittings to ensure correct installation.

5 When checking vacuum hoses, be sure to include any plastic T-fittings in the check. Inspect the fittings for cracks and the hose where it fits over the fitting for distortion, which could cause leakage.

6 A small piece of vacuum hose (1/4-inch inside diameter) can be used as a stethoscope to detect vacuum leaks. Hold one end of the hose to your ear and probe around vacuum hoses and fittings, listening for the "hissing" sound characteristic of a vacuum

12.5f On belts that don't have an idler pulley, adjust the driven component by loosening the lock bolt (A) and turning the adjusting bolt (B)

leak. **Warning:** *When probing with the vacuum hose stethoscope, be very careful not to come into contact with moving engine components such as the drivebelt, cooling fan, etc.*

Fuel hose

Warning: *Gasoline is extremely flammable, so take extra precautions when you work on any part of the fuel system. Don't smoke or allow open flames or bare light bulbs near the work area, and don't work in a garage where a gas-type appliance (such as a water heater or clothes dryer) is present. Since gasoline is carcinogenic, wear fuel-resistant gloves when there's a possibility of being exposed to fuel, and, if you spill any fuel on your skin, rinse it off immediately with soap and water. Mop up any spills immediately and do not store fuel-soaked rags where they could ignite. When you perform any kind of work on the fuel system, wear safety glasses and have a Class B type fire extinguisher on hand. The fuel system is under pressure, so if any lines must be disconnected, the pressure in the system must be relieved first (see Chapter 4 for more information).*

7 Check all rubber fuel lines for deterioration and chafing. Check especially for cracks in areas where the hose bends and just before fittings, such as where a hose attaches to the fuel filter and fuel injection unit.

8 High quality fuel line, specifically designed for high-pressure fuel injection applications, must be used for fuel line replacement. Never, under any circumstances, use regular fuel line, unreinforced vacuum line, clear plastic tubing or water hose for fuel lines.

9 Spring-type (pinch) clamps are commonly used on fuel lines. These clamps often lose their tension over a period of time, and can be "sprung" during removal. Replace all spring-type clamps with screw clamps whenever a hose is replaced.

Metal lines

10 Sections of metal line are routed along the frame, between the fuel tank and the engine. Check carefully to be sure the line has not been bent or crimped and no cracks have started in the line.

11 If a section of metal fuel line must be replaced, only seamless steel tubing should be used, since copper and aluminum tubing don't have the strength necessary to withstand normal engine vibration.

12 Check the metal brake lines where they enter the master cylinder and brake proportioning unit for cracks in the lines or loose fittings. Any sign of brake fluid leakage calls for an immediate and thorough inspection of the brake system.

14 Cooling system check (every 6000 miles or 6 months)

Refer to illustration 14.4

1 Many major engine failures can be attributed to a faulty cooling system. If the vehicle is equipped with an automatic transmission, the cooling system also cools the transmission fluid and thus plays an important role in prolonging transmission life.

2 The cooling system should be checked with the engine cold. Do this before the vehicle is driven for the day or after it has been shut off for at least three hours.

3 Remove the radiator cap by turning it to the counterclockwise until it reaches a stop. If you hear any hissing sounds (indicating there is still pressure in the system), wait until it stops, then depress the cap and continue turning until it can be removed. Thoroughly clean the cap, inside and out, with clean water. Also clean the filler neck on the radiator. All traces of corrosion should be removed. The coolant inside the radiator (and coolant reservoir) should be relatively transparent. If it is rust colored, the system should be drained and refilled (see Section 29). If the coolant level is low, add additional antifreeze/coolant mixture (see Section 4).

4 Carefully check the large upper and lower radiator hoses along with any smaller diameter heater hoses that run from the engine to the firewall. Inspect each hose along its entire length, replacing any hose that is cracked, swollen or shows signs of deterioration. Cracks may become more apparent if the hose is squeezed **(see illustration)**.

5 Make sure all hose connections are tight. A leak in the cooling system will usually show up as white or rust-colored deposits on the areas adjoining the leak. If spring-type clamps are used at the ends of the hoses, it may be wise to replace them with more secure, screw-type clamps.

6 Use compressed air or a soft brush to remove bugs, leaves, etc. from the front of the radiator or air conditioning condenser. Be careful not to damage the delicate cooling

Chapter 1 Tune-up and routine maintenance

Check for a chafed area that could fail prematurely.

Check for a soft area indicating the hose has deteriorated inside.

Overtightening the clamp on a hardened hose will damage the hose and cause a leak.

Check each hose for swelling and oil-soaked ends. Cracks and breaks can be located by squeezing the hose.

14.4 Hoses, like drivebelts, have a habit of failing at the worst possible time - to prevent the inconvenience of a blown radiator or heater hose, inspect them carefully as shown here

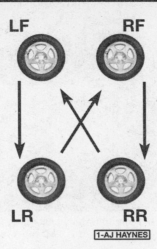

15.2 The recommended four-tire rotation pattern for non-directional radial tires

16.2a Inspect the muffler (A) for signs of deterioration, and all hangers (B)

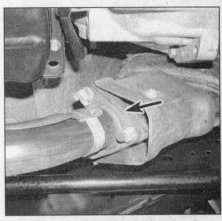

16.2b Inspect all flanged joints for signs of exhaust gas leakage

fins or cut yourself on them.

7 Every other inspection, or at the first indication of cooling system problems, have the cap and system pressure tested. If you don't have a pressure tester, most gas stations and repair shops will do this for a minimal charge.

15 Tire rotation (every 6000 miles or 6 months)

Refer to illustration 15.2

1 The tires should be rotated at the specified intervals and whenever uneven wear is noticed.
2 Tires must be rotated in the recommended pattern **(see illustration)**.
3 Refer to the information in *Jacking and towing* at the front of this manual for the proper procedures to follow when raising the vehicle and changing a tire. If the brakes are to be checked, don't apply the parking brake as stated. Make sure the tires are blocked to prevent the vehicle from rolling as it's raised. Before raising the vehicle, loosen the wheel lug nuts slightly.
4 Preferably, the entire vehicle should be raised at the same time. This can be done on a hoist or by jacking up each corner and then lowering the vehicle onto jackstands placed under the frame rails. Always use four jackstands and make sure the vehicle is safely supported.
5 After rotation, check and adjust the tire pressures as necessary. Tighten the lug nuts to the torque listed in this Chapter's Specifications.

16 Exhaust system check (every 6,000 miles or 6 months)

Refer to illustrations 16.2a and 16.2b

1 With the engine cold (at least three hours after the vehicle has been driven), check the complete exhaust system from the manifold to the end of the tailpipe. Be careful around the catalytic converter, which may be hot even after three hours. The inspection should be done with the vehicle on a hoist to permit unrestricted access. If a hoist isn't available, raise the vehicle and support it securely on jackstands.
2 Check the exhaust pipes and connections for signs of leakage and/or corrosion indicating a potential failure. Make sure that all brackets and hangers are in good condition and tight **(see illustrations)**.
3 Inspect the underside of the body for holes, corrosion, open seams, etc. which may allow exhaust gasses to enter the passenger compartment. Seal all body openings with silicone sealant or body putty.
4 Rattles and other noises can often be traced to the exhaust system, especially the hangers, mounts and heat shields. Try to move the pipes, mufflers and catalytic converter. If the components can come in contact with the body or suspension parts, secure the exhaust system with new brackets and hangers.

17 Differential lubricant level check (every 15,000 miles or 12 months)

Refer to illustrations 17.2a and 17.2b
Note: *4WD vehicles have two differentials - one in the center of each axle. 2WD vehicles have one differential - in the center of the rear axle. On 4WD vehicles, be sure to check the lubricant level in both differentials.*

1 The check/filler plug on the differential(s) is a threaded metal type and can be loosened with a 1/2-inch drive ratchet or breaker bar. If the vehicle is raised to gain access to the plug, be sure to support it safely on jackstands - DO NOT crawl under the vehicle when it's supported only by the jack. Be sure the vehicle is level or the check may not be accurate.
2 Remove the plug from the filler hole in

17.2a Remove the rear axle filler plug (A) to check the differential lubricant level - (B) is the drain plug

17.2b Front differential filler plug (A) and drain plug (B) - 4WD models

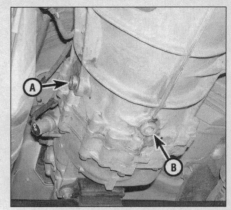

18.2 Manual transmission filler plug (A) and drain plug (B)

the differential housing or cover **(see illustrations)**.
3   The lubricant level should be up to the bottom of the filler hole. If not, use a pump or squeeze bottle to add the recommended lubricant until it just starts to run out of the opening.
4   Install the plug in the filler hole and tighten it to the torque listed in this Chapter's Specifications.

18 Manual transmission lubricant level check (every 15,000 miles or 12 months)

Refer to illustration 18.2
1   The manual transmission has a filler plug which must be removed to check the lubricant level. If the vehicle is raised to gain access to the plug, be sure to support it safely on jackstands - DO NOT crawl under a vehicle that is supported only by a jack! Be sure the vehicle is level or the check may be inaccurate.
2   Using the appropriate wrench, unscrew the plug from the transmission **(see illustration)**.
3   Use your little finger to reach inside the housing to feel the lubricant level. The level should be at or near the bottom of the plug hole. If it isn't, add the recommended lubricant through the plug hole with a syringe or squeeze bottle.
4   Install and tighten the plug. Check for leaks after the first few miles of driving.

19 Transfer case lubricant level check (4WD models) (every 15,000 miles or 12 months)

Refer to illustration 19.1
1   The transfer case lubricant level is checked by removing the upper plug located at the rear of the case **(see illustration)**.
2   After removing the plug, reach inside the hole. The lubricant level should be just at the bottom of the hole. If not, add the appropriate lubricant through the opening.

19.1 Transfer case filler plug (A) and drain plug (B, on other side)

20 Chassis lubrication (every 15,000 miles or 12 months)

Refer to illustration 20.1
1   Refer to *Recommended lubricants and fluids* at the front of this Chapter to obtain the necessary grease, etc. You'll also need a grease gun **(see illustration)**. If a suspension component has no grease fitting in place, this indicates the part is sealed and doesn't require periodic lubrication.
2   Look under the vehicle and locate the grease fittings. Occasionally, plugs may be installed rather than grease fittings. If so, grease fittings will have to be purchased and installed (they're available at auto parts stores).
3   For easier access under the vehicle, raise it with a jack and place jackstands under the frame. Make sure it's safely supported by the stands. If the wheels are to be removed at this interval for tire rotation or brake inspection, loosen the lug nuts slightly while the vehicle is still on the ground.
4   Before beginning, force a little grease out of the nozzle to remove any dirt from the end of the gun. Wipe the nozzle clean with a rag.
5   With the grease gun and plenty of clean rags, crawl under the vehicle and begin lubri-

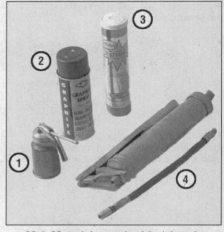

20.1 Materials required for chassis and body lubrication

1   **Engine oil** - Light engine oil in a can like this can be used for door and hood hinges
2   **Graphite spray** - Used to lubricate lock cylinders
3   **Grease** - Grease, in a variety of types and weights, is available for use in a grease gun. Check the Specifications for your requirements
4   **Grease gun** - A common grease gun, shown here with a detachable hose and nozzle, is needed for chassis lubrication. After use, clean it thoroughly!

cating the components.
6   Wipe one of the grease fittings clean and push the nozzle firmly over it. Pump the gun until the component is completely lubricated. On balljoints, stop pumping when the rubber seal is firm to the touch. Do not pump too much grease into the fitting as it could rupture the seal. For all other suspension and steering components, continue pumping grease into the fitting until it oozes out of the joint between the two components. If it escapes around the grease gun nozzle, the fitting is clogged or the nozzle is not completely seated on the fitting. Resecure the

Chapter 1 Tune-up and routine maintenance

gun nozzle to the fitting and try again. If necessary, replace the fitting with a new one.
7   Wipe the excess grease from the components and the grease fitting. Repeat the procedure for the remaining fittings.
8   Clean the fitting and pump grease into the driveline universal joints until the grease can be seen coming out of the contact points. The other U-joints are sealed and do not require lubrication. **Note:** *Most replacement driveshaft U-joints aren't permanently sealed, and are sold with grease fittings. If your U-joints have been replaced, make sure you include them in your routine chassis lubrication.*
9   Also clean and lubricate the parking brake cable guides and levers. **Caution:** *Do not use chassis lubrication on the brake cables themselves. The grease will cause the cable housings to deteriorate.*

21 Fuel system check (every 15,000 miles or 12 months)

Warning: *Gasoline is extremely flammable, so take extra precautions when you work on any part of the fuel system. Don't smoke or allow open flames or bare light bulbs near the work area, and don't work in a garage where a gas-type appliance (such as a water heater or clothes dryer) is present. Since gasoline is carcinogenic, wear fuel-resistant gloves when there's a possibility of being exposed to fuel, and, if you spill any fuel on your skin, rinse it off immediately with soap and water. Mop up any spills immediately and do not store fuel-soaked rags where they could ignite. When you perform any kind of work on the fuel system, wear safety glasses and have a Class B type fire extinguisher on hand. The fuel system is under constant pressure, so, before any lines are disconnected, the fuel system pressure must be relieved (see Chapter 4).*

1   If you smell gasoline while driving or after the vehicle has been sitting in the sun, inspect the fuel system immediately.
2   Remove the fuel filler cap and inspect it for damage and corrosion. The gasket should have an unbroke1n sealing imprint. If the gasket is damaged or corroded, install a new cap.
3   Inspect the fuel feed and return lines for cracks. Make sure that the connections between the fuel lines and the fuel injection system are tight. **Warning:** *Your vehicle is fuel injected, so you must relieve the fuel system pressure before servicing fuel system components. The fuel system pressure relief procedure is described in Chapter 4.*
4   If the fuel injectors are visible, look for signs of fuel leakage (wet spots) around any of the injectors, they may need new O-rings (see Chapter 4).
5   Since some components of the fuel system - the fuel tank and part of the fuel feed and return lines, for example - are underneath the vehicle, they can be inspected more easily with the vehicle raised on a hoist. If that's not possible, raise the vehicle and support it

22.6a Check for signs of fluid leakage at this point on shock absorbers (rear shock shown)

on jackstands.
6   With the vehicle raised and safely supported, inspect the fuel tank and filler neck for punctures, cracks and other damage. The connection between the filler neck and the tank is particularly critical. Sometimes a rubber filler neck will leak because of loose clamps or deteriorated rubber. Inspect all fuel tank mounting brackets and straps to be sure that the tank is securely attached to the vehicle. **Warning:** *Do not, under any circumstances, try to repair a fuel tank (except rubber components). A welding torch or any open flame can easily cause fuel vapors inside the tank to explode.*
7   Carefully check all rubber hoses and metal lines leading away from the fuel tank. Check for loose connections, deteriorated hoses, crimped lines and other damage. Repair or replace damaged sections as necessary (see Chapter 4).
8   The evaporative emissions control system can also be a source of fuel odors. The function of the system is to store fuel vapors from the fuel tank in a charcoal canister until they can be routed to the intake manifold where they mix with incoming air before being burned in the combustion chambers.
9   The most common symptom of a faulty evaporative emissions system is a strong odor of fuel near the charcoal canister, which is mounted in the engine compartment on 1996 Pathfinders, or under the rear of the vehicle on all other models. If a fuel odor has been detected, and you have already checked the areas described above, check the charcoal canister and the hoses connected to it (see Section 27).

22 Suspension, steering and driveaxle boot check (every 15,000 miles or 12 months)

Note: *The steering linkage and suspension components should be checked periodically. Worn or damaged suspension and steering linkage components can result in excessive and abnormal tire wear, poor ride quality and*

22.6b On Pathfinder models, check the front struts for fluid leakage where the piston rod enters the strut body

vehicle handling and reduced fuel economy. For detailed illustrations of the steering and suspension components, refer to Chapter 10.

Shock absorber/strut check

Refer to illustrations 22.6a and 22.6b
Note: *Pathfinder models are equipped with MacPherson struts at the front end and shock absorbers at the rear; all other models use shock absorbers front and rear.*

1   Park the vehicle on level ground, turn the engine off and set the parking brake. Check the tire pressures.
2   Push down at one corner of the vehicle, then release it while noting the movement of the body. It should stop moving and come to rest in a level position within one or two bounces.
3   If the vehicle continues to move up-and-down or if it fails to return to its original position, a worn or weak shock absorber or strut is probably the reason.
4   Repeat the above check at each of the three remaining corners of the vehicle.
5   Raise the vehicle and support it securely on jackstands.
6   Check the shock absorbers and/or struts for evidence of fluid leakage **(see illustrations)**. A light film of fluid is no cause for concern. Make sure that any fluid noted is from the shocks or struts and not from some other source. If leakage is noted, replace the shocks or struts as a set.
7   Check the shocks/struts to be sure they are securely mounted and undamaged. Check the upper mounts for damage and wear. If damage or wear is noted, replace the shocks or struts as a set (front or rear).
8   If the shocks/struts must be replaced, refer to Chapter 10 for the procedure.

Steering and suspension check

Refer to illustrations 22.9a, 22.9b, 22.9c, 22.9d and 22.11
9   Visually inspect the steering and suspension components (front and rear) for damage and distortion. Look for damaged seals,

Chapter 1 Tune-up and routine maintenance

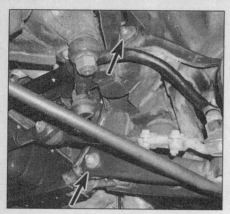

22.9a Check the bushings at the inner ends of the control arms for deterioration

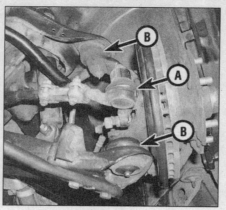

22.9b Inspect the tie-rod ends (A) and the balljoints (B)

22.9c On Pathfinder models, inspect the steering gear boots for signs of cracking or fluid leakage (if fluid leakage is noted, the rack seals are faulty)

boots and bushings and leaks of any kind. Examine the bushings where the control arms meet the chassis **(see illustrations)**.

10   Clean the lower end of the steering knuckle. Have an assistant grasp the lower edge of the tire and move the wheel in-and-out while you look for movement at the steering knuckle-to-control arm balljoint. If there is any movement the suspension balljoint(s) must be replaced.

11   Grasp each front tire at the front and rear edges, push in at the front, pull out at the rear and feel for play in the steering system components. If any freeplay is noted, check the idler arm and the tie-rod ends for looseness **(see illustration)**.

12   Additional steering and suspension system information and illustrations can be found in Chapter 10.

Driveaxle boot check (4WD models)

Refer to illustration 22.14

13   The driveaxle boots are very important because they prevent dirt, water and foreign material from entering and damaging the constant velocity (CV) joints. Oil and grease can cause the boot material to deteriorate prematurely, so it's a good idea to wash the boots with soap and water. Because it constantly pivots back and forth following the steering action of the front hub, the outer CV boot wears out sooner and should be inspected regularly.

14   Inspect the boots for tears and cracks as well as loose clamps **(see illustration)**. If there is any evidence of cracks or leaking lubricant, they must be replaced as described in Chapter 8.

23 Brake system check (every 15,000 miles or 12 months)

Warning: *The dust created by the brake system is harmful to your health. Never blow it out with compressed air and don't inhale any of it. An approved filtering mask should be worn when working on the brakes. Do not, under any circumstances, use petroleum-based solvents to clean brake parts. Use brake system cleaner only! Try to use non-asbestos replacement parts whenever possible.*

Note: *For detailed photographs of the brake system, refer to Chapter 9.*

1   In addition to the specified intervals, the brakes should be inspected every time the

22.9d On Frontier and Xterra models, check the steering gear for signs of lubricant leakage (A), and make sure the Pitman arm nut (B) is tight

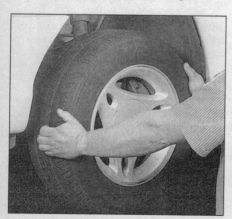

22.11 With the steering wheel in the locked position and the vehicle raised, grasp the front tire as shown and try to move it back-and-forth - if any play is noted, check the steering gear mounts and tie-rod ends for looseness

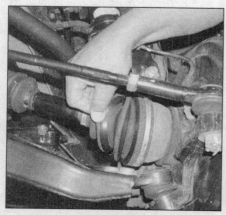

22.14 Inspect the inner and outer driveaxle boots on 4WD models for loose clamps, cracks or signs of leaking lubricant

23.7 With the wheel off, check the thickness of the inner brake pads through the inspection hole

Chapter 1 Tune-up and routine maintenance   1-25

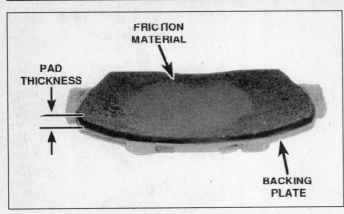

23.9 If a more precise measurement of pad thickness is necessary, remove the pads and measure the remaining friction material

23.11 Check along the brake hoses and at each fitting for seepage, deterioration and cracks

wheels are removed or whenever a defect is suspected.
2   Any of the following symptoms could indicate a potential brake system defect: The vehicle pulls to one side when the brake pedal is depressed; the brakes make squealing or dragging noises when applied; brake pedal travel is excessive; the pedal pulsates; or brake fluid leaks, usually onto the inside of the tire or wheel.
3   Loosen the wheel lug nuts.
4   Raise the vehicle and place it securely on jackstands.
5   Remove the wheels (see *Jacking and towing* at the front of this book, or your owner's manual, if necessary).

Disc brakes

Refer to illustrations 23.7, 23.9 and 23.11

6   There are two pads (an outer and an inner) in each caliper. The pads are visible with the wheels removed.
7   Check the pad thickness by looking at each end of the caliper and through the inspection window in the caliper body **(see illustration)**. If the lining material is less than the thickness listed in this Chapter's Specifications, replace the pads. **Note:** *Keep in mind that the lining material is riveted or bonded to a metal backing plate and the metal portion is not included in this measurement.*
8   If it is difficult to determine the exact thickness of the remaining pad material by the above method, or if you are at all concerned about the condition of the pads, remove the pads for further inspection (refer to Chapter 9).
9   Once the pads are removed from the calipers, clean them with brake cleaner and re-measure them with a ruler or a vernier caliper **(see illustration)**.
10 Measure the disc thickness with a micrometer to make sure that it still has service life remaining. If any disc is thinner than the specified minimum thickness, replace it (see Chapter 9). Even if the disc has service life remaining, check its condition. Look for scoring, gouging and burned spots. If these

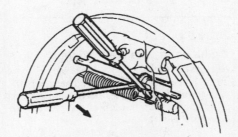

23.15 If the drum binds on the shoes, remove the adjusting hole plug, push the adjusting lever off the star wheel with one screwdriver and turn the star wheel with another screwdriver

conditions exist, remove the disc and have it resurfaced (see Chapter 9).
11 Before installing the wheels, check all brake lines and hoses for damage, wear, deformation, cracks, corrosion, leakage, bends and twists, particularly in the vicinity of the rubber hoses at the calipers **(see illustration)**. Check the clamps for tightness and the connections for leakage. Make sure that all hoses and lines are clear of sharp edges, moving parts and the exhaust system. If any of the above conditions are noted, repair, reroute or replace the lines and/or fittings as necessary (see Chapter 9).

Drum brakes

Refer to illustrations 23.15, 23.17 and 23.19

12 Make sure the parking brake is off, then tap on the outside of the drum with a rubber mallet to loosen it.
13 Remove the brake drums. **Note:** *If the drum won't come off, it may be rusted to the axle at the center hole. Apply a little penetrating oil, allow it to soak in, then try again.*
14 If the drum still cannot be pulled off, the parking brake lever will have to be lifted slightly off its stop. This is done by first removing the small plug from the backing plate.

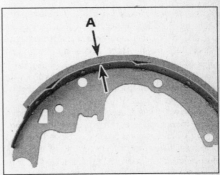

23.17 If the lining is bonded to the brake shoe, measure the lining thickness from the outer surface to the metal shoe, as shown here; if the lining is riveted to the shoe, measure from the lining outer surface to the rivet head

15 With the plug removed, insert a thin screwdriver and lift the adjusting lever off the star wheel, then use an adjusting tool or screwdriver to back off the star wheel several turns **(see illustration)**. This will move the brake shoes away from the drum. If the drum still won't pull off, tap around its inner circumference with a soft-face hammer.
16 With the drums removed, carefully clean the brake assembly with brake system cleaner. **Warning:** *Don't blow the dust out with compressed air and don't inhale any of it (it may contain asbestos, which is harmful to your health).*
17 Note the thickness of the lining material on both front and rear brake shoes. If the material has worn away to within 1/16-inch of the recessed rivets or the metal backing on bonded type shoes, the shoes should be replaced **(see illustration)**. The shoes should also be replaced if they're cracked, glazed (shiny areas), or covered with brake fluid.
18 Make sure all the brake assembly springs are connected and in good condition.
19 Check the brake components for signs of fluid leakage. With your finger or a small screwdriver, carefully pry back the rubber cups on the wheel cylinder located at the top

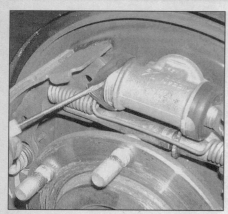

23.19 To check for wheel cylinder leakage, use a small screwdriver to pry the boot away from the cylinder

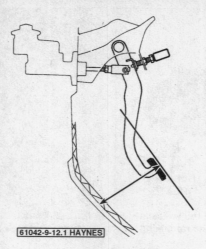

23.30 With the brake pedal fully released, measure the distance from the top of the pad to the floor

23.32 To adjust brake pedal released height, loosen the locknut in front of the brake booster clevis and turn the input rod until free height is correct (this procedure is also used to adjust brake pedal freeplay)

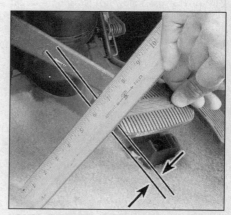

23.36 To measure brake pedal freeplay, press down lightly on the brake pedal and measure the distance that it moves freely before resistance is felt

of the brake shoes (see illustration). Any leakage here is an indication that the wheel cylinders should be replaced immediately (see Chapter 9). Also, check all hoses and connections for signs of leakage.

20 Clean the inside of the drum with brake system cleaner. Again, be careful not to breathe the dust.

21 Check the inside of the drum for cracks, score marks, deep scratches and "hard spots" which will appear as small discolored areas. If imperfections cannot be removed with fine emery cloth, the drum must be taken to an automotive machine shop for resurfacing.

22 Repeat the procedure for the remaining wheel. If the inspection reveals that all parts are in good condition, reinstall the brake drums, install the wheels and lower the vehicle to the ground.

Brake booster check

23 Sit in the driver's seat and perform the following sequence of tests.

24 With the brake fully depressed, start the engine - the pedal should move down a little when the engine starts.

25 With the engine running, depress the brake pedal several times - the travel distance should not change.

26 Depress the brake, stop the engine and hold the pedal in for about 30 seconds - the pedal should neither sink nor rise.

27 Restart the engine, run it for about a minute and turn it off. Then firmly depress the brake several times - the pedal travel should decrease with each application.

28 If your brakes do not operate as described, the brake booster has failed. Refer to Chapter 9 for the replacement procedure.

Parking brake

29 One method of checking the parking brake is to park the vehicle on a steep hill with the parking brake set and the transmission in Neutral (be sure to stay in the vehicle during this check!). If the parking brake cannot prevent the vehicle from rolling, it's in need of adjustment (see Chapter 9).

Brake pedal

Brake pedal released height

Refer to illustrations 23.30 and 23.32

30 Peel back the carpet and insulator pad. With the brake pedal fully released, measure the distance from the top of the pad to the floor (see illustration).

31 If the height is not as listed in the Specifications Section at the beginning of this Chapter it must be adjusted.

32 Loosen the locknut just in front of the power brake booster clevis (see illustration).

33 Turn the booster input rod until the pedal height is correct.

34 Tighten the locknut.

35 After adjusting the pedal height, check the freeplay. It may also be necessary to adjust the brake light switch (see Chapter 9).

Brake pedal freeplay

Refer to illustration 23.36

36 Press down lightly on the brake pedal and measure the distance that it moves freely before resistance is felt (see illustration). The freeplay should be within the specified limits. If it isn't, it must be adjusted.

37 Loosen the locknut for the brake booster clevis (see illustration 23.32).

38 Turn the booster input rod until the pedal freeplay is correct.

39 Tighten the locknut.

Brake pedal depressed height

40 After checking and, if necessary, adjusting the pedal released height and freeplay, the pedal depressed height must be checked.

41 With the engine running, press the brake pedal fully and measure the pedal pad-to-floor distance.

42 If the minimum depressed height is below that listed in the Specifications Section listed at the beginning of this Chapter, check the brake system for leaks or other damage.

24 Brake fluid change (every 30,000 miles or 24 months)

Warning: *Brake fluid can harm your eyes and damage painted surfaces, so use extreme caution when handling or pouring it. Do not use brake fluid that has been standing open or is more than one year old. Brake fluid absorbs moisture from the air. Excess moisture can cause a dangerous loss of braking effectiveness.*

1 At the specified intervals, the brake fluid should be drained and replaced. Since the brake fluid may drip or splash when pouring it, place plenty of rags around the master cylinder to protect any surrounding painted surfaces.

2 Before beginning work, purchase the specified brake fluid (see *Recommended lubricants and fluids* at the beginning of this Chapter).

3 Remove the cap from the master cylinder reservoir.

Chapter 1 Tune-up and routine maintenance

25.1a To remove the air filter element on four-cylinder models, unscrew these two wing nuts . . .

25.1b . . . then remove the cover and take the element out

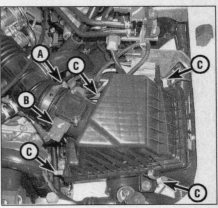

25.1c On models with a 3.3L V6 engine, loosen the intake duct hose clamp (A), unplug the Mass Airflow sensor (B), release the clips (C) . . .

25.1d . . . then lift the cover off and take out the filter element

25.1e On models with a 3.5L V6 engine, release the clips (arrows), pull the filter carrier up and remove the filter element from the carrier

4 Using a hand suction pump or similar device, withdraw the fluid from the master cylinder reservoir.
5 Add new fluid to the master cylinder until it rises to the line indicated on the reservoir.
6 Bleed the brake system as described in Chapter 9 at all four brakes until new and uncontaminated fluid is expelled from the bleeder screw. Be sure to maintain the fluid level in the master cylinder as you perform the bleeding process. If you allow the master cylinder to run dry, air will enter the system.
7 Refill the master cylinder with fluid and check the operation of the brakes. The pedal should feel solid when depressed, with no sponginess. **Warning:** *Do not operate the vehicle if you are in doubt about the effectiveness of the brake system.*

25 Air filter replacement (every 30,000 miles or 24 months)

Refer to illustrations 25.1a, 25.1b, 25.1c, 25.1d and 25.1e

1 The air filter on four-cylinder models is located in a housing in the center of the engine compartment. On V6 models it's located inside a housing at the left side of the engine compartment. To remove the air filter on a four-cylinder model, unscrew the two wing nuts and remove the top of the housing, then remove the top of the housing and take out the filter element **(see illustrations)**. On 3.3L V6 models, release the spring clips that keep the two halves of the air cleaner housing together, then lift the cover up and remove the air filter element **(see illustrations)**. On 3.5L V6 models, release the clips, pull the filter carrier out of the housing and remove the filter element from the carrier **(see illustration)**.
2 Inspect the outer surface of the filter element. If it is dirty, replace it. If it is only moderately dusty, it can be reused by blowing it clean from the back to the front surface with compressed air. Because it is a pleated paper type filter, it cannot be washed or oiled. If it cannot be cleaned satisfactorily with compressed air, discard and replace it. While the cover is off, be careful not to drop anything down into the housing. **Caution:** *Never drive the vehicle with the air cleaner removed. Excessive engine wear could result and backfiring could even cause a fire under the hood.*

3 Wipe out the inside of the air cleaner housing.
4 Place the new filter into the air cleaner housing, making sure it seats properly.
5 Installation of the cover is the reverse of removal.

26 Fuel filter replacement (every 30,000 miles or 24 months)

Refer to illustrations 26.3a and 26.3b

Warning: *Gasoline is extremely flammable, so take extra precautions when you work on any part of the fuel system. Don't smoke or allow open flames or bare light bulbs near the work area, and don't work in a garage where a gas-type appliance (such as a water heater or clothes dryer) is present. Since gasoline is carcinogenic, wear fuel-resistant gloves when there's a possibility of being exposed to fuel, and, if you spill any fuel on your skin, rinse it off immediately with soap and water. Mop up any spills immediately and do not store fuel-soaked rags where they could ignite. The fuel system is under constant pressure, so, if any fuel lines are to be disconnected, the fuel pressure in the system must be relieved first (see Chapter 4 for more information). When you perform any kind of work on the fuel system, wear safety glasses and have a Class B type fire extinguisher on hand.*

1 Relieve the fuel system pressure (see Chapter 4), then disconnect the cable from the negative terminal of the battery.
2 Raise the vehicle and support it securely on jackstands. **Note:** *If you're working on a Frontier or Xterra, raise the front of the vehicle. If you're working on a Pathfinder, raise the rear of the vehicle (be sure block the front wheels to prevent the vehicle from rolling off the stands).*
3 On Frontier models the fuel filter is mounted to the right frame rail, near the fuel tank. On Xterra models it's located on the crossmember just in front of the fuel tank

26.3a Fuel filter location - Xterra models

26.3b Fuel filter location - Pathfinder models

(see illustration). On Pathfinders it's under the vehicle, near the rear differential **(see illustration)**.
4 Clean off any dirt surrounding the fuel inlet and outlet lines.
5 Loosen the clamps on the inlet and outlet hoses, then detach the hoses from the filter. *Note: Have spare rags or a small container to catch or wipe up the gasoline that will spill from the filter.*
6 Unscrew the fuel filter mounting bracket bolt and remove the fuel filter from the bracket.
7 Installation is the reverse of removal. Be sure the filter is installed facing the proper direction. *Note: If spring-type clamps are used, it's a good idea to replace them with screw-type clamps.*

27 Evaporative emissions control system check (every 30,000 miles or 24 months)

Refer to illustrations 27.2a and 27.2b
1 The function of the evaporative emissions control system is to draw fuel vapors from the gas tank and fuel system, store them in a charcoal canister and route them to the intake manifold during normal engine operation.
2 The most common symptom of a fault in the evaporative emissions system is a strong fuel odor. If a fuel odor is detected, inspect the charcoal canister, located in the engine compartment on 1996 Pathfinders, or under the vehicle on all other vehicles **(see illustrations)**. On Frontier and Xterra models, remove the spare tire for access. Check the canister and all hoses for damage and deterioration.
3 The evaporative emissions control system is explained in more detail in Chapter 6.

28 Spark plug replacement (see maintenance schedule for replacement interval)

Refer to illustrations 28.1, 28.4a, 28.4b, 28.5, 28.7, 28.8 and 28.9
1 In most cases, the tools necessary for spark plug replacement include a spark plug socket which fits onto a ratchet (spark plug sockets are padded inside to prevent damage to the porcelain insulators on the new plugs), various extensions and a gap gauge to check and adjust the gaps on the new plugs **(see illustration)**. A special plug wire removal tool is available for separating the wire boots from the spark plugs, but it isn't absolutely necessary. A torque wrench should be used to tighten the new plugs.
2 The best approach when replacing the spark plugs is to purchase the new ones in advance, adjust them to the proper gap and replace them one at a time. When buying the new spark plugs, be sure to obtain the correct plug type for your particular engine. This information can be found in the vehicle owner's manual and the Specifications at the front of this Chapter.
3 Allow the engine to cool completely before attempting to remove any of the plugs. While you're waiting for the engine to cool, check the new plugs for defects and adjust the gaps. *Caution: The manufacturer recommends against checking the gap on platinum-tipped spark plugs (the platinum coating could be scraped off).*
4 The gap is checked by inserting the proper-thickness gauge between the electrodes at the tip of the plug **(see illustration)**. The gap between the electrodes should be the same as the one specified in this Chapter's Specifications. The wire should just slide between the electrodes with a slight amount of drag. If the gap is incorrect, use the adjuster on the gauge body to bend the curved side electrode slightly until the proper gap is obtained **(see illustration)**. If the side electrode is not exactly over the center electrode, bend it with the adjuster until it is. Check for cracks in the porcelain insulator (if any are found, the plug should not be used).

27.2a Check the charcoal canister for damage and the hose connections for cracks and deterioration (Frontier model shown, Xterra similar

27.2b On 1997 and later Pathfinder models the charcoal canister is located behind the left rear wheel

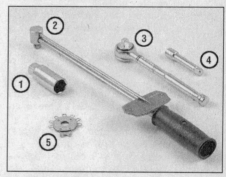

28.1 Tools required for changing spark plugs

1 **Spark plug socket** - *This will have special padding inside to protect the spark plug's porcelain insulator*
2 **Torque wrench** - *Although not mandatory, using this tool is the best way to ensure the plugs are tightened properly*
3 **Ratchet** - *Standard hand tool to fit the spark plug socket*
4 **Extension** - *Depending on model and accessories, you may need special extensions and universal joints to reach one or more of the plugs*
5 **Spark plug gap gauge** - *This gauge for checking the gap comes in a variety of styles. Make sure the gap for your engine is included*

Chapter 1 Tune-up and routine maintenance

28.4a Spark plug manufacturers recommend using a wire-type gauge when checking the gap - if the wire does not slide between the electrodes with a slight drag, adjustment is required

28.4b To change the gap, bend the side electrode only, as indicated by the arrows, and be very careful not to crack or chip the porcelain insulator surrounding the center electrode

28.5 Twist the spark plug boot back-and-forth while pulling up - don't pull on the wire itself

5 With the engine cool, remove the spark plug wire from one spark plug. Pull only on the boot at the end of the wire - do not pull on the wire **(see illustration)**. **Note:** *On 3.5L V6 engines, each individual ignition coil must be removed (see Chapter 5).*

6 If compressed air is available, use it to blow any dirt or foreign material away from the spark plug hole. The idea here is to eliminate the possibility of debris falling into the cylinder as the spark plug is removed.

7 Place the spark plug socket over the plug and remove it from the engine by turning it in a counterclockwise direction **(see illustration)**.

8 Compare the spark plug with the chart on the inside back cover of this manual to get an indication of the general running condition of the engine. Before installing the new plugs, apply a thin coat of anti-seize compound to the threads **(see illustration)**.

9 Thread one of the new plugs into the hole until you can no longer turn it with your fingers, then tighten it with a torque wrench (if available) or the ratchet. It is a good idea to slip a short length of rubber hose over the end of the plug to use as a tool to thread it into place **(see illustration)**. The hose will grip the plug well enough to turn it, but will start to slip if the plug begins to cross-thread in the hole - this will prevent damaged threads and the accompanying repair costs.

10 Attach the plug wire (or ignition coil) to the new spark plug, again using a twisting motion on the boot until it's seated on the spark plug.

11 Repeat the procedure for the remaining spark plugs, replacing them one at a time to prevent mixing up the spark plug wires.

29 Cooling system servicing (draining, flushing and refilling) (every 30000 miles or 24 months)

Warning: *Do not allow antifreeze to come in contact with your skin or painted surfaces of the vehicle. Flush contacted areas immediately with plenty of water. Do not store new coolant or leave old coolant lying around where it is easily accessible to children and pets, because they are attracted by its sweet smell. Ingestion of even a small amount can be fatal. Wipe up the garage floor and drip pan coolant spills immediately. Keep antifreeze containers covered and repair leaks in your cooling system immediately. Antifreeze is flammable - be sure to read the precautions on the container.*

Note: *Non-toxic coolant is available at local auto parts stores. Although the coolant is non-toxic when fresh, proper disposal is still required.*

Draining

Refer to illustrations 29.3a, 29.3b, 29.4a, 29.4b, 29.4c, 29.5a, 29.5b and 29.5c

1 Periodically, the cooling system should be drained, flushed and refilled to replenish the antifreeze mixture and prevent formation of rust and corrosion, which can impair the performance of the cooling system and cause engine damage. When the cooling system is serviced, all hoses and the radiator cap should be checked and replaced if necessary.

28.7 Use a socket and extension to unscrew the spark plugs - various length extensions and perhaps a flex-joint may be required to reach some plugs on V6 models

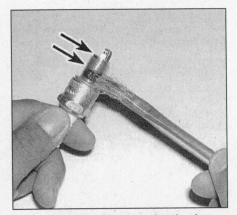

28.8 Apply a thin coat of anti-seize compound to the spark plug threads, being careful not to get any near the lower threads (arrows)

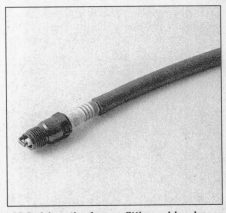

28.9 A length of snug-fitting rubber hose will save time and prevent damaged threads when installing the spark plugs

29.3a Pry out these plastic fasteners that secure this cover under the radiator, then fold the cover back for access to the radiator drain plug

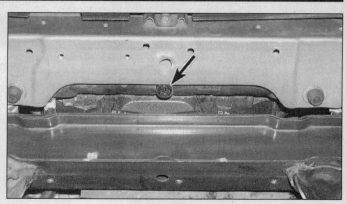

29.3b Loosen the radiator drain plug with a Phillips screwdriver

29.4a The engine block drain plug on four-cylinder engines is on the left side of the block

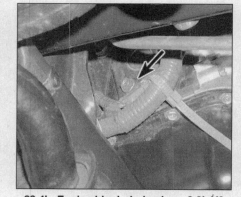

29.4b Engine block drain plug - 3.3L V6 engine (there's another one like it on the other side of the block)

29.4c On the 3.5L V6 engine there's a block drain plug between the air conditioning compressor and the crankshaft pulley (arrow), and one just below the right exhaust manifold

2 Apply the parking brake and block the wheels. **Warning:** *If the vehicle has just been driven, wait several hours to allow the engine to cool down before beginning this procedure.* Turn the ignition key to the On position, then set the heater control to the maximum heat position. Wait at least ten seconds, then turn the ignition Off.

3 Move a large container under the radiator drain to catch the coolant. Remove the plastic fasteners from the cover under the radiator (if equipped), then tuck the cover back over the frame for access to the drain plug, which is located in the radiator's lower tank **(see illustrations)**. Unscrew the drain fitting with a Phillips screwdriver, then remove the radiator cap.

4 After coolant stops flowing out of the radiator, move the container under the engine block drain plug(s) - on four-cylinder engines the plug is on the driver's side of the block **(see illustration)**; on the 3.3L V6 engine there's one on each side of the block **(see illustration)**; on the 3.5L V6 there's one on the front of the engine block, near the water pump cover **(see illustration)**, and one on the right (passenger's) side of the engine block, just below the exhaust manifold. Remove the plugs and allow the coolant in the block to drain. **Note:** *Frequently, the coolant will not drain from the block after the plug is removed. This is due to a rust layer that has built up behind the plug. Insert a Phillips screwdriver into the hole to break the rust barrier.*

5 Also loosen the air relief plug(s) to aid in coolant draining **(see illustrations)**. **Note:** *On 3.5L V6 engines there are two relief plugs;*

29.5a Air relief plug location - four-cylinder engine

29.5b Air relief plug location - 3.3L V6 engine

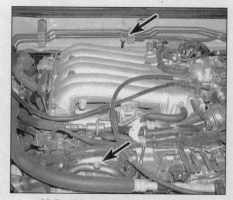

29.5c Air relief plug locations - 3.5L V6 engine

Chapter 1 Tune-up and routine maintenance

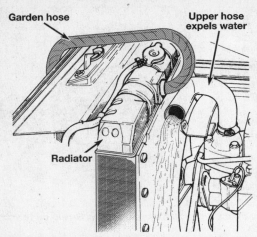

29.10 With the thermostat removed, disconnect the upper radiator hose and flush the radiator and engine block with a garden hose (four-cylinder and 3.3L V6 engines)

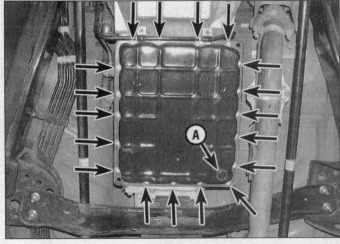

30.5 Transmission pan details - A is the drain plug (the other arrows indicate the pan mounting bolts)

one is threaded into the front water pipe and the other is actually a rubber cap clamped onto a pipe connected to the heater pipe at the rear of the engine.

6   While the coolant is draining, check the condition of the radiator hoses, heater hoses and clamps (refer to Section 14 if necessary). Remove the coolant reservoir (see Chapter 3) and pour the coolant into the drain pan. Rinse out the reservoir with clean water.

7   Once the coolant has drained completely, replace any damaged clamps or hoses. Apply thread sealant to the drain plugs, reinstall them and tighten them securely. Also, temporarily tighten the air relief plug(s).

Flushing

Note 1: *In severe cases of contamination or clogging of the radiator, remove the radiator (see Chapter 3) and have a radiator repair facility clean and repair it if necessary.*
Note 2: *Many deposits can be removed by the chemical action of a cleaner available at auto parts stores. Follow the procedure outlined in the manufacturer's instructions. However, when the coolant is regularly drained and the system refilled with the correct antifreeze/water mixture, there should be no need to use chemical cleaners or descalers.*

8   Make sure your heating system controls are still set to Hot, so that the heater core will be flushed at the same time as the rest of the cooling system.

Four-cylinder and 3.3L V6 engines

Refer to illustration 29.10

9   Once the system is completely drained, remove the thermostat from the engine (see Chapter 3). Then reinstall the thermostat housing without the thermostat. This will allow the system to be thoroughly flushed.
10   Disconnect the upper radiator hose, then place a garden hose in the upper radiator inlet and flush the system until the water runs clear at the upper radiator hose **(see illustration).**

3.5L V6 engine

11   Open the air relief plugs **(see illustration 29.5c)** and fill the cooling system with clean water until water flows from the relief holes. Tighten the relief plugs and re-check to make sure the radiator and coolant reservoir are full of clean water.
12   Start the engine and allow it to reach normal operating temperature, then rev up the engine a few times.
13   Turn the engine off and allow it to cool completely, then drain the system as described earlier.
14   Repeat Steps 11 through 13 until the water being drained is free of contaminants.

Refilling

15   To refill the system, install the thermostat (if removed), reconnect any radiator hoses and install the reservoir and the overflow hose.
16   Place the heater temperature control in the maximum heat position.
17   Loosen the air relief plug(s) **(see illustrations 29.5a, 29.5b and 29.5c).**
18   Make sure to use the proper coolant listed in this Chapter's Specifications. Slowly fill the radiator with the recommended mixture of antifreeze and water until coolant begins flowing from the air relief hole(s).
19   Tighten the air relief plug(s) securely and continue to fill the system until coolant reaches the base of the radiator filler neck. Then add coolant to the reservoir until it reaches the FULL COLD mark. Wait five minutes and recheck the coolant level in the radiator, adding if necessary.
20   Leave the radiator cap off and run the engine in a well-ventilated area until the thermostat opens (coolant will begin flowing through the radiator and the upper radiator hose will become hot).
21   Rev the engine to approximately 2500 rpm for ten seconds then let it idle; do this a few times.
22   Turn the engine off and let it cool. Add more coolant mixture to bring the level back up to the base of the filler neck.
23   Squeeze the upper radiator hose to expel air, then add more coolant mixture if necessary. Reinstall the radiator cap. Add coolant to the reservoir, if necessary.
24   Start the engine, allow it to reach normal operating temperature and check for leaks.

30 Automatic transmission fluid and filter change (every 30,000 miles or 24 months)

Refer to illustrations 30.5, 30.8, 30.9, 30.11 and 30.12

1   At the specified intervals, the transmission fluid should be drained and replaced. Since the fluid will remain hot long after driving, perform this procedure only after the engine has cooled down completely.
2   Before beginning work, purchase the specified transmission fluid (see *Recommended lubricants and fluids* at the front of this Chapter) and a new filter and pan gasket.
3   Other tools necessary for this job include a floor jack, jackstands to support the vehicle in a raised position, a drain pan capable of holding at least eight quarts, newspapers and clean rags.
4   Raise the vehicle and support it securely on jackstands.
5   Place the drain pan underneath the transmission pan. Remove the drain plug and allow the fluid to drain, then reinsert the plug and tighten it securely **(see illustration).** Measure the amount of fluid drained (the same amount will be added to the transmission later).
6   Remove the transmission pan mounting bolts **(see illustration 30.5),** then remove the pan, prying gently if necessary. **Warning:** *There is still some transmission fluid in the pan.*
7   Carefully clean the gasket surface of the transmission and pan to remove all traces of the old gasket and sealant.

1-32 Chapter 1 Tune-up and routine maintenance

30.8 After cleaning the pan and magnet, reinstall the magnet in its proper position

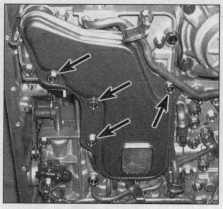

30.9 Remove these bolts and detach the strainer from the valve body

30.11 Install a new O-ring on the strainer

8   Clean the pan with solvent and dry it. **Note:** *Some models are equipped with magnets in the transmission pan to catch metal debris* **(see illustration)**. *Clean the magnet thoroughly. A small amount of metal material is normal at the magnet. If there is considerable debris, consult a dealer or transmission specialist.*
9   Remove the strainer from the valve body inside the transmission **(see illustration)**.
10 Clean the strainer with solvent and dry it with compressed air, if available. If compressed air isn't available, pour some clean automatic transmission fluid through it.
11 Install a new O-ring on the strainer **(see illustration)**.
12 Make sure the gasket surface on the transmission pan is clean, then install a new gasket on the pan **(see illustration)**. Put the pan in place against the transmission and install the bolts. Tighten each bolt a little at a time to the torque listed in this Chapter's Specifications.
13 Lower the vehicle and add the specified type of automatic transmission fluid (the same amount that was drained in Step 5), through the filler tube (see Section 7).
14 With the transmission in Park and the parking brake set, run the engine at a fast idle, but don't race it.
15 Move the gear selector through each

range and back to Park, then let the engine idle for a few minutes. Check the fluid level. It may be low. Add enough fluid to bring the level to the proper mark on the dipstick. Be careful not to overfill.
16 Check under the vehicle for leaks during the first few trips. Check the fluid level again when the transmission is hot (see Section 7).

31 Front hub and wheel bearing check, repack and adjustment (every 30,000 miles or 24 months)

Check and repack

Refer to illustrations 31.1, 31.8a, 31.8b, 31.8c, 31.9a, 31.9b, 31.10, 31.12, 31.16 and 31.20

1   In most cases the front wheel bearings will not need servicing until the brake pads are changed. However, the bearings should be checked whenever the front of the vehicle is raised for any reason. Several items, including a torque wrench and special grease, are required for this procedure **(see illustration)**.
2   With the vehicle securely supported on jackstands, spin each wheel and check for noise, rolling resistance and endplay (wobbling).
3   Grasp the top of each tire with one hand and the bottom with the other. Move the wheel in and out on the spindle. If there's any noticeable movement, the bearings should be checked and then repacked with grease, or replaced if necessary.
4   Remove the wheel.
5   Remove the disc brake caliper (see Chapter 9) and hang it out of the way on a piece of wire. Also remove the caliper mounting bracket.
6   If the vehicle is equipped with a four-wheel Anti-lock Brake System (ABS), remove the front wheel speed sensors.
7   On 4WD models, remove the free-running hub, if so equipped (see Chapter 8). If the vehicle is not equipped with free-running hubs, remove the hub cap, snap-ring and the

drive flange (see Chapter 8, Section 21).
8   On 2WD Frontier models, remove the hub cap **(see illustration)**. Remove the cotter pin from the spindle using a pair of needle nose pliers or cutting pliers to grip the slip-

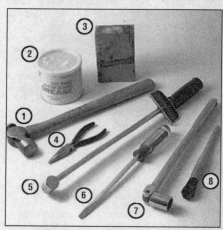

31.1 Tools and materials needed for front wheel bearing maintenance

1 **Hammer** - *A common hammer will do just fine*
2 **Grease** - *High-temperature grease that is formulated specially for front wheel bearings should be used*
3 **Wood block** - *If you have a scrap piece of 2x4, it can be used to drive the new seal into the hub*
4 **Needle-nose pliers** - *Used to straighten and remove the cotter pin in the spindle*
5 **Torque wrench** - *This is used to ensure all play is taken out of the bearing by applying a slight preload before final adjustment*
6 **Screwdriver** - *Used to remove the seal from the hub (a long screwdriver is preferred)*
7 **Socket/breaker bar** - *Needed to loosen the nut on the spindle if it's extremely tight*
8 **Brush** - *Together with some clean solvent, this will be used to remove old grease from the hub and spindle*

30.12 Place the gasket on the transmission pan and install the corner bolts to hold the gasket in place

Chapter 1 Tune-up and routine maintenance

31.8a On 2WD Frontier models, remove the hub cap from the hub . . .

31.8b . . . then remove the cotter pin and nut lock . . .

31.8c . . . and the spindle nut and thrust washer

31.9a On Pathfinders, Xterras and 4WD Frontier models, remove the screws securing the lockwasher . . .

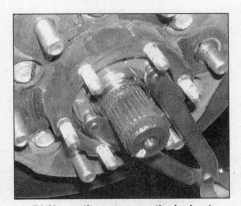

31.9b . . . then unscrew the locknut with a tool like the one shown here, or a special socket

31.10 Pull the hub assembly out slightly, then push it back onto the spindle to disengage the outer wheel bearing

pery surface **(see illustration)**. Remove the nut lock, spindle nut and thrust washer from the spindle **(see illustration)**.

9 On 4WD Frontier models and all Xterras and Pathfinders, remove the screws securing the lockwasher, then remove the lockwasher and unscrew the locknut **(see illustrations)**. **Note:** *A pin spanner or special socket, available at most auto parts stores, is required to unscrew the locknut. The socket type tool, which has pins that engage the holes in the locknut, is preferred (it will enable you to more accurately set the wheel bearing preload during reassembly).*

10 Pull the hub assembly out slightly, then push it back in. This will force the outer bearing and cup off the spindle so it can be removed **(see illustration)**.

11 Pull the hub and disc assembly off the spindle.

12 Use a screwdriver or a seal puller tool to pry the seal out of the rear of the hub assembly **(see illustration)**. Note how the seal is installed.

13 Remove the inner wheel bearing from the hub assembly.

14 Use solvent to remove all traces of the old grease from the bearings, hub and spindle. A small brush may prove helpful; however make sure no bristles from the brush embed themselves inside the bearing rollers.

31.12 Use a screwdriver or seal removal tool to pry out the grease seal

Clean the solvent from the bearings with brake system cleaner and allow them to dry.

15 Carefully inspect the bearings for cracks, heat discoloration, worn rollers, etc. Check the bearing races inside the hub for wear and damage. If the bearing races are defective, drive the bearing races out of the hub using a long brass drift. Drive the new races into the hub using the appropriate size bearing driver (inexpensive bearing driver sets are available at most automotive parts stores). Note that the bearings and races are

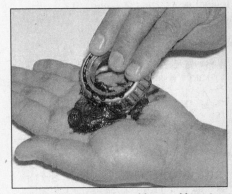

31.16 If a bearing packing tool is not available, work grease into the bearing rollers by pressing it against the palm of your hand

replaced as matched sets; used bearings should never be installed on new races.

16 Use only high-temperature front wheel bearing grease to pack the bearings. Inexpensive bearing packing tools are available at automotive parts stores, but not entirely necessary. If one is not available, pack the grease by hand completely into the bearings, forcing it between the rollers, cone and cage from the large diameter side or the bearing **(see illustration)**.

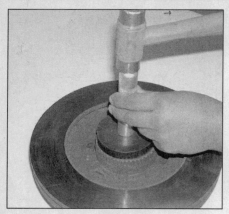

31.20 Install the new grease seal with a seal driver or block of wood

17 Apply a thin coat of grease to the spindle at the outer bearing seat, inner bearing seat, shoulder and seal seat.
18 Apply a coat of grease to the races and to the area in the center of the hub, between the races.
19 Place the grease-packed inner bearing into the rear of the hub and put a little more grease outboard of the bearing.
20 Place a new seal over the inner bearing and tap the seal evenly into place with a hammer and a seal installation tool or block of wood until it's flush with the hub **(see illustration)**.

Adjustment

21 Carefully place the hub assembly onto the spindle and push the grease-packed outer bearing into position.

2WD Frontier models

22 Install the washer and spindle nut. Tighten the nut only slightly (no more than 12 ft-lbs of torque). Spin the hub in a forward direction while tightening the spindle nut to 25 to 29 ft-lbs (34 to 39 Nm) to seat the bearings. Loosen the spindle nut 1/8 turn (45-degrees), then using your hand (not a wrench of any kind), tighten the nut until it is snug.
23 Install the nut lock, then insert a new cotter pin through the hole in the spindle and the slots in the nut lock. If the nut lock slots do not line up, remove the nut lock and turn it slightly until they do. Bend the ends of the cotter pin until they are flattened against the nut. Cut off any extra length that could interfere with the hub cap.
24 Place some grease inside the hub cap, then install the hub cap, tapping it into place with a hammer (be careful not to dent it).

4WD Frontier models and all Xterra and Pathfinder models

Refer to illustration 31.28

25 Install the thrust washer and the locknut on the spindle. Tighten the locknut to 58 to 72 ft-lbs (78 to 98 Nm) while rotating the hub to seat the bearings. Back off the locknut until it can be loosened by hand.
26 Tighten the locknut to 4.3 to 13 in-lbs (0.5

31.28 Using a pull scale, measure the force required to start the hub turning

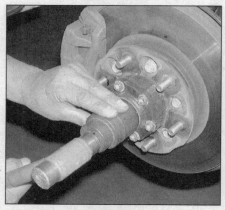

31.34 When installing the hub cap, use a large socket or a piece of pipe that contacts the flange of the cap - don't hit the center of the cap to drive it on

to 1.5 Nm) while rotating the hub assembly.
27 Turn the hub several times, then repeat Step 26. Verify that the hub rotates smoothly and no endplay is present.
28 Using a pull scale, measure the amount of effort required to start the hub rotating **(see illustration)**. Record this measurement.
29 Install the lockwasher and screws. If necessary, the locknut can be tightened 15 to 30-degrees more to allow installation of the lockwasher. Tighten the screws securely.
30 Once again measure the amount of effort required to start the hub turning. Record this figure.
31 Subtract the figure recorded in Step 28 from the figure recorded in Step 30 to calculate the desired wheel bearing preload; it should be within the range listed in this Chapter's Specifications.
32 If the preload is not as specified, remove the lockwasher and loosen or tighten the locknut within 15-degrees and repeat Steps 28 through 31.

All models

Refer to illustrations 31.33 and 31.34

33 The remainder of the installation is the reverse of removal. **Note:** *On models so equipped, lightly grease the internal compo-*

31.33 There are two O-rings on the drive flange (4WD models without free-running hubs)

32.1a On four-cylinder engines the PCV valve is mounted in a grommet on the breather separator (located on the lower timing chain cover on the right side of the engine, below the power steering pump)

nents of the free-running hub assemblies with multi-purpose grease. DO NOT pack the hubs full of grease or poor operation may occur. If you're working on a 4WD model without free-running hubs, install new O-rings on the drive flange **(see illustration)**. *Tighten the free-running hubs or drive flange fasteners to the torque listed in the Chapter 8 Specifications.*
34 When installing the hub cap on an Xterra, Pathfinder or 4WD Frontier model drive it on with a large socket or piece of pipe that contacts the flange of the cover **(see illustration)**. **Note:** *The manufacturer recommends replacing the cover whenever it is removed, but if you were careful when removing it, it can be re-used.*
35 Tighten the wheel lug nuts to the torque listed in this Chapter's Specifications.

32 Positive Crankcase Ventilation (PCV) valve and hose check and replacement (30,000 miles or 24 months)

Refer to illustration 32.1a, 32.1b and 32.1c

1 The PCV valve is located in various loca-

Chapter 1 Tune-up and routine maintenance

32.1b On 3.3L V6 engines the PCV valve is threaded into the upper intake manifold

32.1c On 3.5L V6 engines the PCV valve is mounted in a grommet in the rear of the right valve cover

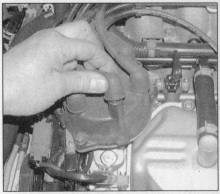

33.8 Detach the wires from the distributor cap one at a time, by pulling only on the boots

tions, depending on the engine **(see illustrations)**.

2   If you're working on a model with a four-cylinder engine, detach the hose and pull the valve out of the breather separator. If you're working on a 3.3L V6 engine, detach the hose and unscrew the valve from the upper intake manifold. If you're working on a 3.5L V6 engine, remove the upper and lower plenums for access to the valve (see Chapter 2C), then pull the valve out of its grommet.

3   Shake the PCV valve, listening for a rattle. If the valve doesn't rattle, replace it with a new one.

4   When purchasing a replacement PCV valve, make sure it's for your particular vehicle and engine size. Compare the old valve with the new one to make sure they're the same.

33 Spark plug wire, distributor cap and rotor check and replacement (every 60,000 miles or 48 months)

Refer to illustrations 33.8, 33.11a, 33.11b, 33.12a and 33.12b

Note: *This procedure applies to four-cylinder and 3.3L V6 engines only.*

1   The spark plug wires should be checked whenever new spark plugs are installed.

2   Begin this procedure by making a visual check of the spark plug wires while the engine is running. In a darkened garage (make sure there is ventilation) start the engine and observe each plug wire. Be careful not to come into contact with any moving engine parts. If there is a break in the wire, you will see arcing or a small spark at the damaged area. If arcing is noticed, make a note to obtain new wires, then allow the engine to cool and check the distributor cap and rotor.

3   The spark plug wires should be inspected one at a time to prevent mixing up the order, which is essential for proper engine operation. Each original plug wire should be numbered to help identify its location. If the number is illegible, a piece of tape can be marked with the correct number and wrapped around the plug wire.

4   Disconnect the plug wire from the spark plug. Grasp the rubber boot, twist the boot half a turn and pull the boot free. Do not pull on the wire itself.

5   Check inside the boot for corrosion, which will look like a white crusty powder.

6   Push the wire and boot back onto the end of the spark plug. It should fit tightly onto the end of the plug. If it doesn't, remove the wire and use pliers to carefully crimp the metal connector inside the wire boot until the fit is snug.

7   Using a clean rag, wipe the entire length of the wire to remove built-up dirt and grease. Once the wire is clean, check for burns, cracks and other damage. Do not bend the wire sharply, because the conductor inside might break.

8   Disconnect the wire from the distributor. Again, pull only on the rubber boot. Check for corrosion and a tight fit **(see illustration)**. Reconnect the wire to the distributor cap.

33.11a The distributor cap is retained by three screws

9   Inspect the remaining spark plug wires, making sure that each one is securely fastened at the distributor and spark plug when the check is complete.

10   If new spark plug wires are required, purchase a set for your specific engine model. Pre-cut wire sets with the boots already installed are available. Remove and replace the wires one at a time to avoid mix-ups in the firing order.

11   Detach the distributor cap by loosening the cap retaining screws. Remove it and look inside for cracks, carbon tracks and worn, burned or loose contacts **(see illustrations)**.

12   Pull the rotor off the distributor shaft and examine it for cracks and carbon tracks. **Note:** *Some rotors are retained by a screw*

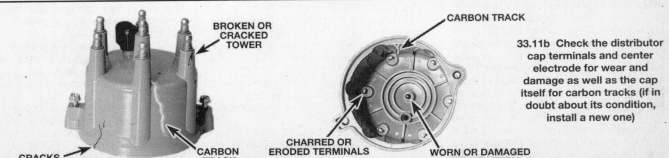

33.11b Check the distributor cap terminals and center electrode for wear and damage as well as the cap itself for carbon tracks (if in doubt about its condition, install a new one)

1-36   Chapter 1 Tune-up and routine maintenance

(see illustration). Replace the cap and rotor if any damage or defects are noted (see illustration).

13 It is common practice to install a new cap and rotor whenever new spark plug wires are installed, but if you wish to continue using the old cap, check the resistance between the spark plug wires and the cap first. If the indicated resistance is more than the maximum value listed in this Chapter's Specifications, replace the cap and/or wires.

14 When installing a new cap, remove the wires from the old cap one at a time and attach them to the new cap in the exact same location - do not simultaneously remove all the wires from the old cap or firing order mix-ups may occur.

34 Exhaust Gas Recirculation (EGR) system check (every 60,000 miles or 48 months)

Refer to illustration 34.2a and 34.2b
Note: *This section does not apply to models with a 3.5L V6 engine.*

1 The EGR valve is usually located on the intake manifold. Most of the time when a problem develops in this emissions system, it's due to a stuck or defective EGR valve.
2 With the engine cold, to prevent burns, push on the EGR valve diaphragm. Using moderate pressure, you should be able to push the diaphragm up into the housing (see illustrations). **Note:** *On four-cylinder engines you'll probably have to remove the air filter housing for access to the EGR valve (see Chapter 4).*
3 If the diaphragm doesn't move or is hard to move, replace the EGR valve with a new one. If in doubt about the condition of the valve, compare the free movement of your EGR valve with a new valve.
4 Refer to Chapter 6 for more information on the EGR system.

35 Manual transmission lubricant change (every 60,000 miles or 48 months)

1 This procedure should be performed after the vehicle has been driven so the lubricant will be warm and therefore will flow out of the transmission more easily. Raise the vehicle and support it securely on jackstands.
2 Move a drain pan, rags, newspapers and wrenches under the transmission.
3 Remove the transmission drain plug at the bottom of the case and allow the lubricant to drain into the pan (see Section 18).
4 After the lubricant has drained completely, reinstall the plug and tighten it to the torque listed in this Chapter's Specifications.
5 Remove the fill plug from the side of the transmission case. Using a hand pump, syringe or squeeze bottle, fill the transmission with the specified lubricant until it just reaches the bottom edge of the hole. Rein-

33.12a Before attempting to pull the rotor off the distributor shaft, check for the presence of a retaining screw (if present, loosen it and remove the rotor)

34.2a EGR valve location - four-cylinder engine (air cleaner housing removed for clarity)

stall the fill plug and tighten it to the torque listed in this Chapter's Specifications.
6 Lower the vehicle.
7 Drive the vehicle for a short distance, then check the drain and fill plugs for leakage.

36 Transfer case lubricant change (4WD models) (every 60,000 miles or 48 months)

1 This procedure should be performed after the vehicle has been driven so the lubricant will be warm and therefore will flow out of the transfer case more easily.
2 Raise the vehicle and support it securely on jackstands.
3 Remove the filler plug from the case (see Section 19).
4 Remove the drain plug from the lower part of the case and allow the lubricant to drain completely.
5 After the case is completely drained, carefully clean and install the drain plug. Tighten the plug to the torque listed in this Chapter's Specifications.
6 Remove the fill plug. Fill the case with the specified lubricant until it is level with the lower edge of the filler hole.
7 Install the filler plug and tighten it to the

33.12b The ignition rotor should be checked for wear and corrosion and burning of the center contact and tip (arrows) and any other cracks or damage (if in doubt about its condition, buy a new one)

34.2b EGR valve location - 3.3L V6 engine

torque listed in this Chapter's Specifications.
8 Drive the vehicle for a short distance and recheck the lubricant level. In some instances a small amount of additional lubricant will have to be added.

37 Differential lubricant change (every 60,000 miles or 48 months)

1 This procedure should be performed after the vehicle has been driven, so the lubricant will be warm and therefore will flow out of the differential more easily.
2 Raise the vehicle and support it securely on jackstands. You'll be draining the lubricant by removing the drain plug, so move a drain pan, rags, newspapers and wrenches under the vehicle.
3 Remove the plug (see Section 17) and allow the lubricant to drain into the pan, then clean and reinstall the drain plug. Tighten the plug to the torque listed in this Chapter's Specifications.
4 Remove the fill plug. Using a hand pump, syringe or squeeze bottle, fill the differential housing with the specified lubricant until it's level with the bottom of the fill-plug hole. Install the plug and tighten it to the torque listed in this Chapter's Specifications.

Chapter 2 Part A
2.4L four cylinder engine

Contents

	Section
Camshafts and lifters - removal, inspection and installation	8
CHECK ENGINE light	See Chapter 6
Crankshaft front oil seal - replacement	13
Crankshaft pulley – removal and installation	12
Cylinder compression check	See Chapter 2D
Cylinder head - removal and installation	11
Drivebelt check, adjustment and replacement	See Chapter 1
Engine - removal and installation	See Chapter 2D
Engine mounts - check and replacement	18
Engine oil and filter change	See Chapter 1
Engine overhaul - general information	See Chapter 2D
Exhaust manifold - removal and installation	10
Flywheel/driveplate - removal and installation	16
General information	1
Intake manifold - removal and installation	9
Oil pan - removal and installation	14
Oil pump and pick-up tube - removal, inspection and installation	15
Rear main oil seal - replacement	17
Repair operations possible with the engine in the vehicle	2
Spark plug replacement	See Chapter 1
Timing chains and sprockets - removal, inspection and installation	7
Top Dead Center (TDC) for number one piston - locating	3
Valve clearance - check and adjustment	5
Valve cover - removal and installation	4
Valve springs, retainers and seals - replacement	6
Valves - servicing	See Chapter 2D
Water pump - removal and installation	See Chapter 3

Specifications

General

Engine designation	KA24DE
Displacement	146 cubic inches (2.4 liters)
Bore	3.50 inches (89 mm)
Stroke	3.78 inches (96 mm)
Cylinder numbers (front to rear)	1-2-3-4
Firing order	1-3-4-2

Cylinder location and distributor rotation

The blackened terminal shown on the distributor cap indicates the Number One spark plug wire position

Warpage limits

Cylinder head-to-block surface	0.004 inch (0.1 mm)

Camshaft

Thrust clearance (endplay)	0.0028 to 0.0058 inch (0.070 to 0.148 mm)
Camshaft journal diameter	1.0998 to 1.1006 inches (27.935 to 27.955 mm)
Camshaft bearing inside diameter	1.1024 to 1.1033 inches (28.000 to 28.025 mm)
Bearing oil clearance	
Standard	0.0018 to 0.0035 inch (0.045 to 0.090 mm)
Service limit	0.0047 inch (0.12 mm)
Runout limit	0.0016 inch (0.04 mm) maximum
Intake lobe height	
2000 and earlier	1.673 to 1.681 inches (42.505 to 42.695 mm)
2001	1.644 to 1.651 inches (41.755 to 41.945 mm)
Exhaust lobe height	
2000 and earlier	1.610 to 1.618 inches (40.905 to 41.095 mm)
2001	1.646 to 1.654 inches (41.815 to 42.005 mm)

Valve clearance (hot)

Intake	0.012 to 0.015 (0.31 to 0.39 mm)
Exhaust	0.013 to 0.016 (0.33 to 0.41 mm)

Chapter 2 Part A 2.4L four cylinder engine

Oil pump
Body-to-outer rotor clearance	0.0059 to 0.0083 inch (0.15 to 0.21 mm)
Inner rotor-to-outer rotor tip clearance	0.0047 inch (0.12 mm) maximum
Inner rotor-to-cover clearance	0.0016 to 0.0039 inch (0.04 to 0.10 mm)
Outer rotor-to-cover clearance	0.0016 to 0.0039 inch (0.04 to 0.10 mm)

Torque specifications

	Ft-lbs (unless otherwise indicated)	Nm
Intake manifold bolts/nuts	144 to 168 in-lbs	16 to 19
Exhaust manifold-to-block bolts/nuts	27 to 35	37 to 48
Crankshaft pulley-to-crankshaft bolt	105 to 112	142 to 152
Flywheel/driveplate bolts	105 to 112	142 to 152
Cylinder head bolts		
Step 1	22	29
Step 2	59	79
Step 3	Loosen all bolts completely	
Step 4	18 to 25	25 to 34
Step 5	55 to 62	75 to 84
Camshaft bearing cap bolts		
Step 1	17 in-lbs	2
Step 2	80 to 104 in-lbs	9 to 12
Camshaft sprocket bolts	123 to 130	167 to 177
Idler sprocket bolt	48 to 61	66 to 83
Engine mount-to-mount bracket nut	30 to 38	41 to 52
Engine mount-to-frame bolts	23 to 31	31 to 42
Engine mount bracket-to-block bolts	23 to 31	31 to 42
Timing chain tensioner bolts	56 to 66 in-lbs	7 to 8
Upper timing chain cover-to-cylinder head bolts	144 to 168 in-lbs	16 to 19
Upper timing chain cover-to-lower timing chain cover bolts	56 to 66 in-lbs	6.5 to 7.5
Lower timing chain cover-to-block bolts		
Two lower left (driver's side) bolts, lower right bolt	56 to 66 in-lbs	6.5 to 7.5
All other bolts	144 to 168 in-lbs	16 to 19
Oil pump-to-front cover bolts	96 to 132 in-lbs	11 to 15
Oil pump-to-front cover screws	52 to 87 in-lbs	6 to 10
Oil pick-up-to-block bolts	144 to 168 in-lbs	16 to 19
Oil pan bolts	56 to 66 in-lbs	7 to 8
Oil pan drain plug	22 to 29	29 to 39
Valve cover bolts	69 to 95 in-lbs	8 to 11

1 General information

This Part of Chapter 2 is devoted to in-vehicle repair procedures for the KA24DE 2.4L Dual Overhead Camshaft (DOHC) four cylinder engine. All information concerning engine removal and installation and engine block and cylinder head overhaul can be found in Chapter 2, Part D.

The following repair procedures are based on the assumption that the engine is installed in the vehicle. If the engine has been removed from the vehicle and mounted on a stand, many of the steps outlined in this Part of Chapter 2 will not apply.

The Specifications included in this Part of Chapter 2 apply only to the procedures contained in this Part. Part D of Chapter 2 contains the Specifications necessary for cylinder head and engine block rebuilding.

2 Repair operations possible with the engine in the vehicle

Many major repair operations can be accomplished without removing the engine from the vehicle.

Clean the engine compartment and the exterior of the engine with some type of degreaser before any work is done. It will make the job easier and help keep dirt out of the internal areas of the engine.

Depending on the components involved, it may be helpful to remove the hood to improve access to the engine as repairs are performed (refer to Chapter 11 if necessary). Cover the fenders to prevent damage to the paint. Special pads are available, but an old bedspread or blanket will also work.

If vacuum, exhaust, oil or coolant leaks develop, indicating a need for gasket or seal replacement, the repairs can generally be made with the engine in the vehicle. The intake and exhaust manifold gaskets, oil pan gasket, crankshaft oil seals and cylinder head gasket are all accessible with the engine in place.

Exterior engine components, such as the intake and exhaust manifolds, the oil pan, the oil pump, the water pump, the starter motor, the alternator, the distributor and the fuel system components can be removed for repair with the engine in place.

Since the cylinder head can be removed without pulling the engine, camshaft and valve component servicing can also be accomplished with the engine in the vehicle. Replacement of the timing chain and sprockets is also possible with the engine in the vehicle.

In extreme cases caused by a lack of necessary equipment, repair or replacement of piston rings, pistons, connecting rods and rod bearings is possible with the engine in the vehicle. However, this practice is not recommended because of the cleaning and preparation work that must be done to the components involved.

3 Top Dead Center (TDC) for number one piston - locating

Refer to illustrations 3.5 and 3.8

1   Top Dead Center (TDC) is the highest point in the cylinder that each piston reaches as it travels up the cylinder bore. Each piston reaches TDC on the compression stroke and again on the exhaust stroke, but TDC generally refers to piston position on the compression stroke.

2   Positioning the piston(s) at TDC is an essential part of many procedures such as valve timing, camshaft and timing chain/sprocket removal.

3   Before beginning this procedure, be sure to place the transmission in Neutral and

Chapter 2 Part A 2.4L four cylinder engine

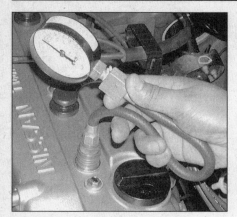

3.5 A compression gauge can be used in the number one plug hole to assist in finding TDC

3.8 Align the red (TDC) mark on the crankshaft pulley with the pointer on the front timing cover – note that the TDC mark is the second mark from the left

apply the parking brake or block the rear wheels. Disable the ignition system by disconnecting the primary electrical connectors at the ignition coil packs and remove the spark plugs (see Chapter 1). Disable the fuel system (see Chapter 4, Section 2).

4 In order to bring any piston to TDC, the crankshaft must be turned using one of the methods outlined below. When looking at the front of the engine, normal crankshaft rotation is clockwise.

   a) The preferred method is to turn the crankshaft with a socket and ratchet attached to the bolt threaded into the front of the crankshaft. Turn the bolt in a clockwise direction.
   b) A remote starter switch, which may save some time, can also be used. Follow the instructions included with the switch. Once the piston is close to TDC, use a socket and ratchet as described in the previous paragraph.
   c) If an assistant is available to turn the ignition switch to the Start position in short bursts, you can get the piston close to TDC without a remote starter switch. Make sure your assistant is out

of the vehicle, away from the ignition switch, then use a socket and ratchet as described in Paragraph (a) to complete the procedure.

5 Install a compression pressure gauge in the number one spark plug hole (**see illustration**). It should be a gauge with a screw-in fitting and a hose at least six inches long.

6 Rotate the crankshaft using one of the methods described above while observing for pressure on the compression gauge. The moment the gauge shows pressure indicates that the number one cylinder has begun the compression stroke.

7 Once the compression stroke has begun, TDC for the compression stroke is reached by bringing the piston to the top of the cylinder.

8 Continue turning the crankshaft until the red notch in the crankshaft damper is aligned with the pointer on the front cover (**see illustration**). At this point, the number one cylinder is at TDC on the compression stroke. If the marks are aligned but there was no compression, the piston was on the exhaust stroke. Continue rotating the crankshaft 360-

degrees (1-turn). **Note:** *If a compression gauge is not available, you can simply place a blunt object (such as the end of a screwdriver handle) over the spark plug hole and listen for compression as the engine is rotated. Once compression at the No.1 spark plug hole is noted the remainder of the Step is the same.*

9 After the number one piston has been positioned at TDC on the compression stroke, TDC for any of the remaining cylinders can be located by turning the crankshaft 180 degrees and following the firing order (refer to the Specifications). Rotating the engine 180 degrees past TDC #1 will put the engine at TDC compression for cylinder #3.

4 Valve cover - removal and installation

Removal

Refer to illustrations 4.2, 4.4 and 4.5

1 Disconnect the cable from the negative terminal of the battery.

2 Remove the spark plug wires from the spark plugs, the crankcase breather hose from the fitting on the valve cover and the throttle cable from its bracket (**see illustration**). Position the throttle cable out of the way to allow valve cover removal.

3 Detach the spark plug wires from the clips on the valve cover.

4 Remove the valve cover retaining bolts/nuts (**see illustration**). If the cover is stuck to the cylinder head, bump the end with a wood block and a hammer to jar it loose. If that doesn't work, try to slip a flexible putty knife between the cylinder head and cover to break the seal. **Caution:** *Don't pry at the cover or housing-to-cylinder head joint or damage to the sealing surfaces may occur, leading to oil leaks after the cover is reinstalled.*

5 Remove the gasket from the valve cover. Check the condition of the spark plug tube seal (**see illustration**), and replace it if necessary.

4.2 Remove the crankcase breather hose (A) and the throttle cable from its bracket (B), then remove the spark plug wires from the spark plugs and the wire retaining bracket (C)

4.4 Valve cover retaining bolts/nuts (arrows)

4.5 The spark plug tubes are sealed by a rubber strip (arrows) - replace if cracked or leaking

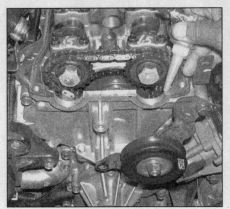

4.7 Apply a thin coat of RTV sealant where the rubber half-circles contact the front cover and the rear of the cylinder head

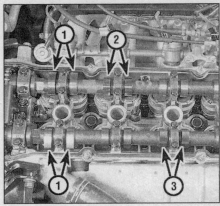

5.6a When the no. 1 piston is at TDC on the compression stroke, the valve clearance for the no. 1 and no. 3 cylinder exhaust valves and the no. 1 and no. 2 cylinder intake valves can be measured

Installation

Refer to illustration 4.7

6   The mating surfaces of the valve cover and cylinder head must be clean when the cover is installed. Use a gasket scraper to remove all traces of sealant from the half-circle areas at the front cover and the rear of the cylinder head, then clean the mating surfaces with lacquer thinner or acetone. If there's residue or oil on the mating surfaces when the cover is installed, oil leaks may develop.
Caution: *Use care when scraping the soft aluminum of the cylinder head or valve cover. It is soft, and deep scratches may lead to oil leaks.*

7   Apply RTV sealant to the half-circle areas at the front cover and the rear of the cylinder head **(see illustration)**. Press the new gasket into the groove on the valve cover and install the valve cover with a new gasket onto the engine.

8   Tighten the bolts/nuts to the torque listed in this Chapter's Specifications, starting with the center fasteners and working out toward each end.

5   Valve clearance - check and adjustment

Refer to illustrations 5.6a, 5.6b, 5.7, 5.9a, 5.9b, 5.9c and 5.10
Note: *The manufacturer recommends adjusting the valve clearance at the specified interval only if the valve train is making excessive noise. The following procedure requires the use of special valve lifter tools. The tools are available from specialty tool manufacturers and many auto parts stores. It is impossible to perform this task without them.*

1   Disconnect the cable from the negative terminal of the battery.
2   Remove the valve cover (see Section 4).
3   On manual transmission vehicles set the parking brake and place the transmission in the neutral position.
4   Remove the spark plugs (see Chapter 1).

5   Position the number 1 piston at TDC on the compression stroke and align the TDC mark with the pointer on the front cover (see Section 3).
6   Measure the clearance of the indicated valves with a feeler gauge **(see illustrations)**. Record each measurement and compare your measurements with the desired valve clearance found in this Chapter's Specifications. Note which are out of specification, this data will be used later to determine the required replacement shims.
7   Turn the crankshaft one complete revolution and realign the TDC mark. Measure and record the clearances of the remaining valves **(see illustration)**.
8   After the clearance of all the valves have been measured, rotate the crankshaft pulley until the camshaft lobe above the first valve which you intend to adjust is pointing up, away from the lifter.
9   Align the notch in the valve lifter with the notch in the cylinder head casting. Place the special valve lifter tool in position as shown, with the upper jaw over the camshaft, next to the lobe and the lower jaw on top of the shim

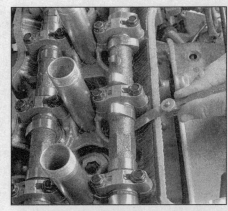

5.6b Measure the clearance for each valve with a feeler gauge of the specified thickness - if the clearance is correct, you should feel a slight drag on the gauge as you pull it out

(see illustration). Depress the valve lifter by squeezing the handles of the valve lifter tool together and rotating the tool away from the

5.7 When the no. 4 piston is at TDC on the compression stroke, the valve clearances for the no. 2 and no. 4 cylinder exhaust valves and the no. 3 and no. 4 cylinder intake valves can be measured

5.9a Install the valve lifter tool as shown - squeeze the handles together and rotate the tool away from the camshaft to depress the valve lifter

5.9b With the small tool wedged between the lifter and the camshaft, pry the shim up with a small screwdriver at the notch . . .

5.9c . . . and remove the shim with a pair of tweezers or a magnet as shown

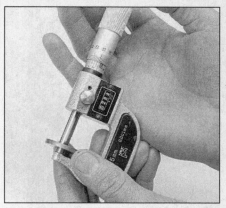

5.10 Measure the shim thickness with a micrometer or a dial caliper

6.5 The air hose adapter threads into the spark plug hole - they're commonly available from auto parts stores

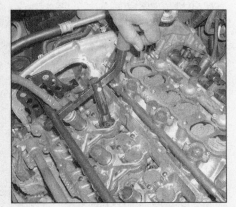

6.7a Compress the valve spring enough to release the valve stem keepers . . .

6.7b . . . and lift them out with a magnet or needle-nose pliers

camshaft. Insert the small tool between the edge of the lifter and the camshaft and release the lifter. Remove the adjusting shim with a small screwdriver and a magnet or a pair of tweezers **(see illustration)**.

10   Measure the thickness of the shim with a micrometer **(see illustration)**. To calculate the correct thickness of a replacement shim that will place the valve clearance within the specified value, use the following formula:

Intake and Exhaust:
N = R + (M − 0.0146-inch [0.37 mm])
R = thickness of the old shim
M = valve clearance measured
N = thickness of the new shim

11   Select a shim with a thickness as close as possible to the valve clearance calculated. Shims, which are available in 37 sizes in increments of 0.0008-inch (0.020 mm). Available shims range in size from 0.0772 inch (1.96 mm) to 0.1055 inch (2.68 mm) (see accompanying chart below). **Note:** *Through careful analysis of the shim sizes needed to bring all the out-of-specification valve clearances within specification, it is often possible to simply move a shim that has to come out anyway to another valve lifter requiring a shim of that particular size, thereby reducing the number of new shims that must be purchased.*

12   Place the special valve lifter tools in position as shown in **illustration 5.9a**, depress the valve lifter and install the new adjusting shim. Measure the clearance with a feeler gauge to make sure that your calculations are correct.

13   Repeat this procedure until all the valves which are out of specification have been corrected.

14   The remainder of installation is the reverse of removal.

6   Valve springs, retainers and seals - replacement

Refer to illustrations 6.5, 6.7a, 6.7b, 6.13a, 6.13b, 6.14 and 6.15

Note: *Broken valve springs and defective valve stem seals can be replaced without removing the cylinder heads. Two special* tools and a compressed air source are normally required to perform this operation, so read through this Section carefully. The universal shaft-type valve spring compressor required for the tight valve spring pockets of this vehicle may not be available at all tool rental yards, so check on the availability before beginning the job.

1   Remove the valve cover (see Section 4).
2   Refer to Section 7 and remove the upper timing chain, then refer to Section 8 and remove the camshafts and lifters from the affected cylinder head.
3   Remove the spark plug from the cylinder that has the defective component. If all of the valve stem seals are being replaced, all of the spark plugs should be removed.
4   Turn the crankshaft until the piston in the affected cylinder is at Top Dead Center on the compression stroke (refer to Section 3). If you're replacing all of the valve stem seals, begin with cylinder number one and work on the valves for one cylinder at a time. Move from cylinder-to-cylinder following the firing order sequence (see this Chapter's Specifications).
5   Thread a long adapter into the spark plug hole and connect an air hose from a compressed air source to it **(see illustration)**. Most auto parts stores can supply the air hose adapter. **Note:** *Because of the length of the spark plug tubes, it will be necessary to use a long spark plug adapter with a length of hose attached (as used on many cylinder compression gauges) utilizing a quick-disconnect fitting to hook to your air source.*
6   Apply compressed air to the cylinder. **Warning:** *The piston may be forced down by the compressed air, causing the crankshaft to turn suddenly. If the wrench used when positioning the number one piston at TDC is still attached to the bolt in the crankshaft nose, it could cause damage or injury when the crankshaft moves.*
7   Stuff shop rags into the cylinder head holes around the valves to prevent parts and tools from falling into the engine, then use a valve spring compressor to compress the spring **(see illustrations)**. Remove the valve stem keepers with small needle-nose pliers

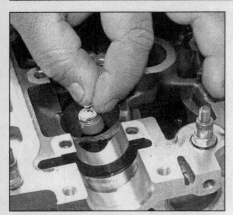

6.13a Lubricate the new seal and slip it over the valve stem, past the valve keeper groove

6.13b Using a deep socket and hammer, gently tap the new seals onto the valve guide only until seated

or a magnet. **Note:** *The valves should be held in place by the air pressure. If the valve faces or seats are in poor condition, leaks may prevent air pressure from retaining the valves. If the valves cannot hold air, the cylinder head should be removed for a valve job at a machine shop.*

8   Remove the spring retainer and valve spring, then remove the valve stem seal.
9   Wrap a rubber band or tape around the top of the valve stem so the valve won't fall into the combustion chamber, then release the air pressure.
10 Inspect the valve stem for damage. Rotate the valve in the guide and check the end for eccentric movement, which would indicate that the valve is bent.
11 Move the valve up-and-down in the guide and make sure it doesn't bind. If the valve stem binds, either the valve is bent or the guide is damaged. In either case, the cylinder head will have to be removed for repair.
12 Reapply air pressure to the cylinder to retain the valve in the closed position, then remove the tape or rubber band from the valve stem.
13 Lubricate the valve stems with engine oil and install a new valve stem seals. Valve stem seals can be installed with a special tool, or a deep socket and hammer - tap the seal only until seated **(see illustrations)**.
14 Install the valve spring in position over the valve, with the more closely-wound spring coils and/or the paint mark toward the cylinder head **(see illustration)**.

15 Install the valve spring retainer. Compress the valve springs and carefully position the keepers in the groove. Apply a small dab of grease to the inside of each keeper to hold it in place **(see illustration)**.
16 Remove the pressure from the spring tool and make sure the keepers are seated.
17 Disconnect the air hose and remove the adapter from the spark plug hole.
18 Refer to Section 8 and install the camshaft and lifters, then refer to Section 7 and install the timing chain.
19 Refer to Section 4 and install the valve covers.
20 Install the spark plugs and the spark plug wires.
21 Start and run the engine, then check for oil leaks and unusual sounds coming from the valve cover area.

7   Timing chains and sprockets - removal, inspection and installation

Refer to illustrations 7.9, 7.10, 7.11, 7.12, 7.17, 7.21a, 7.21b, 7.22a, 7.22b, 7.23 and 7.24

Warning: *Wait until the engine is completely cool before beginning this procedure.*
Note: *The timing chain assembly on this engine consists of a lower chain connecting the crankshaft sprocket to an idler sprocket (attached to the front of the cylinder head),* and an upper chain connecting the idler sprocket to the two camshaft sprockets. The timing chains should last for the life of the engine, and replacement is a difficult procedure, involving removing the oil pan and the cylinder head. The procedure described below involves removing the intake manifold with the cylinder head.

Removal
Upper timing chain
1   Relieve the fuel system pressure (see Chapter 4).
2   Drain the cooling system (see Chapter 1).
3   Remove the air cleaner housing, the air intake duct and the air intake resonator (see Chapter 4).
4   Remove the accessory drivebelts and the engine cooling fan, then remove the radiator and the engine cooling fan shroud as an assembly from the engine compartment (see Chapter 3).
5   Label and disconnect the hoses and electrical connectors attached to the intake manifold and the throttle body, then unbolt the support brace from the intake manifold. (see Section 9, Steps 4 and 5). Also disconnect the fuel feed and return lines from the fuel rail and the electrical connectors from the fuel injectors.
6   Disconnect the main wiring harness from the wiring harness retaining clips and position the main harness out of the way.
7   Position the engine at TDC for number one cylinder (see Section 3).
8   Refer to Section 4 and remove the valve cover.
9   Remove the power steering pump (see Chapter 10). Then remove the power steering pump brackets and the upper idler pulley from the front of the engine **(see illustration)**.
10 Use a large open end-wrench on the hex of the camshaft to hold the camshaft as you loosen the camshaft sprocket bolts **(see illustration)**. Loosen the bolt several turns, but do not remove them at this time.
11 Remove the remaining bolts securing the upper timing chain cover to the cylinder head and lower timing chain cover. Remove the upper timing chain cover **(see illustration)**.

6.15 Apply a small dab of grease to each keeper as shown here before installation - it will hold them in place on the valve stem as the spring is released

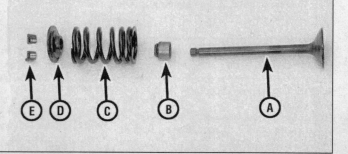

6.14 Arrangement of the valve components

A   Valve
B   Valve stem seal
C   Valve spring
D   Retainer
E   Keepers

Chapter 2 Part A 2.4L four cylinder engine

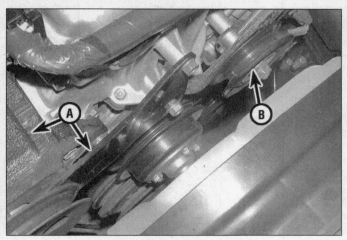

7.9 Remove the power steering pump brackets (A) and the upper idler pulley (B)

7.10 Hold the camshaft with a wrench (arrow) as shown to loosen either camshaft sprocket bolt

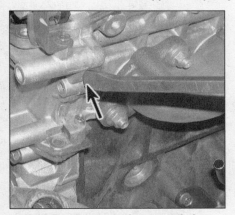

7.11 Pry the upper the upper timing chain cover off the engine at the casting protrusion

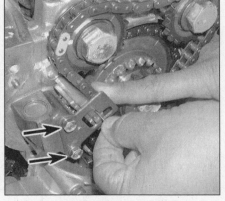

7.12 With the upper timing chain tensioner locked in the retracted position, remove the two bolts (arrows) and the upper timing chain tensioner

7.17 Remove the coolant hoses (A) and the PCV valve hose (B) from the lower timing chain cover

12 Push the upper timing chain tensioner inward and lock it into the retracted position with a paper clip or other suitable tool. Detach the tensioner retaining bolts and remove the upper timing chain tensioner **(see illustration)**.

13 Mark the camshaft sprockets "I" for intake and "E" for exhaust so they can be reinstalled in their original location, then remove the camshaft sprockets and the upper timing chain.

Lower timing chain

14 Remove the upper timing chain as described in Steps 1 through 13.

15 Remove the camshafts and lifters (see Section 8). **Note:** *It is only necessary to follow Steps 7 through 9 in the camshaft removal procedure, since the camshaft sprockets and the upper timing chain are already removed.*

16 Remove the distributor (see Chapter 5).

17 Remove the upper radiator hose and the heater hoses from the thermostat housing (see Chapter 3). Also detach the upper alternator bracket and pivot the alternator out of the way **(see illustration)**.

18 Remove the crankshaft pulley/vibration damper being careful not to rotate the engine from TDC (see Section 12). If the engine rotates off TDC during this step, reposition the engine back to TDC before proceeding. The engine should be left at TDC for the No. 1 piston during this entire procedure.

19 Drain the engine oil, then remove the oil pan and oil pump pick-up (see Section 14). This Step is not absolutely necessary but it will help ensure a leak-proof seal between the oil pan and the lower timing chain cover.

20 Remove the oil pump and the oil pump driveshaft (see Section 15).

21 Remove the lower timing chain cover **(see illustrations)**.

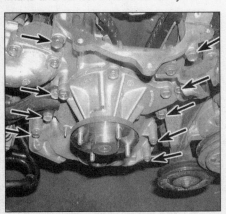

7.21a Lower timing chain cover mounting bolts

7.21b Carefully pry the lower timing chain cover off at the casting protrusion

Chapter 2 Part A 2.4L four cylinder engine

7.22a Compress the lower timing chain tensioner and insert a paper clip in the hole to lock it in place

7.22b The tension arm and the chain guide (arrows) are retained by Torx fasteners

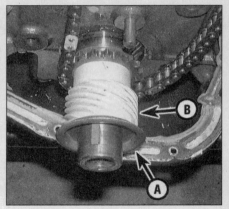

7.23 Remove the oil slinger (A) and the oil pump drive gear (B)

7.24 Idler gear mounting bolt - note that the idler shaft will be removed with the idler gear as the bolt is removed

7.28 Don't forget to inspect the idler shaft and the inside of the idler gear for damage

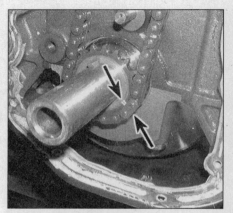

7.31 With the number one cylinder at TDC, align the silver link (arrow) with the dot on the crankshaft sprocket

22 Remove the lower timing chain tensioner, tension arm and the chain guide **(see illustrations)**.
23 Remove the oil slinger and the oil pump drive gear **(see illustration)**
24 Remove the idler sprocket, the idler shaft and the lower timing chain **(see illustration)**. Note: *Since the cylinder head has to be removed anyway, it is easier the cut the head gasket flush with the cylinder head to allow easy removal of the lower timing chain.*
25 Remove the exhaust manifold (see Section 10).
26 Detach the EGR pipe from the EGR valve (see Chapter 6).
27 Remove the cylinder head and the intake manifold as an assembly from the engine (see Section 11). **Note:** *It is because of the cylinder head gasket design that the cylinder head must be removed for the timing chain installation procedure. Since the upper and lower timing chain covers must seal against the cylinder head gasket, the cylinder head gasket must be replaced.*

Inspection

Refer to illustration 7.28

28 Inspect the camshaft sprockets, the idler sprocket and the crankshaft sprockets for wear of the teeth and keyways. Also inspect the oil pump drive gear for wear and damage and the timing chains for cracks or excessive wear on the rollers. Inspect the facing of the chain guides and the idler shaft for excessive wear **(see illustration)**.

Installation

Refer to illustrations 7.31, 7.32a, 7.32b, 7.34a, 7.34b, 7.37, and 7.38

29 Remove all traces of gasket material from the cylinder block, the timing chain covers, the oil pan and the cylinder head mating surfaces.
30 Refer to Section 11 and install the cylinder head with a new gasket. Install the bolts finger tight at this time - do not tighten them! **Caution:** *Make sure the beveled edges of the cylinder head-bolt washers face the bolts, and the flat sides face the cylinder head.*
31 Install the lower timing chain on the crankshaft sprocket with the silver-colored link aligned with the mark on the crankshaft sprocket **(see illustration)**.
32 Feed the chain up through the cylinder head gasket and position the idler sprocket with the idler shaft on the chain, aligning the timing mark with the yellow-colored link. Install the idler sprocket/idler shaft to the cylinder head and tighten the sprocket bolt hand tight **(see illustrations)**.
33 Reinstall the lower timing chain guide and the tensioner arm, then install the lower timing chain tensioner on the block. Remove the tensioner lock pin (installed in Step 22) to set tension on the chain.
34 Install new O-rings in the oil passages on the front of the block **(see illustration)**. Also install the oil pump drive gear and the oil slinger onto the crankshaft. Apply a bead of

7.32a Align the dot on the idler sprocket with the yellow link in the chain (arrow)

Chapter 2 Part A 2.4L four cylinder engine

7.32b Pull the idler gear and the lower timing chain up through the opening in the cylinder head gasket and install it on the engine

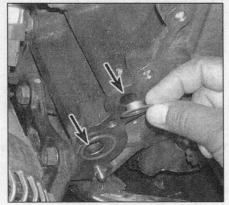

7.34a Install new oil-passage O-rings (arrows) in the block before installing the lower timing cover

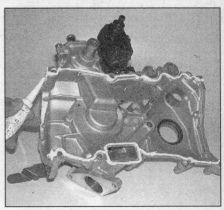

7.34b Apply RTV sealant to the lower cover-to-engine block sealing surface

7.37 Upper timing chain installation details – align the yellow links on the chain with the marks on the sprockets and the number one piston at TDC on the compression stroke

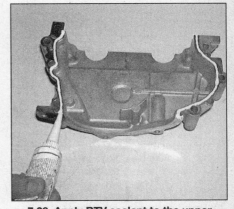

7.38 Apply RTV sealant to the upper cover-to-engine block sealing surface

8.4 With the engine at TDC for the number one cylinder, apply a dab of paint across the timing chain links and the camshaft sprockets - also mark the sprockets "I" for intake and "E" for exhaust

RTV sealant to the lower timing chain cover sealing surface and install the cover on to the engine block **(see illustration)**. Install any brackets that were removed from the cover and tighten the cover bolts to the torque listed in this Chapter's Specifications.

35 Now tighten the cylinder head according to the procedure in Section 11 and reinstall the camshafts and lifters as described in Section 8.

36 Install the exhaust camshaft sprocket and bolt onto the exhaust camshaft hand tight.

37 Install the upper timing chain on the idler sprocket with a yellow-colored link aligned with the dot on the front of the idler gear. Then slip the chain over the exhaust camshaft sprocket, aligning its dot with the second yellow link in the chain. Slip the intake camshaft sprocket under the chain and align the third yellow chain link with the dot on the front of the intake sprocket. Install the intake sprocket onto the intake camshaft and install the bolt **(see illustration)**. Tighten the camshaft and idler sprocket bolts to the torque listed in this Chapter's Specifications. **Note:** *There are exactly seven chain links between each yellow link, therefore it does not matter which yellow link you align with the idler sprocket, just make sure the three yellow links are aligned with the marks on the sprockets with engine at TDC for the number one cylinder.*

38 Install the upper timing chain tensioner and remove the tensioner lock pin (installed in Step 12) to set tension on the upper chain **(see illustration 7.37)**. Apply a bead of RTV sealant to the upper timing chain cover sealing surface and install the upper timing cover on to the engine block **(see illustration)**. Install any brackets that were removed from the upper timing cover and tighten the cover bolts to the torque listed in this Chapter's Specifications.

39 The remainder of the installation is the reverse of the disassembly sequence.

40 Refill the crankcase with oil, fill the cooling system with coolant, then run the engine and check for leaks.

8 Camshaft and lifters - removal, inspection and installation

Note: *The camshafts and lifters should always be thoroughly inspected before installation and camshaft endplay should always be checked prior to camshaft removal. Refer to Chapter 2D for the camshaft and lifter inspection procedures.*

Removal

Refer to illustrations 8.4, 8.7, 8.9a and 8.9b

1 Detach the cable from the negative terminal of the battery.

2 Remove the valve cover (see Section 4) and the air intake duct (see Chapter 4).

3 Position the engine at TDC for number one cylinder (see Section 3).

4 With the TDC marks aligned, apply a dab of paint to the timing chain links and the sprockets to help aid in the installation process. Also mark the sprockets "I" for intake or "E" for exhaust. The sprockets must be reinstalled in their original location **(see illustration)**.

5 Using a wrench to hold the camshaft sprockets from turning, loosen the camshaft sprocket bolts several turns **(see illustration 7.10)**. If the camshaft sprockets have rotated during the bolt loosening process, rotate the engine clockwise until the "TDC" marks on the crankshaft pulley are once again aligned.

8.7 The camshaft bearing caps are numbered and have an arrow that should face the timing chain end of the engine

8.9a Mark the lifters (I for intake, E for exhaust and number their location) and pull them straight up to remove them or use a magnetic retrieval tool

8.9b The lifters and shims can be stored in individually-marked plastic bags, or in a divided, marked box like this one

6 Remove the camshaft sprocket retaining bolts. Disengage the timing chain from the sprockets and remove the camshaft sprockets from the engine, letting the upper timing chain rest in the upper timing chain cover opening. **Note:** *There is a cast-in protrusion inside the upper timing chain cover that will keep the upper timing chain from falling off the (lower) idler gear.*

7 Loosen the camshaft bearing caps in two or three steps, in the reverse of the tightening sequence **(see illustration 8.12)**. They are numbered from 1 to 5, are stamped with an "I" or an "E" to indicate intake or exhaust, and have arrows to indicate which way faces the timing-chain end of the engine **(see illustration)**. **Caution:** *Keep the caps in order. They must go back in the same location they were removed from.*

8 Remove the bearing caps and lift the camshafts straight up and out. Also mark the camshafts I" for intake or "E" for exhaust. If the camshafts are to be reused they must be reinstalled in their original location.

9 Pull the lifters straight up, and store them in numbered plastic bags or a marked box, keeping the proper shim with each lifter **(see illustrations)**.

Installation

Refer to illustration 8.12

10 Apply moly-based engine assembly lubricant to the camshaft lobes and journals and install the camshaft into the cylinder head with the No.1 cylinder camshaft lobes pointing inward toward each other and the dowel pins facing upward. If the old camshafts are being used, make sure they're installed in the exact location from which they came.

11 Install the bearing caps and bolts and tighten them hand tight.

12 Tighten the bearing cap bolts in several equal steps, to the torque listed in this Chapter's Specifications, using the proper tightening sequence **(see illustration)**.

13 Engage the camshaft sprocket teeth with the timing chain links so that the match marks made during removal align with the marks on the sprockets, then position the sprockets over the dowels on the camshaft hubs and install the camshaft sprocket bolts finger tight. **Note:** *It will be necessary to depress the upper timing chain tensioner to create enough slack in the upper timing chain to install the camshaft sprockets onto the dowel pins.*

14 Double check that the sprockets are returned to the proper camshaft and tighten the camshaft sprocket bolts to the torque listed in this Chapter's Specifications.

15 The remainder of installation is the reverse of removal.

9 Intake manifold - removal and installation

Warning: *Wait until the engine is completely cool before beginning this procedure.*

Removal

Refer to illustration 9.4

1 Relieve the fuel system pressure (see Chapter 4). Disconnect the cable from the negative terminal of the battery.

2 Drain the cooling system (see Chapter 1). Remove the power steering pump and set it aside without disconnecting the fluid lines (see Chapter 10).

3 Remove throttle body, the fuel rail and the fuel injectors (see Chapter 4).

4 Label and detach all wire harnesses, control cables, coolant and vacuum hoses connected to the intake manifold **(see illustration)**. Also disconnect the EGR pipes from the EGR valve.

5 Working underneath the vehicle, unbolt the lower brace from the intake manifold.

6 Remove the mounting nuts/bolts, then detach the manifold from the engine **(see illustration 9.9)**.

Installation

Refer to illustration 9.9

7 Use a scraper to remove all traces of old gasket material and sealant from the manifold and cylinder head, then clean the mating surfaces with lacquer thinner or acetone.

8 Install a new gasket, then position the manifold on the cylinder head and install the nuts/bolts.

9 Tighten the nuts/bolts in three or four equal steps to the torque listed in this Chapter's Specifications. Follow the recommended tightening sequence **(see illustration)**.

10 Install the remaining parts in the reverse order of removal. Refill the cooling system (see Chapter 1).

8.12 Camshaft bearing cap TIGHTENING sequence - the sequence is the same for both camshafts

Chapter 2 Part A 2.4L four cylinder engine

11 Before starting the engine, check the throttle linkage for smooth operation.
12 Run the engine and check for coolant and vacuum leaks. Road test the vehicle and check for proper operation of all accessories.

10 Exhaust manifold - removal and installation

Warning: *The engine must be completely cool before beginning this procedure.*

Removal

Refer to illustrations 10.4, 10.5 and 10.7

1 Disconnect the cable from the negative terminal of the battery.
2 Block the rear wheels and set the parking brake. Raise the front of the vehicle and support it securely on jackstands.
3 Remove the engine splash shields (if equipped).
4 Working underneath the vehicle, disconnect the exhaust pipe from the exhaust manifold **(see illustration)**. Note: *Applying penetrating oil to the exhaust manifold fasteners may make removing the nuts/bolts easier.*
5 Working in the engine compartment, remove the lower heat shield from the exhaust manifold **(see illustration)**.
6 Disconnect the electrical connector from the oxygen sensor.
7 Remove the upper heat shield from the exhaust manifold **(see illustration)**.
8 Disconnect the EGR pipe from the exhaust manifold.
9 Remove the exhaust manifold-to-cylinder head nuts, working from the outside toward the middle, and detach the manifold and gaskets **(see illustration 10.13)**.

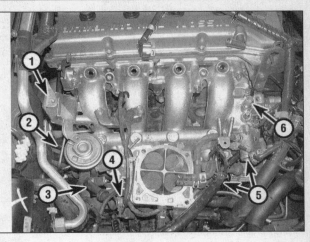

9.4 Label and disconnect the following components required for intake manifold removal

1 Air cleaner support bracket
2 EGR pipe(s)
3 Transmission dipstick tube mounting bolt
4 Heater hose
5 Coolant, vacuum and PCV valve hoses at front of manifold (not visible)
6 Electrical connectors

Installation

Refer to illustration 10.13

10 Use a scraper to remove all traces of old gasket material and carbon deposits from the exhaust manifold and cylinder head mating surfaces.
11 Position the new exhaust manifold gaskets over the cylinder head studs.
12 Install the manifold and thread the mounting nuts into place.

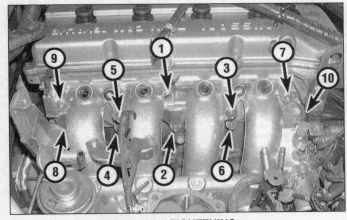

9.9 Intake manifold TIGHTENING sequence

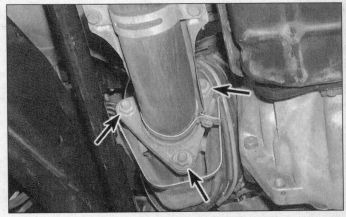

10.4 Remove the nuts (arrows) retaining the exhaust pipe to the exhaust manifold

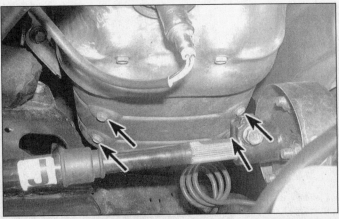

10.5 Remove the bolts retaining the lower heat shield to the exhaust manifold and remove the heat shield

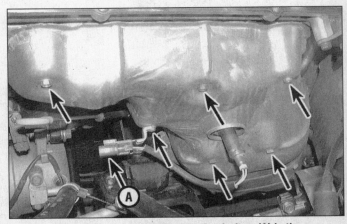

10.7 Upper heat shield mounting bolts - (A) is the oxygen sensor connector

13 Tighten the nuts in the recommended sequence to the torque listed in this Chapter's Specifications **(see illustration)**.
14 Reinstall the remaining parts in the reverse order of removal. Use anti-seize lubricant on the exhaust pipe studs and the EGR pipe nut.
15 Run the engine and check for exhaust leaks.

11 Cylinder head - removal and installation

Caution: *The engine must be completely cool before beginning this procedure.*
Note: *The cylinder head may be removed with the intake manifold still attached, and the manifold can be removed later on the workbench.*

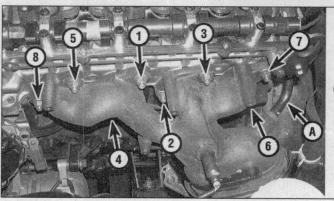

10.13 Exhaust manifold TIGHTENING sequence - (A) is the EGR pipe nut

Removal

Refer to illustrations 11.4a, 11.4b, 11.8a, 11.8b and 11.9

1 Disconnect the cable from the negative battery terminal.
2 Refer to Section 7, Steps 1 through 13 and remove the upper timing chain.
3 Install a timing chain tensioner wedge through the opening in the lower timing chain cover and force the wedge down between the narrowest section of the lower timing chain. **Caution:** *Never force the wedge past the narrowest section of the lower timing chain as damage to the tensioner may occur.* If a timing chain wedge is not available, one may be fabricated using a block of wood that is approximately 1/2-inch (13mm) thick and a piece of wire to pull the wedge out of the cylinder head after installation.
4 Make reference mark(s) on the lower timing chain and the idler sprocket, then remove the idler sprocket bolt **(see illustration)**. The idler gear/idler shaft and the lower timing chain should stay engaged together and rest on the timing chain wedge as the remaining steps of this procedure are performed **(see illustration)**.
5 Remove the camshafts and lifters (see Section 8). **Note:** *It is only necessary to follow

11.4a Make reference mark(s) on the idler gear and the lower timing chain, then remove the idler shaft retaining bolt (arrow) . . .

Steps 7 through 9 in the camshaft removal procedure, since the camshaft sprockets and the upper timing chain are already removed.
6 Remove the exhaust manifold (see Section 10).
7 Label and remove any remaining items attached to the cylinder head, such as coolant fittings, oil dipstick tube, cables, hoses or wiring harness.
8 Using a breaker bar and the appropriate-sized hex bit, loosen the cylinder head bolts in 1/4-turn increments until they can be removed by hand **(see illustrations)**. Loosen

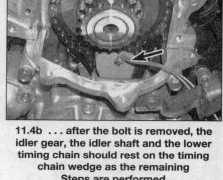

11.4b . . . after the bolt is removed, the idler gear, the idler shaft and the lower timing chain should rest on the timing chain wedge as the remaining Steps are performed

the bolts in the reverse of the tightening sequence **(see illustration 11.20)** to avoid warping or cracking the cylinder head.
9 Lift the cylinder head off the engine block. If it's stuck, very carefully pry up at a casting protrusion, beyond the gasket surface **(see illustration)**. The cylinder head is easiest to remove with the intake manifold in place.
10 Remove all external components from the cylinder head to allow for thorough cleaning and inspection. **Note:** *See Chapter 2, Part D, for cylinder head servicing procedures.*

11.8a Use a hex bit and a long extension to loosen the cylinder head bolts

11.8b Don't overlook this small bolt (arrow) retaining the front of the cylinder head to the lower timing chain cover

11.9 Pry the cylinder head off at a casting protrusion - do not pry under the gasket surface

Chapter 2 Part A 2.4L four cylinder engine

11.12 Remove all traces of old gasket material - the cylinder head and block mating surfaces must be perfectly clean to ensure a good gasket seal

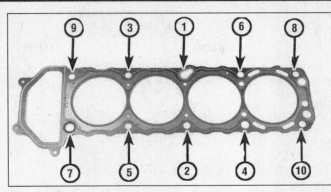

11.20 Cylinder head bolt TIGHTENING sequence

Installation

Refer to illustrations 11.12 and 11.20

11 The mating surfaces of the cylinder head and block must be perfectly clean when the cylinder head is installed.

12 Use a gasket scraper to remove all traces of carbon and old gasket material from the block and the cylinder head being careful not to gouge the aluminum **(see illustration)**. Then clean the mating surfaces with lacquer thinner or acetone. If there's oil on the mating surfaces when the cylinder head is installed, the gasket may not seal correctly and leaks could develop. When working on the block, stuff the cylinders with clean shop rags to keep out debris. Use a vacuum cleaner to remove material that falls into the cylinders.

13 Check the block and cylinder head mating surfaces for nicks, deep scratches and other damage. If damage is slight, it can be removed with a fine file; if it's excessive, machining may be the only alternative.

14 Use a tap of the correct size to chase the threads in the cylinder head bolt holes, then clean the holes with compressed air - make sure that nothing remains in the holes. **Warning:** *Wear eye protection when using compressed air!*

15 Use a wire wheel or brush to remove corrosion and dirt from the bolt threads. Dirt, corrosion, sealant will affect torque readings.

16 Once the cylinder head's gasket surface is clean, check the cylinder head for warpage (see Chapter 2D). Also check the intake and exhaust manifolds for warpage.

17 Install the components that were removed from the cylinder head, including the intake manifold (see Section 9).

18 Position the new cylinder head gasket over the dowel pins in the block and carefully set the cylinder head on the block without disturbing the gasket.

19 Before installing the cylinder head bolts, apply a small amount of clean engine oil to the threads and hardened washers. The chamfered side of the washers must face the bolt heads, and the flat side of the washers must face the cylinder head.

20 Install the cylinder head bolts in their original locations and tighten them in the recommended sequence to the torque listed in this Chapter's Specifications **(see illustration)**.

21 Refer to Section 8 and install the lifters, shims and the camshafts on the cylinder head. If any machine work was done to the cylinder head (a valve job), it will be necessary to check and adjust the valve clearances (see Section 5).

22 Install the idler gear/idler shaft bolt hand tight making sure the reference marks on the lower timing chain and idler gear (made in Step 4) are still aligned, then remove the timing chain wedge from the lower timing chain.

23 Refer to Section 7, Steps 36 through 38 and install the upper timing chain and the upper timing chain cover. Make sure to tighten the small bolt connecting the cylinder head to the lower timing cover to the same specifications as the timing cover bolts.

24 The remaining installation steps are the reverse of removal.

25 Change the engine oil and filter (see Chapter 1).

26 Refill the cooling system (see Chapter 1), then run the engine and check for leaks.

12 Crankshaft pulley – removal and installation

Refer to illustrations 12.4 and 12.5

1 Disconnect the negative cable from the battery. Raise the front of the vehicle and secure it on jackstands.

2 Remove the engine under cover. Remove the lower radiator shroud and remove the engine cooling fan (see Chapter 3).

3 Remove the drivebelts (see Chapter 1).

4 Use a strap wrench around the crankshaft pulley to hold it while using a breaker bar and socket to remove the crankshaft pulley center bolt **(see illustration)**.

5 Wedge a prybar or two screwdrivers behind the crankshaft pulley and carefully pry it off the crankshaft **(see illustration)**. If the pulley is difficult to remove, use a jaw-type puller and pull it off.

6 Installation is the reverse of removal. Tighten the pulley bolt to the torque listed in this Chapter's Specifications.

7 Adjust the drivebelts (see Chapter 1).

13 Crankshaft front oil seal – replacement

Crankshaft front oil seal replacement for the 2.4L four cylinder engine is identical to the crankshaft front oil seal replacement procedure for the 3.5L V6 engine. Refer to Section 12 for the crankshaft pulley removal procedure, then refer to Chapter 2C for the front oil seal removal procedure. Be sure to use the torque figures in this Chapter's Specifications.

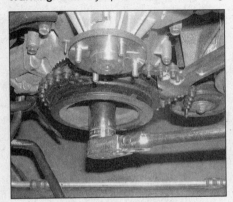

12.4 Use a strap wrench to hold the crankshaft pulley while removing the center bolt (a chain-type wrench may be used if you wrap a section of old drivebelt around the crankshaft pulley first)

12.5 If the pulley is difficult to remove, place a three-jaw puller around the center hub and pull it off

2A-14   Chapter 2 Part A   2.4L four cylinder engine

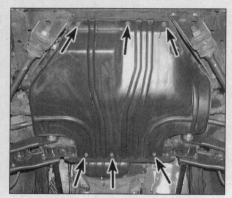

14.3 Remove the engine under cover (if equipped)

14.7 Lower crossmember mounting bolts – 2WD models

14.13 Apply a bead of RTV sealant around the perimeter of the oil pan

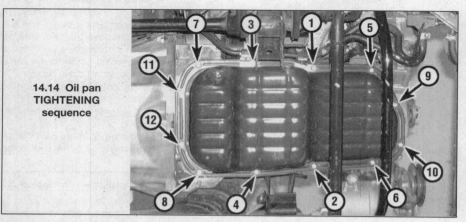

14.14 Oil pan TIGHTENING sequence

15.2 Mark the position of the rotor with the number one cylinder at TDC to help aid the installation process

14   Oil pan - removal and installation

Removal

Refer to illustrations 14.3 and 14.7

1   Disconnect the cable from the negative terminal of the battery.
2   Raise the vehicle and support it securely on jackstands.
3   Remove the engine under-cover (if equipped) **(see illustration)**.
4   Drain the engine oil and remove the oil filter (see Chapter 1).
5   Unbolt and remove front exhaust pipe from the exhaust system (see Chapter 4).
6   Detach the bellhousing cover.
7   On 2WD models, remove the front suspension crossmember from beneath the oil pan **(see illustration)**.
8   On 4WD models, remove the front axle/differential assembly (see Chapter 8).
9   Remove the oil pan bolts, following the reverse of the tightening sequence **(see illustration 14.14)**.
10   Detach the oil pan. Don't pry between the pan and engine block or damage to the sealing surfaces may result and oil leaks could develop. If the pan is stuck, dislodge it with a soft-face hammer.
11   Use a gasket scraper to remove all traces of old gasket material and sealant from the engine block and pan. Clean the mating surfaces with lacquer thinner or acetone.

Installation

Refer to illustrations 14.13 and 14.14

12   Ensure that the threaded holes in the engine block are clean (use a tap to remove any sealant or corrosion from the threads).
13   Apply a continuous 5/32-inch (3.5 mm) bead of RTV sealant to the inner sealing surface of the oil pan **(see illustration)**. **Note:** *Install the oil pan within five minutes of sealant application.*
14   Install the oil pan and tighten the bolts in three or four steps following the sequence shown **(see illustration)** to the torque listed in this Chapter's Specifications.
15   The remaining installation steps are the reverse of removal.
16   Allow at least 30 minutes for the sealant to dry, add oil and a new oil filter, start the engine and check for oil pressure and leaks.

15   Oil pump and pick-up tube - removal, inspection and installation

Removal

Refer to illustrations 15.2, 15.3 and 15.4

1   Position the engine at TDC for number one cylinder (see Section 3).
2   Remove the distributor cap and mark the position of the rotor to the distributor body **(see illustration)**. This mark is necessary to help align the rotor with the number one spark plug terminal on the distributor cap during installation of the oil pump and the oil pump driveshaft.
3   The oil pump is affixed to the right side of the lower timing chain cover. Remove the oil pump mounting bolts and detach the

15.3 The oil pump is located on the outside of the lower timing chain cover - remove the bolts (arrows) and detach the oil pump from the lower timing chain cover – if the oil pump driveshaft does not come out with the oil pump, use needle-nose pliers to remove it

Chapter 2 Part A 2.4L four cylinder engine

15.4 Oil pick-up tube mounting details

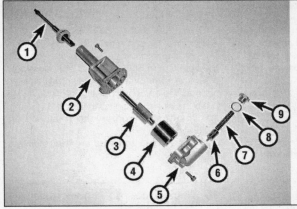

15.7 Oil pump components

1. Oil pump driveshaft
2. Oil pump body
3. Inner rotor
4. Outer rotor
5. Oil pump cover
6. Relief valve
7. Spring
8. Washer
9. Cap

15.8a Check the outer rotor-to-body clearance with a feeler gauge as shown

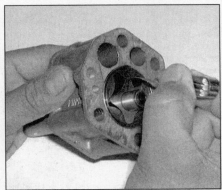

15.8b Check the clearance between the inner and outer rotor tips

15.8c With a precision straightedge placed over the pump body and rotors, check the clearance between the inner and outer rotors and the pump cover with the gasket in place

pump and the oil pump driveshaft from the engine **(see illustration)**. **Note:** *It may be necessary to use a pair of needle-nose pliers to disengage the oil pump driveshaft from the drive gear on the crankshaft. Reach up through the hole in the lower timing chain cover and pull the oil pump driveshaft downward to disengage it.*

4 To remove the oil pump pick-up tube, it will be necessary to remove the oil pan first (see Section 14), then remove the pick-up mounting bolts and detach it from the engine **(see illustration)**.

Inspection

Refer to illustrations 15.7, 15.8a, 15.8b and 15.8c

5 Remove the bolt(s) retaining the oil pump cover to the oil pump body, then lift out the inner and outer rotors. **Caution:** *Be very careful with these components, as the close tolerances are critical in creating the correct oil pressure. Any nicks or damage will require replacement of the complete pump/timing chain cover assembly.*

6 Clean all the components, including the timing chain cover and engine block gasket surfaces, with solvent, then inspect all surfaces for excessive wear and/or damage.

7 Disassemble the relief valve by removing the cap, washer, spring and regulator valve **(see illustration)**. Check the oil pressure regulator valve sliding surface and valve spring. The regulator, when clean and oiled, should slide easily in the valve bore. If either the spring or the valve is damaged, they must be replaced as a set. If no damage is found reassemble the relief valve parts, coating the parts with clean engine oil, and reinstall it in the oil pump cover.

8 Check the oil pump component clearance with a feeler gauge **(see illustrations)** and compare the results to this Chapter's Specifications. If any of the Specifications are exceeded, replace the oil pump.

Installation

Refer to illustration 15.11

9 Assemble oil pump components. Pour a generous amount of clean engine oil into the pump cavity and around the rotors. Install the cover with a new gasket to the pump body and tighten the fasteners to the torque listed in this Chapter's Specifications.

10 Install a new oil pump-to-lower timing chain cover gasket on the pump body and install several of the mounting bolts in the pump to hold the gasket in place.

11 Insert the oil pump driveshaft into the end of the oil pump and align the marks as shown **(see illustration)**. Install the oil pump on the engine with the marks on the oil pump and the oil pump driveshaft aligned. It may be necessary to wiggle the rotor in the distributor slightly, to engage the slot on the end of the distributor shaft with the tab on the end of the oil pump driveshaft.

12 With several of the oil pump mounting bolts installed hand tight, verify that the distributor rotor is aligned with the marks made in Step 2. If the marks on the distributor rotor are not aligned with the oil pump installed, repeat Step 11. When the oil pump is installed properly on the engine, the distributor rotor should be aligned with the number one spark plug terminal on the distributor cap with the number one cylinder is at TDC.

13 Install the remaining oil pump mounting

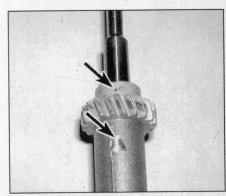

15.11 When installing the oil pump, align the mark on the oil pump driveshaft with the mark on the oil pump body

bolts and tighten the bolts to the torque listed in this Chapter's Specifications.

14 If the oil pick-up tube was removed, install it with a new gasket and reinstall the oil pan (see Section 14). Install a new oil filter and add engine oil to the crankcase (see Chapter 1).

15 Start the engine and check for oil pressure and leaks.

16 Recheck the engine oil level.

16 Flywheel/driveplate - removal and installation

The flywheel/driveplate replacement for 2.4L four cylinder engines is identical to the flywheel/driveplate replacement procedure for the V6 engines. Refer to Chapter 2 Part C for the procedure and use the torque figures in this Chapter's Specifications.

17 Rear main oil seal - replacement

Refer to illustrations 17.3, 17.5 and 17.6

1 Remove the transmission (see Chapter 7).

2 Remove the flywheel/driveplate (see Section 16).

3 Remove the rear oil seal retainer from the block **(see illustration)**.

4 Scrape any sealant or gasket material from the retainer, oil pan and the block.

5 Position the seal and retainer assembly between two wood blocks, to evenly support the aluminum housing, and drive the old seal out with a hammer and punch **(see illustration)**. A screwdriver can also be used to pry the seal out, being careful not to gouge or nick the housing during seal removal.

6 Place the new seal squarely on the retainer and drive it into the retainer with a wood block **(see illustration)** or a section of pipe slightly smaller in diameter than the outside diameter of the seal.

7 Lubricate the crankshaft seal journal and the lip of the new seal with multi-purpose grease.

8 Apply a continuous 1/8-inch bead of RTV sealant around the perimeter of the seal retainer. Also apply sealant to the bottom of the retainer (oil pan mating surface).

9 Slowly and carefully push the seal onto the crankshaft. The seal lip is stiff, so work it onto the crankshaft with a smooth object such as the end of a socket extension as you push the retainer against the block.

10 Install and tighten the retainer bolts securely.

11 The remaining steps are the reverse of removal.

18 Engine mounts - check and replacement

1 Engine mounts seldom require attention, but broken or deteriorated mounts should be replaced immediately or the added strain placed on the driveline components may cause damage or wear.

Check

2 During the check, the engine must be raised slightly to remove the weight from the mounts.

3 Raise the vehicle and support it securely on jackstands and remove the splash shields.

4 Position a jack under the engine oil pan. Place a large wood block between the jack head and the oil pan, then carefully raise the engine *just enough* to take the weight off the mounts. Do not place the wood block under the oil pan drain plug. **Warning:** *DO NOT place any part of your body under the engine when it's supported only by a jack!*

5 Check the mounts to see if the rubber is cracked, hardened or separated from the metal plates. Sometimes the rubber will split right down the center.

6 Check for relative movement between the mount plates and the engine or frame using a large screwdriver or prybar to attempt to move the mounts. If movement is noted, lower the engine and tighten the mount fasteners.

17.3 Unbolt the rear oil seal retainer from the engine block and oil pan

7 Rubber preservative should be applied to the mounts to slow deterioration.

Replacement

Refer to illustration 18.10

8 Disconnect the cable from the negative terminal of the battery, set the parking brake and block the rear wheels.

9 Raise the front of the vehicle and support it securely on jackstands. Remove the splash shields from under the vehicle.

10 Remove the engine mount-to-frame bolts and the engine mount-to-mount bracket retaining nut **(see illustration)**.

11 Attach an engine hoist to the top of the engine for lifting. **Caution:** *Do not use a jack under the oil pan to support the entire weight of the engine or the oil pump pick-up could be damaged.*

12 Raise the engine slightly until the engine mount can be removed from the vehicle.

13 To remove the mount brackets from the engine, simply detach the four retaining bolts securing the mount bracket to each side of the engine.

14 Installation is the reverse of removal. Make sure to reinstall the heat shields over the top of each mount and use thread locking compound on the mount nuts before tightening them to the torque listed in this Chapter's Specifications.

17.5 After removing the retainer from the block, support it between two wood blocks and drive out the old seal with a punch and hammer

17.6 Drive the new seal into the retainer with a wood block or a section of pipe - make sure that you don't cock the seal in the retainer bore

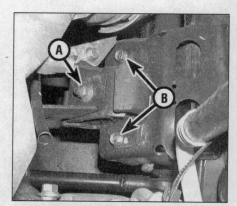

18.10 Engine mounting details

A Engine mount-to-mount bracket nut
B Engine mount-to-frame bolts

Chapter 2 Part B
3.3L V6 engine

Contents

Section		Section
Camshafts, lifters and seals - removal and installation	8	Intake manifold - removal and installation ... 9
Crankshaft front oil seal - replacement	13	Oil pan - removal and installation ... 14
Crankshaft pulley - removal and installation	12	Oil pump - removal, inspection and installation ... 15
Cylinder compression check	See Chapter 2D	Rear main oil seal - replacement ... 17
Cylinder heads - removal and installation	11	Repair operations possible with the engine in the vehicle ... 2
Drivebelt check, adjustment and replacement	See Chapter 1	Rocker arm assembly - removal, inspection and installation ... 5
Engine mounts - check and replacement	18	Spark plug replacement ... See Chapter 1
Engine oil and filter change	See Chapter 1	Timing belt and sprockets - removal and installation ... 7
Engine overhaul - general information	See Chapter 2D	Top Dead Center (TDC) for number one piston - locating ... 3
Engine - removal and installation	See Chapter 2D	Valve covers - removal and installation ... 4
Exhaust manifolds - removal and installation	10	Valve springs, retainers and seals - replacement ... 6
Flywheel/driveplate - removal and installation	16	Valves - servicing ... See Chapter 2D
General information	1	

Specifications

General

Engine designation	VG33E
Displacement	200 cubic inches (3.3 liters)
Bore	3.602 inches (91.5 mm)
Stroke	3.27 inches (83 mm)
Cylinder numbers (front-to-rear)	
Right (passenger) side	1-3-5
Left (driver's) side	2-4-6
Firing order	1-2-3-4-5-6

Cylinder location and distributor rotation diagram

Camshaft and rocker arms

Camshaft endplay	0.0012 to 0.0024 inch (0.03 to 0.06 mm)
Rocker arm shaft diameter	0.7078 to 0.7087 inch (17.979 to 18.000 mm)
Rocker arm bore diameter	0.7089 to 0.7098 inch (18.007 to 18.028 mm)
Rocker arm-to-shaft oil clearance	0.0003 to 0.0019 inch (0.007 to 0.049 mm)

Oil pump

Outer rotor-to-body clearance	0.0045 to 0.0079 inch (0.114 to 0.20 mm)
Outer rotor-to-inner rotor tip clearance	0.0071 inch (0.18 mm) maximum
Cover-to-inner gear side clearance	0.0020 to 0.0035 inch (0.05 to 0.09 mm)
Cover-to-outer gear side clearance	0.0020 to 0.0043 inch (0.05 to 0.11 mm)
Inner rotor ridge-to-body clearance	0.0018 to 0.0036 inch (0.45 to 0.91 mm)

Chapter 2 Part B 3.3L V6 engine

Torque specifications

	Ft-lbs (unless otherwise indicated)	Nm
Camshaft thrust plate bolt	58 to 65	78 to 88
Camshaft sprocket bolt	58 to 65	78 to 88
Crankshaft pulley bolt	141 to 156	191 to 211
Cylinder head bolts (in sequence - **see illustration 11.21**)		
Step one	22	29
Step two	43	59
Step three	Loosen all bolts (in reverse of tightening sequence)	
Step four	22	29
Step five	40 to 47	54 to 64
Driveplate bolts	61 to 69	83 to 93
Exhaust manifold nuts	21 to 25	28 to 32
Exhaust manifold to pre-catalyst bolts/nuts	18 to 22	25 to 29
Intake manifold		
Upper intake plenum bolts	13 to 16	18 to 22
Lower intake manifold bolts/nuts		
1997 and earlier		
Step one	26 to 43 in-lbs	3 to 5
Step two	13 to 16	18 to 22
Step three	Repeat step 2	
1998 and later		
Step one	26 to 43 in-lbs	3 to 5
Step two	70 to 84 in-lbs	8 to 10
Step three	Repeat step 2	
Oil pan bolts/nuts	62 to 70 in-lbs	7 to 8
Oil pan drain plug	22 to 29	29 to 39
Oil pick-up screen mounting bolts	142 to 150 in-lbs	16 to 17
Oil pick-up screen support bracket bolt	62 to 70 in-lbs	7 to 8
Oil pump mounting bolts		
Long	16 to 22	22 to 29
Short	56 to 70 in-lbs	6 to 8
Oil pump cover screws	56 to 70 in-lbs	6 to 8
Rocker arm shaft bolts	13 to 16	18 to 22
Timing belt tensioner nut	32 to 43	43 to 58
Timing belt cover bolts	26 to 44 in-lbs	3 to 5
Rear main oil seal retainer bolts	56 to 70 in-lbs	6 to 8
Engine mount retaining nuts	32 to 41	43 to 55
Engine mount bracket bolts	32 to 41	43 to 55
Valve cover bolts	9 to 26 in-lbs	1 to 3

1 General information

This Part of Chapter 2 is devoted to in-vehicle repair procedures for the VG33E 3.3L Single Overhead Camshaft (SOHC) V6 engine. All information concerning engine removal and installation and engine block and cylinder head overhaul can be found in Chapter 2, Part D.

The following repair procedures are based on the assumption that the engine is installed in the vehicle. If the engine has been removed from the vehicle and mounted on a stand, many of the steps outlined in this Part of Chapter 2 will not apply.

The Specifications included in this Part of Chapter 2 apply only to the procedures contained in this Part. Part D of Chapter 2 contains the Specifications necessary for cylinder head and engine block rebuilding.

2 Repair operations possible with the engine in the vehicle

Many major repair operations can be accomplished without removing the engine from the vehicle.

Clean the engine compartment and the exterior of the engine with some type of degreaser before any work is done. It will make the job easier and help keep dirt out of the internal areas of the engine.

Depending on the components involved, it may be helpful to remove the hood to improve access to the engine as repairs are performed (refer to Chapter 11 if necessary). Cover the fenders to prevent damage to the paint. Special pads are available, but an old bedspread or blanket will also work.

If vacuum, exhaust, oil or coolant leaks develop, indicating a need for gasket or seal replacement, the repairs can generally be made with the engine in the vehicle. The intake and exhaust manifold gaskets, oil pan gasket, crankshaft oil seals and cylinder head gaskets are all accessible with the engine in place.

Exterior engine components, such as the intake and exhaust manifolds, the oil pan, the oil pump, the water pump (see Chapter 3), the starter motor, the alternator, the distributor (see Chapter 5) and the fuel system components (see Chapter 4) can be removed for repair with the engine in place.

Since the cylinder heads can be removed without pulling the engine, valve component servicing can also be accomplished with the engine in the vehicle. Replacement of the camshafts, timing belt and sprockets is also possible with the engine in the vehicle, although the cylinder heads must be removed from the engine to replace the camshafts.

In extreme cases caused by a lack of necessary equipment, repair or replacement of piston rings, pistons, connecting rods and rod bearings is possible with the engine in the vehicle. However, this practice is not recommended because of the cleaning and preparation work that must be done to the components involved.

3 Top Dead Center (TDC) for number one piston - locating

Refer to illustration 3.8

The procedure for finding TDC on 3.3L V6 engines is identical to the procedure for the 3.5L V6 engines. Refer to Chapter 2 Part C for the procedure and use the cylinder location diagram in this Chapter's Specifications.

Chapter 2 Part B 3.3L V6 engine

3.8 Align the zero notch on the crankshaft pulley with the pointer on the timing belt cover (arrow) - the zero notch is the one farthest to the left when facing the front of the engine and is typically yellow in color

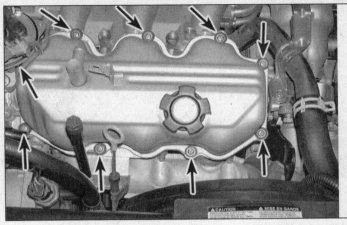

4.6 Valve cover retaining bolts (arrows)

4 Valve covers - removal and installation

Removal

Refer to illustration 4.6

1 Disconnect the cable from the negative terminal of the battery.
2 Label and detach the spark plug wires. Mark them clearly with pieces of masking tape to prevent confusion during installation.
3 If you're removing the left (driver's side) valve cover, remove the upper intake plenum (see Section 9).
4 Remove the breather hose by sliding the hose clamp back and pulling the hose off the fitting on the valve cover.
5 Detach any remaining hoses and wiring which would interfere with valve cover removal.
6 Remove the valve cover bolts and washers **(see illustration)**.
7 Detach the valve cover. *Note: If the cover is stuck to the cylinder head, bump one end with a block of wood and a hammer to jar it loose. If that doesn't work, try to slip a flexible putty knife between the cylinder head and cover to break the gasket seal. Don't pry at the cover-to-cylinder head joint or damage to the sealing surfaces may occur (leading to oil leaks in the future).*

Installation

8 The mating surfaces of each cylinder head and valve cover must be perfectly clean when the covers are installed. Use a gasket scraper to remove all traces of sealant and old gasket material, then clean the mating surfaces with lacquer thinner or acetone. If there's sealant or oil on the mating surfaces when the cover is installed, oil leaks may develop.
9 If necessary, clean the mounting bolt threads with a die to remove any corrosion and restore damaged threads. Make sure the threaded holes in the cylinder head are clean - run a tap into them to remove corrosion and restore damaged threads.
10 The gaskets should be mated to the covers before the covers are installed. Apply a thin coat of RTV sealant to the cover groove, then position the gasket inside the cover and allow the sealant to set up so the gasket adheres to the cover. If the sealant isn't allowed to set, the gasket may fall out of the cover as it's installed on the engine.
11 Carefully position the cover on the cylinder head and install the bolts.
12 Tighten the bolts in three or four steps to the torque listed in this Chapter's Specifications.
13 The remaining installation steps are the reverse of removal.
14 Start the engine and check carefully for oil leaks.

5 Rocker arm assembly - removal, inspection and installation

Removal

Refer to illustrations 5.2, 5.3 and 5.4

1 Remove the valve covers (see Section 4).
2 Loosen the rocker arm shaft retaining bolts **(see illustration)** in two or three stages, working from the ends toward the middle of the shafts. **Caution:** *Some of the valves will be open when you loosen the rocker arm shaft bolts and the rocker arm shafts will be under a certain amount of valve spring pressure. Therefore, the bolts must be loosened gradually. Loosening a bolt all at once near a rocker arm under spring pressure could distort the rocker arm shaft.*
3 Prior to removal, scribe or paint identifying marks on the rockers to ensure they will be installed in their original locations **(see illustration)**.

5.2 Loosen the rocker arm shaft bolts (arrows) a little at a time to avoid distorting the shaft

5.3 Mark the rockers to identify their locations - these are marked LI for Left Intake and LE for Left Exhaust

Chapter 2 Part B 3.3L V6 engine

5.4 The rocker arm shaft assemblies are installed with the large notches (arrows) on the intake manifold side - the small notches on the other shaft must face the exhaust manifold

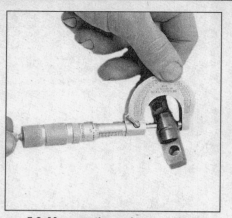

5.6 Measure the rocker arm shaft diameter at each journal where a rocker arm rides on the shaft

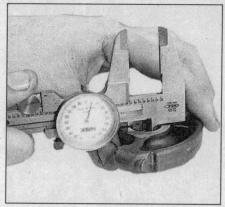

5.7 Measure the inside diameter of each rocker arm bore, subtract the corresponding rocker arm shaft diameter to obtain the clearance and compare the results to Specifications

4 Remove the bolts and lift off the rocker arm shaft assemblies one at a time. Lay them down on a nearby workbench in the same relationship to each other that they're in when installed. They must be reinstalled on the same cylinder head. Note that the shafts with the larger notches go on the intake manifold side (see illustration).

Inspection

Refer to illustrations 5.6 and 5.7

5 Check the rocker arms and shafts for abnormal wear, pits, galling, score marks and rough spots. Don't attempt to restore rocker arms by grinding the pad surfaces.
6 Measure the outside diameter of the rocker arm shaft at each rocker arm journal (see illustration). Compare the measurements to the rocker arm shaft outside diameter specified in this Chapter.
7 Measure the inside diameter of each rocker arm with either an inside micrometer or a dial caliper (see illustration). Compare the measurements to the rocker arm bore diameter specified in this Chapter.
8 Subtract the outside diameter of each rocker arm shaft journal from the corresponding rocker arm bore diameter to compute the clearance between the rocker arm shaft and the rocker arm. Compare the measurements to the clearance specified in this Chapter. If any of them fall outside the specified limits, replace either the rocker arms or the shaft, or both.

Installation

9 Installation is the reverse of the removal procedure. Tighten the rocker arm shaft retaining bolts, in several steps, to the torque listed in this Chapter's Specifications. Work from the ends of the shafts toward the middle.

6 Valve springs, retainers and seals - replacement

Refer to illustrations 6.5, 6.7, 6.13 and 6.15
Note: *Broken valve springs and defective valve stem seals can be replaced without removing the cylinder heads. Two special tools and a compressed air source are normally required to perform this operation, so read through this Section carefully. The universal shaft-type valve spring compressor required for the tight valve spring pockets of*

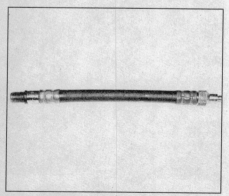

6.5 The air hose adapter threads into the spark plug hole - they're commonly available from auto parts stores

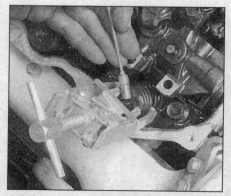

6.7 Compress the valve spring enough to release the keepers and lift them out with a magnet or needle-nose pliers

this vehicle may not be available at all tool rental yards, so check on the availability before beginning the job.
1 Remove the valve cover(s) (see Section 4).
2 Refer to Section 5 and remove the rocker arm assembly, then refer to Section 8 and remove the lifter guide assembly.
3 Remove the spark plug from the cylinder that has the defective component. If all of the valve stem seals are being replaced, all of the spark plugs should be removed.
4 Turn the crankshaft until the piston in the affected cylinder is at Top Dead Center on the compression stroke (refer to Section 3). If you're replacing all of the valve stem seals, begin with cylinder number one and work on the valves for one cylinder at a time. Move from cylinder-to-cylinder following the firing order sequence (see this Chapter's Specifications).
5 Thread a long adapter into the spark plug hole and connect an air hose from a compressed air source to it (see illustration). Most auto parts stores can supply the air hose adapter. **Note:** *Because of the length of the spark plug tubes, it will be necessary to use a long spark plug adapter with a length of hose attached (as used on many cylinder compression gauges) utilizing a quick-disconnect fitting to hook to your air source.*
6 Apply compressed air to the cylinder. **Warning:** *The piston may be forced down by the compressed air, causing the crankshaft to turn suddenly. If the wrench used when positioning the number one piston at TDC is still attached to the bolt in the crankshaft nose, it could cause damage or injury when the crankshaft moves.*
7 Stuff shop rags into the cylinder head holes around the valves to prevent parts and tools from falling into the engine, then use a valve spring compressor to compress the spring (see illustration). Remove the valve keepers with small needle-nose pliers or a magnet. **Note:** *The valves should be held in place by the air pressure. If the valve faces or*

Chapter 2 Part B 3.3L V6 engine

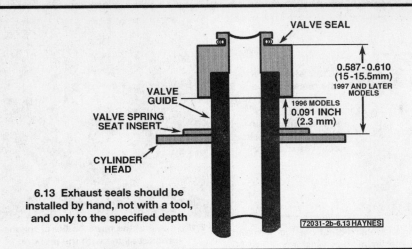

6.13 Exhaust seals should be installed by hand, not with a tool, and only to the specified depth

seats are in poor condition, leaks may prevent air pressure from retaining the valves. If the valves cannot hold air, the cylinder head should be removed for a valve job at a machine shop.

8 Remove the spring retainer, shield and valve spring, then remove the valve stem seal.
9 Wrap a rubber band or tape around the top of the valve stem so the valve won't fall into the combustion chamber, then release the air pressure.
10 Inspect the valve stem for damage. Rotate the valve in the guide and check the end for eccentric movement, which would indicate that the valve is bent.
11 Move the valve up-and-down in the guide and make sure it doesn't bind. If the valve stem binds, either the valve is bent or the guide is damaged. In either case, the cylinder head will have to be removed for repair.
12 Reapply air pressure to the cylinder to retain the valve in the closed position, then remove the tape or rubber band from the valve stem.
13 Lubricate the valve stems with engine oil and install a new valve stem seals. Intake valve seals can be installed with a special tool, or a deep socket and hammer - tap the seal only until seated. Exhaust seals should be installed by hand, not with a tool, and only to the specified depth **(see illustration)**. **Caution:** *Intake and exhaust seals are different, do not mix them up.*
14 Install the inner and outer springs in position over the valve, with the more closely-wound spring coils toward the cylinder head.
15 Install the valve spring retainer. Compress the valve springs and carefully position the keepers in the groove. Apply a small dab of grease to the inside of each keeper to hold it in place **(see illustration)**.
16 Remove the pressure from the spring tool and make sure the keepers are seated.
17 Disconnect the air hose and remove the adapter from the spark plug hole.
18 Refer to Section 8 and install the lifter assembly, then refer to Section 5 and install the rocker arm assembly.
19 Refer to Section 4 and install the valve cover.
20 Install the spark plug(s) and hook up the wire(s).
21 Start and run the engine, then check for oil leaks and unusual sounds coming from the valve cover area.

7 Timing belt and sprockets - removal and installation

Removal

Refer to illustrations 7.9, 7.12, 7.13a, 7.13b, 7.14, 7.15, 7.17 and 7.18

1 Disconnect the cable from the negative terminal of the battery.
2 Raise the front of the vehicle and support it securely on jackstands. Apply the parking brake.
3 Remove the engine under cover and drain the cooling system (see Chapter 1).
4 Remove the engine cooling fan, the cooling fan shroud and the radiator (see Chapter 3).
5 Loosen the water pump pulley nuts, then remove all of the drivebelts (see Chapter 1). Remove the water pump pulley from the water pump.
6 Remove the air conditioning compressor idler pulley and bracket. Also remove any remaining idler pulleys and brackets which would interfere with the removal of the front covers.
7 Detach the breather hoses from the valve covers and the breather pipe from the front of the engine.
8 Disconnect the upper and lower radiator hoses from the engine.
9 Disconnect the hoses and wiring harness at the top of the upper timing belt cover **(see illustration)**.
10 Remove the spark plugs (see Chapter 1). Position the number one piston at TDC on the compression stroke (see Section 3).
11 Remove the crankshaft pulley (see Section 12). **Note:** *Don't allow the crankshaft to rotate during removal of the pulley. If the crankshaft moves, the number one piston will no longer be at TDC.*
12 Remove outer timing belt guide from the crankshaft **(see illustration)**, then remove the bolts securing the timing belt upper and lower covers. Note that various types and sizes of bolts are used. They must be rein-

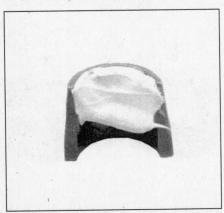

6.15 Apply a small dab of grease to each keeper as shown here before installation - it will hold them in place on the valve stem as the spring is released

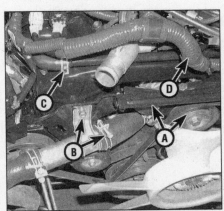

7.9 Remove the air-conditioning drivebelt idler pulley and bracket (A), the lower radiator hose and coolant tube (B) and the crankcase breather pipe (C), then unclip the main harness (D) and position it aside

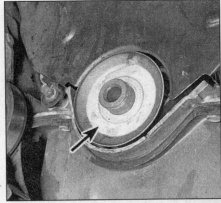

7.12 Remove the outer timing belt guide (arrow) from the crankshaft – be sure to install it with the curved lip facing out

7.13a Make sure the marks on the camshaft sprockets align with the marks on the rear timing belt cover (arrows) . . .

7.13b . . . and the mark on the crankshaft sprocket aligns with the mark on the oil pump housing (arrows)

stalled in their original locations. Mark each bolt or make a sketch to help remember where they go.

13 Confirm that the number one piston is still at TDC on the compression stroke by verifying that the timing marks on the camshaft and crankshaft sprockets are aligned with their respective stationary alignment marks **(see illustrations)**.

14 Relieve tension on the timing belt by loosening the retaining nut one half turn and rotating the timing belt tensioner 70 to 80 degrees clockwise **(see illustration)**.

15 Check to see if the timing belt is marked with an arrow indicating which side faces out **(see illustration)**. If there isn't a mark, paint one on (only if the same belt will be reinstalled). Slide the timing belt off the sprockets. If the belt is cracked, worn or contaminated with oil or coolant, replace it with a new one. Check the belt tensioner and spring for wear and damage, the pulley should rotate smoothly.

16 Make sure the camshaft and crankshaft sprockets are in good condition - if they're worn or damaged, replace them as described in Steps 17 through 19. **Note:** *It is not necessary to remove the camshaft and crankshaft sprockets during the timing belt replacement procedure unless they're damaged or to allow access to and removal of other components such as the cylinder head, oil pump, crankshaft front oil seal or rear timing belt cover.*

17 Insert a screwdriver through a hole in the camshaft sprocket to lock it in place while loosening the mounting bolt **(see illustration)**.

18 Once the bolt is out, the sprocket can be removed by hand. **Note:** *Each sprocket is marked with either an R or L* **(see illustration)**. *If you're removing both camshaft sprockets, don't mix them up. They must be installed on the same cam they were removed from.*

19 To replace the crankshaft sprocket, refer to Section 13.

Installation

Refer to illustrations 7.21, 7.23, 7.25 and 7.27

20 Verify that you have the correct belt for your vehicle. A new factory belt will have three white marks that ease installation by aligning exactly with the two camshaft timing marks and the crankshaft timing marks. Aftermarket belts may or may not have these marks. **Note:** *Also check the tooth design on the camshaft or crankshaft sprockets. Some models may have teeth with a SQUARE edge at the bottom of the sprocket groove, while other models are ROUNDED. The replacement belts are available as either straight-tooth or rounded, and they are NOT interchangeable. Use only a belt that matches the tooth design of your sprockets. Use of the wrong belt will cause whining noise and premature failure.*

21 If the tensioner was removed, reinstall it and make sure the spring is positioned prop-

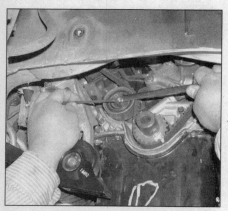

7.14 Loosen the lock nut and rotate the timing belt tensioner 70 to 80 degrees clockwise to relieve belt tension

7.15 The timing belt should be marked (arrow) to indicate which side faces out - if not, use chalk to make an arrow on the belt before removal

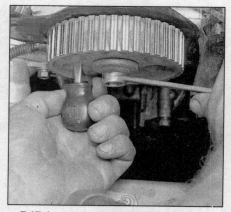

7.17 Insert a screwdriver through the camshaft sprocket to hold it while loosening the bolt

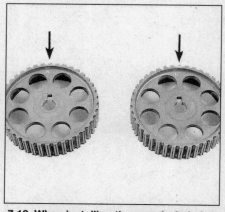

7.18 When installing the camshaft timing belt sprockets, note the R and L marks (arrows) which designate the right cylinder bank and left cylinder bank sprockets - don't mix them up!

Chapter 2 Part B 3.3L V6 engine

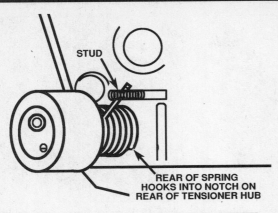

7.21 Belt tensioner spring mounting details (if the stud is removed, use a thread locking compound on the threads during installation)

7.23 When correctly installed, the punch marks on the crankshaft and camshaft sprockets must align with the marks on the rear timing belt covers and the white marks on the belt (arrows) must align with the punch marks on the camshaft sprockets - don't continue with belt installation until you're sure it's correct!

erly **(see illustration)**. Prepare to install the timing belt by turning the tensioner clockwise with an Allen wrench away from the belt and temporarily tightening the locking nut as described in Step 14.

22 Install the timing belt while aligning the belt marks with the punch marks on the sprockets in a counterclockwise direction. Making sure there is no slack, place the belt around the crankshaft sprocket first, then around left camshaft sprocket, over the right camshaft sprocket and finally around the tensioner.

23 Remove any slack from the tensioner side of the belt by hand, loosen the tensioner retaining nut and rotate the tensioner 70 to 80 degrees counterclockwise (this will allow the tensioner to operate and remove all slack from the belt), then retighten the nut. Make sure all three sets of timing marks are properly aligned, and the directional arrow printed on the belt (if present) is pointing away from the engine **(see illustration)**.

24 Slowly turn the crankshaft clockwise two full revolutions, returning the number one piston to TDC on the compression stroke.

Caution: *If excessive resistance is felt while turning the crankshaft, it's an indication that the pistons are coming into contact with the valves. Go back over the procedure to correct the situation before proceeding.*

25 Midway between the right and left camshaft sprockets, push downward on the belt with 22 lbs. (97.8 N) of force and measure the belt deflection, which should be 0.51 to 0.59 inch (13 to 15 mm) **(see illustration)**.

26 If the deflection is as specified, the belt tension is adjusted properly. If not, loosen the tensioner locking nut while keeping the tensioner steady with the Allen wrench. **Note:** *Another person to help with the following procedure will be helpful.*

27 Place a 0.0197-inch (0.5 mm) thick and 1/2-inch (12.7 mm) wide feeler gauge (or a combination of gauges to obtain this thickness) adjacent to the tensioner pulley and slowly turn the crankshaft clockwise until the feeler gauge is between the belt and the tensioner pulley **(see illustration)**.

28 Keeping the tensioner steady against the belt and feeler gauges with the Allen wrench, tighten the tensioner locking nut.

29 Turn the crankshaft to remove the feeler gauge and continue turning for two revolutions and return the number one piston to TDC.

30 Recheck the belt tension as in Step 25, and readjust the belt if necessary. **Note:** *If proper tension with and old belt can't be achieved, installation of a new belt will be necessary.*

31 Install the various components removed during disassembly, referring to the appropriate Sections as necessary.

8 Camshafts, lifters and seals - removal and installation

Lifters

Refer to illustrations 8.2 and 8.3

1 Remove the valve cover (see Section 4) and the rocker arm shaft assemblies (see Section 5).

2 Secure the lifters by raising them slightly and wrapping a rubber band around each

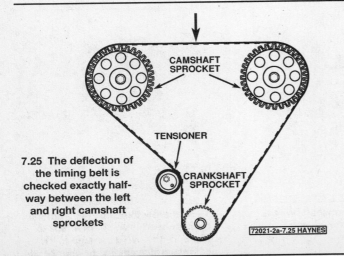

7.25 The deflection of the timing belt is checked exactly halfway between the left and right camshaft sprockets

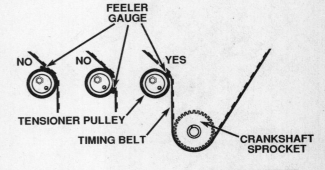

7.27 Position the feeler gauge between the tensioner pulley and the belt, then turn the crankshaft to move the feeler gauge to the point shown here (it must be exact, so work carefully)

8.2 Wrap each lifter with a rubber band so it can't fall out of the lifter guide

8.3 With the lifters retained by rubber bands, the lifter guide assembly can be removed from the cylinder head

8.12 Remove the cover plate bolts (arrows) and gently pry off the cover (left cylinder head shown, right cylinder head similar)

one to prevent them from falling out of the guides **(see illustration)**. **Note:** *If a lifter should fall out of the guide, immediately put it back in its original location.*

3 Remove the lifter guide assembly **(see illustration)**.

4 Remove the lifters from the bores one at a time. Keep them in order. Each lifter must be reinstalled in its original bore. **Caution:** *Do not lay the lifters on their side or upside down, or air can become trapped inside and the lifter will have to be bled as follows. The lifters can be laid on their side only if they are submerged in a pan of clean engine oil until reassembly.*

5 With the lifter in its bore, push down on the plunger at the top of the lifter. If it moves more than 0.040-inch (1 mm), air may be trapped inside the lifter.

6 If you think air is trapped inside a valve lifter, reinstall the rocker arm shaft assemblies and valve cover.

7 Bleed air from the lifters by running the engine at 1,000 rpm under no load for about 10 minutes.

8 Remove the valve cover and rocker arm shaft assemblies again. Repeat the procedure in Step 5 once more. If there's still air in the lifter, replace it with a new one.

9 While the lifters are out of the engine, inspect them for wear. Refer to Chapter 2D for inspection procedures.

10 Installation is the reverse of removal. Be sure to lubricate each lifter with liberal amounts of clean engine oil prior to installation.

Camshafts

Removal

Refer to illustrations 8.12, 8.13, 8.14 and 8.15

11 Remove the cylinder heads from the engine (see Section 11).

12 Remove the bolts **(see illustration)** and gently pry off the camshaft cover plate at the transmission end of the cylinder head.

13 Use the holding lugs to secure the camshaft while loosening the retaining bolt **(see illustration)**. Remove the bolt and the thrust plate.

14 Carefully pry the camshaft oil seal out of the cylinder head with a small screwdriver **(see illustration)**. Don't scratch or nick the camshaft in the process! **Note:** *Pushing on the rear of the camshaft through the thrust plate opening may help facilitate removal of the oil seal.*

15 Carefully pull the camshaft out the front of the cylinder head using a twisting motion **(see illustration)**. **Caution:** *Don't scratch the bearing surfaces with the cam lobes.* Refer to Chapter 2D for camshaft and bearing inspection procedures.

Installation

Refer to illustration 8.19

16 Lubricate the camshaft bearing journals and lobes with moly-based engine assembly lube, then install it carefully in the cylinder head. Don't scratch the bearing surfaces with the camshaft lobes!

17 Install the camshaft thrust plate and retaining bolt at the rear of the cylinder head and tighten it to the torque listed in this Chapter's Specifications.

18 With the camshaft installed in the cylinder head, refer to Chapter 2D and check the camshaft endplay.

19 Endplay outside the specified range requires thrust plate replacement. Measure the old plate **(see illustration)** and obtain a new one that will produce endplay as close to the specification as possible.

20 Install the camshaft oil seal as described in Step 22. The remainder of the cylinder head assembly is the reverse of the disassembly procedure. Refer to Section 11 for cylinder head installation.

8.13 Hold the camshaft lug with pliers or a wrench to prevent the camshaft from moving while loosening the bolt

8.14 Carefully pry the camshaft oil seal out with a small screwdriver

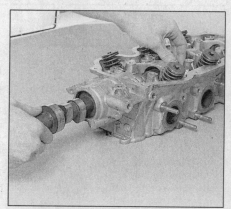

8.15 Withdraw the camshaft from the cylinder head, using both hands to support it to avoid damage to the bearing surfaces in the cylinder head

Chapter 2 Part B 3.3L V6 engine

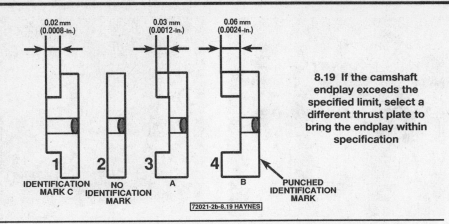

8.19 If the camshaft endplay exceeds the specified limit, select a different thrust plate to bring the endplay within specification

8.22 Use a seal driver to press the new camshaft seal squarely into place

Camshaft oil seals

Refer to illustration 8.22

21 In the course of replacing a camshaft, the old seal is removed. If you are replacing the seals only, the seals may be replaced in vehicle without removal of the cylinder head and camshafts, by removing the timing belt and camshaft sprockets and prying the seal outward. When performing this procedure be extremely careful not to nick or gouge the camshaft journal in the process (see illustration 8.14).

22 After the camshaft has been installed, use a seal installation tool (see illustration 13.7a), deep socket or piece of pipe of the appropriate diameter to press the new seal squarely into the cylinder head (see illustration). Press the seal in only until the seal bottoms.

9 Intake manifold - removal and installation

Warning: *The engine must be completely cool before beginning this procedure.*

Upper intake manifold (plenum)

Refer to illustrations 9.4 and 9.7

1 Disconnect the cable from the negative terminal of the battery.
2 Refer to Chapter 4 and remove the air intake duct.
3 Disconnect the accelerator cable, cruise control cable (if equipped), hoses and electrical connectors from the throttle body. Be prepared for some coolant spillage and plug the hoses immediately to prevent excessive coolant loss.
4 Detach the spark plug wires from the spark plugs and remove the spark plug wires from the retainers on the plenum. Remove the distributor cap from the distributor and position the distributor cap and spark plug wires aside (see illustration). Label and disconnect the hoses and electrical connectors attached to the plenum.
5 Remove the ground strap attached to the upper plenum and loosen the upper intake manifold bolts in the reverse order of the tightening sequence (see illustration 9.7). Remove the upper plenum with the throttle body attached.
6 To install the upper manifold, clean the mounting surfaces of the intake manifold, and the upper plenum with lacquer thinner and remove all traces of the old gasket material or sealant.
7 Install the new gasket over the intake manifold studs, then install the upper plenum onto the intake manifold and tighten the bolts in the proper sequence (see illustration) to the torque listed in this Chapter's Specifications. The remainder of installation is the reverse of removal. Check the coolant level and add some, if necessary (see Chapter 1).

Lower intake manifold

Refer to illustrations 9.14 and 9.19

8 Relieve the fuel pressure (see Chapter 4), then disconnect the cable from the negative terminal of the battery.
9 Remove the upper intake manifold (see Steps 1 through 5).
10 Drain the cooling system (see Chapter 1).
11 Disconnect the electrical connectors from the coolant temperature switch and the coolant temperature sensor. Also remove the radiator hose from the thermostat housing.
12 Label and detach any vacuum hoses which would interfere with the removal of the lower intake manifold.
13 Refer to Chapter 4 and remove the fuel rail and injectors from the lower intake manifold.
14 Remove the bolts and disconnect the coolant pipes at the transmission end of the intake manifold (see illustration).
15 Loosen the manifold mounting bolts/nuts

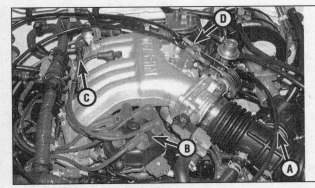

9.4 Disconnect the air intake duct (A) and the accelerator and cruise control cables (B), then detach the spark plug wires (C) and the distributor cap (D) and position them aside

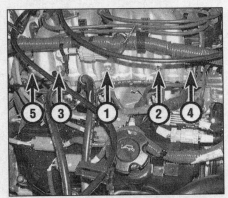

9.7 Upper plenum TIGHTENING sequence

9.14 Remove the bolts and disconnect the coolant pipes (arrow) from the rear of the lower intake manifold

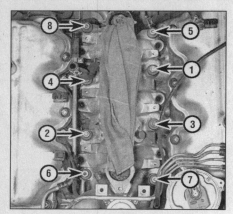

9.19 Lower intake manifold TIGHTENING sequence

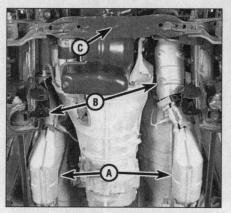

10.3 On California models, detach the front exhaust pipes (A) and the pre-catalytic converters (B) – (C is the front suspension crossmember)

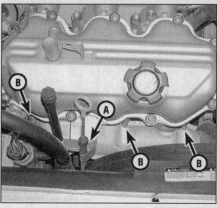

10.6 Unbolt the dipstick bracket (A) and remove the upper row of exhaust manifold nuts (B)

in 1/4-turn increments until they can be removed by hand in the reverse order of the tightening sequence **(see illustration 9.19)**.
16 The manifold will probably be stuck to the cylinder heads and force may be required to break the gasket seal. **Caution:** *Don't pry between the manifold and the heads or damage to the gasket sealing surfaces may occur, leading to vacuum leaks.*
17 Carefully use a scraper to remove all traces of old gasket material and sealant from the manifold and cylinder heads, then clean the mating surfaces with lacquer thinner or acetone.
18 Install new gaskets, then position the lower manifold on the engine. Make sure the gaskets and manifolds are aligned over the dowels in the cylinder heads and install the nuts.
19 Following the recommended tightening sequence, tighten the nuts/bolts, in three equal steps, to the torque listed in this Chapter's Specifications **(see illustration)**.
20 The remainder of the installation is the reverse of the removal procedure. Refill the cooling system and change the engine oil (see Chapter 1). Run the engine and check for fuel, vacuum and coolant leaks.

10 Exhaust manifolds - removal and installation

Caution: *The engine must be completely cool before beginning this procedure.*

Removal

Refer to illustrations 10.3, 10.6 and 10.7
1 Disconnect the cable from the negative terminal of the battery. Raise the front of the vehicle and support it on jackstands.
2 Spray penetrating oil on the exhaust manifold fasteners and allow it to soak in.
3 Working under the vehicle, disconnect the front exhaust pipes from the vehicle (see Chapter 4). **Note:** *On California models, detach the heat shields from the pre-catalytic converter, disconnect the oxygen sensor connectors and remove the pre-cat from the exhaust manifold* **(see illustration)**. If both exhaust manifolds are being removed both pre-catalytic converters or front exhaust pipes must be removed. On 2WD models, remove the front suspension crossmember first, to help facilitate to removal of the front exhaust pipes.

Left manifold

4 Remove the air intake duct and the top half of air cleaner housing as an assembly from the throttle body (see Chapter 4).
5 Disconnect the EGR flare tube from the exhaust manifold.
6 Remove the bracket holding the engine oil dipstick. Remove the heat shield from the manifold **(see illustration)** to provide access to the manifold fasteners, and remove the upper row of manifold nuts. Work from the ends toward the middle when removing the manifold fasteners.
7 Working from below, remove the lower row of manifold fasteners and remove the manifold from the engine **(see illustration)**.

Right manifold

8 Working from above, remove the manifold heat shield. Remove the upper row of manifold nuts. Work from the ends toward the middle when removing the manifold fasteners. Working from below, remove the lower row of manifold fasteners and remove the manifold from the engine.

Installation

9 Carefully inspect the manifolds and fasteners for cracks and damage.
10 Use a scraper to remove all traces of old gasket material and carbon deposits from the manifold and cylinder head mating surfaces. If the gasket was leaking, have the manifold checked for warpage at an automotive machine shop and resurfaced if necessary.
11 Position new gaskets over the cylinder head studs.
12 Install the manifold and thread the mounting nuts into place.
13 Working from the center out, tighten the nuts to the torque listed in this Chapter's Specifications in three or four equal steps.
14 Reinstall the remaining parts in the reverse order of removal. Use new gaskets when connecting the exhaust pipe flange connections.
15 Run the engine and check for exhaust leaks.

11 Cylinder heads - removal and installation

Warning: *Allow the engine to cool completely before beginning this procedure.*

Removal

1 Relieve the fuel system pressure (see Chapter 4), then disconnect the cable from the negative terminal of the battery. Refer to Chapter 1 and drain the cooling system.
2 Detach the spark plug wires from the spark plugs and remove the distributor cap and spark plug wires as an assembly from the engine.
3 Remove the intake manifold (see Section 9).
4 Remove the timing belt, camshaft sprocket(s) and the rear timing belt cover (see Section 7).
5 Remove the rocker arm components (see Section 5) and lifters (see Section 8).

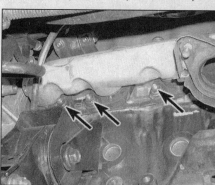

10.7 Remove the lower row of manifold nuts (arrows) from below

Chapter 2 Part B 3.3L V6 engine

11.14 Remove the old gaskets and clean the engine block and cylinder head mating surfaces thoroughly

11.18 Position the gasket over the dowel pins (arrows) so that all the holes line up

6 Remove the exhaust manifold(s) (see Section 10).

7 Label and disconnect any remaining hoses and electrical harness connectors at the timing belt end of the engine. Also remove the bolts holding the coolant pipes to the rear of each cylinder head.

8 Remove the air conditioning compressor from the engine without disconnecting the refrigerant hoses (see Chapter 3) and set it aside. Secure the compressor to the vehicle with rope or wire to make sure it doesn't hang by its hoses.

9 Remove the power steering pump and position it aside without disconnecting the fluid lines (see Chapter 10). Also remove the alternator (see Chapter 5).

10 Remove the air conditioning compressor bracket, the power steering pump bracket and alternator bracket from the engine.

11 Remove the distributor (see Chapter 5).

12 Loosen the cylinder head bolts with a hex drive tool in 1/4-turn increments until they can be removed by hand. Be sure to loosen the bolts in the reverse order of the tightening sequence, removing the small bolt, located outside of the cylinder head first **(see illustration 11.21)**.

13 Lift the cylinder head off the engine block. If resistance is felt, dislodge the cylinder head by striking it with a wood block and hammer. If prying is required, be very careful not to damage the cylinder head or engine block!

Installation

Refer to illustrations 11.14, 11.18 and 11.21

14 Remove the old cylinder head gaskets **(see illustration)**. The mating surfaces of the cylinder heads and engine block must be perfectly clean when the heads are installed.

15 Use a gasket scraper to remove all traces of carbon and old gasket material, then clean the mating surfaces with lacquer thinner or acetone. If there's oil on the mating surfaces when the heads are installed, the gaskets may not seal correctly and leaks may develop. Use a vacuum cleaner to remove any debris that falls into the cylinders.

16 Check the engine block and cylinder head mating surfaces for nicks, deep scratches and other damage. If damage is slight, it can be removed with a file - if it's excessive, machining may be the only alternative.

17 Use a tap of the correct size to chase the threads in the cylinder head bolt holes. Dirt, corrosion, sealant and damaged threads will affect torque readings. Ensure that the threaded holes in the engine block are clean and dry.

18 Position the new gaskets over the dowel pins in the engine block **(see illustration)**.

19 Carefully position the heads on the engine block without disturbing the gaskets.

20 Lightly oil the threads of the cylinder head bolts and install them in the proper locations **(see illustration 11.21)**. Tighten them finger tight. **Caution:** *Make sure the washers are in place on the bolts - the chamfered side of the washer must be against the bolt head, which means the flat side must be against the cylinder head surface.*

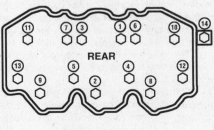

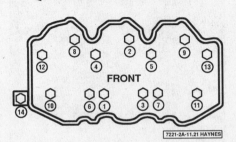

11.21 Cylinder head bolt TIGHTENING sequence - note that bolts 4, 5, 12 and 13 in the sequence are 5.0 inches long, the rest (except no. 14) are 4.17 inches in length; make sure the bolts are installed in the correct locations

21 Follow the recommended sequence and tighten the bolts (except the small bolt on the outside of the cylinder head) in five steps to the torque specified in this Chapter **(see illustration)**. **Caution:** *Bolts 4, 5, 12 and 13 in the sequence are longer than the others - be sure all bolts are in their proper locations!*

22 Tighten the small bolt on the outside of the cylinder head to 78 to 104 in-lbs (9 to 12 Nm).

23 The remaining installation steps are the reverse of removal.

24 Add coolant and change the engine oil and filter (see Chapter 1), then start the engine and check carefully for oil and coolant leaks.

12 Crankshaft pulley - removal and installation

Crankshaft pulley replacement for the 3.3L V6 engine is identical to the crankshaft pulley replacement procedure for the 2.4L four-cylinder engine. Refer to Chapter 2A for the procedure and use the torque figures in this Chapter's Specifications.

13 Crankshaft front oil seal - replacement

Refer to illustrations 13.5, 13.6, 13.7a and 13.7b

1 Remove the crankshaft pulley (see Section 12), timing belt covers and timing belt (see Section 7).

2 Carefully remove the crankshaft sprocket with a prybar or two screwdrivers, being very careful not to damage the oil pump body.

3 If the sprocket cannot be pried off, drill and tap two holes into the face of the sprocket and use a bolt-type puller to pull it off the crankshaft. **Caution:** *Do not reuse a drilled sprocket - replace it.*

4 Remove the inner timing belt guide, noting the side facing out (mark it if necessary).

5 Carefully pry the oil seal out with a screwdriver **(see illustration)**. Don't scratch or nick the crankshaft in the process!

13.5 Pry the seal out very carefully with a seal removal tool or screwdriver - if the crankshaft is nicked or otherwise damaged, the new seal will leak!

6 Before installation, apply a thin coat of multi-purpose grease to the inside of the seal **(see illustration)**.
7 Fabricate a seal installation tool with a short length of pipe of equal or slightly smaller outside diameter than the seal itself. File the end of the pipe that will bear down on the seal until it's free of sharp edges. You'll also need a long bolt of the same thread pitch as the crankshaft pulley bolt and a large washer, slightly larger in diameter than the pipe, on which the bolt head can seat **(see illustration)**. Install the oil seal by pressing it into position with the seal installation tool **(see illustration)**. When the seal is bottomed in the housing, don't turn the bolt any more or you'll damage the seal.
8 Install the inner timing belt guide onto the nose of the crankshaft.
9 Make sure the Woodruff key is in place in the crankshaft.
10 Apply a thin coat of assembly lube to the inside of the timing belt sprocket and slide it onto the crankshaft.
11 Installation of the remaining components is the reverse of removal. Refer to Section 7 for the timing belt installation and adjustment procedure. Tighten all bolts to the torque listed in this Chapter's Specifications.

14 Oil pan - removal and installation

Removal

Refer to illustrations 14.6a and 14.6b

1 Disconnect the cable from the negative terminal of the battery.
2 Raise the vehicle and support it securely on jackstands.
3 Remove the engine under-cover (if equipped).
4 Drain the engine oil and remove the oil filter (see Chapter 1).
5 Unbolt and remove front exhaust pipes from the exhaust system (see Chapter 4). Also remove the front stabilizer bar and the relay rod (see Chapter 10).
6 On vehicles equipped with automatic

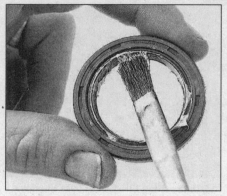

13.6 Apply multi-purpose grease or clean engine oil to the lips of the new seal before installing it (if you apply a small amount of grease to the outer edge, it will be easier to press into the bore)

transmissions detach the bellhousing cover **(see illustration)**. On vehicles equipped with manual transmissions detach the transmission support brace from the left side of the engine **(see illustration)**.
7 On 2WD models, remove the front suspension crossmember from beneath the oil pan.
8 On 4WD models, remove the front differential (see Chapter 8).
9 Support the engine/transmission securely from above with a hoist. **Warning:** *Be absolutely certain the engine/transmission is securely supported! DO NOT place any part of your body under the engine/transmission - it could crush you if the hoist fails!* Unbolt the left and right engine mounts (see Section 18) from the frame rail brackets and lift the engine enough to allow clearance for removal of the oil pan. Work carefully and be sure not to wedge any parts such as the accelerator cable against any surrounding components.
10 Remove the oil pan bolts, following the reverse of the tightening sequence **(see illustration 14.15)**.
11 Detach the oil pan. Don't pry between the pan and engine block or damage to the sealing surfaces may result and oil leaks

13.7a Fabricate a seal installation tool from a piece of pipe, a long bolt and a large washer - the outside diameter of the pipe must be the same size or slightly smaller than the outer diameter of the seal

could develop. If the pan is stuck, dislodge it with a soft-face hammer.

Installation

Refer to illustration 14.15

12 Use a gasket scraper to remove all traces of old gasket material and sealant from the engine block and pan. Clean the mating surfaces with lacquer thinner or acetone. Also ensure that the threaded holes in the engine block are clean (use a tap to remove any sealant or corrosion from the threads).
13 Apply RTV sealant to the ends of the seals and position them on the oil pump and rear seal housings.
14 Apply a continuous 5/32-inch (3.5 mm) bead of RTV sealant to the inner sealing surface of the oil pan. **Note:** *Install the oil pan within five minutes of sealant application.*
15 Install the oil pan and tighten the bolts in three or four steps following the sequence shown **(see illustration)** to the torque listed in this Chapter's Specifications.
16 The remaining installation steps are the reverse of removal.
17 Allow at least 30 minutes for the sealant to dry, add oil and a new oil filter, start the engine and check for oil pressure and leaks.

13.7b Install the seal installation tool and press the seal into the bore by tightening the bolt

14.6a On automatic transmission equipped vehicles, loosen the bolts and detach the bellhousing cover

14.6b On manual transmission equipped vehicles detach the transmission support brace (arrow)

Chapter 2 Part B 3.3L V6 engine

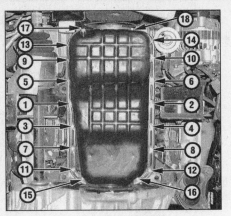

14.15 Oil pan bolt TIGHTENING sequence

15.3 Remove the power steering pump bracket bolts (arrows)

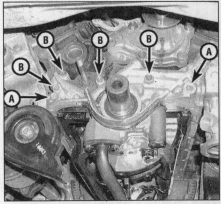

15.4 Remove the oil pump mounting bolts (A indicates the two long bolts, B the shorter bolts) and detach the pump from the engine

15 Oil pump - removal, inspection and installation

Removal

Refer to illustrations 15.3 and 15.5

1 Remove the timing belt and the crankshaft sprocket (see Section 7). Remove the oil pan (see Section 14).
2 Refer to Chapter 5 and remove the alternator and the alternator brackets from the engine.
3 Unbolt the power steering pump (see Chapter 10) and without disconnecting the hoses, position it aside. Remove the power steering pump bracket **(see illustration)**.
4 Detach the oil pick-up tube support bracket from the bottom of the engine, then remove the bolts securing the pick-up tube to the oil pump and remove the oil pick-up tube.
5 Remove the oil pump-to-engine block bolts from the front of the engine **(see illustration)**. **Note:** *The oil filter adapter can remain attached to the pump at this time if desired. One of the two long oil pump mounting bolts is the bolt removed in Step 2 that held the alternator adjusting bar on Frontier and Xterra models.*
6 Pry the oil pump outward to break the oil pump gasket seal and remove it from the dowels on the engine block.
7 Use a scraper to remove old gasket material and sealant from the oil pump and engine block mating surfaces. Clean the mating surfaces with lacquer thinner or acetone.

Inspection

Refer to illustrations 15.8, 15.10, 15.11a, 15.11b, 15.11c and 15.11d

8 Use a large Phillips screwdriver to remove the screws holding the rear cover on the oil pump **(see illustration)**.
9 Clean all components with solvent, then inspect them for wear and damage.
10 Remove the oil pressure regulator cap, washer, spring and valve **(see illustration)**. Check the oil pressure regulator valve sliding surface and valve spring. If either the spring or the valve is damaged, they must be replaced as a set.

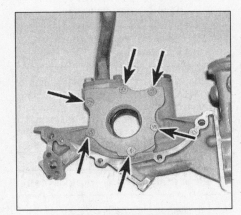

15.8 Remove the screws and lift the cover off

11 Check the clearance of the following oil pump components with a feeler gauge **(see illustrations)** and compare the measurements to the clearance listed in this Chapter's Specifications:

a) Rotor tooth tip clearance
b) Outer rotor-to-body clearance
c) Cover-to-outer rotor clearance

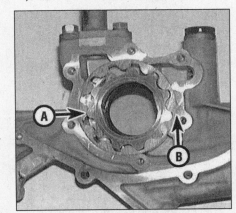

15.11a Use feeler gauges to measure the rotor tooth tip clearance (A) and the outer rotor-to-body clearance (B)

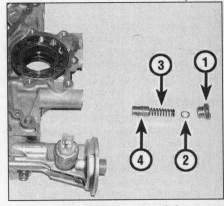

15.10 Oil pressure relief valve components

1 Cap
2 Washer
3 Spring
4 Relief valve

d) Cover-to-inner rotor clearance
e) Inner rotor ridge clearance

If any clearance is excessive, replace the entire oil pump assembly.

12 **Note:** *Pack the pump with petroleum*

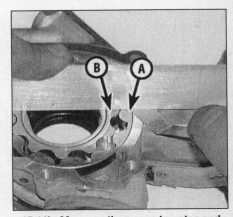

15.11b Measure the cover-to-rotor end clearance with a straightedge and feeler gauges – measure (A) above the outer rotor and (B) above the inner rotor

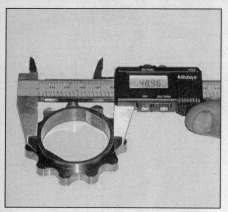

15.11c Use calipers to measure the diameter of the inner rotor ridge (the part of the rotor that rides in the pump housing) . . .

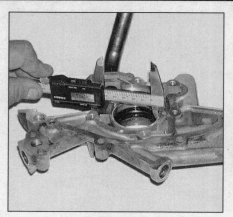

15.11d . . . and subtract the inner rotor ridge diameter from the opening in the pump body where the inner rotor rides to obtain the inner rotor ridge-to body clearance

15.13 There is a flat surface (arrow) on each side of the crankshaft - align them with the flats in the inner rotor

jelly to prime it. Assemble the oil pump and tighten the screws securely. Install the oil pressure regulator valve, spring and washer, then tighten the oil pressure regulator valve cap.

Installation

Refer to illustrations 15.13 and 15.14

13 Install the oil pump-to-block gasket over the dowels on the engine block. Align the flats on the crankshaft **(see illustration)** with the flats on the oil pump rotor, position the oil pump on the engine block and install the mounting bolts. Tighten all fasteners to the torque listed in this Chapter's Specifications.

14 Replace the O-ring on the flange of the oil pick-up tube **(see illustration)** and reinstall the tube. Tighten the pick-up tube bolts to the torque listed in this Chapter's Specifications.

15 The remainder of the installation is the reverse of removal. Be sure to reinstall the rubber end seals between the oil pump and the lower timing belt cover.

16 Flywheel/driveplate - removal and installation

The flywheel/driveplate replacement for 3.3L V6 engines is identical to the flywheel/driveplate replacement procedure for the 3.5L V6 engines. Refer to Chapter 2 Part C for the procedure and use the torque figures in this Chapter's Specifications.

17 Rear main oil seal - replacement

Refer to illustrations 17.2 and 17.3

1 The transmission must be removed from the vehicle for this procedure (see Chapter 7). **Warning:** *The engine must be supported from above with an engine hoist or three-bar support fixture before working underneath the vehicle with the transmission removed.* Remove the flywheel/driveplate (see Section 16).

2 Carefully pry out the old seal out of the retainer with a seal removal tool or screwdriver **(see illustration)**.

3 Apply multi-purpose grease to the crankshaft seal journal and the lip of the new seal. Preferably, a seal installation tool should be used to press the new seal into place. If the proper seal installation tool is unavailable, use a large socket, section of pipe or a blunt tool and carefully drive the new seal into place **(see illustration)**. The lip is stiff so carefully work it onto the seal journal of the crankshaft. Don't rush it or you may damage the seal. **Note:** *Install the seal squarely and only until flush with the back of the seal plate, no further.*

4 The remaining steps are the reverse of removal.

18 Engine mounts - check and replacement

Engine mount replacement for the 3.3L V6 engine is identical to the engine mount replacement procedure for the 3.5L V6 engine. Refer to Chapter 2 Part C for the procedure and use the torque figures in this Chapter's Specifications.

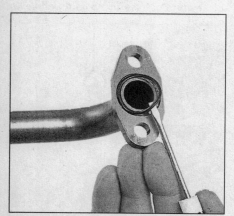

15.14 Remove the oil pick-up tube bolts and replace the rubber O-ring

17.2 Pry the seal out very carefully with a seal removal tool or screwdriver - if the crankshaft is damaged the new seal will leak!

17.3 If you don't have a seal installation tool, use a blunt tool (such as a brass punch) to carefully work the edge of the seal evenly into the bore and around the crankshaft

Chapter 2 Part C
3.5L V6 engine

Contents

Section		Section
Camshafts and lifters - removal and installation	8	Intake manifold - removal and installation ... 9
Crankshaft front oil seal - replacement	13	Oil pan - removal and installation ... 14
Crankshaft pulley - removal and installation	12	Oil pump - removal, inspection and installation ... 15
Cylinder compression check	See Chapter 2D	Rear main oil seal - replacement ... 18
Cylinder head - removal and installation	11	Repair operations possible with the engine in the vehicle ... 2
Drivebelt check, adjustment and replacement	See Chapter 1	Spark plug replacement ... See Chapter 1
Engine mounts - check and replacement	19	Timing chains and sprockets - removal, inspection and installation ... 7
Engine oil and filter change	See Chapter 1	Top Dead Center (TDC) for number one piston - locating ... 3
Engine oil cooler and adapter - general information and replacement	16	Valve clearance - check and adjustment ... 5
Engine overhaul - general information	See Chapter 2D	Valve covers - removal and installation ... 4
Engine - removal and installation	See Chapter 2D	Valve springs, retainers and seals - replacement ... 6
Exhaust manifold - removal and installation	10	Valves - servicing ... See Chapter 2D
Flywheel/driveplate - removal and installation	17	Water pump - removal and installation ... See Chapter 3
General information	1	

Specifications

General

Engine designation	VQ35DE
Displacement	213.45 cubic inches (3.5 liters)
Bore	3.76 inches (95.5 mm)
Stroke	3.205 inches (81.4 mm)
Cylinder numbers (front to rear)	
Right (passenger) side	1-3-5
Left (driver's) side	2-4-6
Firing order	1-2-3-4-5-6

Cylinder location diagram

Camshaft endplay

0.0045 to 0.0074 inch (0.115 to 0.188 mm)

Valve clearance (cold)

Intake	0.010 to 0.013 inch (0.26 to 0.34 mm)
Exhaust	0.011 to 0.015 inch (0.29 to 0.37 mm)

Oil pump

Outer gear-to-body clearance	0.0045 to 0.0079 inch (0.114 to 0.200 mm)
Inner gear-to-outer gear tip clearance	0.0071 inch max (0.18 mm)
Inner gear-to-housing side clearance	0.0012 to 0.0028 inch (0.03 to 0.07 mm)
Outer gear-to-housing side clearance	0.0020 to 0.0043 inch (0.05 to 0.11 mm)
Inner rotor hub-to-housing clearance	0.0018 to 0.0036 inch (0.045 to 0.091 mm)

Torque specifications

	Ft-lbs (unless otherwise indicated)	Nm
Camshaft sprocket bolts	65 to 72	89 to 98
Camshaft bearing cap bolts (see illustration 8.13a)		
Step one (tighten the No.1 bearing cap bolts [13 through 16] first, then follow the tightening sequence)	17 in-lbs	1.96
Step two	52 in-lbs	6
Step three	80 to 104 in-lbs	9 to 11.8
Crankshaft pulley bolt		
Step 1	29 to 36	39 to 49
Step 2	Tighten an additional 60 degrees	
Cylinder head bolts (in sequence; see illustration 11.20)		
Step one	72	98
Step two	Loosen all bolts (in reverse of tightening sequence)	
Step three	25 to 33	34 to 44
Step four	Tighten all bolts an additional 90 to 95 degrees	
Step five	Tighten all bolts an additional 90 to 95 degrees	
Valve cover bolts		
Step one	9 to 26 in-lbs	1 to 3
Step two	61 to 78 in-lbs	6.9 to 8.8
Variable valve timing cover bolts	87 to 112 in-lbs	10 to 12
Driveplate bolts	61 to 69	83 to 93
Exhaust manifold nuts	21 to 24	28 to 32
Exhaust manifold heat shield bolts	45 to 57 in-lbs	5 to 6.5
Intake manifold		
Upper plenum	13 to 15	18 to 21
Lower plenum	13 to 15	18 to 21
Intake manifold bolts/nuts		
Step one	44 to 86 in-lbs	5 to 10
Step two	20 to 23	26 to 31
Step three	Repeat Step 2 several more times	
Oil cooler retaining bolt	26 to 32	34 to 44
Oil cooler adapter bolts	12 to 15	17 to 21
Oil pan bolts		
Aluminum oil pan	15 to 16	20 to 22
Steel oil pan	72 to 83 in-lbs	8.2 to 9.4
Oil pan to transmission	22 to 28	29 to 39
Oil pan drain plug	22 to 28	30 to 39
Oil pick-up tube mounting bolts	15 to 16	20 to 22
Oil pressure switch	108 to 144 in-lbs	13 to 17
Oil pump mounting bolts	75 to 95 in-lbs	8.4 to 10.8
Oil pump cover screws	52 to 70 in-lbs	6 to 8
Front timing chain cover bolts		
6 mm	105 to 120 in-lbs	11.8 to 13.7
8 mm	19 to 23	26 to 31
Rear timing chain cover bolts	106 to 115 in-lbs	12 to 13
Upper timing chain guide(s) bolts	61 to 82 in-lbs	6.9 to 9.3
Main timing chain tensioner bolts	61 to 82 in-lbs	6.9 to 9.3
Main timing chain guide pivot bolt	108 to 168 in-lbs	13 to 19
Secondary timing chain tensioner bolts	61 to 82 in-lbs	6.9 to 9.3
Rear main oil seal retainer bolts	72 to 82 in-lbs	8.2 to 9.3
Lower engine mount nuts	32 to 41	43 to 55
Upper engine mount nuts	43 to 58	59 to 78
Engine mount bracket bolts	32 to 41	43 to 55

1 General information

This Part of Chapter 2 is devoted to in-vehicle repair procedures for the VQ35DE 3.5L Dual Overhead Camshaft (DOHC) V6 engine. All information concerning engine removal and installation and engine block and cylinder head overhaul can be found in Part D of this Chapter.

The following repair procedures are based on the assumption that the engine is installed in the vehicle. If the engine has been removed from the vehicle and mounted on a stand, many of the steps outlined in this Part of Chapter 2 will not apply.

The Specifications included in this Part of Chapter 2 apply only to the procedures contained in this Part. Part D of Chapter 2 contains the Specifications necessary for cylinder head and engine block rebuilding.

2 Repair operations possible with the engine in the vehicle

Many major repair operations can be accomplished without removing the engine from the vehicle.

Clean the engine compartment and the exterior of the engine with some type of degreaser before any work is done. It will make the job easier and help keep dirt out of

Chapter 2 Part C 3.5L V6 engine

3.5 A compression gauge can be used in the number one plug hole to assist in finding TDC

3.8 Align the TDC notch on the crankshaft pulley with the pointer on the timing chain cover (arrow) - the TDC notch is the one farthest to the left when facing the front of the engine and is typically yellow in color

the internal areas of the engine.

Depending on the components involved, it may be helpful to remove the hood to improve access to the engine as repairs are performed (refer to Chapter 11 if necessary). Cover the fenders to prevent damage to the paint. Special pads are available, but an old bedspread or blanket will also work.

If vacuum, exhaust, oil or coolant leaks develop, indicating a need for gasket or seal replacement, the repairs can generally be made with the engine in the vehicle. The intake and exhaust manifold gaskets, oil pan gasket, crankshaft oil seals and cylinder head gaskets are all accessible with the engine in place.

Exterior engine components, such as the intake and exhaust manifolds, the oil pan, the oil pump, the water pump (see Chapter 3), the starter motor, the alternator and the fuel system components (see Chapter 4) can be removed for repair with the engine in place.

Since the cylinder heads can be removed without pulling the engine, valve component servicing can also be accomplished with the engine in the vehicle. Replacement of the camshafts, timing chains and sprockets are also possible with the engine in the vehicle.

In extreme cases caused by a lack of necessary equipment, repair or replacement of piston rings, pistons, connecting rods and rod bearings is possible with the engine in the vehicle. However, this practice is not recommended because of the cleaning and preparation work that must be done to the components involved.

3 Top Dead Center (TDC) for number one piston - locating

Refer to illustrations 3.5 and 3.8

1 Top Dead Center (TDC) is the highest point in the cylinder that each piston reaches as it travels up the cylinder bore. Each piston reaches TDC on the compression stroke and again on the exhaust stroke, but TDC generally refers to piston position on the compression stroke.

2 Positioning the piston(s) at TDC is an essential part of many procedures such as valve timing, camshaft and timing chain/sprocket removal.

3 Before beginning this procedure, be sure to place the transmission in Park (automatic) or Neutral (manual) and apply the parking brake or block the rear wheels. Disable the ignition system by disconnecting the primary electrical connectors at the ignition coil packs, then remove the coil packs and spark plugs (see Chapter 1). Disable the fuel system (see Chapter 4, Section 2).

4 In order to bring any piston to TDC, the crankshaft must be turned using one of the methods outlined below. When looking at the front of the engine, normal crankshaft rotation is clockwise.

  a) The preferred method is to turn the crankshaft with a socket and ratchet attached to the bolt threaded into the front of the crankshaft. Turn the bolt in a clockwise direction.

  b) A remote starter switch, which may save some time, can also be used. Follow the instructions included with the switch. Once the piston is close to TDC, use a socket and ratchet as described in the previous paragraph.

  c) If an assistant is available to turn the ignition switch to the Start position in short bursts, you can get the piston close to TDC without a remote starter switch. Make sure your assistant is out of the vehicle, away from the ignition switch, then use a socket and ratchet as described in Paragraph (a) to complete the procedure.

5 Install a compression pressure gauge in the number one spark plug hole (refer to Chapter 2D). It should be a gauge with a screw-in fitting and a hose at least six inches long (see illustration).

6 Rotate the crankshaft using one of the methods described above while observing for pressure on the compression gauge. The moment the gauge shows pressure indicates that the number one cylinder has begun the compression stroke.

7 Once the compression stroke has begun, TDC for the compression stroke is reached by bringing the piston to the top of the cylinder.

8 Continue turning the crankshaft until the TDC notch in the crankshaft damper is aligned with the pointer on the front cover **(see illustration)**. At this point, the number one cylinder is at TDC on the compression stroke. If the marks are aligned but there was no compression, the piston was on the exhaust stroke. Continue rotating the crankshaft 360-degrees (1-turn). **Note:** *If a compression gauge is not available, you can simply place a blunt object (such as the end of a screwdriver handle) over the spark plug hole and listen for compression as the engine is rotated. Once compression at the No.1 spark plug hole is noted the remainder of the Step is the same.*

9 After the number one piston has been positioned at TDC on the compression stroke, TDC for any of the remaining cylinders can be located by turning the crankshaft 120 degrees and following the firing order (refer to the Specifications). Rotating the engine 120 degrees past TDC #1 will put the engine at TDC compression for cylinder #2.

4 Valve covers - removal and installation

Removal

Refer to illustrations 4.2 and 4.5

1 Disconnect the cable from the negative terminal of the battery.

2 Remove the engine cover **(see illustration)**, then refer to section 9 and remove the upper and lower intake manifold plenums.

3 Remove the ignition coils from the valve cover (see Chapter 5). If both valve covers are being removed, remove all six of the ignition coils.

4.2 Remove the fasteners (arrows) and detach the engine cover

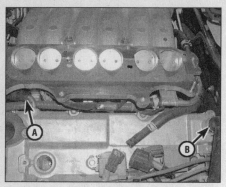

4.5 Remove the breather hose (A) and the PCV hose (B) and position aside any wiring harness that would interfere with the removal of the valve cover

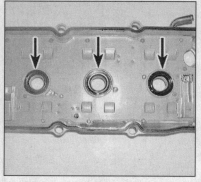

4.10 Make sure to install new spark plug tube seals (arrows) into the valve cover

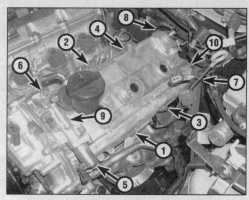

4.13 Valve cover TIGHTENING sequence

4 Remove the breather hose by sliding the hose clamp back and pulling the hose off the fitting on the valve cover.

5 Detach PCV hose and any wiring which would interfere with valve cover removal **(see illustration)**.

6 Remove the valve cover bolts and washers in the reverse order of the tightening sequence **(see illustration 4.13)**.

7 Detach the valve cover. **Note:** *If the cover is stuck to the cylinder head, bump one end with a block of wood and a hammer to jar it loose. If that doesn't work, try to slip a flexible putty knife between the cylinder head and cover to break the gasket seal. Don't pry at the cover-to-cylinder head joint or damage to the sealing surfaces may occur (leading to oil leaks in the future).*

Installation

Refer to illustrations 4.10 and 4.13

8 The mating surfaces of each cylinder head and valve cover must be perfectly clean when the covers are installed. Use a gasket scraper to remove all traces of sealant and old gasket material, then clean the mating surfaces with lacquer thinner or acetone. If there's sealant or oil on the mating surfaces when the cover is installed, oil leaks may develop.

9 If necessary, clean the mounting bolt threads with a wire wheel to remove any corrosion. Make sure the threaded holes in the cylinder head are clean - run a tap into them to remove corrosion and restore damaged threads.

10 Inspect and replace if necessary the spark plug tube sealing washers **(see illustration)**.

11 The valve cover gaskets should be mated to the covers before the covers are installed. Apply a thin coat of RTV sealant to the cover groove and to the corners on the front camshaft journal cap, then position the gasket inside the cover and allow the sealant to set up so the gasket adheres to the cover. If the sealant isn't allowed to set, the gasket may fall out of the cover as it's installed on the engine.

12 Carefully position the cover on the cylinder head and install the bolts.

13 Following the recommended tightening sequence, tighten the bolts, in two equal steps, to the torque listed in this Chapter's Specifications **(see illustration)**.

14 The remaining installation steps are the reverse of removal.

15 Start the engine and check carefully for oil leaks.

5 Valve clearance - check and adjustment

Refer to illustrations 5.6a, 5.6b, 5.7, 5.8, 5.10a, 5.10b, 5.10c and 5.11

Note: *The manufacturer recommends adjusting the valve clearance at the specified interval only if the valve train is making excessive noise. The following procedure requires the use of special valve lifter tools. The tools are available from specialty tool manufacturers and many auto parts stores. It is impossible to perform this task without them.*

1 Disconnect the cable from the negative terminal of the battery.

2 Remove the valve cover (see Section 4).

3 On manual transmission vehicles set the parking brake and place the transmission in the neutral position.

4 Remove the spark plugs (see Chapter 1).

5 Position the number 1 piston at TDC on the compression stroke and align the timing marks (see Section 3).

5.6a When the no. 1 piston is at TDC on the compression stroke, the valve clearance for the no. 1 and no. 6 cylinder intake valves and the no. 2 and no. 3 cylinder exhaust valves can be measured

6 Measure the clearance of the indicated valves with a feeler gauge **(see illustrations)**. Record each measurement and compare your measurements with the desired valve clearance found in this Chapter's Specifications. Note which are out of specification, as this data will be used later to determine the required replacement shims.

7 Turn the crankshaft 240 degrees clockwise and position the number 3 cylinder at TDC on the compression stroke. Measure

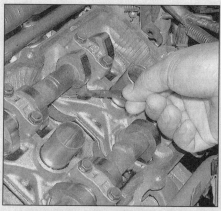

5.6b Measure the clearance for each valve with a feeler gauge of the specified thickness - if the clearance is correct, you should feel a slight drag on the gauge as you pull it out

Chapter 2 Part C 3.5L V6 engine

5.7 When the no. 3 piston is at TDC on the compression stroke, the valve clearance for the no. 2 and no. 3 cylinder intake valves and the no. 4 and no. 5 cylinder exhaust valves can be measured

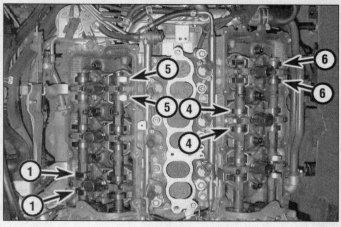

5.8 When the no. 5 piston is at TDC on the compression stroke, the valve clearance for the no. 4 and no. 5 cylinder intake valves and the no. 1 and no. 6 cylinder exhaust valves can be measured

and record the clearances of the indicated valves **(see illustration)**.

8 Turn the crankshaft an additional 240 degrees clockwise and position the number 5 cylinder at TDC on the compression stroke. Measure and record the clearances of the indicated valves **(see illustration)**.

9 After the clearance of all the valves has been measured, rotate the crankshaft pulley until the camshaft lobe above the first valve which you intend to adjust is pointing up, away from the lifter.

10 Rotate the hole in the valve shim toward the center of cylinder head casting. Place the special valve lifter tool in position as shown, with the upper jaw over the camshaft, next to the lobe and the lower jaw on top of the shim **(see illustration)**. Depress the valve lifter by squeezing the handles of the valve lifter tool together and rotating the tool away from the camshaft. Insert the small tool between the edge of the lifter and the camshaft and release the lifter. Remove the adjusting shim with a small screwdriver and a magnet or a pair of tweezers **(see illustrations)**. **Note:** *Blowing compressed air into the valve shim hole may help facilitate removal of the shim.*

11 Measure the thickness of the shim with a micrometer **(see illustration)**. To calculate the correct thickness of a replacement shim that will place the valve clearance within the specified value, use the following formula:

Intake side: $N = R + (M - 0.0118\text{-inch}\ [0.30\ mm])$

Exhaust side: $N = R + (M - 0.0130\text{-inch}\ [0.33\ mm])$

R = thickness of the old shim
M = valve clearance measured
N = thickness of the new shim

12 Select a shim with a thickness as close as possible to the valve clearance calculated. Shims are available in 64 sizes, in increments of 0.0004-inch (0.010 mm). Available shims range in sizes from 0.0913 inch (2.32 mm) to 0.1161 inch (2.95 mm). **Note:** *Through careful analysis of the shim sizes needed to bring all the out-of-specification valve clearances*

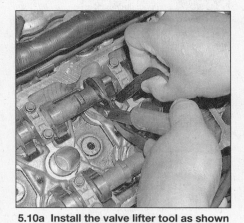

5.10a Install the valve lifter tool as shown - squeeze the handles together and rotate the tool away from the camshaft to depress the valve lifter

within specification, it is often possible to simply move a shim that has to come out anyway to another valve lifter requiring a shim of that particular size, thereby reducing the number of new shims that must be purchased.

13 Place the special valve lifter tools in position as shown in **illustration 5.10a**, depress the valve lifter and install the new

5.10c ... and remove the shim with a pair of tweezers or a magnet as shown

5.10b With the small tool wedged between the lifter and the camshaft pry the shim up with a small screwdriver at the hole in the shim ...

adjusting shim. Measure the clearance with a feeler gauge to make sure that your calculations are correct.

14 Repeat this procedure until all the valves which are out of specification have been corrected.

15 The remainder of installation is the reverse of removal.

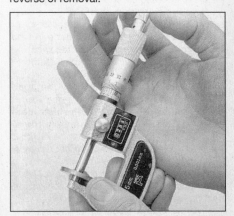

5.11 Measure the shim thickness with a micrometer or a dial caliper

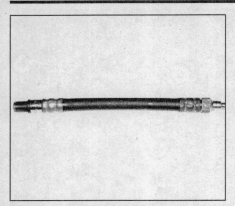

6.5 The air hose adapter threads into the spark plug hole - they're commonly available from auto parts stores

6.7a Compress the valve spring enough to release the valve stem keepers . . .

6.7b . . . and lift them out with a magnet or needle-nose pliers

6 Valve springs, retainers and seals - replacement

Refer to illustrations 6.5, 6.7a, 6.7b, 6.13 and 6.15

Note: *Broken valve springs and defective valve stem seals can be replaced without removing the cylinder heads. Two special tools and a compressed air source are normally required to perform this operation, so read through this Section carefully. The universal shaft-type valve spring compressor required for the tight valve spring pockets of this vehicle may not be available at all tool rental yards, so check on the availability before beginning the job.*

1 Remove the upper and lower intake plenum (see Section 9) and the valve cover(s) (see Section 4).
2 Refer to Section 7 and remove the timing chain, then refer to Section 8 and remove the camshafts and lifters from the affected cylinder head.
3 Remove the spark plug from the cylinder that has the defective component. If all of the valve stem seals are being replaced, all of the spark plugs should be removed.
4 Turn the crankshaft until the piston in the affected cylinder is at Top Dead Center on the compression stroke (refer to Section 3). If you're replacing all of the valve stem seals, begin with cylinder number one and work on the valves for one cylinder at a time. Move from cylinder-to-cylinder following the firing order sequence (see this Chapter's Specifications).
5 Thread a long adapter into the spark plug hole and connect an air hose from a compressed air source to it **(see illustration)**. Most auto parts stores can supply the air hose adapter. **Note:** *Because of the length of the spark plug tubes, it will be necessary to use a long spark plug adapter with a length of hose attached (as used on many cylinder compression gauges) utilizing a quick-disconnect fitting to hook to your air source.*
6 Apply compressed air to the cylinder. **Warning:** *The piston may be forced down by the compressed air, causing the crankshaft to turn suddenly. If the wrench used when positioning the number one piston at TDC is still attached to the bolt in the crankshaft nose, it could cause damage or injury when the crankshaft moves.*
7 Stuff shop rags into the cylinder head holes around the valves to prevent parts and tools from falling into the engine, then use a valve spring compressor to compress the spring **(see illustrations)**. Remove the keepers with small needle-nose pliers or a magnet. **Note:** *The valves should be held in place by the air pressure. If the valve faces or seats are in poor condition, leaks may prevent air pressure from retaining the valves. If the valves cannot hold air, the cylinder head should be removed for a valve job at a machine shop.*
8 Remove the spring retainer and valve spring, then remove the valve stem seal.
9 Wrap a rubber band or tape around the top of the valve stem so the valve won't fall into the combustion chamber, then release the air pressure.
10 Inspect the valve stem for damage. Rotate the valve in the guide and check the end for eccentric movement, which would indicate that the valve is bent.
11 Move the valve up-and-down in the guide and make sure it doesn't bind. If the valve stem binds, either the valve is bent or the guide is damaged. In either case, the

6.13 Using a deep socket and hammer, gently tap the new seals onto the valve guide only until seated

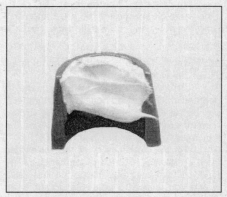

6.15 Apply a small dab of grease to each keeper as shown here before installation - it will hold them in place on the valve stem as the spring is released

cylinder head will have to be removed for repair.
12 Reapply air pressure to the cylinder to retain the valve in the closed position, then remove the tape or rubber band from the valve stem.
13 Lubricate the valve stems with engine oil and install a new valve stem seals. Valve stem seals can be installed with a special tool, or a deep socket and hammer - tap the seal only until seated **(see illustration)**.
14 Install the valve spring in position over the valve, with the more closely-wound spring coils and the paint mark toward the cylinder head.
15 Install the valve spring retainer. Compress the valve springs and carefully position the keepers in the groove. Apply a small dab of grease to the inside of each keeper to hold it in place **(see illustration)**.
16 Remove the pressure from the spring tool and make sure the keepers are seated.
17 Disconnect the air hose and remove the adapter from the spark plug hole.
18 Refer to Section 8 and install the camshaft and lifters, then refer to Section 7 and install the timing chain.
19 Refer to Section 4 and install the valve covers.
20 Install the spark plug(s), ignition coils and the upper and lower intake plenum refer-

Chapter 2 Part C 3.5L V6 engine

7.14a Remove the mounting bolts (A) and the power steering pump bracket (B) . . .

7.14b . . . then remove fan pulley bracket (arrows) from the engine

7.16a Detach the vacuum gallery . . .

7.16b . . . and the water by-pass pipe

ring to the appropriate sections as necessary.
21 Start and run the engine, then check for oil leaks and unusual sounds coming from the valve cover area.

7 Timing chains and sprockets - removal, inspection and installation

Caution: *The engine must be completely cool before beginning this procedure.*

Removal

Refer to illustrations 7.14a, 7.14b, 7.16a, 7.16b, 7.19, 7.20, 7.21, 7.22, 7.23a, 7.23b, 7.24, 7.26 and 7.27

1 Relieve the system fuel pressure (see Chapter 4).
2 Disconnect the cable from the negative terminal of the battery.
3 Remove the engine cover **(see illustration 4.2)**.
4 Remove the drivebelts (see Chapter 1) and the idler pulley brackets.
5 Remove the spark plugs (see Chapter 1). Position the number one piston at TDC on the compression stroke (see Section 3).
6 Block the rear wheels and set the parking brake. Raise the front of the vehicle and support it securely on jackstands.
7 Drain the cooling system and the engine oil (see Chapter 1).
8 Remove the crankshaft sensor (see Chapter 6).
9 Remove the crankshaft pulley (see Section 12). **Note:** *Don't allow the crankshaft to rotate during removal of the pulley. If the crankshaft moves, the number one piston will no longer be at TDC.*
10 Remove the air conditioning compressor (see Chapter 3) and position it aside without disconnecting the refrigerant lines. Also remove the air conditioning compressor bracket from the engine.
11 Remove the upper and lower oil pans (see Section 14), then lower the vehicle.
12 Remove the upper and lower intake manifold plenums (see Section 9). Remove the valve covers (see Section 4).
13 Remove the cooling fan and the radiator (see Chapter 3).
14 Remove the power steering pump (see Chapter 10) and position it aside without disconnecting the fluid lines. Also remove the fan pulley bracket and the power steering pump bracket **(see illustrations)**.
15 Remove the alternator (see Chapter 5).
16 Remove the water by-pass pipe from the front of the engine. Using masking tape and a pen, mark the location of the coolant and vacuum hoses at the front of the engine, them remove any hoses that would interfere with the removal of the water by-pass pipe **(see illustrations)**.
17 Remove the camshaft sensor and the variable valve timing (VVT) sensors from the front timing chain cover (see Chapter 6). Remove the variable valve timing actuator covers. Loosen the bolts in the reverse order of the tightening sequence **(see illustration 7.41c)** and pull the covers straight out to disengage them from the intake camshaft sprocket actuator assemblies.
18 Detach the wiring harness from the brackets at the top of the timing chain cover. Remove the bolts securing the front timing chain cover in the reverse order of the tightening sequence **(see illustration 7.40b)**. Note that various types and sizes of bolts are used. They must be reinstalled in their original locations. Mark each bolt or make a sketch to help remember where they go.
19 Remove the front timing chain cover **(see illustration)**.
20 Confirm that the number one piston is still at TDC on the compression stroke by verifying that the intake and exhaust camshaft lobes on the number one cylinder are pointing upward **(see illustration)**.
21 Relieve tension on the primary timing

7.19 Insert a screwdriver into the notch at the top of the timing cover and pry the front timing cover off the engine

7.20 Verify the engine is at TDC by confirming that the intake and exhaust camshaft lobes on the number one cylinder are pointing upward

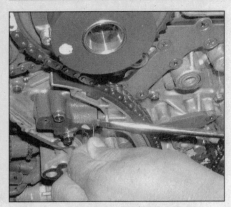

7.21 An ordinary paper clip can be straightened and used to lock the timing chain tensioner(s) in place

chain. Depress the primary tensioner inward and lock it into place by inserting a suitable stopper pin into the hole on the front of the tensioner **(see illustration)**. **Note:** *The 3.5L DOHC engine utilizes three timing chains to produce proper valve timing. The primary timing chain runs around the crankshaft sprocket, the water pump and around two intake camshaft sprockets. This chain synchronizes the valve timing with the crankshaft and pistons, while two secondary timing chains run around the rear of the intake sprockets and separate exhaust camshaft sprockets to synchronize the intake and exhaust camshaft events.*

22 Remove the primary timing chain tensioner, the tensioner pivot arm/chain guide and the upper timing chain guides from the primary timing chain **(see illustration)**.

23 Depress the secondary timing chain tensioners and lock the tensioners in place by inserting a suitable stopper pin into the hole on the front of each tensioner **(see illustrations)**.

24 Remove the camshaft sprocket bolts **(see illustration)**.

25 Disengage the primary timing chain from the teeth on the chain sprockets and remove it from the engine.

26 Mark the camshaft sprockets with either an R or L to indicate the right or left side, then remove the camshaft sprockets and the secondary timing chains from the engine. Don't mix the sprockets up. They must be installed on the same camshaft from which they were removed **(see illustration)**. **Caution:** *The intake camshaft sprockets are identified by the variable valve timing actuator and sensor ring which is fastened to the front of the sprocket. Be extremely careful not to damage or place a magnetic object of any kind near the sensor ring or a no start condition may occur after installation. Do Not disassemble the variable valve timing actuator assembly from the intake camshaft sprocket for any reason.*

27 Remove the crankshaft sprocket and the lower timing chain guide **(see illustration)**.

Inspection

Refer to illustrations 7.28a and 7.28b

28 Inspect the camshaft, water pump and crankshaft sprockets for wear on the teeth and keyways. Inspect the chains for cracks or excessive wear of the rollers. Inspect the facing of the chain guides and the secondary

7.22 Remove the primary timing chain tensioner mounting bolts (A), the tensioner arm/chain guide pivot bolt (B) and the upper chain guides (C)

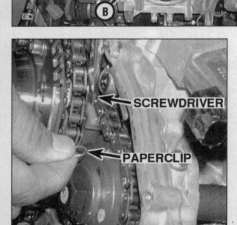

7.23b ... then depress the secondary tensioners with a screwdriver and lock them in place by inserting a paper clip into the hole on the side of each tensioner - note that the secondary tensioner on the right bank points downward, while the secondary tensioner on the left bank points upward

7.24 Hold the lug on the camshaft with a wrench to keep it from rotating as the sprocket bolts are loosened (arrows indicate two of four bolts)

7.23a Bend two paper clips so that they're long enough to protrude out past the camshaft sprocket once they're installed ...

7.26 Note that the left intake camshaft sprocket has a sensor ring (arrow) which is fastened to the front of the sprocket - be extremely careful not to damage or place a magnetic object of any kind near the sensor ring or a no start condition may occur

Chapter 2 Part C 3.5L V6 engine

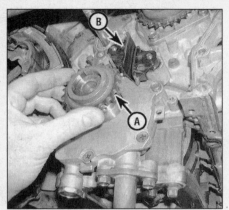

7.27 Remove the crankshaft sprocket (A) and the lower chain guide; make a note that the mark (B) on the chain guide must face up when reinstalling

7.28a Examine the chain guides for deep grooves and excessive wear - replace them if necessary

7.28b Secondary timing chain tensioner mounting bolts (A) - be sure to replace the O-ring (B) before reinstalling the front camshaft bearing cap

timing chain tensioners for excessive wear **(see illustration)**. *Note: If the secondary timing chain tensioners need to be replaced, the front camshaft bearing cap will have to be removed from the affected cylinder head to allow access to the secondary tensioner bolts* **(see illustration)**.

Installation

Refer to illustrations 7.32a, 7.32b, 7.34a, 7.34b, 7.39, 7.40a, 7.40b, 7.41a, 7.41b and 7.41c

29 Install the crankshaft sprocket and the lower timing chain guide with mark facing up **(see illustration 7.27)**.
30 Verify that you have the correct timing chains for your vehicle by counting the number of links each chain has and comparing the new chains with the old chains. Also compare the position of the colored links in the new chains with the position of the colored links in the old chains.
31 If the secondary tensioners were removed, reinstall them and make sure the tensioner spring is locked in place.
32 Make sure the camshafts are positioned with the dowels facing up **(see illustration 8.10)**. Install the secondary timing chains and sprocket assemblies on the camshafts with the timing marks aligned as shown **(see illustrations)**. Then install the camshaft sprocket bolts hand tight.
33 Reconfirm that the secondary camshaft sprocket timing marks are aligned correctly with the light colored links on the secondary timing chains and remove the stopper pins from the secondary chain tensioners **(see illustration 7.32b)**.
34 Install the primary timing chain onto the engine by looping the chain around the crankshaft sprocket and aligning the light colored chain link with the mark on the crankshaft sprocket. Then place the chain around the water pump sprocket and finally around the primary camshaft sprockets making sure the yellow colored links align with their respective marks on the sprockets **(see illustrations)**. *Note: It may be necessary to rotate the camshafts slightly in order to align the yellow colored chain links with the marks on the primary timing chain sprockets.*
35 Install the upper timing chain guides.
36 Install the primary tensioner arm/chain guide and the timing chain tensioner assembly **(see illustration 7.22)**. Reconfirm that the number one piston is still at TDC on the com-

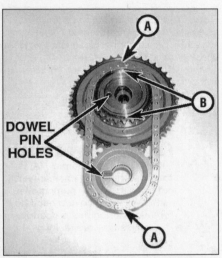

7.32a Align the gold colored links (A) on the secondary chain(s) with the marks on the camshaft sprockets (B) - this secondary timing chain assembly is for the left bank; when aligning the marks for the right bank the dowel pin holes on the sprockets should point to the right

7.32b Note that alignment of exhaust camshaft sprocket mark (B) to the gold colored link (A) can only be viewed from the front

7.34a The light colored or silver link on the timing chain (arrow) aligns with the mark on the crankshaft sprocket

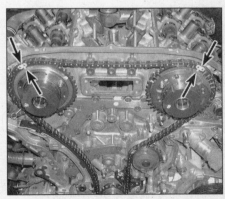

7.34b Make sure the yellow colored links (upper arrows) on the primary timing chain align with the marks (lower arrows) on the intake camshaft sprockets

7.39 Install new O-rings at the indicated area on the right and left side of the rear timing cover case

pression stroke and that the timing marks on the camshaft and crankshaft sprockets are aligned with the colored links on the chain, then remove the stopper pin from the primary timing chain tensioner.

37 Tighten the camshaft sprocket bolts to the torque listed in this Chapter's Specifications.

38 Remove all traces of old sealant from the front timing chain cover, the cover bolts and the rear cover bolt holes.

39 Install new O-rings in the variable valve timing oil control orifice of the rear timing cover **(see illustration)**.

40 Apply an 1/8-inch (3 mm) diameter bead of RTV sealant to the timing cover sealing surfaces **(see illustration)**. Place the front timing cover in position on the engine and install the bolts in their original locations. Following the recommended tightening sequence, tighten the bolts to the torque listed in this Chapter's Specifications **(see illustration)**. **Note:** *It will also be necessary to install the power steering pump bracket and the fan pulley bracket in order to tighten the bolts in the proper sequence* **(see illustrations 7.9a and 7.9b)**.

41 Install new O-rings in the VVT orifices of the front timing chain cover and on the VVT actuator covers. Then apply a 1/8-inch (3 mm) diameter bead of RTV sealant to the sealing surface of variable valve timing actuator covers **(see illustrations)**. Place the VVT covers in position over the dowels on the front timing cover and install the bolts in their original locations. Following the recommended tightening sequence, tighten the bolts to the torque listed in this Chapter's Specifications **(see illustration)**.

42 The remainder of the installation is the reverse of removal. Be sure to follow the sealant manufacturers recommendations for assembly and sealant curing times. Allow all sealant to fully cure before starting the engine.

43 Install a new oil filter, then refill the crankcase with oil and the cooling system with coolant (see Chapter 1).

44 Start the engine and check for leaks.
Note: *Timing chain noise may be apparent after performing this procedure. This noise is normal and should only last until the air has bled out of the high pressure chamber of the primary timing chain tensioner. If after several minutes the noise is still apparent, simply run the engine at 3,000 rpm with the transmission in neutral or park until the noise subsides.*

8 Camshafts and lifters - removal and installation

Note: *The camshafts and lifters should always be thoroughly inspected before installation and camshaft endplay should always be checked prior to camshaft removal. Refer to Chapter 2D for the camshaft and lifter inspection procedures.*

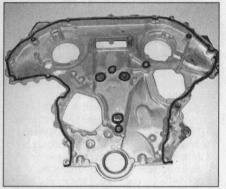

7.40a Apply RTV sealant to the front timing chain cover at the areas shown - be sure to wipe off any excess sealant

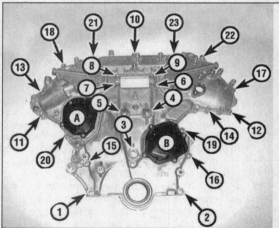

7.40b Front timing chain cover TIGHTENING sequence - (A) is the primary timing chain cover and (B) is the water pump cover; neither of these covers need be removed during this procedure unless they are leaking

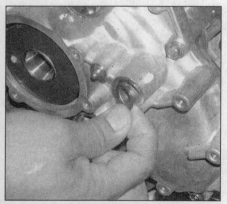

7.41a Install new O-rings in both of the VVT orifices on the front timing chain cover

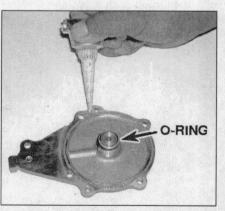

7.41b Apply a light film of engine oil to the new O-rings and install them in the groove on the VVT actuator covers, then apply an 1/8-inch (3 mm) diameter bead of RTV sealant to the indicated areas

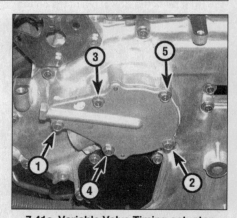

7.41c Variable Valve Timing actuator cover TIGHTENING sequence

Removal

Refer to illustrations 8.3, 8.4, 8.5, 8.6a and 8.6b

1 Detach the cable from the negative terminal of the battery.
2 Remove the upper and lower intake plenums (see section 9) and the valve covers (see Section 4).
3 Remove the timing chains and the camshaft sprockets (see Section 7). Remove the variable valve timing valve(s) from the top of the number one camshaft journal(s) **(see illustration)**.
4 Mark the camshaft bearing caps from 1 to 4, and with an "I" or an "E", to indicate intake or exhaust. Also mark arrows indicating the front of the engine **(see illustration)**. Loosen the camshaft bearing caps in two or three steps, in the reverse order of the tightening sequence **(see illustration 8.13a and 8.13b)**. **Caution:** *Keep the caps in order. They must go back in the same location they were removed from.*
5 Remove the bearing caps and the camshafts. Make a note of the camshaft markings to ensure correct installation **(see illustration)**.
6 Remove the lifters and shims from the cylinder head, keeping the proper shim with each lifter **(see illustrations)**. **Caution:** *Keep the lifters and shims in order. They must go back in the same location they were removed from.*
7 Inspect the camshaft and lifters as described in Chapter 2D.

Installation

Refer to illustrations 8.10, 8.11a, 8.11b, 8.13a and 8.13b

8 Install the lifters and shims into their original locations.
9 Apply moly-based engine assembly lubricant to the camshaft lobes and journals.
10 Install the camshafts in their original positions with the dowel pins facing up (180 degrees from the cylinder head mating surface) and inline with the cylinder bank **(see illustration)**.
11 Apply a bead of RTV sealant to the sealing surfaces of the No. 1 bearing cap(s) and the cylinder head. Install new O-rings on the secondary timing chain tensioner(s) and the VVT oil control orifice on the No. 1 bearing cap(s) **(see illustrations)**.

8.3 Remove the variable valve timing valve(s) from the No.1 camshaft bearing cap(s)

8.4 The camshaft bearing caps should be marked with a number and letter stamp or a marker to ensure correct reinstallation

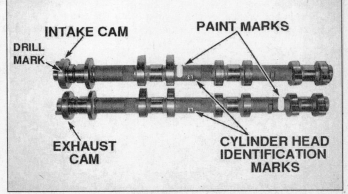

8.5 The ID mark in the center of each camshaft identifies which cylinder head the camshaft belongs to; "L" for left (front) and "R" for right (rear) - paint marks between the No. 2 and No. 3 journals indicate that it is an intake camshaft, while paint marks between the No. 3 and No. 4 journals indicate that it is an exhaust camshaft

8.6a Pull straight up to remove each lifter and shim

8.6b The lifters and shims can be stored in individually marked plastic bags or a divided box as shown

8.10 Install the camshafts with the dowel pins facing up (180 degrees from the cylinder head mating surface) and inline with the cylinder bank

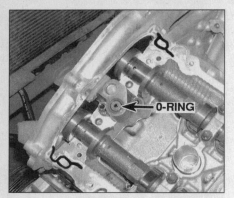

8.11a Apply RTV sealant to the cylinder head at the areas shown and install the secondary tensioner O-ring(s) - be sure to wipe off any excess sealant

8.11b Apply a small dab of grease to the VVT oil control orifice O-ring (arrow) to hold it in place on the No.1 bearing cap, then apply RTV sealant to the areas shown

12 Install the bearing caps and bolts and tighten them hand tight.
13 Tighten the bearing cap bolts in several steps, to the torque listed in this Chapter's Specifications, using the proper tightening sequence **(see illustrations)**.
14 Install the camshaft sprockets and timing chain (see Section 7). Hold the camshafts with a suitable wrench as you tighten the sprocket bolts to the specified torque.
15 The remainder of installation is the reverse of removal. If any part of the valve train was replaced, check and adjust the valve clearance (see Section 5).

9 Intake manifold - removal and installation

Upper intake plenum

Refer to illustrations 9.4, 9.7a and 9.7b

1 Relieve the fuel pressure (see Chapter 4).
2 Disconnect the cable from the negative terminal of the battery.
3 Refer to Chapter 4 and remove the air intake duct. Also remove the engine cover **(see illustration 4.2)**.
4 Label and disconnect the hoses and electrical connectors attached to the plenum and throttle body **(see illustration)**.
5 Loosen the upper intake plenum bolts in the reverse order of the tightening sequence **(see illustration 9.7b)** and remove the upper plenum with the throttle body attached.
6 To install the upper plenum, clean the mounting surfaces of the lower and the upper plenum with lacquer thinner and remove all traces of the old gasket material or sealant.
7 Install the new gaskets over the studs on the lower plenum with the marks (if equipped) facing forward **(see illustration)**, then install the upper plenum onto the lower plenum and tighten the bolts in the recommended tightening sequence **(see illustration)** to the torque listed in this Chapter's Specifications.

8.13a Camshaft bearing cap TIGHTENING sequence

8.13b After the camshaft bearing caps have been tightened in order, tighten the remaining bolts (arrows) securing the rear timing chain cover to the No.1 bearing cap(s)

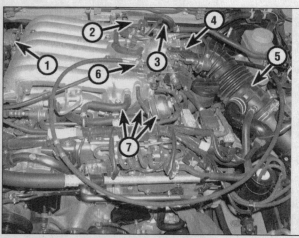

9.4 Label and disconnect the following components required for upper intake plenum removal

1 Electrical connector bracket
2 Power brake booster hose
3 EVAP purge control solenoid
4 Electrical connectors and vacuum hoses from throttle body
5 Air intake duct
6 Accelerator/cruise control cables
7 Vacuum hoses at front of manifold

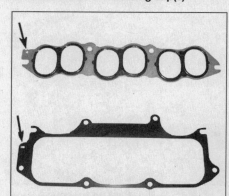

9.7a The upper plenum gaskets must be positioned with the marks (arrows) facing the front of the engine

Chapter 2 Part C 3.5L V6 engine

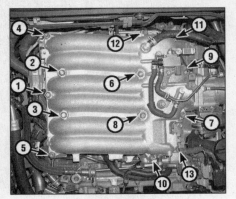

9.7b Upper plenum TIGHTENING sequence

9.10 Detach the PCV hose at the rear of the lower plenum

9.11a Lower plenum front support bracket bolts (arrows)

8 The remainder the installation is the reverse of removal.

Lower intake plenum

Refer to illustrations 9.10, 9.11a, 9.11b and 9.14

9 Follow Steps 1 through 5 and remove the upper intake plenum.
10 Label and disconnect any hoses and electrical connectors attached to the lower plenum **(see illustration)**.
11 Remove the plenum support brackets from the rear of the lower intake plenum **(see illustrations)**.
12 Loosen the lower intake plenum bolts in the reverse order of the tightening sequence **(see illustration 9.14)** and remove the lower plenum.
13 To install the lower plenum, clean the mounting surfaces of the intake manifold and the lower plenum with lacquer thinner and remove all traces of the old gasket material or sealant.
14 Install the new gaskets over the studs on the lower plenum with the marks (if equipped) facing forward, then install the lower plenum onto the intake manifold and tighten the bolts in the recommended tightening sequence **(see illustration)** to the torque listed in this Chapter's Specifications.
15 The remainder the installation is the reverse of removal.

Intake manifold

Refer to illustration 9.23

16 Remove the upper and lower intake plenum (see Steps 1 through 5 and 10 through 12).
17 Label and detach any remaining hoses which would interfere with the removal of the lower intake manifold.
18 Refer to Chapter 4 and remove the fuel rail and injectors from the lower intake manifold.
19 Loosen the manifold mounting bolts/nuts in 1/4-turn increments until they can be removed by hand in the reverse order of the tightening sequence **(see illustration 9.23)**.
20 The manifold will probably be stuck to the cylinder heads and force may be required to break the gasket seal. **Caution:** *Don't pry between the manifold and the heads or damage to the gasket sealing surfaces may occur, leading to vacuum leaks.*
21 Carefully use a scraper to remove all traces of old gasket material and sealant from the manifold and cylinder heads, then clean the mating surfaces with lacquer thinner or acetone.
22 Install new gaskets, then position the lower manifold on the engine. Make sure the gaskets and manifolds are aligned over the studs in the cylinder heads and install the nuts.

23 Following the recommended tightening sequence, tighten the nuts/bolts, in several steps, to the torque listed in this Chapter's Specifications **(see illustration)**.
24 The remainder of the installation is the reverse of the removal procedure. Run the engine and check for fuel, vacuum and coolant leaks.

10 Exhaust manifold - removal and installation

Warning: *The engine must be completely cool before beginning this procedure.*

Removal

Refer to illustrations 10.4 and 10.6

1 Disconnect the cable from the negative terminal of the battery.
2 Block the rear wheels and set the parking brake.
3 Raise the front of the vehicle and support it securely on jackstands, then remove the lower splash shield from below the engine (if equipped).
4 Refer to Chapter 4 and remove the front exhaust pipe from the vehicle. Unbolt the lower heat shields and remove the catalytic converters from the exhaust manifolds **(see illustration)**.

9.11b Lower plenum rear support bracket bolts (arrows) - access is difficult so be patient!

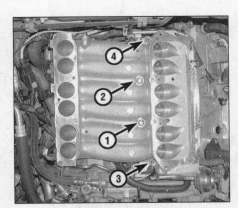

9.14 Lower plenum TIGHTENING sequence

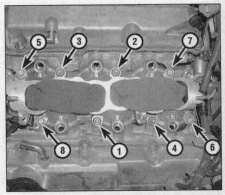

9.23 Intake manifold TIGHTENING sequence

Chapter 2 Part C 3.5L V6 engine

10.4 Detach the heat shields and remove the catalytic converters (arrows)

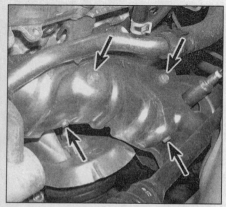

10.6 Left exhaust manifold heat shield mounting bolts (arrows) - right exhaust manifold heat shield similar

5 Disconnect the oxygen sensor connectors.
6 Remove the heat shield from the manifold(s) **(see illustration)**. **Note:** *When removing the right exhaust manifold it will be necessary to remove the coolant tube from the right side of the engine (see illustration 11.6). Also remove the power steering pump reservoir from the inner fenderwell and position it aside without disconnecting the fluid lines to allow access to the manifold.*
7 Remove the manifold-to-head nuts/bolts and detach the manifold and gaskets.

Installation

8 Use a scraper to remove all traces of old gasket material and carbon deposits from the manifold and cylinder head mating surfaces.
9 Position the new exhaust manifold gaskets over the studs on the cylinder head.
10 Install the manifold and thread the mounting nuts/bolts into place.
11 Working from the center out, tighten the nuts/bolts to the torque listed in this Chapter's Specifications in three or four equal steps.
12 Reinstall the remaining parts in the reverse order of removal.
13 Run the engine and check for exhaust leaks.

11 Cylinder head - removal and installation

Caution: *The engine must be completely cool before beginning this procedure.*
Note: *Because the coolant pipe which attaches to the rear of both cylinder heads is impossible to remove with the engine in the vehicle, it will be necessary to remove the engine from the vehicle before the cylinder heads can be removed from the engine (see Chapter 2D).*

Removal

Refer to illustrations 11.6 and 11.8

1 Refer to Section 7, Steps 1 through 26 and remove the timing chain(s) and sprockets. Disregard the Steps that do not apply, since the engine will be removed from the vehicle. **Caution:** *Be careful not to disturb the crankshaft from TDC on the compression stroke of the No.1 cylinder during the remainder of this procedure.*
2 Remove the rear timing chain cover bolts in the reverse order of the tightening sequence **(see illustration 11.22c)**.
3 Detach the rear timing cover from the engine. **Note:** *If the cover is stuck to the cylinder head or engine block, bump one end with a block of wood and a hammer to jar it loose. If that doesn't work, try to slip a flexible putty knife between the cover and the engine to break the gasket seal. Don't pry at the cover-to-cylinder head joint or damage to the sealing surfaces may occur (leading to oil leaks in the future.*
4 Remove the lower intake manifold (see Section 9) and the exhaust manifolds (see Section10).
5 Remove the camshafts and lifters from the cylinder head (see Section 8).
6 Label and remove any remaining items attached to the cylinder head, such as coolant fittings, tubes, cables, hoses, wires or brackets **(see illustration)**.
7 Using a breaker bar and the appropriate sized Allen-head socket, loosen the cylinder head bolts in 1/4-turn increments until they can be removed by hand. Loosen the bolts in the reverse order of the tightening sequence **(see illustration 11.20)** to avoid warping or cracking the head.

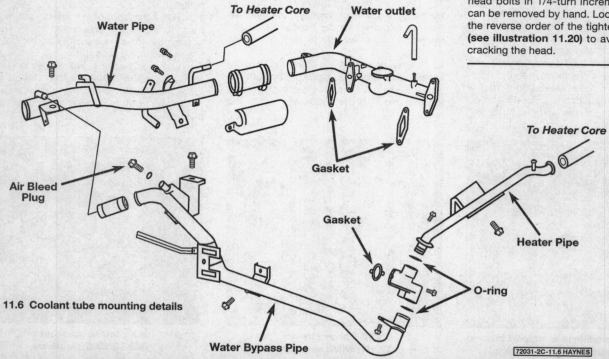

11.6 Coolant tube mounting details

Chapter 2 Part C 3.5L V6 engine

11.8 Pry on a casting protrusion to break the head loose

11.11 Carefully remove all traces of old gasket material from the sealing surfaces

8 Lift the cylinder head off the engine block. If it's stuck, very carefully pry up at a casting protrusion, beyond the gasket surface, **(see illustration)**.

9 Remove all external components from the head to allow for thorough cleaning and inspection. **Note:** *See Chapter 2, Part D, for cylinder head inspection and servicing procedures.*

Installation

Refer to illustrations 11.11, 11.14, 11.20, 11.22a, 11.22b, 11.22c and 11.22d

10 The mating surfaces of the cylinder head and block must be perfectly clean when the head is installed.

11 Use a gasket scraper to remove all traces of carbon and old gasket material from the cylinder head and engine block, then clean the mating surfaces with lacquer thinner or acetone **(see illustration)**. If there's oil on the mating surfaces when the head is installed, the gasket may not seal correctly and leaks could develop. When working on the block, stuff the cylinders with clean shop rags to keep out debris. Use a vacuum cleaner to remove material that falls into the cylinders.

12 Check the block and head mating surfaces for nicks, deep scratches and other damage. If damage is slight, it can be removed with a file; if it's excessive, machining may be the only alternative.

13 Use a tap of the correct size to chase the threads in the head bolt holes, then clean the holes with compressed air - make sure that nothing remains in the holes. **Warning:** *Wear eye protection when using compressed air!*

14 Measure each cylinder head bolt for stretching **(see illustration)**. If the diameter of the bolt threads at point A and the diameter of the bolt threads at point B differ more than 0.0043 inch (0.11 mm), the bolts have exceeded the maximum amount of stretch and will need to be replaced.

15 Check the cylinder head for warpage (see Chapter 2D). Check the head gasket, intake and exhaust manifold surfaces.

16 Install the components that were removed from the head.

17 Position the new cylinder head gasket over the dowel pins on the block noting which direction on the gasket faces up.

18 Carefully set the head on the block without disturbing the gasket.

19 Before installing the head bolts, apply a small amount of clean engine oil to the threads and hardened washers (if equipped). The chamfered side of the washers must face the bolt heads.

20 Install the bolts in their original locations and tighten them finger tight. Then tighten all the bolts in several steps, following the proper sequence **(see illustration)**, to the torque listed in this Chapter's Specifications.

21 Remove all traces of old sealant from the rear timing chain cover and the cover bolts.

22 Apply a bead of RTV sealant to the rear timing chain cover sealing surfaces **(see illustration)**. Install new O-rings in the front of the engine block and in the variable valve timing oil control orifices in the cylinder head **(see illustrations)**. Place the rear timing chain cover in position over the dowels on the engine and install the bolts in their original locations. Following the recommended tightening sequence, tighten the bolts to the torque listed in this Chapter's Specifications **(see illustration)**.

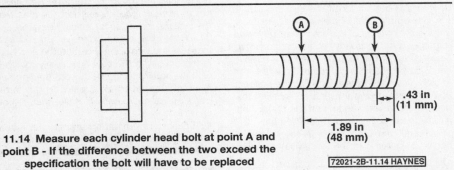

11.14 Measure each cylinder head bolt at point A and point B - If the difference between the two exceed the specification the bolt will have to be replaced

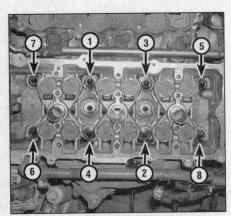

11.20 Cylinder head TIGHTENING sequence

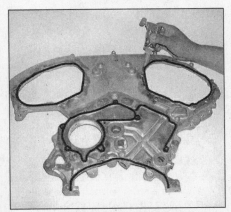

11.22a Apply RTV sealant to the rear timing chain cover at the areas shown - be sure to wipe off any excess sealant

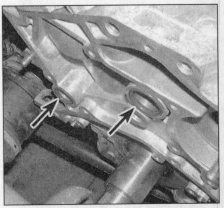

11.22b Install new O-rings (arrows) in the front of the engine block . . .

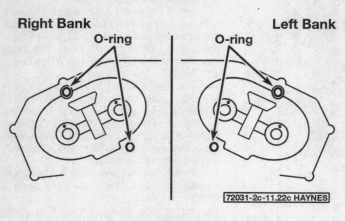

11.22c ...and in the variable valve timing oil control orifices (arrows) in the cylinder head

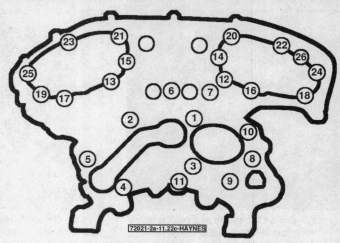

11.22d Rear timing chain cover TIGHTENING sequence

23 Install the camshafts as described in Section 8, then install the timing chains and sprockets as described in Section 7. The remaining installation steps are the reverse of removal. If any part of the valve train was replaced, check and adjust the valve clearance (see Section 5).
24 Install the engine in the vehicle.
25 Refill the cooling system, and change the engine oil and filter (see Chapter 1).
26 Start the engine and check for oil and coolant leaks.

12 Crankshaft pulley - removal and installation

Refer to illustrations 12.4 and 12.7

1 Disconnect the cable from the negative terminal of the battery.
2 Block the rear wheels and set the parking brake. Raise the front of the vehicle and support it securely on jackstands.
4 Remove the lower splash shield if equipped **(see illustration)**.
3 Remove the engine cooling fan and the fan shroud (see Chapter 3).
5 Remove the drivebelts (see Chapter 1).
6 Remove the crankshaft position sensor from the front timing chain cover (see Chapter 6).
7 Use a strap wrench around the crankshaft pulley to hold it while using a breaker bar and socket to remove the crankshaft pulley center bolt **(see illustration)**.
8 Wedge a prybar or two screwdrivers behind the crankshaft pulley and carefully pry it off the crankshaft. If the pulley is difficult to remove, place a two jaw type puller into the openings in the center of the hub and pull it off. **Caution:** *DO NOT place the puller jaws around the outside of the crankshaft pulley or damage to the pulley will occur.* **Note:** *Depending on the type of puller you have it may also be necessary to remove the radiator to gain sufficient clearance to use the puller.*
9 To install the crankshaft pulley align the pulley groove with the key on the crankshaft and slide the pulley onto the crankshaft.
10 Install the crankshaft pulley retaining bolt and tighten it to the torque listed in this Chapter's Specifications.
11 The remainder of installation is the reverse of the removal.

13 Crankshaft front oil seal - replacement

Refer to illustrations 13.2, 13.4 and 13.5

1 Remove the crankshaft pulley from the engine (see Section 12).
2 Carefully pry the seal out of the cover with a seal removal tool or a large screwdriver **(see illustration)**. **Caution:** *Be careful not to scratch, gouge or distort the area that the seal fits into or an oil leak will develop.*
3 Clean the bore to remove any old seal material and corrosion. Position the new seal in the bore with the seal lip (usually the side with the spring) facing IN (toward the engine). A small amount of oil applied to the outer edge of the new seal will make installation easier.
4 Drive the seal into the bore with a seal driver or a large socket and hammer until it's completely seated **(see illustration)**. Select a

12.4 Lower splash shield mounting details (Pathfinder model shown, other models similar)

12.7 Use a strap wrench to hold the crankshaft pulley while removing the center bolt (a chain-type wrench may be used if you wrap a section of old drivebelt around the crankshaft pulley first)

13.2 Pry the seal out very carefully with a seal removal tool or screwdriver, being careful not to nick or gouge the seal bore or the crankshaft

Chapter 2 Part C 3.5L V6 engine

2C-17

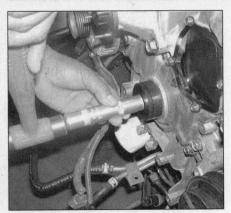

13.4 Use a large socket or seal driver to drive the new seal into the cover

13.5 If the sealing surface of the damper hub has a wear groove from contact with the seal, repair sleeves are available at most auto parts stores

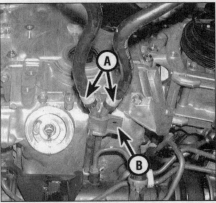

14.14 If equipped with an automatic transmission, disconnect the transmission hoses (A) and remove the clamps (B) securing the lines to the bottom of the oil pan

socket that's the same outside diameter as the seal and make sure the new seal is pressed into place until it bottoms against the cover flange.

5 Check the surface of the damper that the oil seal rides on. If the surface has been grooved from long-time contact with the seal, a press-on sleeve may be available to renew the sealing surface **(see illustration)**. This sleeve is pressed into place with a hammer and a block of wood and is commonly available from most auto parts stores.

6 Lubricate the seal lips with engine oil and reinstall the crankshaft pulley. Install the crankshaft pulley retaining bolt and tighten it to the torque listed in this Chapter's Specifications.

7 The remainder of installation is the reverse of the removal. Run the engine and check for oil leaks.

14 Oil pan - removal and installation

Removal

Refer to illustrations 14.14, 14.19, 14.22 and 14.24

1 Disconnect the cable from the negative terminal of the battery.
2 Set the parking brake and block the rear wheels.
3 Raise the front of the vehicle and sup-

port it securely on jackstands.
4 Remove the splash shields under the engine (if equipped) **(see illustration 12.4)**.
5 Drain the engine oil and remove the oil filter (see Chapter 1).
6 Disconnect the front exhaust pipe from the vehicle (see Chapter 4). Remove the catalytic converter heat shields (see Section 10).
7 Remove the starter motor (see Chapter 5).
8 Remove the brace from the suspension crossmember **(see illustration 14.19)**.
9 On 4WD models, refer to Chapter 8 and remove the front differential. Detach the front driveaxles from the front differential only. It is not necessary to disconnect the driveaxles from the steering knuckles.
10 Working at the front of the engine, remove the engine drivebelts and the lower idler pulley bracket.
11 Remove the alternator (see Chapter 5).
12 Remove the power steering pump (see Chapter 10) and set it aside without disconnecting the fluid lines. Remove the power steering pump bracket **(see illustration 7.14a)**.
13 Remove the crankshaft position sensors from the front timing chain cover and the transmission bellhousing (see Chapter 6).
14 On vehicles equipped with automatic transmissions, disconnect the transmission

lines from the bottom of the oil pan **(see illustration)**. Also disconnect the hoses from the engine oil cooler (see Section 16).
15 Disconnect the electrical connector from the oil pressure switch.
16 Loosen the steering gear lower pinch bolt (see Chapter 10).
17 Support the engine/transmission securely from above with a hoist or a three-bar engine support fixture. **Warning:** *Be absolutely certain the engine/transmission are securely supported from above - it could crush you if the support fails!*
18 Remove the lower engine mount retaining nuts (see Section 19).
19 Place a floor jack under the front suspension brace and loosen the retaining bolts **(see illustration)**. Lower the brace and the steering gear until clearance for removal of the oil pan is adequate.
20 If equipped with an automatic transmission, position the transmission lines aside.
21 Remove the lower steel pan from the upper aluminum section of the oil pan. **Caution:** *Do not pry between the steel pan and the aluminum flange or damage to the sealing surface may result.*
22 Remove the transmission mounting bolts from the upper aluminum section of the oil pan **(see illustration)**.

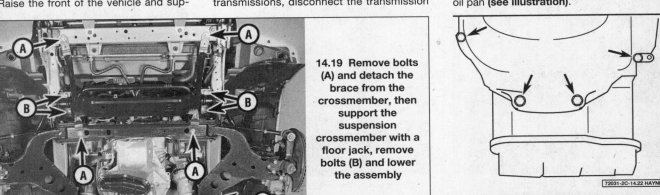

14.19 Remove bolts (A) and detach the brace from the crossmember, then support the suspension crossmember with a floor jack, remove bolts (B) and lower the assembly

14.22 Remove the transmission mounting bolts (arrows)

Chapter 2 Part C 3.5L V6 engine

23 Remove the bolts attaching the upper aluminum section of the oil pan to the block, working from the ends toward the center in the reverse order of the tightening sequence to prevent warpage.
24 To loosen the aluminum section of the oil pan, wedge a flathead screwdriver or pry bar into the notch on the side of the oil pan being careful not to damage the sealing surfaces of the oil pan and engine block (see illustration).

Installation

Refer to illustrations 14.28, 14.29 and 14.32

25 Use a scraper to remove all traces of old gasket material and sealant from the upper aluminum section of the oil pan, the lower steel pan and the engine block. Clean the mating surfaces with lacquer thinner or acetone. **Caution:** *Be careful not to scratch or gouge the gasket surface of the block or oil pan. A leak could develop after the repairs have been completed.*
26 Make sure the threaded bolt holes in the block and aluminum section of the oil pan are clean.
27 Apply a bead of RTV sealant to the ends of the front timing cover gasket and the rear oil seal retainer gasket, then place the gaskets in position on the oil pan. Apply a bead of RTV sealant around the upper aluminum oil pan flange. **Note:** *The oil pan must be installed within 15 minutes once the sealant has been applied.*
28 Install new O-rings in the engine block and the oil pump body (see illustration).
29 Carefully position the upper aluminum section of the oil pan on the engine block and install the bolts. Following the recommended sequence, tighten the fasteners in three or four steps to the torque listed in this Chapter's Specifications (see illustration).
30 Install the transmission mounting bolts (see illustration 14.22).
31 Check the lower steel oil pan flange for distortion, particularly around the bolt holes. If necessary, place the pan on a wood block and use a hammer to flatten and restore the gasket surface.

14.24 Insert a flathead screwdriver or small pry bar into the notch on the side of the oil pan to break it loose - be careful not to damage the sealing surfaces!

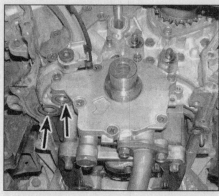

14.28 Install new O-rings in the block and the oil pump housing (arrows)

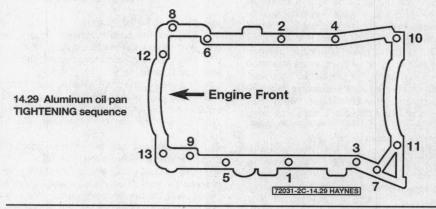

14.29 Aluminum oil pan TIGHTENING sequence

32 Apply a bead of RTV sealant around the steel oil pan flange and install the steel oil pan. **Note:** *The oil pan must be installed within 15 minutes once the sealant has been applied.* Following the recommended sequence, tighten the fasteners in several steps to the torque listed in this Chapter's Specifications (see illustration).
33 The remainder of installation is the reverse of removal. Be sure to install a new oil filter (see Chapter 1) and wait at least one hour before adding oil.

15 Oil pump - removal, inspection and installation

Removal

Refer to illustrations 15.2 and 15.3

1 Refer to Section 7 and remove the primary timing chain and the crankshaft sprocket. **Note:** *It is not necessary to remove the camshaft sprockets, the camshaft sprocket bolts, the secondary timing chains*

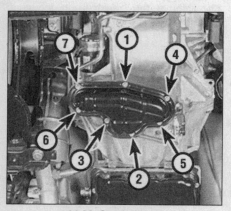

14.32 Steel oil pan TIGHTENING sequence

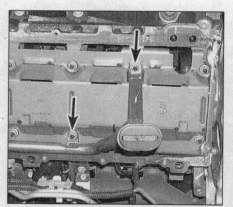

15.2 Remove the bolts (arrows) securing the oil pump pick-up tube support bracket

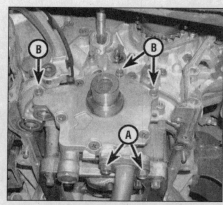

15.3 Remove the mounting bolts (A) and detach the oil pick-up tube from the oil pump, then remove the oil pump housing retaining bolts (B)

Chapter 2 Part C 3.5L V6 engine

15.5 Remove the screws (arrows) and lift the cover off

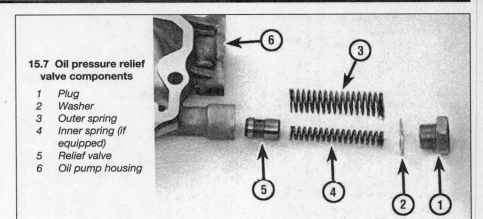

15.7 Oil pressure relief valve components

1. Plug
2. Washer
3. Outer spring
4. Inner spring (if equipped)
5. Relief valve
6. Oil pump housing

or the primary timing chain tensioner pivot arm/chain guide during this procedure. Simply pivot the tensioner arm/chain guide over to the left side to allow removal of the oil pump housing.

2 Remove the oil pans (see Section 14). Remove the oil pump pick-up tube **(see illustration)**.

3 Remove the oil pump-to-engine block bolts from the front of the engine **(see illustration)**.

4 Gently pry the oil pump housing outward enough to clear the dowel pins on the engine block and remove it from the engine.

Inspection

Refer to illustrations 15.5, 15.7, 15.8a, 15.8b, 15.8c, 15.8d and 15.8e

5 Use a large Phillips screwdriver to remove the screws holding the front cover on the oil pump housing **(see illustration)**.

6 Clean all components with solvent, then inspect them for wear and damage.

7 Remove the oil pressure regulator cap, washer, spring(s) and valve **(see illustration)**. Check the oil pressure regulator valve sliding surface and valve spring. If either the spring

15.8a Use feeler gauges to measure the rotor tooth tip clearance . . .

or the valve is damaged, they must be replaced as a set.

8 Check the clearance of the following oil pump components with a feeler gauge **(see illustrations)** and compare the measurements to the clearance listed in this Chapter's Specifications:

a) Rotor tooth tip clearance
b) Outer rotor-to-body clearance
c) Cover-to-inner rotor clearance

15.8b . . . and the outer rotor-to-body clearance

d) Cover-to-outer rotor clearance
e) Inner rotor ridge clearance

If any clearance is excessive, replace the entire oil pump assembly.

9 **Note:** *Pack the pump with petroleum jelly to prime it.* Assemble the oil pump and tighten the screws securely. Install the oil pressure regulator valve, spring and washer, then tighten the oil pressure regulator valve cap.

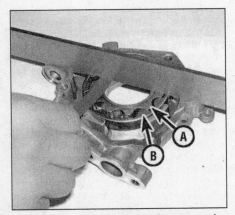

15.8c Measure the cover-to-rotor end clearance with a straightedge and feeler gauge - measure (A) above the inner rotor and (B) above the outer rotor

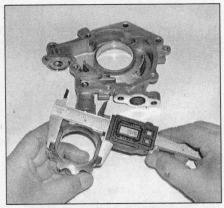

15.8d Use calipers to measure the diameter of the inner rotor ridge (the part of the inner rotor that rides in the pump body) . . .

15.8e . . . and subtract the inner rotor ridge diameter from the opening in the pump body where the inner rotor rides to obtain the inner rotor ridge-to-body clearance

15.10 There is a flat surface (arrow) on each side of the crankshaft - align them with the flats on the inner gear

16.3 Engine oil cooler mounting details - automatic transmission models only

- A Inlet hose
- B Outlet hose
- C Oil cooler retaining bolt
- D Oil pressure sending unit
- E Oil cooler adapter

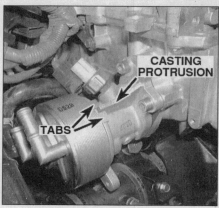

16.8 Align the tabs on the oil cooler with the casting protrusion on the oil cooler adapter

Installation

Refer to illustration 15.10

10 Use new gaskets (where applicable) on all disassembled parts and reverse the removal procedure for installation. Align the flats on the crankshaft **(see illustration)** with the flats on the oil pump gear. Tighten all fasteners to the torque listed in this Chapter's Specifications. **Note:** *Before installing the oil pan, be sure to replace the O-rings on the oil pump housing and engine block* **(see illustration 14.28).**

16 Engine oil cooler and adapter - general information and replacement

General information

1 Engines equipped with automatic transmissions are provided extra engine cooling by an oil cooler which is mounted to an adapter at the front of the oil pan next to the oil filter. The oil filter adapter doubles as a housing to which the oil pressure sending unit and the oil cooler are mounted to. The oil cooler adapter also incorporates an oil pressure relief valve, which redirects oil flow to bypass the oil cooler when pressures are too high. The oil cooler has two hoses connecting the cooler to the engine, where oil temperature is lowered by the radiator coolant. The hose that connects to the passenger side water pipe is the inlet side of the oil cooler and the hose that connects to the main thermostat housing (at the front of the engine) is the outlet side of the oil cooler. Engine oil coolers are not installed on engines with manual transmissions.

Replacement

Refer to illustrations 16.3 and 16.8

Warning: *The engine must be completely cool before beginning this procedure.*

2 Drain the engine oil and the cooling system (see Chapter 1).
3 Detach the hose clamps and remove the inlet and outlet hoses from the oil cooler **(see illustration).**
4 Disconnect the electrical connector from the oil pressure sending unit and remove the oil pressure sending unit from the oil cooler adapter.
5 Loosen the oil cooler retaining bolt and remove the oil cooler and O-rings from the adapter.
6 If it's necessary to remove the oil cooler adapter, simply remove the three retaining bolts and separate the adapter from the upper aluminum section of the oil pan. Be sure to remove the old gasket and thoroughly clean the mating surfaces of the oil pan and the adapter before installing the oil cooler adapter and a new gasket back onto the oil pan. Tighten the adapter bolts to the torque listed in this Chapter's Specifications, then install the oil pressure sending unit back onto the adapter using pipe sealant on the threads.
7 Using a small amount of engine oil, lubricate the oil cooler O-rings. Install the large O-ring in the groove on the oil cooler and the small O-ring over the end of the oil cooler retaining bolt until it seats against the bolt head.
8 Position the oil cooler onto the adapter, so that the casting protrusion on the adapter aligns between the two tabs on the oil cooler and install the retaining bolt hand tight **(see illustration).**
9 Tighten the oil cooler retaining bolt to the torque listed in this Chapter's Specifications. Do not overtighten!
10 Install the oil cooler hoses, then run the engine and check for leaks. Turn off the engine for five minutes and check the oil and coolant levels, adding fluids if necessary.

17 Flywheel/driveplate - removal and installation

Refer to illustration 17.4

1 Raise the vehicle and support it securely on jackstands, then refer to Chapter 7 and remove the transmission. **Warning:** *The engine must be supported from above with an engine hoist or three-bar support fixture before working underneath the vehicle with the transmission removed.*
2 If the vehicle is equipped with a manual transmission, remove the pressure plate and clutch disc (see Chapter 8). Now is a good time to check/replace the clutch components and pilot bushing if necessary. If the vehicle is equipped with an automatic transmission, now would be a good time to check and replace the front pump seal/O-ring.
3 Use paint or a center-punch to make alignment marks on the flywheel/driveplate and crankshaft to ensure correct alignment during reinstallation.
4 Remove the bolts that secure the flywheel/driveplate to the crankshaft **(see illustration).** If the crankshaft turns, hold the flywheel with a pry bar or wedge a screwdriver into the ring gear teeth to jam the flywheel.
5 Remove the flywheel/driveplate from the crankshaft. Since the flywheel is fairly heavy, be sure to support it while removing the last bolt.
6 Clean the flywheel to remove grease and oil. Inspect the surface for cracks, rivet grooves, burned areas and score marks. Light scoring can be removed with emery cloth. Check for cracked and broken ring gear teeth or a loose ring gear. Lay the flywheel on a flat surface and use a straightedge to check for warpage.
7 Clean and inspect the mating surfaces of the flywheel/driveplate and the crankshaft. If the crankshaft rear seal is leaking, replace it before reinstalling the flywheel/driveplate.
8 Position the flywheel/driveplate against the crankshaft. Be sure to align the marks made during removal. Note that some engines have an alignment dowel or staggered bolt holes to ensure correct installation. Before installing the bolts, apply thread locking compound to the threads.
9 Wedge a screwdriver into the ring gear teeth to keep the flywheel/driveplate from turning as you tighten the bolts to the torque

Chapter 2 Part C 3.5L V6 engine

17.4 Hold a lever against a casting protrusion on the engine block or place a screwdriver through a hole in the driveplate to hold the driveplate while the mounting bolts are removed - note the painted marks made at the crank and driveplate for alignment

18.2 Pry the seal out very carefully with a seal removal tool or screwdriver - if the crankshaft is damaged the new seal will leak!

18.3 If you don't have a seal installation tool, use a blunt tool (such as a brass punch) to carefully work the edge of the seal evenly into the bore and around the crankshaft

listed in this Chapter's Specifications.
10 The remainder of installation is the reverse of the removal.

18 Rear main oil seal - replacement

Refer to illustrations 18.2 and 18.3

1 The transmission must be removed from the vehicle for this procedure (see Chapter 7). **Warning:** *The engine must be supported from above with an engine hoist or three-bar support fixture before working underneath the vehicle with the transmission removed.* Remove the flywheel/driveplate (see Section 17).
2 Carefully pry the old seal out of the retainer with a seal removal tool or screwdriver **(see illustration)**.
3 Apply multi-purpose grease to the crankshaft seal journal and the lip of the new seal. Preferably, a seal installation tool should be used to press the new seal into place. If the proper seal installation tool is unavailable, use a large socket, section of pipe or a blunt tool and carefully drive the new seal into place **(see illustration)**. The lip is stiff so carefully work it onto the seal journal of the crankshaft. Don't rush it or you may damage the seal. **Note:** *Install the seal squarely and only until flush with the back of the seal plate, no further.*
4 The remaining steps are the reverse of removal.

19 Engine mounts - check and replacement

1 There are two engine mounts and one transmission mount installed on the vehicles covered by this manual. The two engine mounts are located on the passenger and driver's side of the vehicle attached to the engine block and to each frame rail. The transmission mount is mounted to the rear of the transmission and the transmission crossmember. Refer to Chapter 7B for the transmission mount replacement procedures.

Check

2 During the check, the engine must be raised slightly to remove the weight from the mounts.
3 Raise the vehicle and support it securely on jackstands. Support the engine/transmission from above using a hoist or three bar support fixture.
4 Check the mounts to see if the rubber is cracked, hardened or separated from the bushing in the center of the mount.
5 Check for relative movement between the mounts and the engine or frame (use a large screwdriver or prybar to attempt to move the mounts). If movement is noted, lower the engine and tighten the mount fasteners.
6 Rubber preservative should be applied to the mounts to slow deterioration.

Replacement

Refer to illustrations 19.9 and 19.10

7 Disconnect the cable from the negative terminal of the battery, set the parking brake and block the rear wheels.

8 Raise the front of the vehicle and support it securely on jackstands. Remove the splash shields from under the vehicle.
9 Remove the engine mount-to-frame nuts. There are two nuts on each side securing the mounts to the frame rails **(see illustration)**.
10 Working in the engine compartment, remove the engine mount-to-engine mount bracket nut(s). There is one nut on each side securing the mounts to the engine mount bracket **(see illustration)**.
11 Attach an engine hoist to the top of the engine for lifting. **Caution:** *Do not use a jack under the oil pan to support the entire weight of the engine or the oil pump pick-up could be damaged.*
12 Raise the engine slightly until the engine mount can be removed from the vehicle.
13 To remove the engine mount brackets, simply detach the four retaining bolts securing the mount bracket to each side of the engine.
14 Installation is the reverse of removal. Be sure to reinstall the heat shields over the top of each mount and use thread locking compound on the mount nuts before tightening them to the torque listed in this Chapter's Specifications.

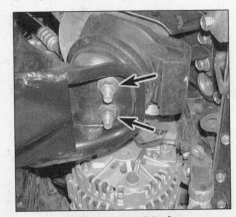

19.9 Engine mount-to-frame rail retaining nuts

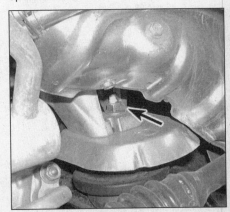

19.10 Engine mount-to-engine bracket retaining nut

Notes

Chapter 2 Part D
General engine overhaul procedures

Contents

	Section		Section
Camshafts, lifters and bearings - inspection	21	Engine removal - methods and precautions	6
Crankshaft - inspection	19	General information - engine overhaul	1
Crankshaft - installation and main bearing oil clearance check	24	Initial start-up and break-in after overhaul	27
Crankshaft - removal	14	Main and connecting rod bearings - inspection and main bearing selection	20
Cylinder compression check	3	Oil pressure check	2
Cylinder head - cleaning and inspection	10	Piston rings - installation	23
Cylinder head - disassembly	9	Pistons/connecting rods - inspection	18
Cylinder head - reassembly	12	Pistons/connecting rods - installation and rod bearing oil clearance check	26
Cylinder honing	17		
Engine - removal and installation	7	Pistons/connecting rods - removal	13
Engine block - cleaning	15	Rear main oil seal installation	25
Engine block - inspection	16	Vacuum gauge diagnostic checks	4
Engine overhaul - disassembly sequence	8	Valves - servicing	11
Engine overhaul - reassembly sequence	22		
Engine rebuilding alternatives	5		

Specifications

General

Engine designation and displacement
  KA24DE four cylinder engine .... 146 cubic inches (2.4 liters)
  VG33E V6 engine .... 200 cubic inches (3.3 liters)
  VQ35DE V6 engine .... 213.45 cubic inches (3.5 liters)
Bore
  2.4 liter .... 3.50 inches (89 mm)
  3.3 liter .... 3.602 inches (91.5 mm)
  3.5 liter .... 3.76 inches (95.5 mm)
Stroke
  2.4 liter .... 3.78 inches (96 mm)
  3.3 liter .... 3.27 inches (83 mm)
  3.5 liter .... 3.205 inches (81.4 mm)
Oil pump pressure (at normal operating temperature)
  2.4 liter
    At idle .... More than 11 psi (78 kPa)
    3000 rpm .... 60 to 70 psi (412 to 481 kPa)
  3.3 liter
    At idle .... More than 9 psi (59 kPa)
    At 2000 rpm .... 60 to 65 psi (412 to 451 kPa)
  3.5 liter
    At idle .... More than 14 psi (98 kPa)
    At 2000 rpm .... 43 psi (294 kPa)
Cylinder compression pressure (at 300 rpm)
  2.4 liter
    Standard .... 178 psi (1,226 kPa)
    Minimum .... 149 psi (1,030 kPa)
    Maximum difference between cylinders .... 14 psi (98 kPa)
  3.3 liter
    Standard .... 173 psi (1,196 kPa)
    Minimum .... 128 psi (883 kPa)
    Maximum difference between cylinders .... 14 psi (98 kPa)
  3.5 liter
    Standard .... 185 psi (1,275 kPa)
    Minimum .... 142 psi (981 kPa)
    Maximum difference between cylinders .... 14 psi (98 kPa)

Cylinder head

Warpage limit .... 0.004 inch (0.1 mm)
Cylinder head height
  2.4 liter .... 4.972 to 4.980 inches (126.3 to 126.5 mm)
  3.3 liter .... 4.205 to 4.220 inches (106.8 to 107.2 mm)
  3.5 liter .... 4.972 to 4.980 inches (126.3 to 126.5 mm)

Valves and related components

Rocker arm-to-shaft oil clearance
  3.3 liter .... 0.0003 to 0.0019 inch (0.007 to 0.049 mm)
Valve stem diameter
  2.4 liter
    Intake .... 0.2742 to 0.2748 inch (6.965 to 6.980 mm)
    Exhaust .... 0.2734 to 0.2740 inch (6.945 to 6.960 mm)
  3.3 liter
    Intake .... 0.2742 to 0.2748 inch (6.965 to 6.980 mm)
    Exhaust .... 0.3135 to 0.3138 inch (7.962 to 7.970 mm)
  3.5 liter
    Intake .... 0.2348 to 0.2354 inch (5.965 to 5.980 mm)
    Exhaust .... 0.2341 to 0.2346 inch (5.945 to 5.960 mm)
Valve margin
  2.4 liter
    Intake .... 0.0374 to 0.0492 inch (0.95 to 1.25 mm)
    Exhaust .... 0.0453 to 0.0571 inch (1.15 to 1.45 mm)
    Service limit (minimum) .... 0.020 inch (0.5 mm)
  3.3 liter
    Intake .... 0.0453 to 0.0571 inch (1.15 to 1.45 mm)
    Exhaust .... 0.0531 to 0.0650 inch (1.35 to 1.65 mm)
    Service limit (minimum) .... 0.020 inch (0.5 mm)
  3.5 liter
    Intake .... 0.0453 to 0.0571 inch (1.15 to 1.45 mm)
    Exhaust .... 0.0571 to 0.0689 inch (1.45 to 1.75 mm)
    Service limit (minimum) .... 0.020 inch (0.5 mm)
Valve stem-to-guide clearance
  2.4 liter
    Intake .... 0.0008 to 0.0021 inch (0.020 to 0.053 mm)
    Exhaust .... 0.0016 to 0.0029 inch (0.040 to 0.073 mm)
    Service limit (maximum)
      Intake .... 0.0031 inch (0.08 mm)
      Exhaust .... 0.004 inch (0.10 mm)
  3.3 liter
    Intake .... 0.0008 to 0.0021 inch (0.020 to 0.053 mm)
    Exhaust .... 0.0012 to 0.0019 inch (0.030 to 0.049 mm)
    Service limit (maximum) .... 0.0039 inch (0.10 mm)
  3.5 liter
    Intake .... 0.0008 to 0.0021 inch (0.020 to 0.053 mm)
    Exhaust .... 0.0016 to 0.0029 inch (0.040 to 0.073 mm)
    Service limit (maximum)
      Intake .... 0.0031 inch (0.08 mm)
      Exhaust .... 0.0040 inch (0.10 mm)
Valve spring free length
  2.4 liter .... 1.9831 inches (50.3 mm)
  3.3 liter
    Outer .... 2.016 inches (51.2 mm)
    Inner .... 1.736 inches (44.1 mm)
  3.5 liter .... 1.8543 inches (47.10 mm)
Valve spring out-of-square limit
  2.4 liter .... 0.087 inch (2.2 mm)
  3.3 liter
    Outer .... 0.087 inch (2.2 mm)
    Inner .... 0.075 inch (1.9 mm)
  3.5 liter .... 0.079 inch (2.0 mm)

Valve lifters

Lifter outside diameter
  2.4 liter .... 1.3370 to 1.3376 inch (33.960 to 33.975 mm)
  3.3 liter .... 0.6278 to 0.6282 inch (15.947 to 15.957 mm)
  3.5 liter .... 1.3764 to 1.3770 inch (34.960 to 34.975 mm)

Chapter 2 Part D General engine overhaul procedures

Lifter guide inside diameter
    2.4 liter .... 1.3386 to 1.3394 inch (34.000 to 34.021 mm)
    3.3 liter .... 0.6299 to 0.6304 inch (16.000 to 16.013 mm)
    3.5 liter .... 1.3780 to 1.3788 inch (35.000 to 35.021 mm)
Lifter-to-guide clearance
    2.4 liter .... 0.0010 to 0.0024 inch (0.025 to 0.061 mm)
    3.3 liter .... 0.0017 to 0.0026 inch (0.043 to 0.066 mm)
    3.5 liter .... 0.0010 to 0.0024 inch (0.025 to 0.061 mm)

Camshaft

Inner diameter of camshaft bearing
    2.4 liter
        Journal No.1 through 5 .... 1.1024 to 1.1033 inches (28.000 to 28.025 mm)
    3.3 liter
        Journal A .... 1.8504 to 1.8514 inches (47.000 to 47.025 mm)
        Journal B .... 1.6732 to 1.6742 inches (42.500 to 42.525 mm)
        Journal C .... 1.8898 to 1.8907 inches (48.000 to 48.025 mm)
    3.5 liter
        Journal No.1 .... 1.0236 to 1.0244 inches (26.000 to 26.021 mm)
        Journal No. 2, 3 and 4 .... 0.9252 to 0.9260 inches (23.500 to 23.521 mm)
Outer diameter of camshaft journal
    2.4 liter
        Journal No.1 through 5 .... 1.0998 to 1.1006 inches (27.935 to 27.955 mm)
    3.3 liter
        Journal A .... 1.8472 to 1.8480 inches (46.920 to 46.940 mm)
        Journal B .... 1.6701 to 1.6709 inches (42.420 to 42.440 mm)
        Journal C .... 1.8866 to 1.8874 inches (47.920 to 47.940 mm)
    3.5 liter
        Journal No.1 .... 1.0211 to 1.0218 inches (25.935 to 25.955 mm)
        Journal No. 2, 3 and 4 .... 0.9230 to 0.9238 inches (23.445 to 23.465 mm)
Camshaft bearing oil clearance
    2.4 liter
        Standard .... 0.0018 to 0.0035 inch (0.045 to 0.090 mm)
        Service limit (maximum) .... 0.0047 inch (0.12 mm)
    3.3 liter
        Standard .... 0.0024 to 0.0041 inch (0.060 to 0.105 mm)
        Service limit (maximum) .... 0.0059 inch (0.15 mm)
    3.5 liter
        Journal No.1 .... 0.0018 to 0.0034 inch (0.045 to 0.086 mm)
        Journal No. 2, 3 and 4 .... 0.0014 to 0.0030 inch (0.035 to 0.076 mm)
        Service limit (maximum) .... 0.0059 inch (0.15 mm)
Camshaft endplay
    2.4 liter .... 0.0028 to 0.0058 inch (0.070 to 0.148 mm)
    3.3 liter .... 0.0012 to 0.0024 inch (0.03 to 0.06 mm)
    3.5 liter .... 0.0045 to 0.0074 inch (0.115 to 0.188 mm)
Camshaft runout (total indicator reading)
    2.4 liter
        Standard .... 0.0008 inch (0.02 mm)
        Service limit .... 0.0016 inch (0.04 mm)
    3.3 liter
        Standard .... 0.0016 inch (0.04 mm)
        Service limit .... 0.004 inch (0.10 mm)
    3.5 liter
        Standard .... 0.0008 inch (0.02 mm)
        Service limit .... 0.0020 inch (0.05 mm)
Camshaft lobe height
    2.4 liter
        Intake
            2000 and earlier .... 1.673 to 1.681 inches (42.505 to 42.695 mm)
            2001 .... 1.644 to 1.651 inches (41.755 to 41.945 mm)
        Exhaust
            2000 and earlier .... 1.610 to 1.618 inches (40.905 to 41.095 mm)
            2001 .... 1.646 to 1.654 inches (41.815 to 42.005 mm)
        Service limit (lobe lift loss) .... 0.008 inch (0.20 mm)
    3.3 liter
        Intake
            1997 and earlier .... 1.5450 to 1.5524 inches (39.242 to 39.432 mm)
            1998 and later .... 1.5332 to 1.5407 inches (38.943 to 39.133 mm)
        Exhaust .... 1.5332 to 1.5407 inches (38.943 to 39.133 mm)
        Service limit (lobe lift loss) .... 0.0059 inch (0.15 mm)

2D-4　Chapter 2 Part D　General engine overhaul procedures

Camshaft (continued)
Camshaft lobe height (continued)
  3.5 liter
    Intake and exhaust.. 1.7506 to 1.7581 inches (44.465 to 44.655 mm)
    Service limit (lobe lift loss)....................................... 0.0080 inch (0.20 mm)

Engine block
Deck warpage limit
  All engines (maximum)... 0.0039 inch (0.10 mm)
Cylinder bore diameter
  2.4 liter
    Standard... 3.5039 to 3.5051 inches (89.000 to 89.030 mm)
    Wear limit... 0.008 inch (0.20 mm)
  3.3 liter
    Standard... 3.6024 to 3.6041 inches (91.500 to 91.545 mm)
    Wear limit... 0.008 inch (0.20 mm)
  3.5 liter
    Standard... 3.7598 to 3.7610 inches (95.500 to 95.530 mm)
    Wear limit... 0.008 inch (0.20 mm)
Cylinder out-of-round and taper limit
  All engines (maximum)... 0.0006 inch (0.015 mm)
Main journal bore diameter
  2.4 liter... not available
  3.3 liter... 2.6238 to 2.6249 inches (66.645 to 66.672 mm)
  3.5 liter... 2.5194 to 2.5203 inches (63.993 to 64.017 mm)

Pistons and rings
Piston skirt diameter
  2.4 liter... 3.5027 to 3.5039 inches (88.970 to 89.500 mm)
  3.3 liter... 3.6010 to 3.6027 inches (91.465 to 91.510 mm)
  3.5 liter... 3.7590 to 3.7602 inches (95.480 to 95.510 mm)
Piston-to-cylinder clearance
  2.4 liter... 0.0008 to 0.0016 inch (0.020 to 0.040 mm
  3.3 liter... 0.0010 to 0.0015 inch (0.025 to 0.038 mm)
  3.5 liter... 0.0004 to 0.0012 inch (0.01 to 0.03 mm)
Piston ring side clearance
  Top compression ring
    All engines
      Standard... 0.0016 to 0.0031 inch (0.040 to 0.080 mm)
      Service limit (maximum)..................................... 0.004 inch (0.10 mm)
  Second compression ring
    All engines
      Standard... 0.0012 to 0.0028 inch (0.030 to 0.070 mm)
      Service limit (maximum)..................................... 0.004 inch (0.10 mm)
  Oil ring
    2.4 liter
      Standard... 0.0026 to 0.0053 inch (0.065 to 0.135 mm)
      Service limit (maximum)..................................... 0.0053 inch (0.135 mm)
    3.3 liter
      Standard... 0.0006 to 0.0073 inch (0.015 to 0.185 mm)
      Service limit (maximum)..................................... 0.0073 inch (0.185 mm)
    3.5 liter
      Standard... 0.0006 to 0.0020 inch (0.015 to 0.050 mm)
      Service limit (maximum)..................................... 0.002 inch (0.050 mm)
Piston ring end gap
  2.4 liter
    Top compression ring
      Standard... 0.0110 to 0.0205 inch (0.28 to 0.52mm)
      Service limit (maximum)..................................... 0.039 inch (1.0 mm)
    Second compression ring
      Standard... 0.0177 to 0.0272 inch (0.45 to 0.69 mm)
      Service limit (maximum)..................................... 0.039 inch (1.0 mm)
    Oil rail
      Standard... 0.0079 to 0.0272 inch (0.20 to 0.69 mm)
      Service limit (maximum)..................................... 0.039 inch (1.0 mm)
  3.3 liter
    Top compression ring
      1997 and earlier
        Standard... 0.0083 to 0.0157 inch (0.21 to 0.40 mm)
        Service limit (maximum)............................... 0.0213 inch (0.54 mm)

Chapter 2 Part D General engine overhaul procedures

    1998 and later
        Standard .... 0.0083 to 0.0122 inch (0.21 to 0.31 mm)
        Service limit (maximum) .... 0.0169 inch (0.43 mm)
    Second compression ring
        1997 and earlier
            Standard .... 0.0197 to 0.0272 inch (0.50 to 0.69 mm)
            Service limit (maximum) .... 0.0315 inch (0.80 mm)
        1998 and later
            Standard .... 0.0197 to 0.0236 inch (0.50 to 0.60 mm)
            Service limit (maximum) .... 0.0272 inch (0.69 mm)
    Oil rail
        1997 and earlier
            Standard .... 0.0079 to 0.0272 inch (0.20 to 0.69 mm)
            Service limit (maximum) .... 0.0374 inch (0.95 mm)
        1998 and later
            Standard .... 0.0079 to 0.0236 inch (0.20 to 0.60 mm)
            Service limit (maximum) .... 0.0331 inch (0.84 mm)
  3.5 liter
    Top compression ring
        Standard .... 0.0091 to 0.0130 inch (0.23 to 0.33 mm)
        Service limit (maximum) .... 0.0213 inch (0.54 mm)
    Second compression ring
        Standard .... 0.0130 to 0.0189 inch (0.33 to 0.48 mm)
        Service limit (maximum) .... 0.0315 inch (0.80 mm)
    Oil rail
        Standard .... 0.0079 to 0.0315 inch (0.20 to 0.80 mm)
        Service limit (maximum) .... 0.0374 inch (0.95 mm)

Crankshaft

Main journal diameter
    2.4 liter .... 2.3603 to 2.3612 inches (59.951 to 59.975 mm)
    3.3 liter .... 2.4784 to 2.4793 inches (62.951 to 62.975 mm)
    3.5 liter .... 2.3603 to 2.3612 inches (59.951 to 59.975 mm)
Rod journal diameter
    2.4 liter .... 1.9668 to 1.9675 inches (49.956 to 49.974 mm)
    3.3 liter .... 1.9667 to 1.9675 inches (49.955 to 49.974 mm)
    3.5 liter .... 2.0445 to 2.0462 inches (51.956 to 51.974 mm)
Crankshaft journal out-of-round and taper limit
    2.4 liter
        Main journal .... 0.0004 inch (0.01 mm)
        Rod journal .... 0.0002 inch (0.005 mm)
    3.3 liter .... 0.0002 inch (0.005 mm)
    3.5 liter .... 0.0001 inch (0.002 mm)
Endplay
    2.4 liter
        Standard .... 0.0020 to 0.0071 inch (0.05 to 0.18 mm)
        Service limit (maximum) .... 0.0120 inch (0.30 mm)
    3.3 liter
        Standard .... 0.0020 to 0.0067 inch (0.05 to 0.17 mm)
        Service limit (maximum) .... 0.0118 inch (0.30 mm)
    3.5 liter
        Standard .... 0.0039 to 0.0098 inch (0.10 to 0.25 mm)
        Service limit (maximum) .... 0.0118 inch (0.30 mm)
Main bearing oil clearance
    2.4 liter
        Standard .... 0.0008 to 0.0019 inch (0.020 to 0.047 mm)
        Service limit (maximum) .... 0.0040 inch (0.10 mm)
    3.3 liter
        2000 and earlier
            Standard .... 0.0011 to 0.0022 inch (0.028 to 0.055 mm)
            Service limit (maximum) .... 0.0035 inch (0.09 mm)
        2001
            Journal No.1
                Standard .... 0.0012 to 0.0019 inch (0.030 to 0.048 mm)
                Service limit (maximum) .... 0.0024 inch (0.060 mm)
            Journal No. 2, 3 and 4
                Standard .... 0.0015 to 0.0026 inch (0.038 to 0.065 mm)
                Service limit (maximum) .... 0.0031 inch (0.080 mm)

Crankshaft (continued)

Main bearing oil clearance (continued)
  3.5 liter
    Standard .... 0.0014 to 0.0018 inch (0.035 to 0.045 mm)
    Service limit (maximum) .... 0.0026 inch (0.065 mm)
Rod bearing oil clearance
  2.4 liter
    Standard .... 0.0004 to 0.0014 inch (0.010 to 0.035 mm)
    Service limit (maximum) .... 0.0035 inch (0.09 mm)
  3.3 liter
    2000 and earlier
      Standard .... 0.0006 to 0.0021 inch (0.014 to 0.054 mm)
      Service limit (maximum) .... 0.0035 inch (0.09 mm)
    2001
      Standard .... 0.0009 to 0.0025 inch (0.024 to 0.064 mm)
      Service limit (maximum) .... 0.0035 inch (0.09 mm)
  3.5 liter
    Standard .... 0.0013 to 0.0023 inch (0.034 to 0.059 mm)
    Service limit (maximum) .... 0.0028 inch (0.07 mm)
Connecting rod side clearance (endplay)
  2.4 liter
    Standard .... 0.008 to 0.016 inch (0.20 to 0.40 mm)
    Service limit (maximum) .... 0.024 inch (0.60 mm)
  3.3 liter
    Standard .... 0.0079 to 0.0138 inch (0.20 to 0.35 mm)
    Service limit (maximum) .... 0.0157 inch (0.40 mm)
  3.5 liter
    Standard .... 0.0079 to 0.0138 inch (0.20 to 0.35 mm)
    Service limit (maximum) .... 0.0157 inch (0.40 mm)

Torque specifications*

	Ft-lbs (unless otherwise indicated)	Nm
Rear main oil seal retainer bolts		
2.4 liter	56 to 66 in-lbs	6.4 to 7.5
3.3 liter	55 to 73 in-lbs	6.3 to 8.3
3.5 liter	72 to 82 in-lbs	8.2 to 9.3
Connecting rod nuts		
2.4 liter		
Step one	120 to 144 in-lbs	14 to 16
Step two	28 to 33	38 to 44
3.3 liter		
Step one	120 to 144 in-lbs	14 to 16
Step two	28 to 33	38 to 44
3.5 liter		
Step one	14 to 15	19 to 21
Step two	Turn an additional 90 to 95 degrees	
Main bearing cap bolts		
2.4 liter	34 to 41	46 to 56
3.3 liter	64 to 74	90 to 100
3.5 liter		
Step one	24 to 28	32 to 38
Step two	Turn an additional 90 to 95 degrees	

*Refer to Part A, B or C for additional torque specifications

1 General information - engine overhaul

Included in this portion of Chapter 2 are the general overhaul procedures for the cylinder head and internal engine components.

The information ranges from advice concerning preparation for an overhaul and the purchase of replacement parts to detailed, step-by-step procedures covering removal and installation of internal engine components and the inspection of parts.

The following Sections have been written based on the assumption that the engine has been removed from the vehicle. For information concerning in-vehicle engine repair, as well as removal and installation of the external components necessary for the overhaul, see Chapter 2A, 2B or 2C.

The Specifications included in this Part are only those necessary for the inspection and overhaul procedures which follow. Refer to Chapter 2, Part A, Part B or Part C for additional Specifications.

It's not always easy to determine when, or if, an engine should be completely overhauled, as a number of factors must be considered.

High mileage is not necessarily an indication that an overhaul is needed, while low mileage doesn't preclude the need for an overhaul. Frequency of servicing is probably the most important consideration. An engine that's had regular and frequent oil and filter changes, as well as other required mainte-

Chapter 2 Part D General engine overhaul procedures

2.2a Oil pressure sending unit location - 2.4L

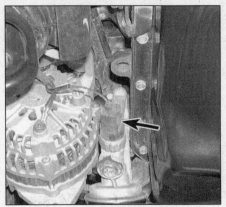

2.2b Oil pressure sending unit location - 3.3L

2.2c Oil pressure sending unit location - 3.5L

nance, will most likely give many thousands of miles of reliable service. Conversely, a neglected engine may require an overhaul very early in its life.

Excessive oil consumption is an indication that piston rings, valve seals and/or valve guides are in need of attention. Make sure that oil leaks aren't responsible before deciding that the rings and/or guides are bad. Perform a cylinder compression check to determine the extent of the work required (see Section 3). Also check the vacuum readings under various conditions (see Section 4).

Loss of power, rough running, knocking or metallic engine noises, excessive valve train noise and high fuel consumption rates may also point to the need for an overhaul, especially if they're all present at the same time. If a complete tune-up doesn't remedy the situation, major mechanical work is the only solution.

An engine overhaul involves restoring the internal parts to the specifications of a new engine. During an overhaul, the piston rings are replaced and the cylinder walls are reconditioned (re-bored and/or honed). If a re-bore is done by an automotive machine shop, new oversize pistons will also be installed. The main bearings, connecting rod bearings and camshaft bearings are generally replaced with new ones and, if necessary, the crankshaft may be reground to restore the journals. Generally, the valves are serviced as well, since they're usually in less-than-perfect condition at this point. While the engine is being overhauled, other components, such as the distributor (if equipped), the starter and alternator, can be rebuilt as well. The end result should be a like new engine that will give many trouble free miles. **Note:** *Critical cooling system components such as the hoses, drivebelts, thermostat and water pump should be replaced with new parts when an engine is overhauled. The radiator should be checked carefully to ensure that it isn't clogged or leaking (see Chapter 3). If you purchase a rebuilt engine or short block, some rebuilders will not warranty their engines unless the radiator has been professionally flushed. Also, we don't recommend overhauling the oil pump - always install a new one when an engine is rebuilt.*

Before beginning the engine overhaul, read through the entire procedure to familiarize yourself with the scope and requirements of the job. Overhauling an engine isn't difficult, but it is time-consuming. Plan on the vehicle being tied up for a minimum of two weeks, especially if parts must be taken to an automotive machine shop for repair or reconditioning. Check on availability of parts and make sure that any necessary special tools and equipment are obtained in advance. Most work can be done with typical hand tools, although a number of precision measuring tools are required for inspecting parts to determine if they must be replaced. Often an automotive machine shop will handle the inspection of parts and offer advice concerning reconditioning and replacement. **Note:** *Always wait until the engine has been completely disassembled and all components, especially the engine block, have been inspected before deciding what service and repair operations must be performed by an automotive machine shop.* Since the block's condition will be the major factor to consider when determining whether to overhaul the original engine or buy a rebuilt one, never purchase parts or have machine work done on other components until the block has been thoroughly inspected. As a general rule, time is the primary cost of an overhaul, so it doesn't pay to install worn or substandard parts.

As a final note, to ensure maximum life and minimum trouble from a rebuilt engine, everything must be assembled with care in a spotlessly clean environment.

2 Oil pressure check

Refer to illustrations 2.2a, 2.2b and 2.2c

1 Low engine oil pressure can be a sign of an engine in need of rebuilding. A "low oil pressure" indicator (often called an "idiot light") is not a test of the oiling system. Such indicators only come on when the oil pressure is dangerously low. Even a factory oil pressure gauge in the instrument panel is only a relative indication, although much better for driver information than a warning light. A better test is with a mechanical (not electrical) oil pressure gauge. When used in conjunction with an accurate tachometer, an engine's oil pressure performance can be compared to factory Specifications for that year and model.

2 Locate the oil pressure sending unit **(see illustrations)**.

3 Remove the oil pressure sending unit and install a fitting which will allow you to directly connect your hand-held, mechanical oil pressure gauge. Use Teflon tape or sealant on the threads of the adapter and the fitting on the end of your gauge's hose.

4 Connect an accurate tachometer to the engine, according to the tachometer manufacturer's instructions.

5 Check the oil pressure with the engine running (normal operating temperature) at the specified engine speed, and compare it to this Chapter's Specifications. If it's extremely low, the bearings and/or oil pump are probably worn out.

3 Cylinder compression check

Refer to illustration 3.6

1 A compression check will tell you what mechanical condition the upper end of your engine (pistons, rings, valves, cylinder head gaskets) are in. Specifically, it can tell you if the compression is down due to leakage caused by worn piston rings, defective valves and seats or a blown cylinder head gasket. **Note:** *The engine must be at normal operating temperature and the battery must be fully charged for this check.*

2 Begin by cleaning the area around the spark plugs before you remove them (compressed air should be used, if available). The idea is to prevent dirt from getting into the cylinders as the compression check is being done.

3 Relieve the fuel system pressure (see Chapter 4). The fuel pump must remain disabled throughout this procedure.

4 Remove all of the spark plugs from the

engine (see Chapter 1). Block the throttle wide open.

5 On models with a distributor, disconnect the ignition coil electrical connector at the distributor (see Chapter 5). On models without a distributor, unplug the electrical connector from each ignition coil.

6 Install the compression gauge in the spark plug hole **(see illustration)**.

7 Crank the engine over at least seven compression strokes and watch the gauge. The compression should build up quickly in a healthy engine. Low compression on the first stroke, followed by gradually increasing pressure on successive strokes, indicates worn piston rings. A low compression reading on the first stroke, which doesn't build up during successive strokes, indicates leaking valves or a blown cylinder head gasket (a cracked cylinder head could also be the cause). Deposits on the undersides of the valve heads can also cause low compression. Record the highest gauge reading obtained.

8 Repeat the procedure for the remaining cylinders and compare the results to this Chapter's Specifications.

9 Add some engine oil (about three squirts from a plunger-type oil can) to each cylinder, through the spark plug hole, and repeat the test.

10 If the compression increases after the oil is added, the piston rings are definitely worn. If the compression doesn't increase significantly, the leakage is occurring at the valves or cylinder head gasket. Leakage past the valves may be caused by burned valve seats and/or faces or warped, cracked or bent valves.

11 If two adjacent cylinders have equally low compression, there's a strong possibility that the cylinder head gasket between them is blown. The appearance of coolant in the combustion chambers or the crankcase would verify this condition.

12 If one cylinder is slightly lower than the others, and the engine has a slightly rough idle, a worn lobe on the camshaft could be the cause.

13 If the compression is unusually high, the combustion chambers are probably coated with carbon deposits. If that's the case, the cylinder head(s) should be removed and decarbonized.

14 If compression is way down or varies greatly between cylinders, it would be a good idea to have a leak-down test performed by an automotive repair shop. This test will pinpoint exactly where the leakage is occurring and how severe it is.

4 Vacuum gauge diagnostic checks

Refer to illustrations 4.4 and 4.6

A vacuum gauge provides valuable information about what is going on in the engine at a low-cost. You can check for worn rings or cylinder walls, leaking cylinder head

3.6 A compression gauge with a threaded fitting for the spark plug hole is preferred over the type that requires hand pressure to maintain the seal

4.4 A simple vacuum gauge can be very handy in diagnosing engine condition and performance

or intake manifold gaskets, incorrect carburetor adjustments, restricted exhaust, stuck or burned valves, weak valve springs, improper ignition or valve timing and ignition problems.

Unfortunately, vacuum gauge readings are easy to misinterpret, so they should be used in conjunction with other tests to confirm the diagnosis.

Both the absolute readings and the rate of needle movement are important for accurate interpretation. Most gauges measure vacuum in inches of mercury (in-Hg). The following references to vacuum assume the diagnosis is being performed at sea level. As elevation increases (or atmospheric pressure decreases), the reading will decrease. For every 1,000 foot increase in elevation above approximately 2000 feet, the gauge readings will decrease about one inch of mercury.

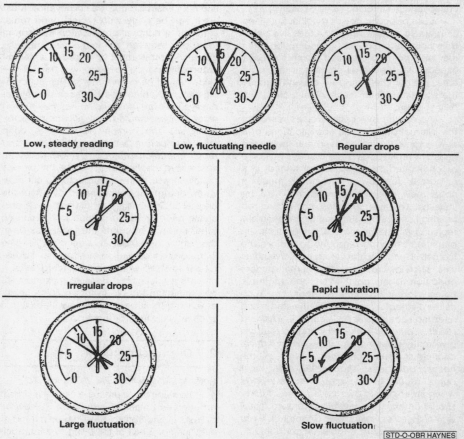

4.6 Typical vacuum gauge readings

Chapter 2 Part D General engine overhaul procedures

Connect the vacuum gauge directly to intake manifold vacuum, not to ported (throttle body) vacuum **(see illustration)**. Be sure no hoses are left disconnected during the test or false readings will result.

Before you begin the test, allow the engine to warm up completely. Block the wheels and set the parking brake. With the transmission in Park, start the engine and allow it to run at normal idle speed. **Warning:** *Keep your hands and the vacuum gauge clear of the fans.*

Read the vacuum gauge; an average, healthy engine should normally produce about 17 to 22 in-Hg of vacuum with a fairly steady needle. Refer to the following vacuum gauge readings and what they indicate about the engine's condition **(see illustration):**

1 A low steady reading usually indicates a leaking gasket between the intake manifold and cylinder head(s) or throttle body, a leaky vacuum hose, late ignition timing or incorrect camshaft timing. Check ignition timing with a timing light and eliminate all other possible causes, utilizing the tests provided in this Chapter before you remove the timing chain cover to check the timing marks.

2 If the reading is three to eight inches below normal and it fluctuates at that low reading, suspect an intake manifold gasket leak at an intake port or a faulty fuel injector.

3 If the needle has regular drops of about two-to-four in-Hg at a steady rate, the valves are probably leaking. Perform a compression check or leak-down test to confirm this.

4 An irregular drop or down-flick of the needle can be caused by a sticking valve or an ignition misfire. Perform a compression check or leak-down test and read the spark plugs.

5 A rapid vibration of about four in-Hg vibration at idle combined with exhaust smoke indicates worn valve guides. Perform a leak-down test to confirm this. If the rapid vibration occurs with an increase in engine speed, check for a leaking intake manifold gasket or cylinder head gasket, weak valve springs, burned valves or ignition misfire.

6 A slight fluctuation, say one inch up and down, may mean ignition problems. Check all the usual tune-up items and, if necessary, run the engine on an ignition analyzer.

7 If there is a large fluctuation, perform a compression or leak-down test to look for a weak or dead cylinder or a blown cylinder head gasket.

8 If the needle moves slowly through a wide range, check for a clogged PCV system, incorrect idle fuel mixture, carburetor/throttle body or intake manifold gasket leaks.

9 Check for a slow return after revving the engine by quickly snapping the throttle open until the engine reaches about 2,500 rpm and let it shut. Normally the reading should drop to near zero, rise above normal idle reading (about 5 in-Hg over) and then return to the previous idle reading. If the vacuum returns slowly and doesn't peak when the throttle is snapped shut, the rings may be worn. If there is a long delay, look for a restricted exhaust system (often the muffler or catalytic converter). An easy way to check this is to temporarily disconnect the exhaust ahead of the suspected part and repeat the test.

5 Engine rebuilding alternatives

The do-it-yourselfer is faced with a number of options when performing an engine overhaul. The decision to replace the engine block, piston/connecting rod assemblies and crankshaft depends on a number of factors, with the number one consideration being the condition of the block. Other considerations are cost, access to machine shop facilities, parts availability, time required to complete the project and the extent of prior mechanical experience on the part of the do-it-yourselfer.

Some of the rebuilding alternatives include:

Individual parts - If the inspection procedures reveal that the engine block and most engine components are in reusable condition, purchasing individual parts may be the most economical alternative. The block, crankshaft and piston/connecting rod assemblies should all be inspected carefully. Even if the block shows little wear, the cylinder bores should be surface-honed.

Crankshaft kit - This rebuild package consists of a reground crankshaft and a matched set of pistons and connecting rods. The pistons will already be installed on the connecting rods. Piston rings and the necessary bearings will be included in the kit. These kits are commonly available for standard cylinder bores, as well as for engine blocks which have been bored to a regular oversize.

Short block - A short block consists of an engine block with renewed crankshaft and piston/connecting rod assemblies already installed. All new bearings are incorporated and all clearances will be correct. The existing cylinder head(s), camshaft, valve train components and external parts can be bolted to the short block with little or no machine shop work necessary.

Long block - A long block consists of a short block plus an oil pump, oil pan, cylinder heads, valve covers, camshaft and valve train components, timing sprockets, timing chain and timing cover. All components are installed with new bearings, seals and gaskets incorporated throughout. The installation of manifolds and external parts is all that is necessary.

Used engine assembly - While overhaul provides the best assurance of a like-new engine, used engines available from wrecking yards and importers are often a very simple and economical solution. Many used engines come with warranties, but always give any engine a thorough diagnostic check-out before purchase. Check compression, vacuum and also for signs of oil leakage. If possible, have the seller run the engine, ether in the vehicle or on a test stand so you can be sure it runs smoothly with no knocking or other noises.

Give careful thought to which alternative is best for you and discuss the situation with local automotive machine shops, auto parts dealers or parts store countermen before ordering or purchasing replacement parts.

6 Engine removal - methods and precautions

If you've decided that an engine must be removed for overhaul or major repair work, several preliminary steps should be taken.

Locating a suitable place to work is extremely important. Adequate work space, along with storage space for the vehicle, will be needed. If a shop or garage isn't available, at the very least a flat, level, clean work surface made of concrete or asphalt is required.

Cleaning the engine compartment and engine before beginning the removal procedure will help keep tools clean and organized.

An engine hoist or A-frame will also be necessary. Make sure the equipment is rated in excess of the combined weight of the engine and transmission. Safety is of primary importance, considering the potential hazards involved in lifting the engine out of the vehicle.

If the engine is being removed by a novice, a helper should be available. Advice and aid from someone more experienced would also be helpful. There are many instances when one person cannot simultaneously perform all of the operations required when lifting the engine out of the vehicle.

Plan the operation ahead of time. Arrange for or obtain all of the tools and equipment you'll need prior to beginning the job. Some of the equipment necessary to perform engine removal and installation safely and with relative ease are (in addition to an engine hoist) a heavy duty floor jack, complete sets of wrenches and sockets as described in the front of this manual, wooden blocks and plenty of rags and cleaning solvent for mopping up spilled oil, coolant and gasoline. If the hoist must be rented, make sure that you arrange for it in advance and perform all of the operations possible without it beforehand. This will save you money and time.

Plan for the vehicle to be out of use for quite a while. A machine shop will be required to perform some of the work which the do-it-yourselfer can't accomplish without special equipment. These shops often have a busy schedule, so it would be a good idea to consult them before removing the engine in order to accurately estimate the amount of time required to rebuild or repair components that may need work.

Always be extremely careful when removing and installing the engine. Serious injury can result from careless actions. Plan ahead, take your time and a job of this nature, although major, can be accomplished successfully.

2D-10 Chapter 2 Part D General engine overhaul procedures

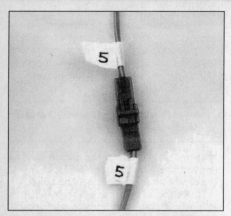

7.6 Label the hoses (arrow) and wires to ensure proper assembly

7.13a On 2.4L engines, remove the distributor cap (A) with the wires attached, disconnect all the electrical connectors (B indicates one) and pull the main electrical harness (C) away from the engine

7 Engine - removal and installation

Warning: *The models covered by this manual are equipped with Supplemental Restraint Systems (SRS), more commonly known as airbags. Always disable the airbag system before working in the vicinity of any airbag system components to avoid the possibility of accidental deployment of the airbag(s), which could cause personal injury (see Chapter 12).*

Removal

Refer to illustrations 7.6, 7.13a, 7.13b, 7.13c, 7.17 and 7.22

1 Relieve the fuel system pressure (see Chapter 4).
2 Disconnect the cable from the negative terminal of the battery (see Chapter 1).
3 Place protective covers on the fenders and cowl and remove the hood (see Chapter 11).
4 Remove the air cleaner assembly (see Chapter 4).
5 Raise the vehicle and support it securely on jackstands. Remove the engine under cover (if equipped) and drain the cooling system, transmission and engine oil and remove the drivebelts (see Chapter 1).
6 Remove the engine cover (if equipped). Clearly label, then disconnect all vacuum lines, coolant and emissions hoses, wiring harness connectors and ground straps. Masking tape and/or a touch up paint applicator work well for marking items **(see illustration)**. Take instant photos or sketch the locations of components and brackets.
7 Disconnect the radiator hoses and heater hoses from the engine. Remove the cooling fans and the radiator (see Chapter 3).
8 Release the residual fuel pressure in the tank by removing the gas cap, then disconnect the fuel lines from the fuel rail (see Chapter 4). Plug or cap all open fittings.
9 Refer to Chapter 3 and unbolt and set aside the air conditioning compressor, without disconnecting the refrigerant lines.
10 Disconnect the accelerator cable (and cruise control cable, if equipped) from the engine (see Chapter 4).
11 Unbolt the power steering pump. Tie the pump aside without disconnecting the hoses (see Chapter 10). Remove the alternator (see Chapter 5).
12 On 2.4L and 3.3L engines, refer to Chapter 5 and remove the distributor cap with the spark plug wires attached.
13 Label and disconnect the main engine electrical harnesses at each end of the engine **(see illustrations)**. Refer to Chapter 6 for the location of the engine control system sensors.
14 On V6 engines, remove the upper intake manifold to make engine removal easier. Be sure to label and disconnect all hoses, connectors, and any ground straps.
15 Disconnect the front exhaust pipe(s) from the exhaust manifold(s) and at the converter, then remove the pipe (see Chapter 4). Remove the crankshaft position sensor from the transmission bellhousing, if equipped (see Chapter 6).
16 Support the transmission with a jack. Position a block of wood between the jack and transmission to prevent damage to the transmission. Special transmission jacks with safety chains are available - use one if possible.
17 Attach an engine sling or a length of chain to the lifting brackets on the engine **(see illustration). Caution:** *DO NOT lift the engine by the intake manifold. Lift the engine*

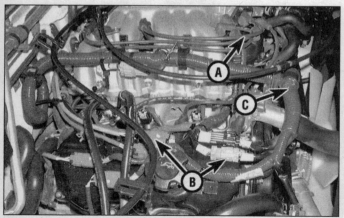

7.13b On 3.3L engines, remove the distributor cap (A) with the wires attached, disconnect all the electrical connectors (B indicates some) and pull the main electrical harness (C) away from the engine

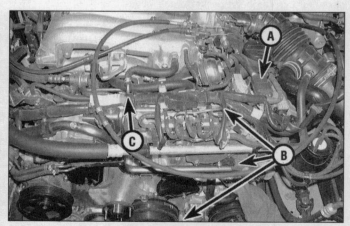

7.13c On 3.5L engines, disconnect the electrical connectors from the main harness (A), then disconnect the electrical connectors from the engine control sensors (B indicates some - After the sensors and harness brackets have been disconnected or detached, position the main harness (C) aside

Chapter 2 Part D General engine overhaul procedures

7.17 Attach the chain or sling to the lifting eyes (if equipped)

7.22 With the chain or sling attached securely to the engine, raise the engine enough to remove the mounts, then pull the engine forward as far as possible to clear the transmission and the cowl and lift the engine from the engine compartment

by the lifting brackets, the engine block or the cylinder head(s) only.

18 Roll the hoist into position and connect the sling to it. Take up the slack in the sling or chain, but don't lift the engine. **Warning:** *DO NOT place any part of your body under the engine when it's supported only by a hoist or other lifting device.*

19 Remove the transmission-to-engine block bolts.

20 Remove the engine mount retaining nuts and bolts (see Chapter 2A, 2b or 2C for engine mount removal procedures).

21 Recheck to be sure nothing is still connecting the engine to the transmission or vehicle. Disconnect anything still remaining.

22 Raise the engine slightly. Carefully work it forward to separate it from the transmission. If you're working on a vehicle with an automatic transmission, be sure the torque converter stays in the transmission (clamp a pair of vise-grips to the housing to keep the converter from sliding out). If you're working on a vehicle with a manual transmission, the input shaft must be completely disengaged from the clutch. Slowly raise the engine out of the engine compartment **(see illustration)**. Check carefully to make sure nothing is hanging up.

23 Remove the flywheel/driveplate and mount the engine on an engine stand.

Installation

24 Install the flywheel/driveplate on the engine (see Chapter 2A, 2B or 2C). Check the engine and transmission mounts. If they're worn or damaged, replace them.

25 If you're working on a vehicle with a manual transmission, install the clutch and pressure plate onto the flywheel (see Chapter 7A). Now is a good time to install a new clutch.

26 Carefully lower the engine into the engine compartment - make sure the engine mounts line up.

27 If you're working on a vehicle with an automatic transmission, guide the torque converter into the crankshaft following the procedure outlined in Chapter 7B.

28 If you're working on a vehicle with a manual transmission, apply a dab of high-temperature grease to the input shaft and guide it into the crankshaft pilot bearing until the bellhousing is flush with the engine block.
Note: *It may be necessary to place the transmission into first gear, then turn the output shaft on the transmission until the splines on the input shaft align with the splines on the clutch disc.*

29 Install the transmission-to-engine bolts and tighten them securely. **Caution:** *DO NOT use the bolts to force the transmission and engine together!*

30 Reinstall the remaining components in the reverse order of removal.

31 Add coolant, oil, power steering and transmission fluid as needed.

32 Run the engine and check for leaks and proper operation of all accessories, then install the hood and test drive the vehicle.

8 Engine overhaul - disassembly sequence

1 It's much easier to disassemble and work on the engine if it's mounted on a portable engine stand. A stand can often be rented quite cheaply from an equipment rental yard. Before the engine is mounted on a stand, the flywheel/driveplate and engine rear plate (if equipped) should be removed from the engine.

2 If a stand isn't available, it's possible to disassemble the engine with it blocked up on the floor. Be extra careful not to tip or drop the engine when working without a stand.

3 If you're going to obtain a rebuilt engine, all external components must come off first, to be transferred to the replacement engine, just as they will if you're doing a complete engine overhaul yourself. These include:

Alternator and brackets
Emissions control components
Distributor and spark plug wires (2.4L and 3.3L engines)
Spark plugs
Thermostat and housing cover
Water pump
EFI components
Intake/exhaust manifolds
Oil filter
Engine mounts
Flywheel/driveplate
Engine rear plate

Note: *When removing the external components from the engine, pay close attention to details that may be helpful or important during installation. Note the installed position of gaskets, seals, spacers, pins, brackets, washers, bolts and other small items.*

4 If you're obtaining a short block, which consists of the engine block, crankshaft, pistons and connecting rods all assembled, then the cylinder heads, oil pan and oil pump will have to be removed as well. See *Engine rebuilding alternatives* for additional information regarding the different possibilities to be considered.

5 If you're planning a complete overhaul, the engine must be disassembled and the internal components removed in the following general order:

2.4L engine

Valve cover
Oil pan and pick-up tube
Oil pump and oil pump driveshaft
Upper timing chain cover
Lower timing chain cover
Timing chain and sprockets
Camshaft(s) and lifters
Cylinder head
Piston/connecting rod assemblies
Crankshaft rear oil seal retainer
Crankshaft and main bearings
Piston oil jets

3.3L engine

Valve covers
Rocker arm assemblies
Valve lifters and guides
Timing covers
Timing belt and sprockets
Cylinder heads
Camshafts
Oil pan and pick-up
Oil pump
Piston/connecting rod assemblies
Crankshaft and main bearings

3.5L engine

Valve covers
Steel oil pan
Aluminum oil pan
Variable valve timing covers
Front timing cover
Timing chains and sprockets
Rear timing cover
Camshaft and lifters
Cylinder heads
Oil pump and pick-up
Piston/connecting rod assemblies
Crankshaft and main bearings
Piston oil jets

9.2 A small plastic bag, with an appropriate label, can be used to store the valve train components so they can be kept together and reinstalled in the original position

9.3 Use a valve spring compressor to compress the spring, then remove the valve stem keepers from the valve stem

9.4 If the valve won't pull through the guide, deburr the edge of the stem end and the area around the top of the valve stem keeper groove with a fine file or whetstone

6 Before beginning the disassembly and overhaul procedures, make sure the following items are available. Also, refer to Engine overhaul - reassembly sequence for a list of tools and materials needed for engine reassembly.

    Common hand tools
    Small cardboard boxes or plastic bags for storing parts
    Gasket scraper
    Ridge reamer
    Vibration damper puller
    Micrometers
    Telescoping gauges
    Dial indicator set
    Valve spring compressor
    Cylinder surfacing hone
    Piston ring groove cleaning tool
    Electric drill motor
    Tap and die set
    Wire brushes
    Oil gallery brushes
    Cleaning solvent

9 Cylinder head - disassembly

Refer to illustrations 9.2, 9.3 and 9.4

Note: *New and rebuilt cylinder heads are commonly available for most engines at dealerships and auto parts stores. Due to the fact that some specialized tools are necessary for the disassembly and inspection procedures, and replacement parts may not be readily available, it may be more practical and economical for the home mechanic to purchase replacement heads rather than taking the time to disassemble, inspect and recondition the originals.*

1 Cylinder head disassembly involves removal of the intake and exhaust valves and related components. If they're still in place, remove the rocker arms (3.3L only) and the lifters and camshafts (see Chapter 2A, 2B or 2C) from the cylinder head. Label the parts or store them separately so they can be reinstalled in their original locations. Refer to Section 21 for camshaft and lifter inspection procedures. **Caution:** *Do not lay the lifters on 3.3L engines on their side or upside down, or air can become trapped inside and the lifter will have to be bled (see Chapter 2B). The lifters can be laid on their side only if they are submerged in a pan of clean engine oil until reassembly.*

2 Before the valves are removed, arrange to label and store them, along with their related components, so they can be kept separate and reinstalled in the same valve guides they are removed from **(see illustration)**.

3 Compress the springs on the first valve with a spring compressor and remove the valve stem keepers **(see illustration)**. Carefully release the valve spring compressor and remove the retainer, the spring and the spring seat (if used).

4 Pull the valve out of the cylinder head, then remove the oil seal from the guide. If the valve binds in the guide (won't pull through), push it back into the cylinder head and deburr the area around the valve stem keeper groove with a fine file or whetstone **(see illustration)**.

5 Repeat the procedure for the remaining valves. Remember to keep all the parts for each valve together so they can be reinstalled in the same locations.

6 Once the valves and related components have been removed and stored in an organized manner, the cylinder head should be thoroughly cleaned and inspected. If a complete engine overhaul is being done, finish the engine disassembly procedures before beginning the cylinder head cleaning and inspection process.

10 Cylinder head - cleaning and inspection

1 Thorough cleaning of the cylinder heads and related valve train components, followed by a detailed inspection, will enable you to decide how much valve service work must be done during the engine overhaul. **Note:** *If the engine was severely overheated, the cylinder head(s) are probably warped (see Step 12).*

Cleaning

2 Scrape all traces of old gasket material and sealing compound off the cylinder head gasket, intake manifold and exhaust manifold sealing surfaces. Be very careful not to gouge the cylinder head. Special gasket removal solvents that soften gaskets and make removal much easier are available at auto parts stores.

3 Remove all built up scale from the coolant passages.

4 Run a stiff wire brush through the various holes to remove deposits that may have formed in them.

5 Run an appropriate-size tap into each of the threaded holes to remove corrosion and thread sealant that may be present. If compressed air is available, use it to clear the holes of debris produced by this operation. **Warning:** *Wear eye protection when using compressed air!*

6 Clean the combustion chambers with a brass wire brush and solvent if carbon has accumulated.

7 Clean the cylinder head with solvent and dry it thoroughly. Compressed air will speed the drying process and ensure that all holes and recessed areas are clean. **Note:** *Decarbonizing chemicals are available and may prove very useful when cleaning cylinder heads and valve train components. They are very caustic and should be used with caution. Be sure to follow the instructions on the container.*

8 On 3.3L engines clean the rocker arms and shafts with solvent and dry them thoroughly (don't mix them up during the cleaning process). Compressed air will speed the drying process and can be used to clean out the oil passages.

9 Clean all the valve springs, spring seats, valve stem keepers and retainers with solvent and dry them thoroughly. Do the components from one valve at a time to avoid mixing up the parts.

10 Scrape off any heavy deposits that may have formed on the valves, then use a motorized wire brush to remove deposits from the valve heads and stems. Again, make sure the valves don't get mixed up.

Chapter 2 Part D General engine overhaul procedures   2D-13

10.12 Check the cylinder head gasket surface for warpage by trying to slip a feeler gauge under the straightedge (see the Specifications for the maximum warpage allowed and use a feeler gauge of that thickness)

Inspection

Note: *Be sure to perform all of the following inspection procedures before concluding that machine shop work is required. Make a list of the items that need attention.*

Cylinder head

Refer to illustrations 10.12 and 10.14

11 Inspect the heads very carefully for cracks, evidence of coolant leakage and other damage. If cracks are found, check with an automotive machine shop concerning repair. If repair isn't possible, a new cylinder head should be obtained.

12 Using a straightedge and feeler gauge, check the cylinder head gasket mating surface for warpage **(see illustration)**. If the warpage exceeds the limit specified in this Chapter, it can be resurfaced at an automotive machine shop. **Note:** *The cylinder heads have a specific MINIMUM height, measured from the cylinder head gasket surface to the valve cover surface. If the cylinder head will fall below the minimum height (see Specifications) after it is machined, a new cylinder head will have to be purchased.*

13 Examine the valve seats in each of the combustion chambers. If they're pitted, cracked or burned, the cylinder head will

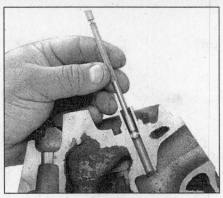

10.14 Use a small hole gauge to determine the inside diameter of the valve guides (the gauge is then measured with a micrometer)

require valve service that's beyond the scope of the home mechanic.

14 Check the valve stem-to-guide clearance with a small hole gauge and micrometer **(see illustration)**, then measure the valve stem diameter with a micrometer and subtract it from the valve guide inside diameter to obtain the stem to guide clearance. When using a small hole gauge or telescoping snap gauge, insert the gauge to the middle portion of the valve guide (where wear should be minimal) and tighten the gauge. Move the gauge up and down in the guide. If the guide isn't worn the clearance should be equal from top to bottom. Loose areas indicate that the guide is tapered. If the measurement exceeds the stem-to-guide clearance limit found in this Chapter's Specifications, the valve guides should be replaced. After this is done, if there's still some doubt regarding the condition of the valve guides they should be checked by an automotive machine shop (the cost should be minimal).

Valves

Refer to illustrations 10.15 and 10.16

15 Carefully inspect each valve face for uneven wear, deformation, cracks, pits and burned areas **(see illustration)**. Check the valve stem for scuffing and galling and the

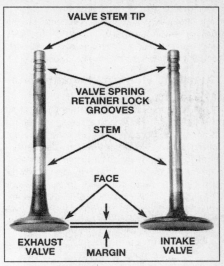

10.15 Check for valve wear at the points shown here

neck for cracks. Rotate the valve and check for any obvious indication that it's bent. Look for pits and excessive wear on the end of the stem. The presence of any of these conditions indicates the need for valve service by an automotive machine shop.

16 Measure the margin width on each valve **(see illustration)**. Any valve with a margin narrower than specified in this Chapter will have to be replaced with a new one.

Valve components

Refer to illustrations 10.17 and 10.18

17 Check each valve spring for wear (on the ends) and pits. Measure the free length and compare it to the Specifications in this Chapter **(see illustration)**. Any springs that are shorter than specified have sagged and should not be reused. The tension of all springs should be checked with a special fixture before deciding that they're suitable for use in a rebuilt engine (take the springs to an automotive machine shop for this check).

18 Stand each spring on a flat surface and check it for squareness **(see illustration)**. If any of the springs are distorted or sagged, replace all of them with new parts.

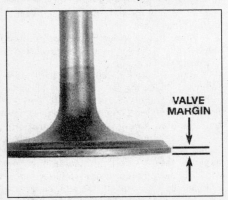

10.16 The margin width on each valve must be as specified (if no margin exists, the valve cannot be reused)

10.17 Measure the free length of each valve spring with a dial or vernier caliper

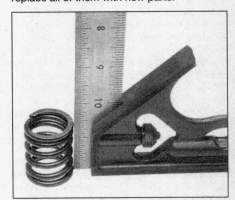

10.18 Check each valve spring for squareness

19   Check the spring retainers and valve stem keepers for obvious wear and cracks. Any questionable parts should be replaced with new ones, as extensive damage will occur if they fail during engine operation.

Camshaft, lifters, rocker arms and shafts

20   Refer to Section 21 of this Chapter for the camshaft, lifters and bearing inspection procedures. Be sure to inspect the camshaft bearing journals before the cylinder head is sent to the machine shop to have the valves serviced. If the journals are gouged or scored the cylinder head will have to be replaced regardless of the condition of the valves and related components. If you're working on a 3.3L engine refer to Chapter 2B and also inspect the rocker arms and shafts.

21   Any damaged or excessively worn parts must be replaced with new ones.

22   If the inspection process indicates that the valve components are in generally poor condition and worn beyond the limits specified, which is usually the case in an engine that's being overhauled, reassemble the valves in the cylinder head and refer to Section 11 for valve servicing recommendations.

11 Valves - servicing

1   Because of the complex nature of the job and the special tools and equipment needed, servicing of the valves, the valve seats and the valve guides, commonly known as a valve job, should be done by a professional.

2   The home mechanic can remove and disassemble the heads, do the initial cleaning and inspection, then reassemble and deliver them to an automotive machine shop for the actual service work. Doing the inspection will enable you to see what condition the cylinder head and valvetrain components are in and will ensure that you know what work and new parts are required when dealing with an automotive machine shop.

3   The machine shop will remove the valves and springs, recondition or replace the valves and valve seats, recondition or replace the valve guides, check and replace the valve springs, spring retainers and valve stem keepers (as necessary), replace the valve seals with new ones, reassemble the valve components and make sure the installed spring height is correct. The cylinder head gasket surface will also be resurfaced if it's warped.

4   After the valve job has been performed by a professional, the cylinder head will be in like-new condition. When the cylinder head is returned, be sure to clean it again before installation on the engine to remove any metal particles and abrasive grit that may still be present from the valve service or cylinder head resurfacing operations. Use compressed air, if available, to blow out all the oil holes and passages.

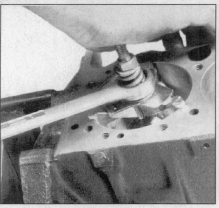

13.1 A ridge reamer is required to remove the ridge from the top of each cylinder - do this before removing the pistons!

12 Cylinder head - reassembly

1   Regardless of whether or not the cylinder head was sent to an automotive repair shop for valve servicing, make sure it is clean before beginning reassembly.

2   If the cylinder head was sent out for valve servicing, the valves and related components will already be in place. Begin the reassembly procedure with Step 5.

3   Install the valves, with light oiling on the stems into valve guides. Install the valve spring seat(s) in place on the cylinder head. **Note:** *3.3L engines are equipped with an inner and outer valve spring seat while 2.4L and 3.5L engines are equipped with a single valve spring seat. Be sure to install the outer valve spring seat first, then install the inner valve spring seat second on 3.3L engines.* After the valve spring seat(s) have been positioned on the cylinder head correctly, install the valve seals over the top of the valves tips by hand. Using the stem of the valves as a guide, slide the seals down to the top of each valve guide. Gently tap each seal into place until it is properly seated onto the guide to specified depth (see Chapter 2A, 2B or 2C). **Caution:** *Do not hammer on the guide seal once it is seated or you may damage the seal. Do not twist or cock the seals during installation or they will not seal properly on the valve stems.* **Note:** *On 3.3L engines it will only be necessary to use a seal installation tool on the intake valve guide. The exhaust valve seals should simply slide into place on the valve guide by hand.*

4   Slip the valve spring(s) in place on the cylinder head, then use a spring compressor to install the springs, retainers and valve stem keepers (see Chapter 2A, 2B or 2C). **Note:** *3.3L engines are equipped with an inner and outer valve spring while 2.4L and 3.5L engines are equipped with a single valve spring. Be sure to install the inner valve spring into the outer valve spring and install the springs as a set on 3.3L engines. On all engines, always install the end of the outer valve spring with the more closely wound*

13.3 Check the connecting rod side clearance with a feeler gauge as shown

coils or paint marks towards the cylinder head.

5   On 2.4L and 3.5L engines refer to Chapter 2A or 2C and install the lifters and the camshafts.

6   On 3.3L engines refer to Chapter 2B and install the camshafts, camshaft oil seals, the lifter guide assembly and the rocker arms.

7   Always install old cylinder head components in their original locations.

13 Pistons/connecting rods - removal

Refer to illustrations 13.1, 13.3 and 13.6
Note: *Prior to removing the piston/connecting rod assemblies, remove the cylinder heads, the oil pan, the windage tray (3.5L engine only) and the oil pump pick-up by referring to the appropriate Sections in Chapter 2A, 2B or 2C, if not already removed.*

1   Use your fingernail to feel if a ridge has formed at the upper limit of ring travel (about 1/4-inch down from the top of each cylinder). If carbon deposits or cylinder wear have produced ridges, they must be completely removed with a special tool **(see illustration)**. Follow the manufacturer's instructions provided with the tool. Failure to remove the ridges before attempting to remove the piston/connecting rod assemblies may result in piston breakage.

2   After the cylinder ridges have been removed, turn the engine upside-down so the crankshaft is facing up.

3   Before the connecting rods are removed, check the side clearance (endplay) with feeler gauges. Slide them between the first connecting rod and the crankshaft throw until the play is removed **(see illustration)**. The side clearance is equal to the thickness of the feeler gauge(s). If the side clearance exceeds the service limit, new connecting rods will be required. If new rods (or a new crankshaft) are installed, the side clearance may fall under the specified minimum (if it does, the rods will have to be machined to restore it - consult an automotive machine

Chapter 2 Part D General engine overhaul procedures 2D-15

13.6 To prevent damage to the crankshaft journals and cylinder walls, slip sections of rubber or plastic hose over the connecting rod bolts before removing the pistons

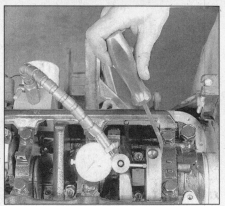

14.1 Checking crankshaft endplay with a dial indicator

14.3 Crankshaft endplay can also be measured with a feeler gauge at the crankshaft thrust bearing

shop for advice if necessary). Repeat the procedure for the remaining connecting rods.

4   Check the connecting rods and caps for identification marks. If they aren't plainly marked, use a small center punch to make the appropriate number of indentations on each rod and cap (1, 2, 3, etc.).

5   Loosen each of the connecting rod cap nuts 1/2-turn at a time until they can be removed by hand. Remove the number one connecting rod cap and bearing insert. Don't drop the bearing insert out of the cap.

6   Slip a short length of plastic or rubber hose over each connecting rod cap bolt to protect the crankshaft journal and cylinder wall as the piston is removed (see illustration). Note: *On 3.5L engines the rod bolts are removed with the caps, so it may be helpful to make a set of connecting rod guides. Find several bolts that fit the rods at your local hardware store. or purchase a couple of new rod bolts. Cut the heads of the bolts, off with a hacksaw or similar tool and slip short length of hose over the end of each bolt to make a pair connecting rod guides. Screw the guides into the connecting rods during removal and installation of the piston/connecting rod from the engine block.*

7   Remove the bearing insert and push the connecting rod/piston assembly out through the top of the engine. Use a wooden hammer handle to push on the upper bearing surface in the connecting rod. If resistance is felt, double-check to make sure that all of the ridge was removed from the cylinder.

8   Repeat the procedure for the remaining cylinders.

9   After removal, reassemble the connecting rod caps and bearing inserts in their respective connecting rods and install the cap nuts finger tight. Leaving the old bearing inserts in place until reassembly will help prevent the connecting rod bearing surfaces from being accidentally nicked or gouged.

10 Don't separate the pistons from the connecting rods (see Section 18 for additional information).

14 Crankshaft - removal

Refer to illustrations 14.1, 14.3, 14.4 and 14.7
Note: *The crankshaft can be removed only after the engine has been removed from the vehicle. It's assumed that the flywheel/driveplate, crankshaft pulley, timing belt (3.3L engine) or timing chain (2.4L and 3.5L engine), camshaft and crankshaft sprockets, oil pan, oil pump and piston/connecting rod assemblies have already been removed. The rear main oil seal retainer must also be unbolted and separated from the block before proceeding with crankshaft removal.*

1   Before the crankshaft is removed, check the endplay. Mount a dial indicator with the stem in line with the crankshaft and just touching one of the crank throws (see illustration).

2   Push the crankshaft all the way to the rear and zero the dial indicator. Next, pry the crankshaft to the front as far as possible and check the reading on the dial indicator. The distance that it moves is the endplay. If it's

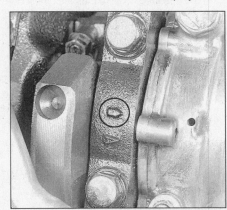

14.4 The main bearing cap assembly typically has cast-in arrows which point toward the drivebelt (front) end of the engine - If no arrows exist use a number stamp or punch to mark the location and direction of the cap assembly

greater than specified in this Chapter, check the crankshaft thrust surfaces for wear. If no wear is evident, new main bearings should correct the endplay.

3   If a dial indicator isn't available, feeler gauges can be used. Gently pry or push the crankshaft all the way to the front of the engine. Slip feeler gauges between the crankshaft and the front face of the crankshaft thrust bearing to determine the clearance (see illustration).

4   Loosen the main bearing cap bolts in the reverse of the tightening sequence (see Section 24) 1/4-turn at a time, until the bearing cap(s) and brace assembly (if equipped) can be removed by hand. The main bearing cap assembly typically have cast-in arrows, which points to the drivebelt end of the engine (see illustration). If the caps do not have markings, use a number stamp or punch to indicate the direction and location of the main bearing caps. **Note:** *On 3.3L engines it will only be necessary to mark the direction of the front of the bearing cap brace, since the bearing caps and the brace are one piece. On 3.5L engines, mark the front of the bearing cap brace and remove it, then inspect the main bearing caps for installation marks (marking them if necessary) and remove the main bearing caps from the engine. 2.4L engines are not equipped with a main bearing cap brace.*

5   Gently tap the cap assembly with a soft-face hammer, then separate it from the engine block. If necessary, use the bolts as levers to remove the cap assembly. Try not to drop the bearing inserts if they come out with the cap assembly.

6   Carefully lift the crankshaft out of the engine. It may be a good idea to have an assistant available, since the crankshaft is quite heavy. With the bearing inserts in place in the engine block and main bearing caps, return the cap assembly to it's location on the engine block and tighten the bolts finger tight.

7   On 3.5L engines inspect the main bearing cap bolts for excessive stretching (see

2D-16 Chapter 2 Part D General engine overhaul procedures

illustration). If there is a difference of 0.0043 inch (0.11 mm) or larger in the diameter of the bolts between the indicated areas, the bolts must be replaced. Many times it may be cheaper to purchase a complete set of bolts rather than a number of bolts individually.

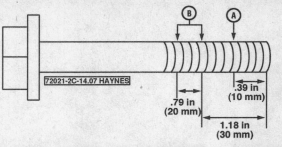

14.7 On 3.5L engines inspect the main bearing cap bolts for stretch at the indicated areas (A and B)

15 Engine block - cleaning

Refer to illustrations 15.1a, 15.1b, 15.4, 15.8 and 15.10

Caution: *The core plugs (also known as freeze or soft plugs) may be difficult or impossible to retrieve if they're driven completely into the block coolant passages.*

1 Using the blunt end of a punch, tap in on the outer edge of the core plug to turn the plug sideways in the bore. Then use pliers to pull the core plug from the engine block **(see illustrations)**.

2 Using a gasket scraper, remove all traces of gasket material from the engine block. Be very careful not to nick or gouge the gasket sealing surfaces.

3 Remove the main bearing cap assembly and separate the bearing inserts from the caps and the engine block. Label the bearings, indicating which cylinder they were removed from and whether they were in the cap or the block, then set them aside.

4 Remove all of the threaded oil gallery plugs from the block. The plugs are usually very tight - they may have to be drilled out and the holes retapped. Use new plugs when the engine is reassembled. Where applicable, it will also be necessary to remove the oil spray jets from the engine block before it is cleaned **(see illustration)**.

5 If the engine is extremely dirty, it should be taken to an automotive machine shop to be steam cleaned or hot tanked.

6 After the block is returned, clean all oil holes and oil galleries one more time. Brushes specifically designed for this purpose are available at most auto parts stores. Flush the passages with warm water until the water runs clear, dry the block thoroughly

15.1a The core plugs can be removed by tapping in one edge until the plug turns sideways ...

15.1b ... then remove the core plug with pliers

and wipe all machined surfaces with a light, rust-preventive oil. If you have access to compressed air, use it to speed the drying process and to blow out all the oil holes and galleries. **Warning:** *Wear eye protection when using compressed air!*

7 If the block isn't extremely dirty or sludged up, you can do an adequate cleaning job with hot soapy water and a stiff brush. Take plenty of time and do a thorough job. Regardless of the cleaning method used, be sure to clean all oil holes and galleries very thoroughly, dry the block completely and coat all machined surfaces with light oil.

8 The threaded holes in the block must be clean to ensure accurate torque readings during reassembly. Run the proper size tap into each of the holes to remove rust, corrosion, thread sealant or sludge and restore damaged threads **(see illustration)**. If possible, use compressed air to clear the holes of debris produced by this operation. Now is a good time to clean the threads on the cylinder head bolts and the main bearing cap bolts as well.

9 Reinstall the main bearing caps and tighten the bolts finger tight.

10 After coating the sealing surfaces of the new core plugs with core plug sealant, install them in the engine block **(see illustration)**. Make sure they're driven in straight and seated properly or leakage could result. Spe-

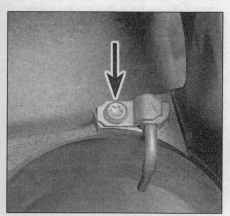

15.4 On 2.4L and 3.5L engines, it will also be necessary to remove the oil jets from the block before cleaning it

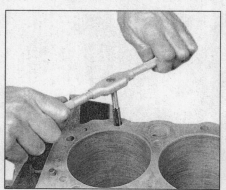

15.8 All bolt holes in the block - particularly the main bearing cap and cylinder head bolt holes - should be cleaned and restored with a tap (be sure to remove debris from the holes after this is done)

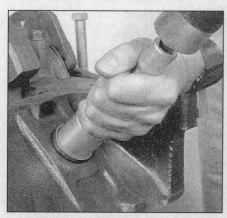

15.10 A large socket on an extension can be used to drive the new core plugs into the bores

Chapter 2 Part D General engine overhaul procedures   2D-17

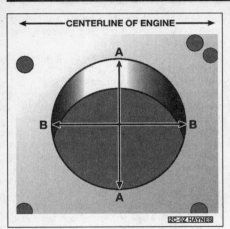

16.4a Measure the diameter of each cylinder at a right angle to the engine centerline (A), and parallel to engine centerline (B) - out-of-round is the distance between A and B; taper is the difference between A and B at the top of the cylinder and A and B at the bottom of the cylinder

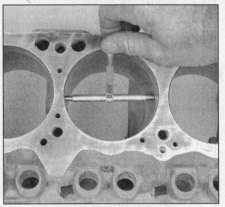

16.4b Use a telescoping gauge to measure the bore - the ability to "feel" when it is at the correct point will be developed over time, so work slowly and repeat the check until you're satisfied the bore measurement is accurate

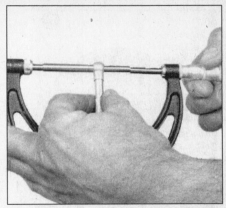

16.4c The gauge is then measured with a micrometer to determine the bore size

cial tools are available for this purpose, but a large socket, with an outside diameter that will just slip into the core plug, a 1/2-inch drive extension and a hammer will work just as well. **Note:** *Make sure the socket only contacts the inside of the core plug, not the rim.*
11   Apply non-hardening thread sealant to the new oil gallery plugs and thread them into the holes in the block. Make sure they're tightened securely.
12   If the engine isn't going to be reassembled right away, cover it with a large plastic trash bag to keep it clean.

16 Engine block - inspection

Refer to illustrations 16.4a, 16.4b, 16.4c and 16.11

1   Before the block is inspected, it should be cleaned as described in Section 15.
2   Visually check the block for cracks, rust and corrosion. Look for stripped threads in the threaded holes. It's also a good idea to have the block checked for hidden cracks by an automotive machine shop that has the special equipment to do this type of work. If defects are found, have the block repaired, if possible, or replaced.
3   Check the cylinder bores for scuffing and scoring.
4   Check the cylinders for taper and out-of-round conditions as follows **(see illustrations)**.
5   Measure the diameter of each cylinder at the top (just under the ridge area), center and bottom of the cylinder bore, parallel to the crankshaft axis.
6   Next measure each cylinder's diameter at the same three locations perpendicular to the crankshaft axis.
7   The taper of the cylinder is the difference between the bore diameter at the top of the cylinder and the diameter at the bottom. The out-of-round specification of the cylinder bore is the difference between the parallel and perpendicular readings. Compare your results to those listed in this Chapter's Specifications.
8   Repeat the procedure for the remaining pistons and cylinders.
9   If the cylinder walls are badly scuffed or scored, or if they're out-of-round or tapered beyond the limits given in this Chapter's Specifications, have the engine block rebored and honed at an automotive machine shop. If a rebore is done, oversize pistons and rings will be required.
10   If the cylinders are in reasonably good condition and not worn to the outside of the limits, and if the piston-to-cylinder clearances can be maintained properly, then they don't have to be rebored. Honing is all that's necessary (see Section 17).
11   Using a precision straightedge and feeler gauge, check the block deck (the surface that mates with the cylinder head) for distortion **(see illustration)**.

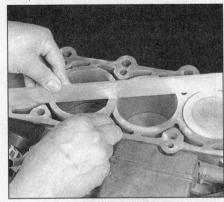

16.11 Check the flatness of the top of the block with a straightedge and feeler gauge - if distortion exceeds Specifications, the block deck will have to be machined

17 Cylinder honing

Refer to illustrations 17.3a and 17.3b

1   Prior to engine reassembly, the cylinder bores must be honed so the new piston rings will seat correctly and provide the best possible combustion chamber seal. **Note:** *If you don't have the tools or don't want to tackle the honing operation, most automotive machine shops will do it for a reasonable fee.*
2   Before honing the cylinders, install the main bearing caps and tighten the bolts to the torque specified in this Chapter.
3   Two types of cylinder hones are commonly available - the flex hone or "bottle brush" type and the more traditional surfacing hone with spring-loaded stones. Both will do the job, but for the less experienced mechanic the "bottle brush" hone will probably be easier to use. You'll also need some kerosene or honing oil, rags and an electric drill motor. Proceed as follows:

   a) Mount the hone in the drill motor, compress the stones and slip it into the first cylinder **(see illustration)**. Be sure to wear safety goggles or a face shield!
   b) Lubricate the cylinder with plenty of honing oil, turn on the drill and move the

17.3a A "bottle brush" hone is the easiest type of hone to use

2D

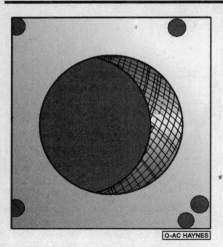

17.3b The cylinder hone should leave a smooth, crosshatch pattern with the lines intersecting at approximately a 60-degree angle

hone up-and-down in the cylinder at a pace that will produce a fine crosshatch pattern on the cylinder walls. Ideally, the crosshatch lines should intersect at approximately a 60-degree angle (see illustration). Be sure to use plenty of lubricant and don't take off any more material than is absolutely necessary to produce the desired finish. **Note:** *Piston ring manufacturers may specify a smaller crosshatch angle than the traditional 60-degrees - read and follow any instructions included with the new rings.*

c) *Don't withdraw the hone from the cylinder while it's running. Instead, shut off the drill and continue moving the hone up-and-down in the cylinder until it comes to a complete stop, then compress the stones and withdraw the hone. If you're using a "bottle brush" type hone, stop the drill motor, then turn the chuck in the normal direction of rotation while withdrawing the hone from the cylinder.*

d) *Wipe the oil out of the cylinder and repeat the procedure for the remaining cylinders.*

4 After the honing job is complete, chamfer the top edges of the cylinder bores with a small file so the rings won't catch when the pistons are installed. Be very careful not to nick the cylinder walls with the end of the file.

5 The entire engine block must be washed again very thoroughly with warm, soapy water to remove all traces of the abrasive grit produced during the honing operation. **Note:** *The bores can be considered clean when a lint-free white cloth - dampened with clean engine oil - used to wipe them out doesn't pick up any more honing residue, which will show up as gray areas on the cloth. Be sure to run a brush through all oil holes and galleries and flush them with running water.*

6 After rinsing, dry the block and apply a coat of light rust-preventive oil to all machined surfaces. Wrap the block in a plastic trash bag to keep it clean and set it aside until reassembly.

18 Pistons/connecting rods - inspection

Refer to illustrations 18.4a, 18.4b, 18.10, 18.11a and 18.11b

1 Before the inspection process can be carried out, the piston/connecting rod assemblies must be cleaned and the original piston rings removed from the pistons. **Note:** *Always use new piston rings when the engine is reassembled.*

2 Using a piston ring installation tool, carefully remove the rings from the pistons. Be careful not to nick or gouge the pistons in the process.

3 Scrape all traces of carbon from the top of the piston. A hand-held brass wire brush or a piece of fine emery cloth can be used (with solvent) once the majority of the deposits have been scraped away. Do not, under any circumstances, use a wire brush mounted in a drill motor to remove deposits from the pistons. The piston material is soft and may be eroded away by the wire brush.

4 Use a piston ring groove-cleaning tool to remove carbon deposits from the ring grooves. If a tool isn't available, a piece broken off the old ring will do the job. Be very careful to remove only the carbon deposits - don't remove any metal and do not nick or scratch the sides of the ring grooves **(see illustrations)**.

5 Once the deposits have been removed, clean the piston/rod assemblies with solvent and dry them with compressed air (if available). Make sure the oil return holes in the back sides of the ring grooves are clear.

6 If the pistons and cylinder walls aren't damaged or worn excessively, and if the engine block is not rebored, new pistons won't be necessary. Normal piston wear appears as even vertical wear on the piston thrust surfaces and slight looseness of the top ring in its groove. New piston rings, however, should always be used when an engine is rebuilt.

7 Carefully inspect each piston for cracks around the skirt, at the pin bosses and at the ring lands.

8 Look for scoring and scuffing on the thrust faces of the skirt, holes in the piston crown and burned areas at the edge of the crown. If the skirt is scored or scuffed, the engine may have been suffering from overheating and/or abnormal combustion, which caused excessively high operating temperatures. The cooling and lubrication systems should be checked thoroughly. A hole in the piston crown is an indication that abnormal combustion (preignition) was occurring. Burned areas at the edge of the piston crown are usually evidence of spark knock (detonation). If any of the above problems exist, the causes must be corrected or the damage will occur again. The causes may include intake air leaks, incorrect fuel/air mixture, incorrect ignition timing and EGR system malfunctions.

9 Corrosion of the piston, in the form of small pits, indicates that coolant is leaking into the combustion chamber and/or the crankcase. Again, the cause must be corrected or the problem may persist in the rebuilt engine.

10 Measure the piston ring side clearance by laying a new piston ring in each ring groove and slipping a feeler gauge in beside it **(see illustration)**. Check the clearance at three or four locations around each groove. Be sure to use the correct ring for each groove - they are different. If the side clearance is greater than specified in this Chapter,

18.4a The piston ring grooves can be cleaned with a special tool, as shown here . . .

18.4b . . . or a section of a broken ring

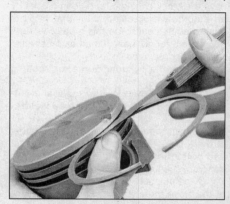

18.10 Check the ring side clearance with a feeler gauge at several points around the groove

Chapter 2 Part D General engine overhaul procedures

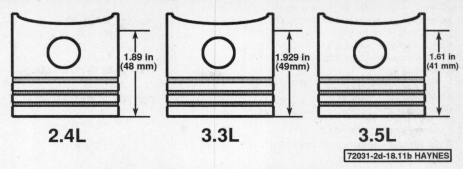

18.11b . . . at the indicated area for your type of engine

18.11a Measure the piston diameter at a 90-degree angle to the piston pin . . .

new pistons will have to be used.

11 Check the piston-to-bore clearance by measuring the bore (see Section 16) and the piston diameter. Make sure the pistons and bores are correctly matched. Measure the piston across the skirt, at a 90-degree angle to the piston pin at the specified distance from the bottom or top of the piston **(see illustrations)**. Subtract the piston diameter from the bore diameter to obtain the clearance. If it's greater than specified in this Chapter, the block will have to be rebored and new pistons and rings installed.

12 Check the piston-to-rod clearance by twisting the piston and rod in opposite directions. Any noticeable play indicates excessive wear, which must be corrected. The piston/connecting rod assemblies should be taken to an automotive machine shop to have the pistons and rods resized and new pins installed.

13 If the pistons must be removed from the connecting rods for any reason, they should be taken to an automotive machine shop. While they are there have the connecting rods checked for bend and twist, since automotive machine shops have special equipment for this purpose. **Note:** *Unless new pistons and/or connecting rods must be installed, do not disassemble the pistons and connecting rods.*

14 Check the connecting rods for cracks and other damage. Temporarily remove the rod caps, lift out the old bearing inserts, wipe the rod and cap bearing surfaces clean and inspect them for nicks, gouges and scratches. After checking the rods, replace the old bearings, slip the caps into place and tighten the nuts finger tight. **Note:** *If the engine is being rebuilt because of a connecting rod knock, be sure to install new or rebuilt rods.*

19 Crankshaft - inspection

Refer to illustrations 19.1, 19.2, 19.5 and 19.7

1 Remove all burrs from the crankshaft oil holes with a stone, file or scraper **(see illustration)**.
2 Clean the crankshaft with solvent and dry it with compressed air (if available). Be

19.1 The oil holes should be chamfered so sharp edges don't gouge or scratch the new bearings

sure to clean the oil holes with a stiff brush **(see illustration)** and flush them with solvent.
3 Check the main and connecting rod bearing journals for uneven wear, scoring, pits and cracks.
4 Check the rest of the crankshaft for cracks and other damage. It should be magnafluxed to reveal hidden cracks - an automotive machine shop will handle the procedure.
5 Using a micrometer, measure the diameter of the main and connecting rod journals **(see illustration)** and compare the results to the Specifications in this Chapter. By measuring the diameter at a number of points around each journal's circumference, you'll

19.2 Use a wire or stiff plastic bristle brush to clean the oil passages in the crankshaft

be able to determine whether or not the journal is out-of-round. Take the measurement at each end of the journal, near the crank throws, to determine if the journal is tapered.
6 If the crankshaft journals are damaged, tapered, out-of-round or worn beyond the limits given in the Specifications in this Chapter, have the crankshaft reground by an automotive machine shop. Be sure to use the correct size bearing inserts if the crankshaft is reconditioned.
7 Check the oil seal journals at each end of the crankshaft for wear and damage. If the seal has worn a groove in the journal, or if it's nicked or scratched **(see illustration)**, the

19.5 Measure the diameter of each crankshaft journal at several points to detect taper and out-of-round conditions

19.7 If the seals have worn grooves in the crankshaft journals, or if the seal contact surfaces are nicked or scratched, the new seals will leak

new seal may leak when the engine is reassembled. In some cases, an automotive machine shop may be able to repair the journal by pressing on a thin sleeve. If repair isn't feasible, a new or different crankshaft should be installed.

8 Refer to Section 20 and examine the main and rod bearing inserts.

20 Main and connecting rod bearings - inspection and main bearing selection

Inspection

Refer to illustration 20.1

1 Even though the main and connecting rod bearings should be replaced with new ones during the engine overhaul, the old bearings should be retained for close examination, as they may reveal valuable information about the condition of the engine **(see illustration)**.

2 Bearing failure occurs because of lack of lubrication, the presence of dirt or other foreign particles, overloading the engine and corrosion. Regardless of the cause of bearing failure, it must be corrected before the engine is reassembled to prevent it from happening again.

3 When examining the bearings, remove them from the engine block, the main bearing caps, the connecting rods and the rod caps and lay them out on a clean surface in the same general position as their location in the engine. This will enable you to match any bearing problems with the corresponding crankshaft journal.

4 Dirt and other foreign particles get into the engine in a variety of ways. It may be left in the engine during assembly, or it may pass through filters or the PCV system. It may get into the oil, and from there into the bearings. Metal chips from machining operations and normal engine wear are often present. Abrasives are sometimes left in engine components after reconditioning, especially when parts are not thoroughly cleaned using the proper cleaning methods. Whatever the source, these foreign objects often end up embedded in the soft bearing material and are easily recognized. Large particles will not embed in the bearing and will score or gouge the bearing and journal. The best prevention for this cause of bearing failure is to clean all parts thoroughly and keep everything spotlessly clean during engine assembly. Frequent and regular engine oil and filter changes are also recommended.

5 Lack of lubrication (or lubrication breakdown) has a number of interrelated causes. Excessive heat (which thins the oil), overloading (which squeezes the oil from the bearing face) and oil leakage or throw off (from excessive bearing clearances, worn oil pump or high engine speeds) all contribute to lubrication breakdown. Blocked oil passages, which usually are the result of misaligned oil holes in a bearing shell, will also oil starve a

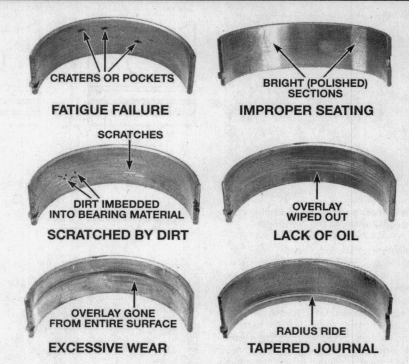

bearing and destroy it. When lack of lubrication is the cause of bearing failure, the bearing material is wiped or extruded from the steel backing of the bearing. Temperatures may increase to the point where the steel backing turns blue from overheating.

6 Driving habits can have a definite effect on bearing life. Low-speed operation in too-high a gear (lugging the engine) puts very high loads on bearings, which tends to squeeze out the oil film. These loads cause the bearings to flex, which produces fine cracks in the bearing face (fatigue failure). Eventually the bearing material will loosen in pieces and tear away from the steel backing. Short-trip driving leads to corrosion of bearings because insufficient engine heat is produced to drive off the condensed water and corrosive gases. These products collect in the engine oil, forming acid and sludge. As the oil is carried to the engine bearings, the acid attacks and corrodes the bearing material.

7 Incorrect bearing installation during engine assembly will lead to bearing failure as well. Tight-fitting bearings leave insufficient bearing oil clearance and will result in oil starvation. Dirt or foreign particles trapped behind a bearing insert result in high spots on the bearing which lead to failure.

Selection

8 If the original bearings are worn or damaged, or if the oil clearances are incorrect (see Sections 24 or 26), new bearings will have to be purchased. It is rare during a thorough rebuild of an engine with many miles on it that new replacement bearings would not be employed. However, if the crankshaft has

20.1 Typical bearing failures

21.1 Inspect the cam bearing surfaces in each cylinder head for pits, score marks and abnormal wear - if wear or damage is noted, the cylinder head must be replaced

been reground, new undersize bearings must be installed.

9 The automotive machine shop that reconditions the crankshaft will provide or help you select the correct size bearings. Depending on how much material has to be ground from the crankshaft to restore it, different undersize bearings are required. Crankshafts are normally ground in increments of 0.010-inch. Sometimes the amount of material machined on a crankshaft will differ between the mains and rod journals, especially if a rod journal was damaged. Markings on most reground crankshafts indicate how much was machined, such as "10-10", meaning that 0.010-inch was removed from both the rod and main journals. Such a crankshaft would require 0.010-inch undersize bearings, a common replacement bearing size.

Chapter 2 Part D General engine overhaul procedures

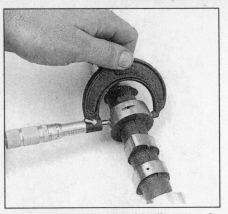

21.2a Measure the outside diameter of each camshaft journal and the inside diameter of each bearing to determine the oil clearance measurement

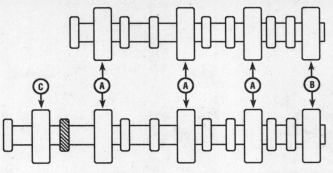

21.2b 3.3L engine camshaft journal designations - B are the rearmost journals, while C is the journal ahead of the distributor drive gear

10 Regardless of how the bearing sizes are determined, use the oil clearance, measured with Plastigage, as the final guide to ensure the bearings are the right size. If you have any questions or are unsure which bearings to use, get help from your machine shop or a dealer parts or service department.

21 Camshafts, lifters and bearings - inspection

Refer to illustrations 21.1, 21.2a, 21.2b, 21.4, 21.5a, 21.5b, 21.6 and 21.7

1 Visually check the camshaft bearing surfaces for pitting, score marks, galling and abnormal wear. If the bearing surfaces are damaged, the cylinder head will have to be replaced **(see illustration)**.

2 Measure the outside diameter of each camshaft bearing journal and record your measurements **(see illustrations)**. Compare them to the journal outside diameter specified in this Chapter, then measure the inside diameter of each corresponding camshaft bearing and record the measurements. Subtract each cam journal outside diameter from its respective cam bearing bore inside diameter to determine the oil clearance for each bearing. Compare the results to the specified journal-to-bearing clearance. If any of the measurements fall outside the standard specified wear limits in this Chapter, either the camshaft or the cylinder head, or both, must be replaced.

3 Check camshaft runout by placing the camshaft between two V-blocks or back into the cylinder head and set up a dial indicator on the center journal. Zero the dial indicator. Turn the camshaft slowly and note the dial indicator readings. Record your readings and compare them with the specified runout in this Chapter. If the measured runout exceeds the runout specified in this Chapter, replace the camshaft.

4 Check the camshaft lobe height by measuring each lobe with a micrometer **(see illustration)**. Compare the measurement to

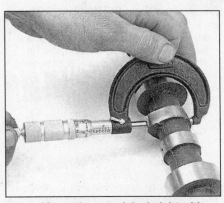

21.4 Measuring cam lobe height with a micrometer; make sure you move the micrometer to get the highest reading (top of cam lobe)

the cam lobe height specified in this Chapter. Then subtract the measured cam lobe height from the specified height to compute wear on the cam lobes. Compare it to the specified wear limit. If it's greater than the specified wear limit, replace the camshaft.

5 Inspect the contact and sliding surfaces of each lifter for wear and scratches **(see illustrations). Note:** *If the lifter pad is worn, it's a good idea to check the corresponding camshaft lobe or adjusting shim on 2.4L and 3.5L engines, because it will probably be worn too.* **Caution:** *On 3.3L engines do not lay the lifters on their side or upside down, or*

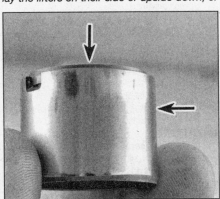

21.5b On 2.4L and 3.5L engines, inspect the valve lifter at the areas shown – don't forget to also inspect the valve adjusting shims for wear

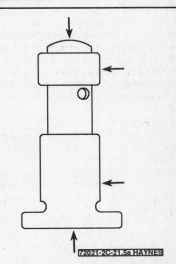

21.5a On 3.3L engines, check the contact and sliding surfaces of each lifter for wear and damage at the areas shown

air can become trapped inside and the lifter will have to be bled (see Chapter 2B). The lifters can be laid on their side only if they are submerged in a pan of clean engine oil until reassembly.

6 Measure the outside diameter of each lifter with a micrometer **(see illustration)** and compare it to the Specifications in this Chapter. If any lifter is worn beyond the specified

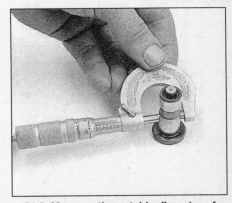

21.6 Measure the outside diameter of each lifter with a micrometer . . .

2D-22 Chapter 2 Part D General engine overhaul procedures

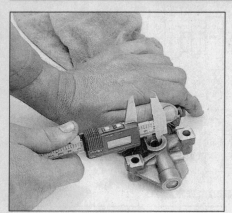

21.7 ... and the inside diameter of each lifter bore - subtract the lifter diameter from the lifter bore diameter to obtain the lifter-to-guide clearance (3.3L shown, on 2.4L and 3.5L engines the procedure is the same except that the lifter bore diameter is an integral part of the cylinder head)

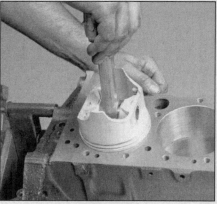

23.3 When checking piston ring end gap, the ring must be square in the cylinder bore (this is done by pushing the ring down with the top of a piston as shown)

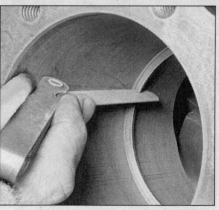

23.4 With the ring square in the cylinder, measure the end gap with a feeler gauge

limit, replace it.

7   Check each lifter bore diameter **(see illustration)** and compare the results to the Specifications in this Chapter. If any lifter bore is worn beyond the specified limit, the lifter guide assembly (3.3L engines) or the cylinder head (2.4L and 3.5L engines) must be replaced.

8   Subtract the outside diameter of each lifter from the inside diameter of the lifter bore and compare the difference to the clearance specified in this Chapter. If both the lifter and the bore are within acceptable limits, this measurement should fall within tolerance as well. However, if you purchase new parts such as the lifters, the lifter guide assembly on 3.3L engines or the cylinder head has been changed on 2.4L and 3.5L engines, you may find that this clearance no longer falls within the specified limit.

8   On 3.3L engines, inspect the rocker arms and shafts as described in Chapter 2B. If the pads are worn the rocker arms must be replaced. Don't under any circumstances attempt to restore rocker arms by grinding the pad surfaces.

9   In any case make sure all the parts new or old have been thoroughly inspected before reassembly.

22 Engine overhaul - reassembly sequence

1   Before beginning engine reassembly, make sure you have all the necessary new parts (including new cylinder head bolts), gaskets and seals as well as the following items on hand:

   Common hand tools
   A 1/2-inch drive torque wrench
   Piston ring installation tool
   Piston ring compressor
   Short lengths of rubber or plastic hose
      to fit over connecting rod bolts
   Plastigage
   Feeler gauges
   A fine-tooth file
   New engine oil
   Engine assembly lube or moly-base
      grease
   Gasket sealant
   Thread locking compound

2   In order to save time and avoid problems, engine reassembly must be done in the following general order:

2.4L engines

   Piston rings
   Piston oil jets
   Crankshaft and main bearings
   Rear main oil seal and retainer
   Piston/connecting rod assemblies
   Cylinder head
   Camshafts and lifters
   Timing chains and sprockets
   Lower timing cover
   Upper timing cover
   Oil pump and oil pump driveshaft
   Oil pump pick-up tube
   Oil pan
   Crankshaft pulley
   Intake and exhaust manifolds
   Valve covers
   Engine rear plate
   Flywheel/driveplate

3.3L engines

   Piston rings
   Crankshaft and main bearings
   Rear main oil seal and retainer
   Piston/connecting rod assemblies
   Oil pump and pick up tube
   Oil pan
   Cylinder heads, camshafts, lifters and
      rocker arms
   Timing belt and sprockets
   Timing belt covers
   Crankshaft pulley
   Intake and exhaust manifolds
   Valve covers
   Engine rear plate
   Flywheel/driveplate

3.5L engines

   Piston rings
   Piston oil jets
   Crankshaft, main bearings and main
      bearing cap brace
   Rear main oil seal and retainer
   Piston/connecting rod assemblies
   Oil pump and pick up tube
   Cylinder heads
   Camshaft and lifters
   Rear timing cover
   Timing chains and sprockets
   Front timing cover
   Variable valve timing covers
   Crankshaft pulley
   Aluminum oil pan
   Steel Oil pan
   Intake and exhaust manifolds
   Valve covers
   Engine rear plate
   Flywheel/driveplate

23 Piston rings - installation

Refer to illustrations 23.3, 23.4, 23.9a, 23.9b and 23.12

1   Before installing the new piston rings, the ring end gaps must be checked. It's assumed that the piston ring side clearance has been checked and verified correct (Section 18).

2   Lay out the piston/connecting rod assemblies and the new ring sets so the ring sets will be matched with the same piston and cylinder during the end gap measurement and engine assembly.

3   Insert the top (number one) ring into the first cylinder and square it up with the cylinder walls by pushing it in with the top of the piston **(see illustration)**. The ring should be near the bottom of the cylinder, at the lower limit of ring travel.

4   To measure the end gap, slip feeler gauges between the ends of the ring until a gauge equal to the gap width is found **(see illustration)**. The feeler gauge should slide between the ring ends with a slight amount of drag. Compare the measurement to the Specifications in this Chapter. If the gap is

23.9a Installing the spacer/expander in the oil control ring groove

23.9b DO NOT use a piston ring installation tool when installing the oil ring side rails

23.12 Installing the compression rings with a ring expander - the mark (arrow) must face up

larger or smaller than specified, double-check to make sure you have the correct rings before proceeding.

5 If the gap is too small, you may have to file the rings to fit or exchange the set. The type of ring set you buy, and the material the rings are faced with, determine whether they can be filed. Carefully read the instructions with the ring set.

6 Excess end gap isn't as critical as too little gap, unless the gap is greater than 0.040-inch. Compare your measurements to this Chapter's Specifications for maximum end gap. Again, double-check to make sure you have the correct rings for your engine. If you do file the ring gaps, mount a file in a vise, lubricate the tops of the jaws, and slide the ring back and forth across the file, resting the ring against the top of the jaws and with even pressure on both sides of the ring gap. File a little, then recheck that ring's end gap in the bore before filing any more. When the correct gap is achieved, use a whetstone or fine file to deburr the edges that have been filed.

7 Repeat the procedure for each ring that will be installed in the first cylinder and for each ring in the remaining cylinders. Remember to keep rings, pistons and cylinders matched up.

8 Once the ring end gaps have been checked/corrected, the rings can be installed on the pistons.

9 The oil control ring (lowest one on the piston) is usually installed first. It's composed of three separate components. Slip the spacer/expander into the groove **(see illustration)**. If an anti-rotation tang is used, make sure it's inserted into the drilled hole in the ring groove. Next, install the lower side rail. Don't use a piston ring installation tool on the oil ring side rails, as they may be damaged. Instead, place one end of the side rail into the groove between the spacer/expander and the ring land, hold it firmly in place and slide a finger around the piston while pushing the rail into the groove **(see illustration)**. Next, install the upper side rail in the same manner.

10 After the three oil ring components have been installed, check to make sure that both the upper and lower side rails can be turned smoothly in the ring groove.

11 The number two (middle) ring is installed next. It's usually stamped with a mark which must face up, toward the top of the piston. **Note:** *Always follow the instructions printed on the ring package or box - different manufacturers may require different approaches. Do not mix up the top and middle rings, as they have different cross sections.*

12 Use a piston ring installation tool and make sure the identification mark is facing the top of the piston, then slip the ring into the middle groove on the piston **(see illustration)**. Don't expand the ring any more than necessary to slide it over the piston.

13 Install the number one (top) ring in the same manner. Make sure the mark is facing up. Be careful not to confuse the number one and number two rings.

14 Repeat the procedure for the remaining pistons and rings.

24 Crankshaft - installation and main bearing oil clearance check

Refer to illustrations 24.5, 24.11, 24.13a, 24.13b, 24.13c and 24.15

1 Crankshaft installation is the first step in engine reassembly. It's assumed at this point that the engine block and crankshaft have been cleaned, inspected and repaired or reconditioned.

2 Position the engine on the stand with the crankcase facing up.

3 Remove the main bearing cap bolts and lift out the bearing cap and brace assembly.

4 If they're still in place, remove the original bearing inserts from the block and the main bearing caps. Wipe the bearing surfaces of the block and caps with a clean, lint-free cloth. They must be kept spotlessly clean.

Main bearing oil clearance check

5 Clean the back sides of the new main bearing inserts and lay one in each main bearing saddle in the block. If any of the bearing inserts have a large groove in it, make sure the grooved inserts are installed in the block. Lay the other bearings from the set in the corresponding main bearing caps. Make sure the tab on the bearing insert fits into the recess in the block or cap. **Caution:** *The oil holes in the block must line up with the oil holes in the bearing inserts* **(see illustration)**. *Do not hammer the bearing into place and don't nick or gouge the bearing faces. No lubrication should be used at this time.*

6 The flanged thrust bearing must be installed in the fourth (rear) cap and saddle on 3.3L engines. On 2.4L and 3.5L engines the thrust bearing must be installed in the third cap and saddle.

7 Clean the faces of the bearings in the block and the crankshaft main bearing journals with a clean, lint-free cloth.

8 Check or clean the oil holes in the crankshaft, as any dirt here can go only one way - straight through the new bearings.

9 Once you're certain the crankshaft is clean, carefully lay it in position in the main bearings. Do not lubricate the crankshaft with oil at this time.

10 Before the crankshaft can be permanently installed, the main bearing oil clearance must be checked.

11 Cut several pieces of the appropriate size Plastigage (they must be slightly shorter than the width of the main bearings) and place one

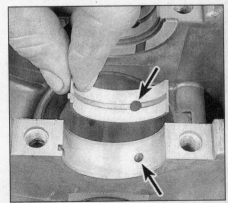

24.5 Make sure the oil holes in the bearings are aligned with the oil holes in the block (arrows)

Chapter 2 Part D General engine overhaul procedures

24.11 Lay the Plastigage strips (arrow) on the main bearing journals, parallel to the crankshaft centerline

24.13a Main bearing cap TIGHTENING sequence (2.4L engine)

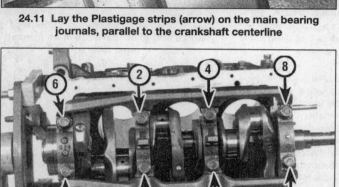

24.13b Main bearing cap TIGHTENING sequence (3.3L engine)

24.13c Main bearing cap TIGHTENING sequence (3.5L engine)

piece on each crankshaft main bearing journal, parallel with the journal axis (see illustration).

12 Clean the faces of the bearings in the cap and install the bearing caps and brace assembly with the arrows pointing toward the drivebelt end of the engine. Don't disturb the Plastigage.

13 Starting with the center main and working out toward the ends (see illustrations), tighten the main bearing cap assembly bolts, in three steps, to the torque specified in this Chapter. Don't rotate the crankshaft at any time during this operation. Note: *Make sure the main bearing cap bolts have been thoroughly inspected as described in Section 14 before reusing any main bearing cap bolts or the torque readings may be incorrect.*

14 Remove the bolts and carefully lift off the main bearing caps and brace assembly if equipped. Don't disturb the Plastigage or rotate the crankshaft.

15 Compare the width of the crushed Plastigage on each journal to the scale printed on the Plastigage envelope to obtain the main bearing oil clearance (see illustration). Check the Specifications in this Chapter to make sure it's correct.

16 If the clearance is not as specified, the bearing inserts may be the wrong size (which means different ones will be required). Before deciding that different inserts are needed, make sure that no dirt or oil was between the bearing inserts and the caps or block when the clearance was measured. If the Plastigage

was wider at one end than the other, the journal may be tapered (refer to Section 19).

17 Carefully scrape all traces of the Plastigage material off the main bearing journals and/or the bearing faces. Use your fingernail or the edge of a credit card - don't nick or scratch the bearing faces.

Final crankshaft installation

18 Carefully lift the crankshaft out of the engine.

19 Clean the bearing faces in the block, then apply a thin, uniform layer of moly-base grease or engine assembly lube to each of the bearing surfaces. Be sure to coat the thrust faces as well as the journal face of the thrust bearing.

20 Make sure the crankshaft journals are clean, then lay the crankshaft back in place in the block.

21 Clean the faces of the bearings in the caps, then apply lubricant to them.

22 Install the bearing caps and brace assembly with the arrows pointing toward the drivebelt end of the engine.

23 Tighten the bearing cap bolts to 10-to-12 ft-lbs.

24 Gently tap the ends of the crankshaft forward and backward with a lead or brass hammer to line up the main bearing and crankshaft thrust surfaces.

25 Retighten all main bearing cap bolts to the specified torque, starting with the center main and working out toward the ends (see illustration 24.13a, 24.13b and 24.13c).

26 Rotate the crankshaft a number of times by hand to check for any obvious binding.

27 The final step is to check the crankshaft endplay with a feeler gauge or a dial indicator as described in Section 14. The endplay should be correct if the crankshaft thrust faces aren't worn or damaged and new bearings have been installed.

28 Refer to Section 25 and install the new seal, then bolt the retainer to the block.

24.15 Compare the width of the crushed Plastigage to the scale on the envelope to determine the main bearing oil clearance (always take the measurement at the widest point of the Plastigage); be sure to use the correct scale - standard and metric ones are included

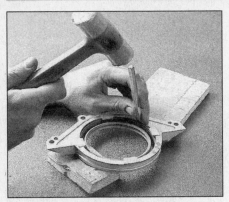

25.3 Place the retainer between two blocks of wood and drive the seal out of the retainer from the rear

25.4 Drive the new seal into the retainer with a block of wood or a section of pipe - make sure that you don't cock the seal in the bore

25 Rear main oil seal installation

Refer to illustrations 25.3 and 25.4

1 All models are equipped with a one-piece seal that fits into a housing (retainer) attached to the transmission end of the block. The crankshaft must be installed first and the main bearing caps bolted in place, then the new seal should be installed in the retainer and the retainer bolted to the block.

2 Check the seal contact surface very carefully for scratches and nicks that could damage the new seal lip and cause oil leaks. If the crankshaft is damaged, the only alternative is a new or different crankshaft.

3 The old seal can be removed from the retainer with a hammer and punch by driving it out from the back side **(see illustration)**. Be sure to note how far it's recessed into the retainer bore before removing it; the new seal will have to be recessed an equal amount. Be very careful not to scratch or otherwise damage the bore in the retainer or oil leaks could develop.

4 Make sure the retainer is clean, then apply a thin coat of engine oil to the outer edge of the new seal. The seal must be pressed squarely into the retainer bore, so hammering it into place is not recommended. If you don't have access to a press, sandwich the retainer and seal between two smooth pieces of wood and press the seal into place with the jaws of a large vise. The pieces of wood must be thick enough to distribute the force evenly around the entire circumference of the seal. Work slowly and make sure the seal enters the bore squarely **(see illustration)**.

5 The seal lips must be lubricated with clean engine oil or multi-purpose grease before the seal/retainer is slipped over the crankshaft and bolted to the block. Apply RTV sealant to retainer housing to block surface - and make sure the dowel pins are in place before installing the retainer.

6 Tighten the screws a little at a time until the torque specified in this Chapter is reached.

26 Pistons/connecting rods - installation and rod bearing oil clearance check

Refer to illustrations 26.5, 26.11, 26.13 and 26.17

1 Before installing the piston/connecting rod assemblies, the cylinder walls must be perfectly clean, the top edge of each cylinder must be chamfered, and the crankshaft must be in place.

2 Remove the cap from the end of the number one connecting rod (refer to the marks made during removal). Remove the original bearing inserts and wipe the bearing surfaces of the connecting rod and cap with a clean, lint-free cloth. They must be kept spotlessly clean.

Connecting rod bearing oil clearance check

3 Clean the back side of the new upper bearing insert, then lay it in place in the connecting rod. Make sure the tab on the bearing fits into the recess in the rod. Don't hammer the bearing insert into place and be very careful not to nick or gouge the bearing face. Don't lubricate the bearing at this time.

4 Clean the back side of the other bearing insert and install it in the rod cap. Again, make sure the tab on the bearing fits into the recess in the cap, and don't apply any lubricant. It's critically important that the mating surfaces of the bearing and connecting rod are perfectly clean and oil free when they're assembled for clearance checking.

5 Position the piston ring gaps at the specified intervals around the piston **(see illustration)**.

6 Slip a section of plastic or rubber hose over each connecting rod cap bolt. On 3.5L engines install the connecting rod guides that were fabricated previously.

7 Lubricate the piston and rings with clean engine oil and attach a piston ring compressor to the piston. Leave the skirt protruding about 1/4-inch to guide the piston into the cylinder. The rings must be compressed until they're flush with the piston.

8 Rotate the crankshaft until the number one connecting rod journal is at BDC (bottom dead center) and apply a coat of engine oil to the cylinder walls.

9 With the notch on top of the piston facing the drivebelt end of the engine, gently insert the piston/connecting rod assembly into the number one cylinder bore and rest the bottom edge of the ring compressor on the engine block.

10 Tap the top edge of the ring compressor to make sure it's contacting the block around its entire circumference.

11 Gently tap on the top of the piston with the end of a wooden hammer handle **(see illustration)** while guiding the end of the connecting rod into place on the crankshaft journal. The piston rings may try to pop out of the

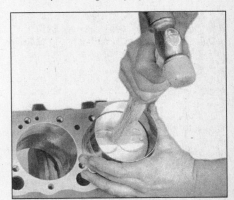

26.11 The piston can be driven (gently) into the cylinder bore with the end of a wooden hammer handle - make sure the notch on top of the piston is facing the front of the engine

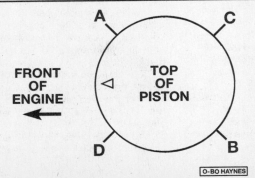

26.5 Stagger the ring end gaps as shown

A Oil ring expander
B Oil ring lower rail
C Top ring gap and upper oil ring rail
D Second ring gap

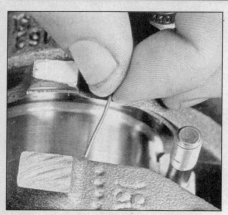

26.13 Lay the Plastigage strips on each rod bearing journal, parallel to the crankshaft centerline

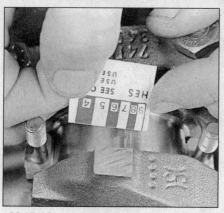

26.17 Measure the width of the crushed Plastigage to determine the rod bearing oil clearance (be sure to use the correct scale - standard and metric ones are included)

ring compressor just before entering the cylinder bore, so keep some downward pressure on the ring compressor. Work slowly, and if any resistance is felt as the piston enters the cylinder, stop immediately. Find out what's hanging up and fix it before proceeding. Do not, for any reason, force the piston into the cylinder - you might break a ring and/or the piston.

12 Once the piston/connecting rod assembly is installed, the connecting rod bearing oil clearance must be checked before the rod cap is permanently bolted in place.

13 Cut a piece of the appropriate size Plastigage slightly shorter than the width of the connecting rod bearing and lay it in place on the number one connecting rod journal, parallel with the journal axis (see illustration).

14 Clean the connecting rod cap bearing face, remove the protective hoses from the connecting rod bolts and install the rod cap. Make sure the mating mark on the cap is on the same side as the mark on the connecting rod.

15 Install the nuts and tighten them to the torque specified in this Chapter (work up to it in three steps). Note: Use a thin-wall socket to avoid erroneous torque readings that can result if the socket is wedged between the rod cap and nut. If the socket tends to wedge itself between the nut and the cap, lift up on it slightly until it no longer contacts the cap. Do not rotate the crankshaft at any time during this operation.

16 Remove the nuts and detach the rod cap, being very careful not to disturb the Plastigage.

17 Compare the width of the crushed Plastigage to the scale printed on the Plastigage envelope to obtain the oil clearance (see illustration). Compare it to the Specifications in this Chapter to make sure the clearance is correct.

18 If the clearance is not as specified, the bearing inserts may be the wrong size (which means different ones will be required). Before deciding that different inserts are needed, make sure that no dirt or oil was between the bearing inserts and the connecting rod or cap when the clearance was measured. Also, recheck the journal diameter. If the Plastigage was wider at one end than the other, the journal may be tapered (refer to Section 19).

Final connecting rod installation

19 Carefully scrape all traces of the Plastigage material off the rod journal and/or bearing face. Be very careful not to scratch the bearing - use your fingernail or the edge of a credit card.

20 Make sure the bearing faces are perfectly clean, then apply a uniform layer of clean moly-base grease or engine assembly lube to both of them. You'll have to push the piston into the cylinder to expose the face of the bearing insert in the connecting rod - be sure to slip the protective hoses over the rod bolts first.

21 Slide the connecting rod back into place on the journal, remove the protective hoses from the rod cap bolts, install the rod cap and tighten the nuts to the specified torque. Again, work up to the torque in three steps.

22 Repeat the entire procedure for the remaining pistons/connecting rods.

23 The important points to remember are:
 a) Keep the back sides of the bearing inserts and the insides of the connecting rods and caps perfectly clean when assembling them.
 b) Make sure you have the correct piston/rod assembly for each cylinder.
 c) The "W" mark on the piston and rod must face the drivebelt end of the engine.
 d) Lubricate the cylinder walls with clean oil.
 e) Lubricate the bearing faces when installing the rod caps after the oil clearance has been checked.

24 After all the piston/connecting rod assemblies have been properly installed, rotate the crankshaft a number of times by hand to check for any obvious binding.

25 As a final step, the connecting rod endplay must be checked. Refer to Section 13 for this procedure.

26 Compare the measured endplay to the Specifications to make sure it's correct. If it was correct before disassembly and the original crankshaft and rods were reinstalled, it should still be right. If new rods or a new crankshaft were installed, the endplay may be inadequate. If so, the rods will have to be removed and taken to an automotive machine shop for resizing. If the endplay is too great, new rods may be required.

27 Initial start-up and break-in after overhaul

Warning: *Have a fire extinguisher ready when starting the engine for the first time.*

1 Once the engine has been installed in the vehicle, double-check the engine oil and coolant levels.

2 With the spark plugs out of the engine and the ignition and fuel systems disabled (see Section 3), crank the engine until oil pressure registers on the gauge.

3 Install the spark plugs, and restore the ignition and fuel system functions.

4 Start the engine. It may take a few moments for the fuel system to build up pressure, but the engine should start without a great deal of effort. Note: *If backfiring occurs through the throttle body, recheck the valve timing and ignition timing.*

5 After the engine starts, it should be allowed to warm up to normal operating temperature. While the engine is warming up, make a thorough check for fuel, oil and coolant leaks. Also check the automatic transmission fluid level (if equipped).

6 Shut the engine off and recheck the engine oil and coolant levels.

7 Drive the vehicle to an area with no traffic, accelerate from 30 to 50 mph, then allow the vehicle to slow rapidly to 30 mph with the throttle closed. Repeat the procedure 10 or 12 times. This will load the piston rings and cause them to seat properly against the cylinder walls. Check again for oil and coolant leaks.

8 Drive the vehicle gently for the first 500 miles (no sustained high speeds) and keep a constant check on the oil level. It is not unusual for an engine to use oil during the break-in period.

9 At approximately 500 to 600 miles, change the oil and filter.

10 For the next few hundred miles, drive the vehicle normally. Do not pamper it or abuse it.

11 After 2000 miles, change the oil and filter again and consider the engine broken in.

Chapter 3
Cooling, heating and air conditioning systems

Contents

	Section		Section
Air conditioning and heating system - check and maintenance	12	Cooling system servicing (draining, flushing and refilling)	See Chapter 1
Air conditioning compressor - removal and installation	14	Drivebelt check, adjustment and replacement	See Chapter 1
Air conditioning condenser - removal and installation	15	Engine cooling fan - check and replacement	4
Air conditioning receiver/drier - removal and installation	13	General information	1
Antifreeze - general information	2	Heater and air conditioning control assembly - removal and installation	10
Blower motor - removal and installation	9		
Blower motor circuit - check	8	Heater core - replacement	11
CHECK ENGINE light	See Chapter 6	Radiator and coolant reservoir - removal and installation	5
Coolant level check	See Chapter 1	Thermostat - check and replacement	3
Coolant temperature gauge sending unit - check and replacement	7	Underhood hose check and replacement	See Chapter 1
		Water pump - check and replacement	6

Specifications

General

Coolant capacity	See Chapter 1
Drivebelt tension	See Chapter 1
Radiator pressure cap rating	11 to 14 psi
Thermostat rating	
Four-cylinder engine	
Valve opens	170-degrees F (77-degrees C)
Fully open	194-degrees F (90-degrees C)
3.3L V6 engine	
Valve opens	180-degrees F (83-degrees C)
Fully open	203-degrees F (96-degrees C)
3.5L V6 engine	
Thermostat	
Valve opens	170-degrees F (77-degrees C)
Fully open	194-degrees F (90-degrees C)
Water control valve	
Valve opens	203-degrees F (96-degrees C)
Fully open	226-degrees F (108-degrees C)
Refrigerant type (all models)	R-134a

Torque specifications

	In-lbs	Nm
Thermostat housing cover bolts		
Four-cylinder engine	56 to 75	6.5 to 8.5
3.3L V6 engine	144 to 180	16 to 21
3.5L V6 engine	75 to 99	8.4 to 11.2
Water outlet housing bolts (3.5L V6 engine)	156 to 204	18 to 23
Water pump retaining bolts		
Four-cylinder engine	144 to 180	16 to 21
3.3L V6 engine	144 to 180	16 to 21
3.5L V6 engine	75 to 95	8.5 to 10.7
Steering column retaining nuts/bolts	See Chapter 10	

1 General information

Engine cooling system

All vehicles covered by this manual employ a pressurized engine cooling system with thermostatically controlled coolant circulation. An impeller-type water pump mounted on the engine block pumps coolant through the engine. The coolant flows around each cylinder and toward the rear of the engine. Cast-in coolant passages direct coolant around the intake and exhaust ports, near the spark plug areas and in close proximity to the exhaust valve guides.

A wax-pellet type thermostat controls engine coolant temperature. During warm up, the closed thermostat prevents coolant from circulating through the radiator. As the engine nears normal operating temperature, the thermostat opens and allows hot coolant to travel through the radiator, where it's cooled before returning to the engine.

The cooling system is sealed by a pressure-type radiator cap, which raises the boiling point of the coolant and increases the cooling efficiency of the radiator. If the system pressure exceeds the cap pressure relief value, the excess pressure in the system forces the spring-loaded valve inside the cap off its seat and allows the coolant to escape through the overflow tube into a coolant reservoir. When the system cools the excess coolant is automatically drawn from the reservoir back into the radiator.

The coolant reservoir serves as both the point at which fresh coolant is added to the cooling system to maintain the proper fluid level and as a retaining tank for overheated coolant. This type of cooling system is known as a closed design because coolant that escapes past the pressure cap is saved and reused.

Heating system

The heating system consists of a blower fan and heater core located in the heater unit, the hoses connecting the heater core to the engine cooling system and the heater/air conditioning control panel on the dashboard. Hot engine coolant is circulated through the heater core. When the heater mode is activated, a flap door opens to expose the heater unit to the passenger compartment. A fan switch on the control head activates the blower motor, which forces air through the core, heating the air.

Air conditioning system

The air conditioning system consists of a condenser mounted in front of the radiator, an evaporator mounted adjacent to the heater core, a compressor mounted on the engine, a receiver-drier which contains a high pressure relief valve and the plumbing connecting all of the above components.

A blower fan forces the warmer air of the passenger compartment through the evaporator core (sort of a radiator-in-reverse), transferring the heat from the air to the refrigerant. The liquid refrigerant boils off into low pressure vapor, taking the heat with it when it leaves the evaporator. **Warning:** *The models covered by this manual are equipped with Supplemental Restraint Systems (SRS), more commonly known as airbags. Always disable the airbag system before working in the vicinity of any airbag system components to avoid the possibility of accidental deployment of the airbag(s), which could cause personal injury* (see Chapter 12).

2 Antifreeze - general information

Refer to illustration 2.4

Warning: *Do not allow antifreeze to come in contact with your skin or painted surfaces of the vehicle. Rinse off spills immediately with plenty of water. Antifreeze is highly toxic if ingested. Never leave antifreeze lying around in an open container or in puddles on the floor; children and pets are attracted by its sweet smell and may drink it. Check with local authorities about disposing of used antifreeze. Many communities have collection centers, which will see that antifreeze is disposed of safely. Never dump used anti-freeze on the ground or into drains.*

The cooling system should be filled with a water/ethylene glycol based antifreeze solution, which will prevent freezing down to at least -20-degrees F, or lower if local climate requires it. It also provides protection against corrosion and increases the coolant boiling point.

The cooling system should be drained, flushed and refilled at the specified intervals (see Chapter 1). Old or contaminated antifreeze solutions are likely to cause damage and encourage the formation of rust and scale in the system. Use distilled water with the antifreeze.

Before adding antifreeze, check all hose connections, because antifreeze tends to leak through very minute openings. Engines don't normally consume coolant, so if the level goes down, find the cause and correct it.

The exact mixture of antifreeze-to-water that you should use depends on the relative weather conditions. The mixture should contain at least 50-percent antifreeze, but should never contain more than 70-percent antifreeze. Consult the mixture ratio chart on the antifreeze container before adding coolant. Hydrometers are available at most auto parts stores to test the ratio of antifreeze to water **(see illustration)**. Antifreeze test strips are available at some auto parts stores instead of the hydrometer gauge. Use antifreeze that meets the vehicle manufacturer's specifications.

2.4 The condition of your coolant can easily be checked with this type of hydrometer, available at auto parts stores

Chapter 3 Cooling, heating and air conditioning systems

3.6 A thermostat can be accurately checked by heating it in a container of water with a thermometer and observing the opening and fully open temperature

3.12a On four-cylinder engines, detach the power steering pump support bracket and bolts (arrows)

3 Thermostat - check and replacement

Warning: *Do not attempt to remove the radiator cap, coolant or thermostat until the engine has cooled completely.*

Note: *3.5L V6 engines are equipped with two thermostats! The first or main thermostat is mounted at the front of the engine at the water inlet. The second thermostat is referred to as the "water control valve" and is located in the water outlet pipe that runs between the cylinder heads at the rear of the engine.*

Check

Refer to illustration 3.6

1 Before assuming the thermostat is responsible for a cooling system problem, check the coolant level (see Chapter 1), drivebelt tension (see Chapter 1) and temperature gauge (or light) operation.
2 If the engine takes a long time to warm up (as indicated by the temperature gauge or heater operation), the thermostat is probably stuck open. Replace the thermostat with a new one.
3 If the engine runs hot or overheats, a thorough test of the thermostat should be performed.
4 Testing of the thermostat can only be made when it is removed from the vehicle (see *Replacement*). If the thermostat is stuck in the open position at room temperature, it is faulty and must be replaced. **Caution:** *Do not drive the vehicle without a thermostat. The computer may stay in open loop and emissions and fuel economy will suffer.*
5 To properly test a thermostat, suspend the (closed) thermostat on a length of string or wire in a container of cold water, with a thermometer (cooking type that reads beyond 212-degrees F [100-degrees C]).
6 Heat the water on a stove while observing the temperature and the thermostat. Neither should contact the sides of the container **(see illustration)**.
7 Note the temperature when the thermostat begins to open and when it is fully open. Compare the temperatures to the Specifications in this Chapter. The number stamped into the thermostat is generally the fully open temperature. Some manufacturers provide Specifications for the beginning-to-open temperature, the fully open temperature, and sometimes the amount the valve should open.
8 If the thermostat doesn't open and close as specified, or sticks in any position, replace it.

Replacement

Refer to illustrations 3.12a, 3.12b, 3.13a, 3.13b, 3.13c, 3.14a, 3.14b, 3.15, 3.16, 3.18a and 3.18b

9 Disconnect the cable from the negative terminal of the battery. Remove the lower engine cover (if equipped) from below the vehicle.
10 Drain the coolant from the radiator and the engine block (see Chapter 1).
11 Remove the engine cooling fan (see Section 4).

Four-cylinder engine

12 Remove the air cleaner and the air intake duct from the engine compartment. Also remove the power steering pump support bracket **(see illustration)**. Remove the thermostat housing cover from the engine with the lower radiator hose attached **(see illustration)**. Be prepared for some coolant to spill as the gasket seal is broken.

3.3L V6 engine

13 Remove the upper radiator hose from the coolant outlet at the intake manifold and the radiator. Remove the air conditioning compressor drivebelt and the upper idler pulley bracket **(see illustration)**. Unbolt the coolant tube from the front of the engine, then loosen the hose clamp and detach the lower radiator hose from the thermostat housing cover. Remove the thermostat housing cover from the engine **(see illustrations)**. Be prepared for some coolant to spill as the gasket seal is broken.

3.12b Thermostat housing cover bolts - four-cylinder engine

3.13a On 3.3L engines, remove the upper idler pulley bracket bolts (arrows) . . .

3-4 Chapter 3 Cooling, heating and air conditioning systems

3.5L V6 engine

Thermostat

14 Remove the drivebelts and the water pump drain plug from the front of the block **(see illustration)**. Detach the lower radiator hose and the coolant by-pass hose from the thermostat housing cover. Remove the thermostat housing cover from the engine **(see illustration)**. Be prepared for some coolant to spill as the gasket seal is broken.

Water control valve

15 Remove the intake manifold from the engine (see Chapter 2C). Detach the hose from the water outlet housing and remove the water outlet cover from the engine **(see illustration)**.

All models

16 Remove the thermostat or the water control valve, noting the direction in which it was installed **(see illustration)**.
17 Scrape off any old gasket or sealant on the thermostat housing and the thermostat cover, then clean them with lacquer thinner.
18 Apply a bead of RTV sealant around the

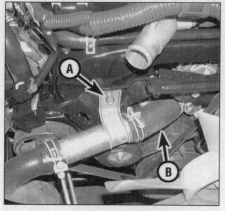

3.13b ... the upper radiator hose, the coolant tube retaining bolt (A) and detach the lower radiator hose (B) from the thermostat housing cover

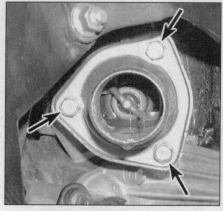

3.13c Thermostat housing cover bolts - 3.3L V6 engine

perimeter of the cover, install the new thermostat with the jiggle valve UP **(see illustrations)** and bolt the cover in place within 5 minutes of applying the sealant. **Note:** *The water control valve on 3.5L engines uses a rubber sealing ring around the circumference of the thermostat and does not need any RTV sealant to seal the cover. Always replace the rubber sealing ring when replacing the water control valve.*

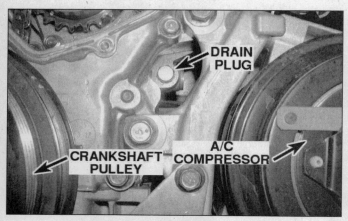

3.14a Water pump drain bolt - 3.5L V6 engines

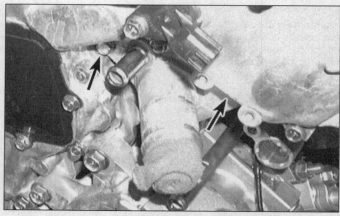

3.14b Thermostat housing cover bolts - 3.5L V6 engine (arrows indicate two of the three bolts)

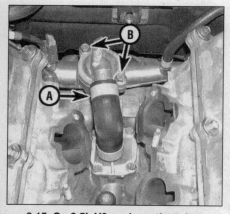

3.15 On 3.5L V6 engines, there is a secondary thermostat (called the water control valve) that can be accessed only after removal of the intake manifold - remove the coolant hose (A) and the water outlet cover bolts (B)

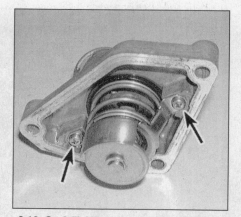

3.16 On 3.5L V6 engines, the thermostat is installed in the thermostat cover and is retained by two screws (arrows) - on other engines the thermostat is installed in the thermostat housing on the engine block

3.18a On all engines, install the thermostat with the jiggle valve (arrow) UP (four-cylinder engine shown)

Chapter 3 Cooling, heating and air conditioning systems

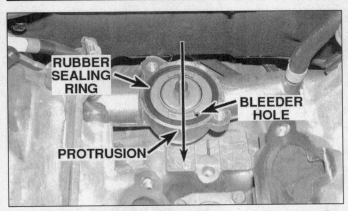

3.18b When installing a water control valve on 3.5L engines, the bleed hole (arrow) must face the front of the engine and the valve stopper must align with the casting protrusions on the water outlet housing

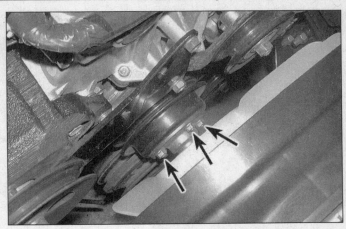

4.6 Water pump/fan pulley mounting nuts (arrows indicate three of four nuts) (2.4L shown, all others similar)

19 Installation is the reverse of removal. Tighten the thermostat cover or the water control valve cover fasteners to the torque listed in this Chapter's Specifications, then reinstall the hoses.
20 Wait at least a half-hour for the sealant to cure. Refill and bleed the cooling system (see Chapter 1). Run the engine and check for leaks and proper operation.

4 Engine cooling fan - check and replacement

Warning: *To avoid possible injury or damage, DO NOT operate the engine with a damaged fan. Do not attempt to repair fan blades. Always replace a damaged fan with a new one.*

Check

1 Disconnect the cable from the negative terminal of the battery and rock the fan back and forth by hand to check for excessive bearing play.
2 With the engine cold (and not running), turn the fan blades by hand. The fan should turn freely.
3 Visually inspect for substantial fluid leakage from the clutch assembly. If problems are noted, replace the clutch assembly.
4 With the engine completely warmed up, turn off the ignition switch and disconnect the negative battery cable from the battery. Turn the fan by hand. Some drag should be evident. If the fan turns easily, replace the fan clutch.

Replacement

Refer to illustrations 4.6, 4.8a and 4.8b

5 Disconnect the cable from the negative terminal of the battery.
6 Loosen the fan/water pump pulley nuts several turns **(see illustration)**.
7 Remove any drivebelts which would interfere with the removal of the engine cooling fan drivebelt, then remove the cooling fan drivebelt (see Chapter 1).
8 Working under the vehicle, detach the lower half of the fan shroud **(see illustrations)**. (If necessary, raise the vehicle and support it securely on jackstands.)
9 Remove the fan/water pump pulley nuts and pull the fan/clutch assembly out from the bottom.
10 Carefully inspect the fan blades for damage and defects. Replace it if necessary.
11 At this point, the fan may be unbolted from the clutch, if necessary. Be sure to re-install the fan blade with the "F" mark facing the front of the engine.
12 Installation is the reverse of removal. Be sure to tighten the fan and clutch mounting nuts evenly and securely.

5 Radiator and coolant reservoir - removal and installation

Warning: *Wait until the engine is completely cool before beginning this procedure.*

Coolant reservoir

Refer to illustration 5.2

1 The coolant reservoir is mounted adjacent to the battery in the right front corner of the engine compartment on Xterra and Frontier models. On Pathfinder models, the coolant reservoir is mounted on the opposite side of the engine compartment in the left front corner.
2 On Xterra and Frontier models, detach the hose from the reservoir, remove the reservoir retaining bolts and lift the reservoir

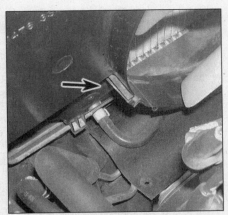

4.8a Unclip the lower fan shroud (arrow) at each side ...

4.8b ... then pull it downward to remove it

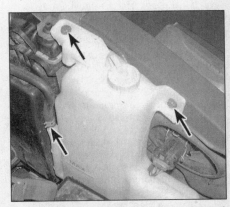

5.2 Detach the coolant hose (lower arrow) and the retaining bolts (upper arrows) and lift the coolant reservoir straight up out of its bracket

5.10 If equipped with an automatic transmission remove the cooler lines (arrow indicates one of two lines)

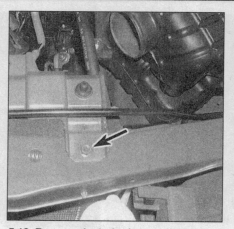

5.13 Remove the bolts (arrow) that attach the upper radiator mounts to the radiator support

6.3 The water pump weep hole (arrow) is located on the underside of the water pump (four-cylinder engine shown, 3.3L V6 engine similar)

straight up out of the bracket (see illustration).

3 On Pathfinder models, refer to Chapter 11 and remove the radiator grille. Then refer to Chapter 12 and remove the left headlight housing. Detach the hose from the reservoir, remove the reservoir retaining bolts and remove the reservoir from the vehicle.

4 Pour the coolant into a container.

5 After washing the reservoir inside and out (use a household "bottle" brush to clean inside), inspect the reservoir for cracks and chafing. If it's damaged or so obscured by age as to make reading the water level difficult, replace it. **Warning:** *If you use a brush to clean the coolant reservoir, never again use it for cleaning drinking glasses or bottles.*

6 Installation is the reverse of removal.

Radiator

Refer to illustrations 5.10 and 5.13

7 Disconnect the cable from the negative terminal of the battery.

8 Set the parking brake and block the rear wheels. Raise the front of the vehicle and support it securely on jackstands.

9 Drain the cooling system (see Chapter 1). If the coolant is relatively new or in good condition, save it and reuse it. Read the **Warning** in Section 2.

10 Disconnect the automatic transmission cooler lines from the radiator if equipped (see illustration). Use a drip pan to catch spilled fluid and plug the lines and fittings.

11 Loosen the hose clamps, then detach the radiator hoses from the radiator. If they're stuck, grasp each hose near the end with a pair of slip joint pliers and twist it to break the seal, then pull it off - be careful not to damage the radiator fittings! If the hoses are old or deteriorated, cut them off and install new ones. Also disconnect the small hose to the coolant reservoir at the radiator filler neck.

12 Refer to Section 4 and remove the engine cooling fan.

13 Unbolt the small brackets that attach the top of the radiator to the radiator support (see illustration).

14 Carefully lift out the radiator with the upper fan shroud attached. Don't spill coolant on the vehicle or scratch the paint.

15 Inspect the radiator for leaks and damage. If it needs repair, have a radiator shop or dealer service department perform the work as special techniques are required.

16 Bugs and dirt can be removed from the radiator by spraying with a garden hose nozzle from the back side. The radiator should be flushed out with a garden hose before reinstallation.

17 Check the radiator mounts for deterioration and replace if necessary.

18 Installation is the reverse of the removal procedure. Guide the radiator into the mounts until they seat properly.

19 Refill and bleed the cooling system (see Chapter 1).

20 Start the engine and check for leaks. Allow the engine to reach normal operating temperature, indicated by the upper radiator hose becoming hot. Allow the engine to cool completely, then recheck the coolant level and add more if required.

21 Check and add automatic transmission fluid if needed (see Chapter 1).

6 Water pump - check and replacement

Warning: *Wait until the engine is completely cool before beginning this procedure.*

Check

Refer to illustration 6.3

1 A failure in the water pump can cause serious engine damage due to overheating.

2 There are two ways to check the operation of the water pump while it's installed on the engine. If the pump is found to be defective, it should be replaced with a new or rebuilt unit.

3 Water pumps are equipped with weep (or vent) holes (see illustration). If a failure occurs in the pump seal, coolant will leak from the hole. **Note:** *On 3.3L V6 engines you can remove the timing belt cover and use a flashlight and a small mirror to find the hole on the water pump from underneath to check for leaks. On 3.5L V6 engines, the weep hole directs coolant out from between the timing chain cover and the engine block.*

4 If the water pump shaft bearings fail, there may be a howling sound at the pump while it's running. On four-cylinder and 3.3L V6 engines shaft wear can be felt with the drivebelt removed if the water pump pulley is rocked up and down (with the engine off). On 3.5L V6 engines shaft wear can be felt by relieving the main timing chain tension (see Chapter 2C) and removing the water pump access cover from the front timing chain cover, then rock the water pump sprocket up and down to detect shaft wear. In either case don't mistake drivebelt slippage, which causes a squealing sound, for water pump bearing failure.

5 Even a pump that exhibits no outward signs of a problem, such as noise or leakage, can still be due for replacement. Removal for close examination is the only sure way to tell. Sometimes the fins on the back of the impeller can corrode to the point that cooling efficiency is hampered.

Replacement

Refer to illustrations 6.10, 6.12a, 6.12b, 6.13a, 6.13b, 6.13c and 6.18

6 Disconnect the cable from the negative terminal of the battery.

7 Drain the engine coolant from the block and the radiator (see Chapter 1). If the coolant is relatively new or in good condition, save it and reuse it.

8 Remove the drivebelts (see Chapter 1).

9 Remove the engine cooling fan and the water pump/fan pulley (see Section 4).

10 Detach the upper radiator hose and the fan shroud from the radiator (see illustration).

11 On 3.3L V6 engines, detach the upper idler pulley bracket (see illustration 3.13a). Also remove the crankshaft pulley and timing

Chapter 3 Cooling, heating and air conditioning systems

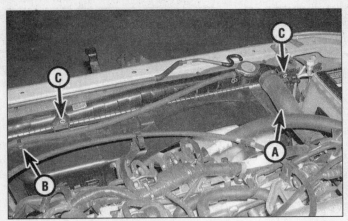

6.10 Detach the upper radiator hose (A) and the overflow tube from the retaining clip (B), then remove the fan shroud retaining bolts (C) and lift the fan shroud out of the engine compartment

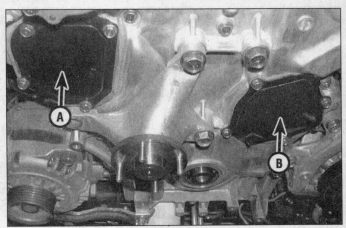

6.12a On 3.5L V6 engines, remove the timing chain tensioner cover (A) and the water pump cover (B) . . .

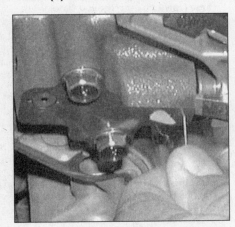

6.12b . . . then depress the timing chain tensioner and lock it into place by inserting a paper clip or another appropriate sized pin into the hole on the front of the tensioner

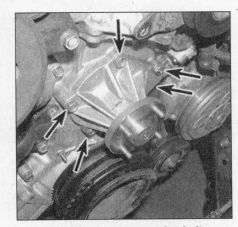

6.13a Water pump mounting bolts - four-cylinder engine

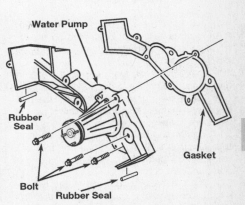

6.13b Water pump mounting bolts - 3.3L V6 engine

belt covers (see Chapter 2B).
12   On 3.5L V6 engines, remove the water pump drain plug from the front of the engine block **(see illustration 3.14a)**. Also remove the timing chain tensioner cover and the water pump cover from the front timing chain cover **(see illustration)**. Press the timing chain tensioner piston inward and insert an appropriate size pin into the tensioner hole to lock the tensioner in place **(see illustration)**. Rotate the crankshaft 20 degrees counterclockwise to loosen the chain from around the water pump sprocket.
13   Remove the bolts and detach the water pump from the engine **(see illustrations)**. Check the impeller on the backside for evidence of corrosion or missing fins.
14   Clean the bolt threads and the threaded holes in the engine to remove corrosion and sealant.
15   Compare the new pump to the old one to make sure they're identical.
16   Remove all traces of old gasket sealant from the engine.
17   Clean the engine and new water pump

mating surfaces with lacquer thinner or acetone.
18   On four-cylinder engines, apply 1/8-inch (3 mm) bead of RTV sealant around the perimeter of the water pump **(see illustration)**. On 3.3L V6 engines apply a thin layer of RTV sealant to the new pump, then carefully set a new gasket in place. On 3.5L V6

6.13c Water pump mounting bolts - 3.5L V6 engine - after the mounting bolts have been removed, install two M8 bolts three to four inches long into the holes designated by the letter (A) and tighten them evenly until the water pump is forced out the engine block

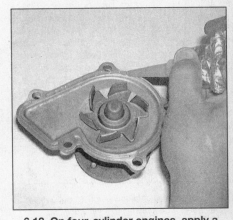

6.18 On four-cylinder engines, apply a 1/8-inch (3 mm) bead of RTV sealant to the sealing surface of the water pump - 3.3L V6 engines use a paper gasket and 3.5L V6 engines use rubber O-rings to seal the water pump to the engine block

7.1a The temperature gauge sending unit (arrow) on four-cylinder engines is located at the front of the engine on the intake manifold

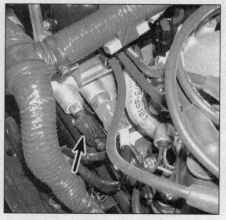

7.1b The temperature gauge sending unit (arrow) on 3.3L V6 engines is located at the front of the engine adjacent to the distributor

7.1c The temperature gauge sending unit (arrow) on 3.5L V6 engines is located at the rear of the engine in the water outlet tube (valve cover removed for clarity)

engines install new O-rings on the pump, then apply a light coat of grease on the O-rings to ease installation into the engine block.

19 Carefully attach the pump to the engine and thread the bolts into the holes finger tight. Use a small amount of RTV sealant on the bolt threads, and make sure that the dowel pins, if used, are in their original locations.

20 Tighten the bolts to the torque listed in this Chapter's Specifications in 1/4-turn increments. Don't overtighten the bolts or the pump may be distorted.

21 On 3.5L V6 engines rotate the crankshaft 20 degrees clockwise to tighten the timing chain around the water pump sprocket. Remove the stopper pin from the timing chain tensioner and install the tensioner cover and water pump access cover after cleaning them and applying RTV sealant to them.

22 Reinstall all parts removed for access to the pump.

23 Wait at least a one hour for the sealant to cure. Refill and bleed the cooling system (see Chapter 1). Run the engine and check for leaks and proper operation. **Note:** *Timing chain noise may be apparent after performing this procedure on 3.5L V6 engines. This noise is normal and should only last until the air has bled out of the high pressure chamber. Simply start the engine and allow it to run at 3,000 rpm with the transmission in neutral or park until the noise subsides.*

7 Coolant temperature gauge sending unit - check and replacement

Warning: *Wait until the engine is completely cool before beginning this procedure.*

Check

Refer to illustrations 7.1a, 7.1b and 7.1c

1 The coolant temperature indicator system consists of a temperature gauge mounted in the instrument panel and a coolant temperature sending unit mounted on the engine **(see illustrations).** There is more than one temperature sensor, but only one is used for the indicator system.

2 If an overheating indication occurs even when the engine is cold, check the wiring between the dash and the sending unit for a short circuit to ground.

3 If the gauge is inoperative, test the circuit by briefly grounding the wire to the sending unit while the ignition is On (engine not running for safety). If the gauge deflects full scale, replace the sending unit.

4 If the gauge doesn't respond in the test outlined in Step 3, check for an open circuit in the gauge wiring or for a defective gauge.

5 To test the sending unit, disconnect the electrical connector and attach an ohmmeter from the pin on top of the sender to an engine ground. With the engine slightly warm (140 degrees F [60-degrees C]) resistance should be 70 to 90 ohms for 1997 and earlier engines or 170 to 210 ohms for 1998 and later engines. When the engine is hot (212-degrees F [100-degrees C]), the resistance should drop to 21 to 24 ohms for 1997 and earlier engines or 47 to 53 ohms for 1998 and later engines. If the sender fails the test, replace it.

Replacement

6 If the sending unit must be replaced, unscrew it from the engine and quickly install the replacement. Use a conductive sealant on the threads (not Teflon tape). Make sure the engine is cool before removing the defective sending unit. There will be some coolant loss as the unit is removed, so be prepared to catch it. Check the coolant level after the replacement part has been installed.

8 Blower motor circuit - check

Refer to illustrations 8.4, 8.6, 8.8, 8.9, 8.12a and 8.12b

Warning: *The models covered by this manual are equipped with Supplemental Restraint Systems (SRS), more commonly known as airbags. Always disable the airbag system before working in the vicinity of any airbag system components to avoid the possibility of accidental deployment of the airbag(s), which could cause personal injury (see Chapter 12).* **Note:** *This procedure applies to vehicles equipped with manual heating and air conditioning systems only. Vehicles equipped with automatic heating and air conditioning systems are very complex and considered beyond the scope of the home mechanic. Vehicles equipped with automatic heating and air conditioning systems should be taken to dealer service department or other qualified repair facility.*

1 If the blower motor speed does not correspond to the setting selected on the blower switch, or the blower motor does not operate at all, the problem could be a bad fuse, relay, switch, blower motor resistor, blower motor or blower motor circuit wiring.

2 Before checking the blower motor or circuit, always check the fuse and the blower relay first (see Chapter 12). The fuse(s) and relay are located in the interior compartment fuse box.

3 Remove the glove compartment and lower dash trim (see Chapter 11) to gain access to the heater case and blower motor.

4 With the ignition key in the ON position, turn the blower switch to the faulty position(s) and, using a test light or voltmeter, check for voltage at the motor electrical connector **(see illustration).**

5 If the motor is receiving voltage but not operating, either the motor ground is bad (on these models the blower switch and resistor are part of the ground circuit) or the motor itself is faulty or the fan is binding.

6 To check the fan, disconnect the electrical connector from the blower motor and connect a jumper wire between the ground wire terminal on the blower motor and a good chassis ground, then connect a fused jumper wire between the battery positive terminal and the positive terminal on the blower motor

Chapter 3 Cooling, heating and air conditioning systems   3-9

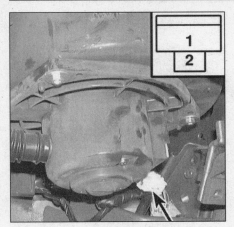

8.4 Test for voltage at the blower motor electrical connector with the ignition key On

8.6 To test the blower motor, disconnect the electrical connector and attach jumper wires (arrows) to apply power and ground - if it does operate, replace the blower motor

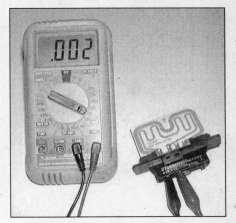

8.8 Check the blower resistor for continuity between all the terminals

(see illustration). If the motor now operates properly, the ground circuit or the blower resistor are bad. If the motor does not operate, the fan is either binding or faulty.

7   If you suspect the blower motor fan is binding, remove the blower motor (see Section 9) to check for free operation of the fan.

8   If the ground circuit to the motor is bad, check the blower resistor for continuity between the terminals (see illustration). If any checks indicate infinite resistance, replace the blower resistor and retest the ground circuit.

9   If the blower resistor is OK, refer to the wiring diagrams at the end of this manual and backprobe the wires at the resistor leading from the blower switch. Using an ohmmeter, check for continuity to ground at the resistor electrical connector while the blower switch is rotated through all the switch position(s). Continuity to ground should be present at terminal No. 1 in all of the switch positions except Off. Continuity to ground should be present at terminal No. 2 when the blower speed switch is placed in the first, second and third positions. Continuity to ground should be present at terminal No. 3 when the blower speed switch is placed in the first and

second positions. Continuity to ground should be present at terminal No. 4 when the blower speed switch is placed in the first position (see illustration).

10 If a ground signal does not exist at the blower resistor connector repair the blower motor switch and or the circuit leading to the switch.

11 To check the blower speed switch refer to Section 10 and pull the heater/air conditioning control unit out from the dash enough to access the rear of the fan switch.

12 Unplug the electrical connector from the blower switch and check the continuity across the indicated switch terminals with an ohmmeter (see illustrations).

13 If continuity isn't as specified, replace the switch.

9 Blower motor - removal and installation

Refer to illustration 9.3
Warning: *The models covered by this manual*

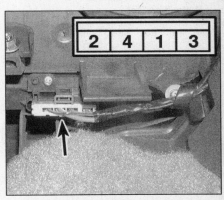

8.9 Terminal guide for the blower motor resistor

are equipped with Supplemental Restraint Systems (SRS), more commonly known as airbags. Always disable the airbag system before working in the vicinity of any airbag system components to avoid the possibility of accidental deployment of the airbag(s), which could cause personal injury (see Chapter 12).

1   Disconnect the cable from the negative terminal of the battery.
2   Remove the glove compartment and

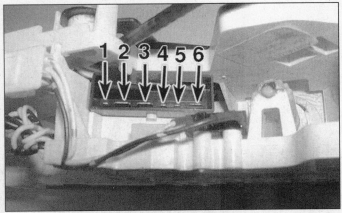

8.12a Terminal guide for the blower motor switch

Terminal	Off	1	2	3	4
1		X	X	X	X
2					X
3				X	
4			X		
5		X			
6		X	X	X	X

8.12b Blower speed control switch continuity chart - there must be continuity between the terminals marked with an X when the switch is in the indicated position

Chapter 3 Cooling, heating and air conditioning systems

9.3 Blower motor mounting details

A Electrical connector   C   Mounting screws
B Vent hose

10.2 Remove the screws (arrows) retaining the control panel to the instrument panel

lower dash trim (see Chapter 11) to gain access to the heater case and blower motor.
3   Detach the blower motor vent hose and disconnect the electrical connector from the blower motor **(see illustration)**.
4   Remove the blower motor mounting screws, and pull the blower motor carefully out of the housing.
5   If you are replacing the motor, detach the fan and transfer it to the new motor.
6   Installation is the reverse of removal. Run the blower and check for proper operation.

10 Heater and air conditioning control assembly - removal and installation

Refer to illustrations 10.2 and 10.3
Warning: *The models covered by this manual are equipped with Supplemental Restraint Systems (SRS), more commonly known as airbags. Always disable the airbag system before working in the vicinity of any airbag system components to avoid the possibility of accidental deployment of the airbag(s), which could cause personal injury (see Chapter 12).*
1   Refer to Chapter 11 for removal of the center trim panels around the control assembly.
2   Remove the four screws retaining the control assembly to the instrument panel **(see illustration)**.
3   Pull the control assembly out of the instrument panel, disconnect the electrical connectors and detach the control cables **(see illustration)**.
4   Refer to Section 8 for electrical checks of the blower motor speed switch. The speed switch, temperature control, and blend-control switch can all be removed from the control panel (manual air conditioning) by pulling the knob off from the front side, then depressing the plastic tabs on the back of the switch to release it from the control panel.
5   Installation is the reverse of the removal procedure.

11 Heater core - replacement

Refer to illustrations 11.4, 11.6, 11.8a, 11.8b, 11.8c, 11.8d, 11.8e, 11.9, 11.11, 11.12, 11.14a and 11.14b
Warning 1: *The models covered by this manual are equipped with Supplemental Restraint Systems (SRS), more commonly known as airbags. Always disable the airbag system before working in the vicinity of any airbag system components to avoid the possibility of accidental deployment of the airbag(s), which could cause personal injury (see Chapter 12).*
Warning 2: *The air conditioning system is under high pressure. DO NOT loosen any fittings or remove any components until after the system has been discharged. Air conditioning refrigerant should be properly discharged into an EPA-approved container at a dealer service department or an automotive air conditioning repair facility. Always wear eye protection when disconnecting air conditioning system fittings.*
Warning 3: *Wait until the engine is completely cool before beginning this procedure.*
Note: *Replacement of the heater core is a difficult procedure for the home mechanic,*

involving removal of the entire dashboard, floor console and many wiring connectors. If you attempt this procedure at home, keep track of the assemblies by taking notes and keeping screws and other hardware in small, marked plastic bags for reassembly.
1   If the vehicle is equipped with air conditioning, have the air conditioning system discharged at a dealer service department or service station.
2   Turn the heater control setting to HOT. Drain the cooling system (see Chapter 1). If the coolant is relatively new, or tests in good condition (see Section 2), save it and re-use it. **Warning:** *Do not allow antifreeze to come in contact with your skin or painted surfaces of the vehicle. Rinse off spills immediately with plenty of water. Antifreeze is highly toxic if ingested. Never leave antifreeze lying around in an open container or in puddles on the floor; children and pets are attracted by it's sweet smell and may drink it. Check with local authorities about disposing of used antifreeze. Many communities have collection centers which will see that antifreeze is disposed of safely. Never dump used anti-freeze on the ground or into drains.*
3   Disconnect the cable from the negative

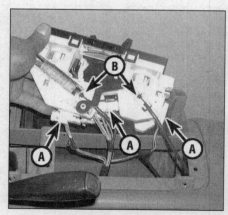

10.3 Unplug the electrical connectors (A) and detach the control cables (B)

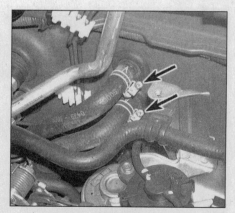

11.4 Loosen the hose clamps and disconnect the heater hoses (arrows) from the heater core tubes at the firewall

Chapter 3 Cooling, heating and air conditioning systems

11.6 Have the air conditioning system discharged (if equipped) then disconnect the refrigerant lines at the firewall

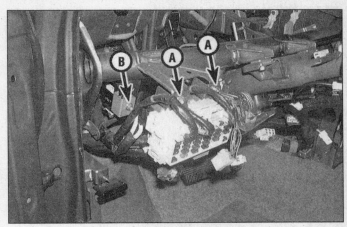

11.8a After removing the instrument panel and lowering the steering column, detach the fuse panel (A), the relay (B) . . .

11.8b . . . and the center braces (arrows) from the reinforcement bar

11.8c Unbolt the reinforcement bar from the driver's side door pillar . . .

11.8d . . . at the center of the vehicle . . .

terminal of the battery.

4   Working in the engine compartment, disconnect the heater hoses where they enter the firewall (see illustration). **Caution:** *If the heater hoses are stuck, it is better to cut off the hoses than to twist them with pliers and risk breaking the heater core tubes.*

5   Remove the rubber grommets where the heater core tubes go through the firewall.

6   Disconnect the air conditioning refrigerant lines and rubber grommet from the evaporator core if the vehicle is so equipped (see illustration). **Warning:** *Always wear eye protection when disconnecting air conditioning system fittings.*

7   Unbolt the steering column and lower it (see Chapter 10), then remove the entire instrument panel (see Chapter 11).

8   Unbolt the fuse panel and any remaining components attached to the instrument panel reinforcement bar (see illustrations). Remove the reinforcement bar from the vehicle (see illustrations). **Note:** *Heater core removal is very difficult. Take your time and don't use excessive force - there may be fasteners you haven't found yet.*

9   Remove the floor heater ducting and

11.8e . . . and from the passenger side door pillar

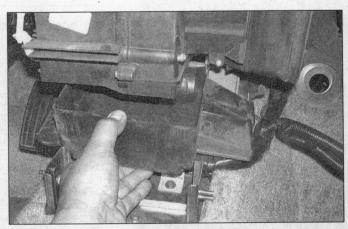

11.9 Remove the floor heater ducting (Frontier and Xterra shown)

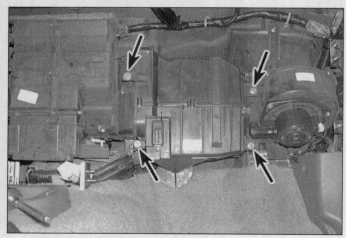

11.11 Evaporator core housing mounting bolts (arrows)

11.12 Heater core housing mounting bolts (arrows)

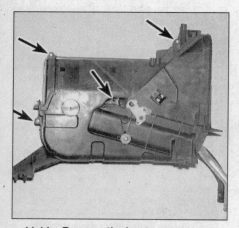

11.14a Remove the heater core cover screws (arrows) . . .

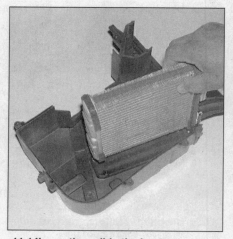

11.14b . . . then slide the heater core out of the case

12.1 Check that the evaporator housing drain tube (arrow) at the firewall is clear of any blockage - the view here is from below the vehicle

any remaining electrical connectors, then refer to chapter 6 to remove the PCM (computer) **(see illustration)**.

10 Remove the defroster ducts from above the heating/cooling unit.

11 Remove the evaporator core housing assembly (if equipped with air conditioning) or passenger side heater duct on models without air conditioning **(see illustration)**.

12 Remove the bolts/nuts securing the heater unit to the firewall. Disconnect the control cables from the unit **(see illustration)**.

13 Pull down and back on the heater unit, making sure that all fasteners have been removed. Some twisting is required to separate the heater unit from the upper ducts.

14 Remove the heater core from the heater unit **(see illustrations)**.

15 Reassemble the heater unit and check the operation of the control flaps. If any parts bind, correct the problem before installation.

16 Reinstall the remaining parts in the reverse order of removal. When attaching the steering column to the support bracket, tighten the nuts or bolts to the torque listed in the Chapter 10 Specifications.

17 Refill the cooling system (see Chapter 1), reconnect the battery and run the engine. Check for leaks and proper operation of the system. Have the air conditioning system recharged if equipped.

12 Air conditioning and heating system - check and maintenance

Air conditioning system

Refer to illustration 12.1

Warning: *The air conditioning system is under high pressure. Do not loosen any hose fittings or remove any components until after the system has been discharged. Air conditioning refrigerant should be properly discharged into an EPA-approved recovery/recycling unit at a dealer service department or an automotive air conditioning repair facility. Always wear eye protection when disconnecting air conditioning system fittings.*

Caution 1: *All models covered by this manual use environmentally friendly R-134a. This refrigerant (and its appropriate refrigerant oils) are not compatible with R-12 refrigerant system components and must never be mixed or the components will be damaged.*

Caution 2: *When replacing entire components, additional refrigerant oil should be added equal to the amount that is removed with the component being replaced. Be sure to read the can before adding any oil to the system, to make sure it is compatible with the R-134a system.*

1 The following maintenance checks should be performed on a regular basis to ensure that the air conditioning continues to operate at peak efficiency.

 a) Inspect the condition of the compressor drivebelt. If it is worn or deteriorated, replace it (see Chapter 1).
 b) Check the drivebelt tension and, if necessary, adjust it (see Chapter 1).
 c) Inspect the system hoses. Look for cracks, bubbles, hardening and deterioration. Inspect the hoses and all fittings for oil bubbles or seepage. If there is any evidence of wear, damage or leakage, replace the hose(s).
 d) Inspect the condenser fins for leaves,

Chapter 3 Cooling, heating and air conditioning systems

bugs and any other foreign material that may have embedded itself in the fins. Use a "fin comb" or compressed air to remove debris from the condenser.
e) Make sure the system has the correct refrigerant charge.
f) If you hear water sloshing around in the dash area or have water dripping on the carpet, check the evaporator housing drain tube **(see illustration)** and insert a piece of wire into the opening to check for blockage.

2 It's a good idea to operate the system for about ten minutes at least once a month. This is particularly important during the winter months because long term non-use can cause hardening, and subsequent failure, of the seals. Note that using the Defrost function operates the compressor.

3 If the air conditioning system is not working properly, proceed to Step 6 and perform the general checks outlined below.

4 Because of the complexity of the air conditioning system and the special equipment necessary to service it, in-depth troubleshooting and repairs beyond checking the refrigerant charge and the compressor clutch operation are not included in this manual. However, simple checks and component replacement procedures are provided in this Chapter. For more complete information on the air conditioning system, refer to the *Haynes Automotive Heating and Air Conditioning Manual*.

5 The most common cause of poor cooling is simply a low system refrigerant charge. If a noticeable drop in system cooling ability occurs, one of the following quick checks will help you determine whether the refrigerant level is low. Should the system lose its cooling ability, the following procedure will help you pinpoint the cause.

Check

Refer to illustration 12.9

6 Warm the engine up to normal operating temperature.

7 Place the air conditioning temperature selector at the coldest setting and put the blower at the highest setting. Open the doors (to make sure the air conditioning system doesn't cycle off as soon as it cools the passenger compartment).

8 After the system reaches operating temperature, feel the pipe exiting the evaporator at the firewall.

9 The outlet pipe should be cold (the tubing that leads back to the compressor). If the evaporator outlet is warm, the system probably needs a charge. Insert a thermometer in the center air distribution duct **(see illustration)** while operating the air conditioning system at its maximum setting - the temperature of the output air should be 35 to 40 degrees F below the ambient air temperature (down to approximately 40 degrees F). If the ambient (outside) air temperature is very high, say 110 degrees F, the duct air temperature may be as high as 60 degrees F, but generally the air conditioning is 35 to 40 degrees F cooler than the ambient air.

10 If the air isn't as cold as it used to be, the system probably needs a charge.

11 If the air is warm and the system doesn't seem to be operating properly check the operation of the compressor clutch.

12 Have an assistant switch the air conditioning On while you observe the front of the compressor. The clutch will make an audible click and the center of the clutch should rotate. If it doesn't, shut the engine off and disconnect the air conditioning system pressure switch **(see illustration 12.22)**. Bridge the terminals of the connector with a jumper wire and turn the air conditioning On again. If it works now, the system pressure is too high or too low. Have your system tested by a dealer service department or air conditioning shop.

13 If the clutch still didn't operate, check the appropriate fuses. Inspect the fuses in the interior fuse panel.

14 Remove the compressor clutch (A/C) relay from the engine compartment relay panel and test it (see Chapter 12). With the relay out and the ignition On, check for battery power at two of the relay terminals (refer to the wiring diagrams for wire color designations to determine which terminals to check). There should be battery power with the key On, at one terminal for the relay control circuit and at one terminal for the power circuit.

15 Using a jumper wire, connect the terminals in the relay box from the relay power circuit to the terminal that leads to the compressor clutch (refer to the wiring diagrams for wire color designations to determine which terminals to connect). Listen for the clutch to click as you make the connection. If the clutch doesn't respond, disconnect the clutch connector at the compressor and check for battery voltage at the compressor clutch connector. Check for continuity to ground from the compressor body to the engine block. If power and ground are available and the clutch doesn't operate when connected, the compressor clutch is defective.

16 If the compressor clutch, relay and related circuits are good and the system is fully charged with refrigerant and the compressor does not operate under normal conditions, have the PCM and related circuits checked by a dealer service department or other properly equipped repair facility.

17 Further inspection or testing of the system is beyond the scope of the home mechanic and should be left to a professional.

Adding refrigerant

Refer to illustrations 12.18, 12.21a, 12.21b and 12.22

Caution: *Make sure any refrigerant, refrigerant oil or replacement component your purchase is designated as compatible with environmentally friendly R-134a systems.*

18 Purchase an R-134a automotive charging kit at an auto parts store **(see illustration)**. A charging kit includes a 12-ounce can of refrigerant, a tap valve and a short section of hose that can be attached between the tap valve and the system low side service valve. Because one can of refrigerant may not be sufficient to bring the system charge up to the proper level, it's a good idea to buy an additional can. **Warning:** *Never add more than two cans of refrigerant to the system.*

19 Hook up the charging kit by following the manufacturer's instructions. **Warning:** *DO NOT hook the charging kit hose to the system high side! The fittings on the charging kit are designed to fit **only** on the low side of the system.*

20 Back off the valve handle on the charging kit and screw the kit onto the refrigerant can, making sure first that the O-ring or rubber seal inside the threaded portion of the kit is in place. **Warning:** *Wear protective eyewear when dealing with pressurized refrigerant cans.*

12.9 Insert a thermometer in the center duct while operating the air conditioning system - the output air should be 35 to 40 degrees F less than the ambient temperature, depending on humidity (but not lower than 40-degrees F)

12.18 A basic charging kit for 134a systems is available at most auto parts stores - it must say 134a (not R-12) and so should the can of refrigerant

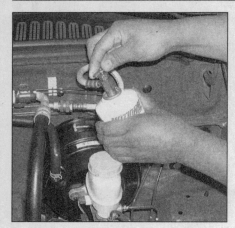

12.21a Attach the refrigerant kit to the low-side charging port - it's near the brake booster on Frontier and Xterra models - the cap should be marked with an "L"

12.21b The low-side charging port on Pathfinder models is located near the right shock tower

12.22 The air conditioning pressure switch (arrow) is located on top of the receiver/drier - if the compressor will not stay engaged, disconnect the connector and bridge it with a jumper wire during the charging procedure

21 Remove the dust cap from the low-side charging port and attach the quick-connect fitting on the kit hose **(see illustrations)**.
22 Warm up the engine and turn On the air conditioning. Keep the charging kit hose away from the fan and other moving parts. **Note:** *The charging process requires the compressor to be running. If the clutch cycles off, you can put the air conditioning switch on High and leave the car doors open to keep the clutch on and compressor working.* **Note:** *The compressor can be kept on during the charging by removing the connector from the dual-pressure switch and bridging it with a paper clip or jumper wire during the procedure* **(see illustration)**.
23 Turn the valve handle on the kit until the stem pierces the can, then back the handle out to release the refrigerant. You should be able to hear the rush of gas. Add refrigerant to the low side of the system, keeping the can upright at all times, but shaking it occasionally. Allow stabilization time between each addition. **Note:** *The charging process will go faster if you wrap the can with a hot-water-soaked shop rag to keep the can from freezing up.*
24 If you have an accurate thermometer, you can place it in the center air conditioning duct inside the vehicle and keep track of the output air temperature **(see illustration 12.9)**. A charged system that is working properly should cool down to approximately 40-degrees F. If the ambient (outside) air temperature is very high, say 110 degrees F, the duct air temperature may be as high as 60 degrees F, but generally the air conditioning is 30-40 degrees F cooler than the ambient air.
25 When the can is empty, turn the valve handle to the closed position and release the connection from the low-side port. Replace the dust cap.
26 Remove the charging kit from the can and store the kit for future use with the piercing valve in the UP position, to prevent inadvertently piercing the can on the next use.

Heating systems

27 If the carpet under the heater core is damp, or if antifreeze vapor or steam is coming through the vents, the heater core is leaking. Remove it (see Section 11) and install a new unit (most radiator shops will not repair a leaking heater core).
28 If the air coming out of the heater vents isn't hot, the problem could stem from any of the following causes:

a) *The thermostat is stuck open, preventing the engine coolant from warming up enough to carry heat to the heater core. Replace the thermostat (see Section 3).*
b) *There is a blockage in the system, preventing the flow of coolant through the heater core. Feel both heater hoses at the firewall. They should be hot. If one of them is cold, there is an obstruction in one of the hoses or in the heater core, or the heater control valve is shut. Detach the hoses and back flush the heater core with a water hose. If the heater core is clear but circulation is impeded, remove the two hoses and flush them out with a water hose.*
c) *If flushing fails to remove the blockage from the heater core, the core must be replaced (see Section 11).*

Eliminating air conditioning odors

Refer to illustration 12.32

29 Unpleasant odors that often develop in air conditioning systems are caused by the growth of a fungus, usually on the surface of the evaporator core. The warm, humid environment there is a perfect breeding ground for mildew to develop.
30 The evaporator core on most vehicles is difficult to access, and factory dealerships have a lengthy, expensive process for eliminating the fungus by opening up the evaporator case and using a powerful disinfectant and rinse on the core until the fungus is gone. You can service your own system at home, but it takes something much stronger than basic household germ-killers or deodorizers.
31 Aerosol disinfectants for automotive air conditioning systems are available in most auto parts stores, but remember when shopping for them that the most effective treatments are also the most expensive. The basic procedure for using these sprays is to start by running the system in the RECIRC mode for ten minutes with the blower on its highest speed. Use the highest heat mode to dry out the system and keep the compressor from engaging by disconnecting the wiring connector at the compressor (see Section 14).
32 The disinfectant can usually comes with a long spray hose. Remove the blower motor resistor (see Section 8), point the nozzle inside the hole and to the left towards the evaporator core, and spray according to the manufacturer's recommendations **(see illustration)**. Try to cover the whole surface of the evaporator core, by aiming the spray up, down and sideways. Follow the manufacturer's recommendations for the length of spray and waiting time between applications.
33 Once the evaporator has been cleaned,

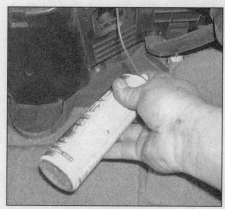

12.32 With the blower motor resistor removed, spray the disinfectant at the evaporator core

Chapter 3 Cooling, heating and air conditioning systems

13.4 Receiver/drier mounting details

A Refrigerant lines
B Clamp bolt
C Pressure switch

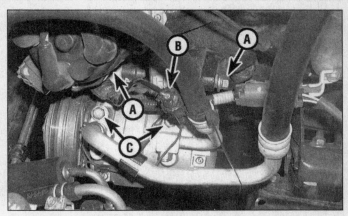

14.6a Air conditioning compressor mounting details - four-cylinder engine

A Upper mounting bolts
B Electrical connector
C Refrigerant lines

the best way to prevent the mildew from coming back again is to make sure your evaporator housing drain tube is clear **(see illustration 12.1)**.

Automatic heating and air conditioning systems

34 Some models are equipped with an optional automatic climate control system. This system has its own computer that receives inputs from various sensors in the heating and air conditioning system. This computer, like the PCM, has self-diagnostic capabilities to help pinpoint problems or faults within the system. Vehicles equipped with automatic heating and air conditioning systems are very complex and considered beyond the scope of the home mechanic. Vehicles equipped with automatic heating and air conditioning systems should be taken to dealer service department or other qualified facility for repair.

13 Air conditioning receiver/drier - removal and installation

Warning: *The air conditioning system is under high pressure. DO NOT loosen any fittings or remove any components until after the system has been discharged. Air conditioning refrigerant should be properly discharged into an EPA-approved container at a dealer service department or an automotive air conditioning repair facility. Always wear eye protection when disconnecting air conditioning system fittings.*

Removal

Refer to illustration 13.4

1 The receiver/drier stores refrigerant and removes moisture from the system. When any major air conditioning component (compressor, condenser, evaporator) is replaced, or the system has been apart and exposed to air for any length of time, the receiver/drier must be replaced.

2 Take the vehicle to a dealer service department or automotive air conditioning shop and have the air conditioning system discharged and the refrigerant recovered (see the **Warning** at the beginning of this Section). Disconnect the cable from the negative terminal of the battery.

3 Disconnect the electrical connector at the compressor clutch cycling switch on top of the receiver/drier. If the receiver/drier is to be replaced with a new one, remove the cycling switch to transfer to the new drier.

4 Disconnect the refrigerant inlet and outlet lines **(see illustration)**. Cap or plug the open lines immediately.

5 Loosen the clamp bolt on the mounting bracket and slide the receiver/drier assembly up and out of the mounting bracket **(see illustration 13.4)**.

Installation

6 If you are replacing the receiver/drier, add two ounces of clean refrigerant oil to the new receiver/drier. This will maintain the correct oil level in the system after the repairs are completed.

7 Place the new a receiver/drier into position, tighten the mounting bracket bolt lightly, still allowing the receiver/drier to be turned to align the line connections.

8 Install the inlet and outlet lines. Lubricate the O-rings using clean refrigerant oil and reconnect the lines. Now tighten the clamp bolt securely and reconnect the electrical connector.

9 Connect the cable to the negative terminal of the battery.

10 Have the system evacuated, recharged and leak tested by a dealer service department or an air conditioning repair facility.

14 Air conditioning compressor - removal and installation

Warning: *The air conditioning system is under high pressure. Do not loosen any hose fittings or remove any components until after the system has been discharged. Air conditioning refrigerant should be properly discharged into an EPA-approved recovery/recycling unit at a dealer service department or an automotive air conditioning repair facility. Always wear eye protection when disconnecting air conditioning system fittings.*
Note: *The receiver/drier should be replaced whenever the compressor is replaced (see Sections 13).*

Removal

Refer to illustrations 14.6a, 14.6b and 14.6c

1 Have the air conditioning system refrigerant discharged and recycled by an air conditioning technician (see **Warning** above).

2 Disconnect the cable from the negative terminal of the battery.

3 Set the parking brake, block the rear wheels and raise the front of the vehicle, supporting it securely on jackstands and remove the splash cover from below the engine.

4 Remove the drivebelt (see Chapter 1).

5 Disconnect the refrigerant lines from the compressor. Plug the open fittings to prevent entry of dirt and moisture .

6 Disconnect the compressor clutch

14.6b Lower two mounting bolts - 2.4L engine

Chapter 3 Cooling, heating and air conditioning systems

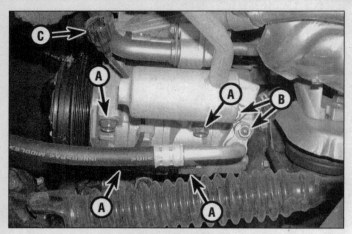

14.6c Typical air conditioning compressor mounting details - 3.5L V6 engine shown, 3.3L V6 engine similar

- A Mounting bolts
- B Refrigerant lines
- C Electrical connector

15.4 Disconnect the condenser lines (arrow indicates one of two lines)

15.5 Remove the two condenser mounting bolts (arrow indicates the left side bolt)

wiring harness. Unbolt the compressor from the mounting bracket and remove it from the vehicle **(see illustrations)**. **Note:** *On some models, the upper mounting bolts may not come all the way out of the compressor - leave them in the compressor until it is removed from the vehicle.*

Installation

7 The clutch may have to be transferred from the old compressor to the new unit.

8 Add the proper amount of refrigerant oil to the new compressor using the following calculations:

 a) Drain the refrigerant oil from the old compressor through the suction fitting and measure it in ounces.
 b) Drain any new oil from the new compressor.
 c) If the amount from the old compressor was 3 to 5 ounces, put that amount of clean, new oil in the new compressor.
 d) If the amount from the old compressor was less than 3 ounces, put 3 ounces of clean, new oil in the new compressor.
 e) If the amount from the old compressor was more than 5 ounces, put 5 ounces of clean, new oil in the new compressor.

9 Installation is the reverse of removal, using new O-rings where the lines attach to the compressor. **Note:** *Remember to slip the two upper mounting bolts into the compressor before installing the compressor in the vehicle.*

10 Have the system evacuated, recharged and leak tested by an air conditioning technician.

15 Air conditioning condenser - removal and installation

Warning: *The air conditioning system is under high pressure. Do not loosen any hose fittings or remove any components until after the system has been discharged. Air conditioning refrigerant should be properly discharged into an EPA-approved recovery/recycling unit at a dealer service department or an automotive air conditioning repair facility. Always wear eye protection when disconnecting air conditioning system fittings.*

Removal

Refer to illustrations 15.4 and 15.5

1 Have the refrigerant discharged and recycled by an air conditioning technician (see **Warning** above).

2 Disconnect the cable from the negative terminal of the battery and drain the cooling system (see Chapter 1).

3 Remove the radiator grille and the hood latch support brace (see Chapter 11). Also loosen the bolts the windshield washer tank bolts.

4 Disconnect the condenser line and discharge line from the condenser **(see illustration)**. Cap the fittings on the condenser and lines to prevent entry of dirt or moisture.

5 Remove the condenser retaining bolts **(see illustration)**.

6 Lean the condenser forward and remove it from the vehicle.

Installation

7 Installation is the reverse of removal. If a new condenser was installed, add 1 to 1.7 ounces of fresh refrigerant oil.

8 Have the system evacuated, charged and leak tested by an air conditioning technician.

Chapter 4
Fuel and exhaust systems

Contents

	Section
Accelerator cable - removal, installation and adjustment	10
Air filter housing - removal and installation	9
Air filter replacement	See Chapter 1
CHECK ENGINE light	See Chapter 6
Engine idle speed and fast idle cam (four-cylinder and 3.3L V6 models) - check and adjustment	16
Exhaust system check	See Chapter 1
Exhaust system servicing - general information	17
Fuel injection system - check	12
Fuel injection system - general information	11
Fuel level sending unit - check and replacement	6
Fuel lines and fittings - replacement	4

	Section
Fuel pressure regulator - removal and installation	14
Fuel pressure relief procedure	2
Fuel pump - removal and installation	5
Fuel pump/fuel pressure - check	3
Fuel rail and injectors - removal and installation	15
Fuel system check	See Chapter 1
Fuel tank - cleaning and repair	8
Fuel tank - removal and installation	7
General information	1
Idle air control system	See Chapter 6
Throttle body - removal and installation	13

Specifications

Engine idle speed	750 rpm
Fuel system pressure	
Key On, engine Off	43 psi
Engine idling	34 psi
Fuel injector resistance	10 to 14 ohms
Fuel level sending unit resistance	
Empty	78 to 85 ohms
1/2	27 to 35 ohms
Full	4 to 6 ohms

Torque specifications

	Ft-lbs (unless otherwise indicated)	Nm
Fuel injector cap screws	26 to 34 in-lbs	3 to 4
Fuel rail mounting bolts		
2.4L four-cylinder		
Step 1	84 to 96 in-lbs	9 to 11
Step 2	15 to 20	21 to 26
3.3L V6		
Step 1	42 to 54 in-lbs	5 to 6
Step 2	96 to 132 in-lbs	11 to 15
3.5L V6		
Step 1	84 to 96 in-lbs	9 to 11
Step 2	15 to 20	21 to 26
Throttle body mounting bolts		
Step 1	84 to 96 in-lbs	9 to 11
Step 2	13 to 16	18 to 22

4-2 Chapter 4 Fuel and exhaust systems

1.1a Fuel injection system and related components - 2.4L four-cylinder models

1. Throttle body
2. Accelerator cable
3. Fuel pressure regulator
4. Air intake duct
5. Fuel rail and fuel injectors
6. Idle speed adjusting screw

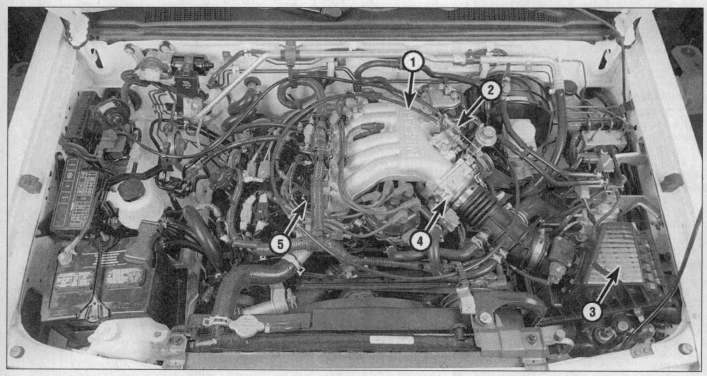

1.1b Fuel injection system and related components - 3.3L V6 models

1. Fuel pressure regulator (behind upper intake manifold)
2. Accelerator cable
3. Air filter housing
4. Throttle body
5. Fuel rail and injectors (under upper intake manifold)

Chapter 4 Fuel and exhaust systems

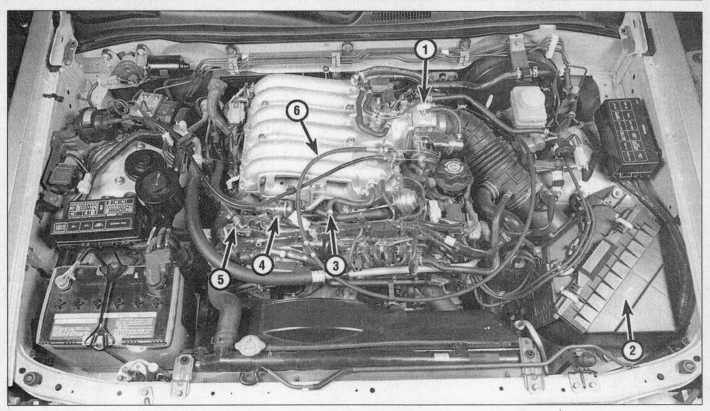

1.1c Fuel injection system and related components - 3.5L V6 models

1 Throttle body
2 Air filter housing
3 Fuel rail and injectors (under upper intake manifold)
4 Fuel damper
5 Fuel pressure regulator
6 Accelerator cable

1 General information

Refer to illustrations 1.1a, 1.1b and 1.1c

The fuel system consists of a fuel tank, an electric fuel pump (located in the fuel tank), a fuel pressure regulator, a fuel pump relay, the fuel rail and fuel injectors, an air filter assembly and a throttle body unit. All models are equipped with a multi-port fuel injection system **(see illustrations)**.

Multi-port fuel injection system

Multi-port fuel injection uses timed impulses to inject the fuel directly into the intake port of each cylinder according to its firing order. The injectors are controlled by the Powertrain Control Module (PCM). The PCM monitors various engine parameters and delivers the exact amount of fuel required into the intake ports. The throttle body serves only to control the amount of air passing into the system. Because each cylinder is equipped with its own injector, much better control of the fuel/air mixture ratio is possible.

Fuel pump and lines

Fuel is circulated from the fuel tank to the fuel injection system, and back to the fuel tank, through a pair of metal lines running along the underside of the vehicle. An electric fuel pump and fuel level sending unit is located inside the fuel tank. A vapor return system routes all vapors back to the fuel tank through a separate return line.

The fuel pump relay is equipped with a primary and secondary voltage circuit. The primary circuit is controlled by the PCM and the secondary circuit is linked directly to battery voltage from the ignition switch. With the ignition switch On (engine not running), the PCM will energize the relay for five seconds. During cranking, the PCM supplies voltage to the fuel pump relay as long as the camshaft position sensor or crankshaft position sensor sends its position signal (see Chapter 6). If there is no signal, the fuel pump will shut off after five seconds.

Exhaust system

The exhaust system includes the exhaust manifold(s), catalytic converter(s), muffler(s) and the exhaust pipes. The catalytic converters are an emission control device added to the exhaust system to reduce pollutants. Refer to Chapter 6 for more information regarding the catalytic converters.

2 Fuel pressure relief procedure

Warning: *Gasoline is extremely flammable, so take extra precautions when you work on any part of the fuel system. Don't smoke or allow open flames or bare light bulbs near the work area, and don't work in a garage where a gas-type appliance (such as a water heater or a clothes dryer) is present. Since gasoline is carcinogenic, wear latex gloves when there's a possibility of being exposed to fuel, and, if you spill any fuel on your skin, rinse it off immediately with soap and water. Mop up any spills immediately and do not store fuel-soaked rags where they could ignite. The fuel system is under constant pressure, so, if any fuel lines are to be disconnected, the fuel pressure in the system must be relieved first. When you perform any kind of work on the fuel system, wear safety glasses and have a Class B type fire extinguisher on hand.*

1 Remove the fuel pump fuse from the passenger compartment fuse panel (see Chapter 12).
2 Attempt to start the engine, it should immediately stall. Crank the engine several more times to ensure the fuel system has been completely relieved. Disconnect the cable from the negative terminal of the bat-

tery before working on the fuel system.

3 The fuel system pressure is now relieved. When you're finished working on the fuel system, install the fuel pump fuse back into the fuse panel and connect the negative cable to the battery.

4 After the fuel pressure has been relieved, it's a good idea to lay a shop towel over any fuel connection to be disassembled, to absorb the residual fuel that may leak out when servicing the fuel system.

3 Fuel pump/fuel pressure - check

Warning: *Gasoline is extremely flammable, so take extra precautions when you work on any part of the fuel system. See the* **Warning** *in Section 2.*

Note: *In order to perform the fuel pressure test, you will need to obtain a fuel pressure gauge capable of measuring high fuel pressure and the necessary fittings to connect the fuel gauge to the fuel line.*

Preliminary check

1 If you suspect insufficient fuel delivery, first inspect all fuel lines to ensure that the problem is not simply a leak in a line. Check that there is adequate fuel in the fuel tank.

2 Set the parking brake and have an assistant turn the ignition switch to the ON position while you listen to the fuel pump (inside the fuel tank). You should hear a "whirring" sound, lasting for a couple of seconds indicating the fuel pump is operating. If the fuel pump is operating, proceed to the pressure check.

3 If there is no sound, check the fuel pump circuit, referring to Chapter 12 and the wiring diagrams. Check the related fuses, the fuel pump relay and the related wiring to ensure power is reaching the fuel pump connector. Check the ground circuit for continuity.

4 If the power and ground circuits are good and the fuel pump does not operate, replace the fuel pump (see Section 5).

Pressure check

Refer to illustration 3.6

5 Relieve the fuel system pressure (see Section 2).

6 Remove the fuel line from the fuel filter and install a T-fitting between the fuel filter and the fuel rail **(see illustration)**. Connect a fuel pressure gauge to the T-fitting. Make sure the hose clamps are securely tightened.

7 Turn the ignition switch On. The fuel pump should run for about five seconds - pressure should register on the gauge and should hold steady. Compare the pressure reading with the key On, engine Off value listed in this Chapter's Specifications.

8 Start the engine and allow it to idle. Compare the pressure reading with the engine running value listed in this Chapter's Specifications. Disconnect and plug the vacuum hose from the fuel pressure regulator (see Section 14) - the pressure should increase to the value recorded in Step 7. If the pressure readings are correct, the system is operating properly.

9 If the fuel pressure is not within specifications, check the following:

a) *If the pressure is higher than specified, check for vacuum to the fuel pressure regulator. Vacuum must fluctuate with the increase or decrease in the engine rpm. If vacuum is present, check for a pinched or clogged fuel return hose or pipe. If the return line is OK, replace the fuel pressure regulator.*

b) *If the pressure is lower than specified, check for a restriction in the fuel filter or fuel line. If the fuel filter and lines are OK, start the engine (if possible) and slowly pinch the return hose shut. If the pressure rises above 43 psi, replace the fuel pressure regulator (see Section 14).* **Warning:** *Don't allow the fuel pressure to exceed 60 psi. If the pressure is still low, replace the fuel pump (see Section 5).*

10 Turn the engine Off and place the ignition switch in the On (engine not running) position. Monitor the pressure on the gauge for approximately five minutes - pressure should decrease slowly. If the pressure decreases rapidly, check the following:

a) *Turn the ignition key On and pinch the fuel feed line shut between the T-fitting and the fuel tank. Turn the ignition key Off - if the pressure decreases rapidly, an injector (or injectors) may be leaking.*

b) *Turn the ignition key On and pinch the fuel feed line shut between the T-fitting and the fuel rail. Turn the ignition key Off - if the pressure decreases rapidly, the in-tank fuel pump check valve may be faulty.*

11 After completing the testing, relieve the fuel pressure (see Section 2) and remove the fuel pressure gauge.

3.6 Using a T-fitting, install the fuel pressure gauge between the fuel filter and the fuel rail

4 Fuel lines and fittings - replacement

Warning: *See the* **Warning** *in Section 2.*

1 Because fuel lines used on fuel-injected vehicles are under high pressure, they require special consideration. Always relieve the fuel pressure before servicing fuel lines or fittings (see Section 2).

2 Metal fuel supply and vapor lines extend from the fuel tank to the engine compartment. The lines are secured to the underbody or frame with retainers. Flexible hose connects the metal lines to the fuel tank, fuel filter and fuel rail. Fuel lines must be occasionally inspected for leaks or damage.

3 In the event of any fuel line damage, metal lines may be repaired with steel tubing of the same diameter, provided the correct fittings are used. Never repair a damaged section of steel line with rubber hose and hose clamps. Rubber fuel hose must be replaced with fuel hose specifically designed for a high pressure fuel injection system; others may fail from the high pressures of this system. Flexible lines with quick-connect fittings must be replaced with factory replacement parts.

4 If evidence of contamination is found in the system or fuel filter during disassembly, the line should be disconnected and blown out. Check the fuel strainer on the fuel pump module for damage and deterioration.

5 Don't route fuel line or hose within four inches of any part of the exhaust system or within ten inches of the catalytic converter. Fuel line must never be allowed to chafe against the engine, body or frame. A minimum of 1/4-inch clearance must be maintained around a fuel line.

6 When replacing a fuel line, remove all fasteners attaching the fuel line to the vehicle body.

Steel tubing

7 If replacement of a steel fuel line or emission line is called for, use steel tubing meeting the manufacturer's specification.

8 Don't use copper or aluminum tubing to replace steel tubing. These materials cannot withstand normal vehicle vibration.

9 Some fuel lines have threaded fittings with O-rings. Any time the fittings are loosened to service or replace components:

a) *Use a flare-nut wrench on the fitting nut and a backup wrench on the stationary portion of the fitting while loosening and tightening the fittings.*

b) *Check all O-rings for cuts, cracks and deterioration. Replace any that appear hardened, worn or damaged.*

c) *If the lines are replaced, always use original equipment parts, or parts that meet the original equipment standards.*

Rubber hose

Refer to illustration 4.10

10 Note the routing of the hose and the orientation of the clamps to assure that replace-

Chapter 4 Fuel and exhaust systems

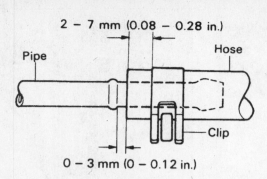

4.10 When attaching a section of rubber hose to a metal fuel line, be sure to overlap the hose as shown, and secure it to the line with a new hose clamp of the proper type

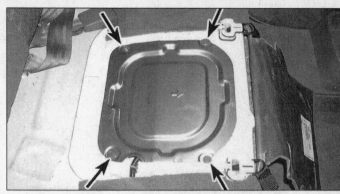

5.3a On Pathfinder and Xterra models, remove the screws (arrows) and the fuel pump/fuel level sending unit access cover located under the rear seat

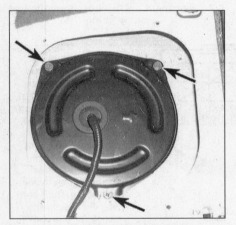

5.3b Remove the tank shield (if equipped) . . .

5.3c . . . disconnect the electrical connectors (arrows) and set the shield and wiring harness aside

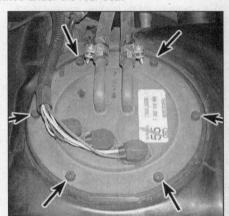

5.4a Disconnect the fuel feed and return lines, then remove the fuel pump/fuel level sending unit mounting screws (arrows)

ment sections are installed in exactly the same manner. Do not kink or twist the hose. When attaching hoses to metal lines, overlap them as shown **(see illustration)**. Tighten the clamp sufficiently to ensure a leak free fit, but do not overtighten the clamp or damage to the rubber hose will result.

Flexible hose with quick-connect fitting

11 Some models may be equipped with flexible hose and quick-connect fittings. There are various methods of disconnecting the fittings, depending upon the type of quick-connect fitting installed on the fuel line. To disconnect a typical quick-connect fitting, push the fitting into the fuel line, squeeze the tabs together and pull the lines apart; do not use any tools to disconnect the fitting. Clean any debris from around the fitting. Disconnect the fitting and carefully remove the fuel line from the vehicle. **Caution:** *The quick-connect fittings are not serviced separately. Do not attempt to repair these types of fuel lines in the event the fitting or line becomes damaged. Replace the entire fuel line as an assembly.*

12 Installation is the reverse of removal with the following additions:

a) *Clean the quick-connect fittings with a lint-free cloth.*

b) *Align the tabs with the openings in the retainer and push the lines together until the tabs click into place.*

c) *After connecting a quick-connect fitting, check the integrity of the connection by attempting to pull the lines apart.*

d) *Cycle the ignition key On and Off several times and check for leaks at the fitting, before starting the engine.*

5 Fuel pump - removal and installation

Warning: *Gasoline is extremely flammable, so take extra precautions when you work on any part of the fuel system. See the* **Warning** *in Section 2.*

Removal

Refer to illustrations 5.3a, 5.3b, 5.3c, 5.4a, 5.4b, 5.5, 5.6, 5.7 and 5.8

1 Relieve the fuel pressure (see Section 2).
2 On Frontier models, remove the fuel tank (see Section 7).
3 On Pathfinder and Xterra models, remove the rear seat (see Chapter 11) and remove the fuel pump/fuel level sending unit access cover **(see illustration)**. Remove the tank shield (if equipped), disconnect the electrical connector from the fuel pump/sending unit assembly and set the cover and wiring harness aside **(see illustrations)**.

4 Detach the fuel feed line and return lines and remove the fuel pump/sending unit mounting screws **(see illustrations)**.

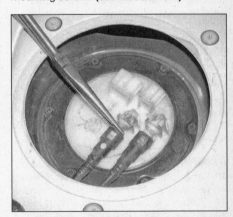

5.4b If the fuel lines are equipped with quick-connect fittings, squeeze the tabs and disconnect the fittings from the fuel tank unit

4-6   Chapter 4   Fuel and exhaust systems

5.5 Carefully angle the fuel pump module out of the fuel tank without damaging the fuel level sending arm and float

5.6 Disconnect the fuel pump electrical connector from the fuel pump

5.7 Disconnect the fuel hose from the fuel pump fitting

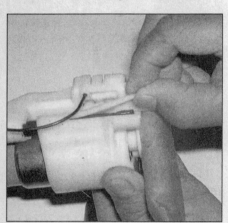

5.8 Depress the locking pawl on the fuel pump bracket and remove the fuel pump and bracket assembly from the fuel level sending unit

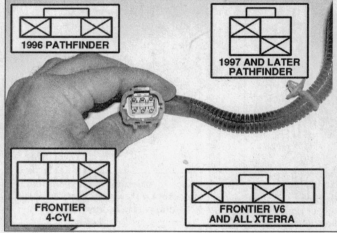

6.2 Connect an ohmmeter to the indicated terminals (marked with an X) of the fuel level sending unit connector and check the resistance of the sending unit with the float positioned down (empty), then move the float up (full) and note the resistance - check for a smooth change in resistance as the float is moved

5   Carefully maneuver the fuel pump/sending unit assembly out of the tank (see illustration). **Caution:** *The fuel level float and sending unit are delicate. Don't bump or bend them during removal or the accuracy of the sending unit may be affected.*
6   Disconnect the fuel pump electrical connector (see illustration).
7   Disconnect the fuel hose from the fitting on the fuel pump (see illustration).
8   Depress the locking pawl and separate the fuel pump and bracket from the fuel level sending unit (see illustration).
9   Check the strainer for contamination and replace it, if necessary.

Installation

10   Installation of the fuel pump to the sending unit assembly is the reverse of removal.
11   Clean the fuel pump mounting flange and the tank mounting surface and seal ring groove.
12   Position a new O-ring around the opening in the fuel tank and guide the fuel pump/sending unit assembly into the tank.
13   Make sure the fuel lines are facing in their original position, then tighten the fuel pump/sending unit mounting screws securely. Connect the fuel lines.
14   On Pathfinder and Xterra models, connect the electrical connector, install the access cover and the rear seat. On Frontier models, install the fuel tank (see Section 7).

6   Fuel level sending unit - check and replacement

Refer to illustration 6.2
Warning: *Gasoline is extremely flammable, so take extra precautions when you work on any part of the fuel system. See the* **Warning** *in Section 2.*
1   Remove the fuel pump/fuel level sending unit assembly (see Section 5).
2   Connect the probes of an ohmmeter to the two indicated terminals of the fuel sending unit electrical connector (see illustration).
3   Position the float in the down (empty) position. Measure the resistance and compare it to the value listed in this Chapter's Specifications.
4   Move the float up to the full position. Measure the resistance and compare it to the value listed in this Chapter's Specifications.
5   If the fuel level sending unit resistance is incorrect or if the resistance does not change smoothly as the float travels from empty to full, the fuel level sending unit assembly is defective.
6   Remove the fuel pump from the sending unit assembly and install it onto the new unit (see Section 5).
7   Install the assembly into the fuel tank with a new O-ring (see Section 5).
8   The remainder of installation is the reverse of removal.

7   Fuel tank - removal and installation

Refer to illustrations 7.6, 7.7a, 7.7b, 7.8a, 7.8b, 7.10a and 7.10b
Warning: *Gasoline is extremely flammable, so take extra precautions when you work on any part of the fuel system. See the* **Warning** *in Section 2.*
1   Relieve the fuel pressure (see Section 2).
2   Disconnect the cable from the negative

Chapter 4 Fuel and exhaust systems

7.6 On Frontier models, disconnect the fuel pump/fuel levels sending unit wiring harness electrical connector

7.7a Remove the fuel tank shield mounting bolts at the front . . .

7.7b . . . and at the rear and remove the shield

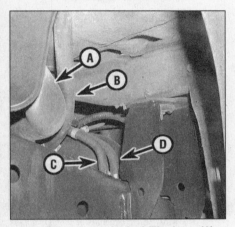

7.8a Disconnect the fuel filler hose (A), vent hose (B), fuel feed hose (C) and fuel return hose (D)

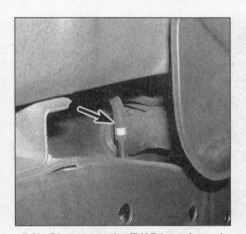

7.8b Disconnect the EVAP hose (arrow) from the fuel tank

7.10a On Frontier and Xterra models, remove the fuel tank mounting bolts (arrows) at the front . . .

terminal of the battery.
3 Siphon the fuel from the fuel tank into an approved fuel container before removing the tank from the vehicle. **Warning:** *DO NOT start the siphoning action by mouth! Use a siphoning kit (available at most auto parts stores).*
4 On Pathfinder and Xterra models, remove the rear seat and the fuel pump/fuel level sending unit access cover. Disconnect the electrical connectors for the fuel pump and fuel level sending unit. Disconnect the fuel feed and return lines (see Section 5).
5 Raise the vehicle and support it securely on jackstands.
6 On Frontier models, disconnect the electrical connector from the fuel pump/fuel level sending unit wiring harness **(see illustration)**.
7 Remove the fuel filler tube protector. Remove the fuel tank shield **(see illustrations)**.
8 Loosen the hose clamps and detach the fuel tank filler hose and vapor hose from the fuel filler neck and the fuel tank. Disconnect the fuel feed and return lines. Disconnect the evaporative emissions canister hose **(see illustrations)**.

9 Place a floor jack under the tank and position a wood plank between the jack pad and the tank. Raise the jack until it's supporting the tank.
10 Remove the fuel tank mounting bolts **(see illustrations)**. On Pathfinder models, remove the mounting straps.
11 Slowly lower the jack while guiding the fuel tank from under the vehicle. Remove the tank from the vehicle.
12 If you're replacing the tank, or having it cleaned or repaired, refer to Section 8.
13 Refer to Section 5 to remove and install the fuel pump/sending unit assembly, if necessary.
14 Installation is the reverse of removal. Clean engine oil can be used as an assembly aid when pushing the fuel filler hose back onto the fuel tank.

8 Fuel tank - cleaning and repair

1 The fuel tank installed in the vehicles covered by this manual is not repairable. If the tank is damaged in any way it must be replaced. If cleaning is required, due to fuel contamination, the process should be carried out by a professional who has experience in this critical and potentially dangerous operation. Even after cleaning and flushing, explosive fumes can remain and ignite.
2 If the fuel tank is removed from the vehicle, it should not be placed in an area where sparks or open flames could ignite the fumes

7.10b . . . and remove the fuel tank mounting bracket bolts (arrows)

Chapter 4 Fuel and exhaust systems

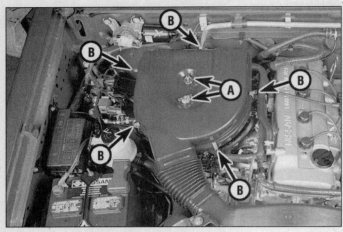

9.1 On four-cylinder models, remove the wing nuts (A), detach the clips (B) and remove the air filter housing cover

9.2 Remove the mounting stud nut (A), disconnect the breather hose (B) and remove the air filter housing base

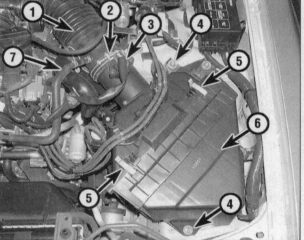

9.4 Air intake duct and air filter housing installation details (3.5L V6 model shown, 3.3L V6 similar)

1. Air intake duct
2. Hose clamp
3. Mass airflow sensor
4. Air filter housing mounting bolts
5. Air filter cover retaining clips
6. Air filter housing
7. Crankcase ventilation hose

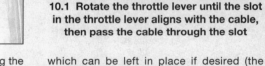

10.1 Rotate the throttle lever until the slot in the throttle lever aligns with the cable, then pass the cable through the slot

coming out of the tank. Be especially careful inside a garage where a gas-type appliance is located, because the flame could cause an explosion.

9 Air filter housing - removal and installation

Four-cylinder engine

Refer to illustrations 9.1 and 9.2

1 Remove the wing nuts, detach the clips and remove the air filter housing cover and air filter element **(see illustration)**.
2 Remove the nut from the mounting stud, disconnect the breather hose and remove the housing base from the throttle body **(see illustration)**. If necessary, remove the nuts/bolts and remove the air intake duct and resonator from the engine compartment.
3 Installation is the reverse of removal.

V6 engines

Refer to illustration 9.4

4 Disconnect the electrical connector from the mass airflow sensor **(see illustration)**.

5 Loosen the retaining clamp securing the air intake duct to the mass airflow sensor. Detach the vacuum hoses from the cover and position the hoses aside.
6 If you're working on a 3.3L V6 model, detach the spring clips and remove the air filter cover and mass airflow sensor as an assembly. On models with a 3.5L V6 engine the filter element rides in a removable carrier,

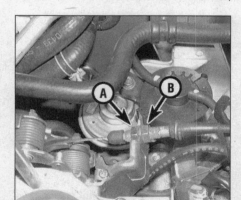

10.2 Loosen the accelerator cable locknut (A) and adjusting nut (B)

which can be left in place if desired (the housing itself is one piece).
7 Remove the air filter housing mounting bolts and remove it from the engine compartment **(see illustration 9.4)**. If necessary, disconnect the crankcase breather hose from the air intake duct, detach the duct from the throttle body and remove the air intake duct.
8 Installation is the reverse of removal.

10 Accelerator cable - removal, installation and adjustment

Note: The adjustment procedure for the accelerator cable and the cruise control cable is similar except where noted below.

Removal

Refer to illustrations 10.1, 10.2 and 10.3

1 Detach the accelerator cable and the cruise control cable (if equipped) from the throttle lever **(see illustration)**.
2 Loosen the cable locknut and adjusting nut, then separate the accelerator cable from the cable bracket **(see illustration)**.
3 From inside the passenger compart-

Chapter 4 Fuel and exhaust systems

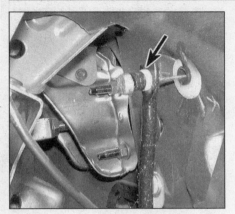

10.3 Working under the dash, pull the cable end from the accelerator pedal recess and pass it through the slot (arrow)

ment, pull the cable end out from the accelerator pedal arm, then pass the cable through the slot in the arm **(see illustration)**. Pull the insulation back and remove the bolts securing the accelerator cable to the firewall.

4 Disconnect any remaining cable clips.
5 Remove the cable through the firewall from the engine compartment side.

Installation

6 Installation is the reverse of removal. Be sure the cable is routed correctly and to fasten all the cable retaining clips.
7 If necessary, at the engine compartment side of the firewall, apply sealant to the accelerator cable bracket where it mates to the firewall to prevent water from entering the passenger compartment.

Adjustment

8 To adjust the accelerator cable:
  a) If equipped with a throttle opener, remove the vacuum hose and apply vacuum to the throttle opener until the throttle opener rod moves off the throttle drum.
  b) Lift up on the cable to remove any slack.
  c) Turn the adjusting nut until the throttle lever just starts to move.
  d) Back off the adjusting nut 1-1/2 to 2 turns.
  e) Tighten the locknut.
  f) Verify that the throttle valve opens all the way when you depress the accelerator pedal to the floor and that it returns to the idle position when you release the accelerator. Verify the cable operates smoothly. It must not bind or stick. Reconnect the vacuum hose to the throttle opener.

9 To adjust the cruise control cable:
  a) Check the accelerator cable for proper adjustment.
  b) Turn the adjusting nut until the throttle lever just starts to move.
  c) Back off the adjusting nut 1/2 to 1 turn.
  d) Tighten the locknut and check for proper operation of the cruise control system.

11 Fuel injection system - general information

All models are equipped with a multi-port fuel injection system. Fuel is delivered into each intake port in sequence with the engine firing order in accordance with engine demand through injectors (one per cylinder) mounted on the intake manifold. On V6 models, the intake manifold incorporates an air intake plenum (upper manifold) to aid in air flow and distribution with a removable throttle body. The air intake plenum bolts to the lower intake manifold, which sits directly in the middle of the engine block.

The multi-port fuel injection system incorporates an on-board electronic engine control computer (known as the Powertrain Control Module - PCM) that accepts inputs from various engine sensors to compute the required fuel flow rate necessary to maintain a prescribed air/fuel ratio throughout the entire engine operational range. The computer then outputs a command to the fuel injectors to meter the quantity of fuel. The system automatically senses and compensates for changes in altitude, load and speed.

The fuel delivery systems include an electric in-tank fuel pump which forces pressurized fuel through a series of metal and rubber lines and an inline fuel filter to the fuel rail assembly. The multi-port fuel injection system uses a single high-pressure pump mounted inside the tank.

The fuel rail assembly incorporates an electrically actuated fuel injector directly above each intake port. When energized, the injectors spray a metered quantity of fuel into the intake air stream.

A constant fuel supply is delivered to the injectors by the fuel rail. The fuel pressure regulator, positioned at the end of the fuel rail, maintains the system fuel pressure. Excess fuel passes through the regulator and returns to the fuel tank through a fuel return line.

Each injector is energized once every other crankshaft revolution in sequence with engine firing order. The period of time that the injectors are energized (known as "on time" or "pulse width") is controlled by the PCM. Air entering the engine is sensed by mass airflow and temperature sensors. The outputs of these, and other, sensors are processed by the PCM. The computer determines the needed injector pulse width and outputs a command to the injector to meter the exact quantity of fuel.

12 Fuel injection system - check

Refer to illustrations 12.7, 12.8 and 12.9
Warning: *Gasoline is extremely flammable, so take extra precautions when you work on any part of the fuel system. See the Warning in Section 2.*
Note: *The following procedure is based on the assumption that the fuel pump is working and the fuel pressure is adequate (see Section 3).*

1 Check all electrical connectors that are related to the system. Loose electrical connectors and poor grounds can cause many problems that resemble more serious malfunctions.
2 Check to see that the battery is fully charged, as the control unit and sensors depend on an accurate supply voltage in order to properly meter the fuel.
3 Check the air filter element - a dirty or partially blocked filter will severely impede performance and economy (see Chapter 1).
4 Check the fuses. If a blown fuse is found, replace it and see if it blows again. If it does, search for a wire shorted to ground in the fuel injection system wiring harness (see Chapter 12 and the wiring diagrams).
5 Check the condition of the vacuum hoses connected to the intake manifold.
6 Remove the air intake duct from the throttle body and check for dirt, carbon or other residue build-up in the throttle body, particularly around the throttle plate. **Caution:** *The throttle body on these models is coated with a sludge-resistant material designed to protect the bore and throttle plate. Do not attempt to clean the interior of the throttle body with carburetor or other spray cleaners. This throttle body is designed to resist sludge accumulation and cleaning may impair the performance of the engine.*
7 With the engine running, place an automotive stethoscope against each injector, one at a time, and listen for a clicking sound, indicating operation **(see illustration)**. If you don't have a stethoscope, you can place the tip of a long screwdriver against the injector

12.7 Use an automotive stethoscope to determine if the injectors are working properly

12.8 Install the fuel injector test light or "noid light" into the fuel injector electrical connector and confirm that it blinks when the engine is cranked or running

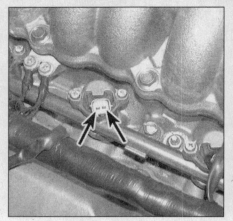

12.9 Measure the resistance across the two terminals of each injector (arrows) - resistance should be within Specifications

13.6a Remove the throttle body mounting bolts (arrows) - four-cylinder models

and listen through the handle.
8 If an injector isn't functioning (not clicking), purchase a special injector test light (sometimes called a "noid" light) and install it into the injector electrical connector **(see illustration)**. Start the engine and check to see if the noid light flashes. If it does, the injector is receiving proper voltage. If it doesn't flash, further diagnosis should be performed by a dealer service department or other properly equipped repair facility.
9 With the engine OFF and the fuel injector electrical connectors disconnected, measure the resistance of each injector **(see illustration)**. Check the Specifications listed in this Chapter for the correct injector resistance.
10 The remainder of the system checks can be found in Chapter 6.

13 Throttle body - removal and installation

Refer to illustrations 13.6a and 13.6b
Warning: *Wait until the engine is completely cool before beginning this procedure.*

1 Disconnect the cable from the negative battery terminal.
2 On four-cylinder models, remove the air filter housing (see Section 9). On V6 models, remove the air intake duct from the throttle body.
3 Disconnect the electrical connectors from the throttle body (see Chapter 6). Also label and detach all vacuum hoses from the throttle body.
4 Detach the accelerator cable (see Section 10) and if equipped, the cruise control cable.
5 Detach the coolant hoses from the throttle body (if equipped). Plug the lines to prevent coolant loss.
6 Remove the throttle body mounting bolts **(see illustrations)**. Remove the throttle body and gasket. Remove all traces of old gasket material from the throttle body and air intake plenum.
7 Installation is the reverse of removal. Be sure to use a new gasket. Adjust the accelerator cable and the cruise control cable (see Section 10). Check the coolant level and add, if necessary (see Chapter 1). On 3.5L V6 models perform the idle air volume relearn procedure (see Chapter 6, Section 16).

14 Fuel pressure regulator - removal and installation

Refer to illustrations 14.4a, 14.4b and 14.4c
Warning: *Gasoline is extremely flammable, so take extra precautions when you work on any part of the fuel system. See the Warning in Section 2.*
1 Relieve the fuel pressure (see Section 2). Disconnect the cable from the negative terminal of the battery.
2 On four cylinder models, remove the air filter housing (see Section 9). On 3.3L V6 models, remove the upper intake manifold (see Chapter 2B). On 3.5L V6 models, remove the engine cover.
3 Clean any dirt from around the fuel pressure regulator. Detach the vacuum hose and the fuel return hose from the fuel pressure regulator.
4 Remove the two screws retaining the fuel pressure regulator and remove the regulator from the fuel rail **(see illustrations)**.
5 Install new O-rings on the pressure regulator and lubricate them with a light coat of oil.
6 Installation is the reverse of removal. Tighten the pressure regulator mounting screws securely.

13.6b Remove the throttle body mounting bolts (arrows) - V6 models

14.4a Remove the fuel pressure regulator screws (arrows) - four-cylinder models

Chapter 4 Fuel and exhaust systems

14.4b Fuel pressure regulator location (arrow) - 3.3L V6 models

14.4c Fuel pressure regulator details - 3.5L V6 models

    A   Fuel pressure regulator   B   Fuel damper

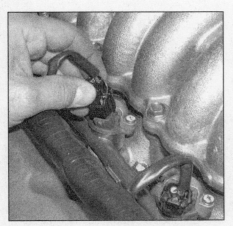

15.4 Disconnect the electrical connectors from the fuel injectors

15.5a If fuel injector service is necessary on a four-cylinder or 3.3L V6 model, remove the cap screws (arrows) and pull the injector(s) from the fuel rail

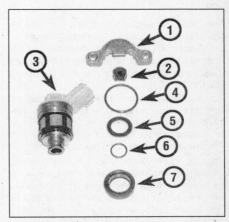

15.5b Fuel injector components - four-cylinder and 3.3L V6 models

1	Cap	5	Center O-ring
2	Insulator	6	Bottom O-ring
3	Fuel injector	7	Fuel rail
4	Top O-ring		insulator

15 Fuel rail and injectors - removal and installation

Refer to illustrations 15.4, 15.5a, 15.5b, 15.6a, 15.6b, 15.6c and 15.9

Warning: *Gasoline is extremely flammable, so take extra precautions when you work on any part of the fuel system. See the* **Warning** *in Section 2.*

Note: *On four-cylinder and 3.3L V6 models, it is not necessary to remove the fuel rail if you're only servicing a fuel injector. Individual fuel injectors can be removed without detaching the fuel rail. On 3.5L V6 models, the fuel rail and injector assembly must be removed as a unit.*

1   Relieve the fuel pressure (see Section 2).
2   Disconnect the cable from the negative terminal of the battery.
3   On V6 models, remove the upper intake manifold from the lower intake manifold (see Chapter 2B or 2C).
4   Disconnect the fuel injector electrical connectors **(see illustration)**.
5   If servicing of the fuel injector(s) is necessary on a four-cylinder or 3.3L V6 model, remove the injector cap screws and cap, then pull the injector(s) from the fuel rail cup **(see illustrations)**. Inspect the injector O-rings for signs of deterioration. Replace as required. Lubricate the new O-rings with light grade oil. Using a light twisting motion, install the injector(s) into the fuel rail cup. Ensure that the injector caps are clean and free of contamination and tighten the cap screws to the torque listed in this Chapter's Specifications.
6   If removal of the fuel rail assembly is necessary, disconnect the fuel feed and return lines. Disconnect the vacuum hose

15.6a Fuel rail mounting bolt locations (arrows) - four-cylinder models

15.6b Fuel rail mounting bolt locations (arrows) - 3.3L V6 models

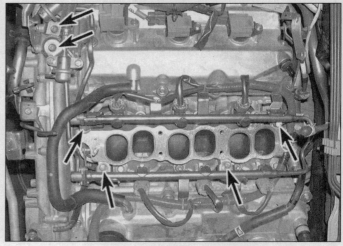

15.6c Fuel rail mounting bolt locations (arrows) - 3.5L V6 models

from the fuel pressure regulator. Remove the fuel rail retaining bolts **(see illustrations)**.

7 Using a rocking, side-to-side motion, carefully lift the fuel rail and the fuel injectors as an assembly from the lower intake manifold.

8 On 3.5L V6 models, remove and discard the retaining clips and withdraw the fuel injectors from the fuel rail.

9 Installation is the reverse of removal with the following exceptions **(see illustration)**:

a) On four-cylinder and 3.3L V6 models, inspect the fuel rail insulators for signs of deterioration and replace as required. Install the fuel rail insulators into the lower intake manifold and lubricate them with a light coat of engine oil before installing the fuel rail assembly onto the lower intake manifold.

b) On 3.5L V6 models, replace the fuel injector O-rings and lubricate them with clean engine oil. Install the fuel injectors onto the fuel rail, aligning the tabs on the injectors with the protrusions on the fuel rail. Install new retaining clips. Make sure the clips are properly seated in the groove on the fuel rail flange.

c) Install the fuel rail and tighten the bolts to the torque listed in this Chapter's Specifications.

16 Engine idle speed and fast idle cam (four-cylinder and 3.3L V6 models) - check and adjustment

Note 1: *Performing the following procedure will set a diagnostic trouble code and illuminate the Check Engine light. Clear the diagnostic trouble code after performing the procedure (see Chapter 6).*
Note 2: *On 3.5L V6 models, the engine idle speed is controlled by the PCM and is not adjustable.*

Engine idle speed

1 Engine idle speed is the speed at which the engine operates when no accelerator pedal pressure is applied, as when stopped at a traffic light. This speed is critical to the performance of the engine itself, as well as many subsystems. Before checking the engine idle speed, check the following items:

a) Check the air filter for restriction.
b) Check the air intake system for leaks.
c) Check the vacuum hoses for leaks.
d) Check the battery and ignition system, including the ignition timing (see Chapter 5).
e) Check the fuel pressure (see Section 3).
f) Check the engine compression (see Chapter 2C).
g) Check the EGR and EVAP systems (see Chapter 6).
h) Check the TPS (see Chapter 6).
i) Check the On Board Diagnostic system for trouble codes (see Chapter 6).

Check

2 Connect a hand-held tachometer in accordance with the tool manufacturer's instructions.

3 Set the parking brake firmly and block the wheels to prevent the vehicle from rolling. Place the transmission in the Neutral position. Make sure all accessories are turned Off.

4 Start the engine and allow it to warm-up to normal operating temperature.

5 Run the engine at around 2000 rpm for two minutes, increasing the speed to over 3000 rpm three times, then allow the engine to idle for one minute.

6 Stop the engine and disconnect the electrical connector from the Throttle Position Sensor (see Chapter 6). Start the engine, increase the engine speed to over 3000 rpm three times and return the engine to idle.

7 Note the idle speed rpm on the tachometer and compare it to that listed on the VECI label or in this Chapter's Specifications. **Note:** *If the idle speed listed on the VECI label is different than that listed in this Chapter's Specifications, use the specification shown on the VECI label.*

Adjustment

Refer to illustrations 16.8a and 16.8b

8 If the idle speed is too low or too high, turn the idle speed adjustment screw to obtain the specified idle speed **(see illustra-

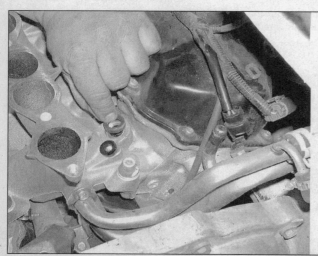

15.9 On four-cylinder and 3.3L V6 models, install the fuel rail insulators into the lower intake manifold and lubricate them with a light coat of oil before installing the fuel rail assembly

Chapter 4 Fuel and exhaust systems

16.8a Idle adjustment screw location (arrow) - four-cylinder models

16.8b Idle adjustment screw location (arrow) - 3.3L V6 models

tions). **Caution:** *Do not attempt to adjust the idle speed with the idle stop screw on the throttle body. The idle stop screw is preset at the factory and should not be tampered with.*

9   Turn the engine off and connect the throttle position sensor electrical connector.

10   Disconnect the tachometer.

Fast idle cam

Refer to illustrations 16.13a and 16.13b

11   A fast idle cam and thermo element is used on four-cylinder and 3.3L V6 models to increase idle speed during cold operating conditions. When the engine temperature is cold the thermo element plunger is in its normal retracted position and the fast idle cam acts on the cam follower lever to lift the throttle lever slightly off the throttle stop, increasing the idle speed. As the engine warms-up, the thermo element plunger extends, rotating the fast idle cam and allowing the throttle lever to return to its base setting.

Check and adjustment

12   Disconnect the electrical connector from the engine coolant temperature sensor and connect an ohmmeter to the two terminals of the sensor (see Chapter 6).

13   Start the engine and monitor the resistance of the coolant temperature sensor. When the sensor resistance is 1,650 to 2,400 ohms, Mark B on the fast idle cam should align with the center of the cam follower lever pin (Mark A) **(see illustrations).**

14   Continue to run the engine. When the coolant temperature resistance is 260 to 390 ohms, Mark C on the fast idle cam should align with the center of the cam follower lever pin (Mark A). If it doesn't, loosen the locknut and turn the adjustment screw until the marks are aligned.

15   If the fast idle cam does not operate as described, or cannot be adjusted properly, replace the throttle body (see Chapter 4).

17 Exhaust system servicing - general information

Refer to illustrations 17.2a, 17.2b and 17.2c

Warning: *Inspection and repair of exhaust system components should be done only after enough time has elapsed after driving the vehicle to allow the system components to cool completely. Also, when working under the vehicle, make sure it is securely supported on jackstands.*

1   The exhaust system consists of the exhaust manifold(s), the catalytic converter(s), the muffler(s), the tailpipe and all connecting pipes, brackets, hangers and clamps. The exhaust system is attached to the body with mounting brackets and rubber hangers. If any of the parts are improperly installed, excessive noise and vibration will be transmitted to the body.

2   Conduct regular inspections of the

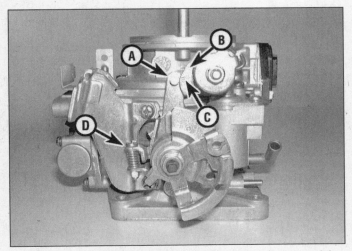

16.13a Fast idle cam adjustment details - four-cylinder models

| A | Mark A (cam follower pin) | C | Mark C |
| B | Mark B | D | Adjusting screw |

16.13b Fast idle cam adjustment details - 3.3L V6 models

| A | Mark A (cam follower pin) | C | Mark C |
| B | Mark B | D | Adjusting screw |

Chapter 4 Fuel and exhaust systems

17.2a Inspect the exhaust system connections for leaks

17.2b Inspect the catalytic converter and heat shield (arrows) for damage

exhaust system to keep it safe and quiet. Look for any damaged or bent parts, open seams, holes, loose connections, excessive corrosion or other defects which could allow exhaust fumes to enter the vehicle **(see illustrations)**. Deteriorated exhaust system components should not be repaired; they should be replaced with new parts.

3   If the exhaust system components are extremely corroded or rusted together, welding equipment will probably be required to remove them. The convenient way to accomplish this is to have a muffler repair shop remove the corroded sections with a cutting torch. If, however, you want to save money by doing it yourself (and you don't have a welding outfit with a cutting torch), simply cut off the old components with a hacksaw. If you have compressed air, special pneumatic cutting chisels can also be used. If you do decide to tackle the job at home, be sure to wear safety goggles to protect your eyes from metal chips and work gloves to protect your hands.

4   Here are some simple guidelines to follow when repairing the exhaust system:
   a) Work from the back to the front when removing exhaust system components.
   b) Apply penetrating oil to the exhaust system component fasteners to make them easier to remove.
   c) Use new gaskets, hangers and clamps when installing exhaust system components.
   d) Apply anti-seize compound to the threads of all exhaust system fasteners during reassembly.
   e) Be sure to allow sufficient clearance between newly installed parts and all points on the underbody to avoid overheating the floor pan and possibly damaging the interior carpet and insulation. Pay particularly close attention to the catalytic converter and heat shield.
   f) Always remove oxygen sensors and connectors before servicing exhaust system components (see Chapter 6).

17.2c Check the condition of the muffler and rubber hangers (arrows) supporting the exhaust system

Chapter 5
Engine electrical systems

Contents

	Section		Section
Alternator - removal and installation	12	General information	1
Battery cables - check and replacement	4	Ignition coil(s) (3.5L V6 models) - removal and installation	7
Battery - check and replacement	3	Ignition system - check	6
Battery - emergency jump starting	2	Ignition system - general information	5
Battery – maintenance and charging	See Chapter 1	Ignition timing - check and adjustment	9
Charging system - check	11	Spark plug replacement	See Chapter 1
Charging system - general information and precautions	10	Starter motor and circuit - in-vehicle check	14
Distributor (four-cylinder and 3.3L V6 models) - removal and installation	8	Starter motor - removal and installation	15
Drivebelt check, adjustment and replacement	See Chapter 1	Starting system - general information and precautions	13

Specifications

Battery voltage
   Engine off .... 12 volts
   Engine running .... 14 to 15 volts
Firing order
   Four-cylinder engine 1-3-4-2
   V6 engines 1-2-3-4-5-6
Ignition coil resistance (at 68-degrees F)
   Four-cylinder and 3.3L V6 engines
      Primary resistance .... 0.5 to 1.0 ohm
      Secondary resistance .... 7 to 13 K-ohms
   3.5L V6 engine .... Not applicable
Ignition timing
   Four-cylinder engine .... 18 to 22 degrees BTDC
   3.3L V6 engine .... 13 to 17 degrees BTDC
   3.5L V6 engine .... Not adjustable
Spark plug wire resistance
   Four-cylinder engine .... 4 to 6 K-ohms per foot
   3.3L V6 engine
      No. 1 .... 6.5 K-ohms
      No. 2 .... 10.1 K-ohms
      No. 3 .... 8.5 K-ohms
      No. 4 .... 12.5 K-ohms
      No. 5 .... 8.5 K-ohms
      No. 6 .... 11.0 K-ohms
   3.5L V6 engine .... Not applicable

Chapter 5 Engine electrical systems

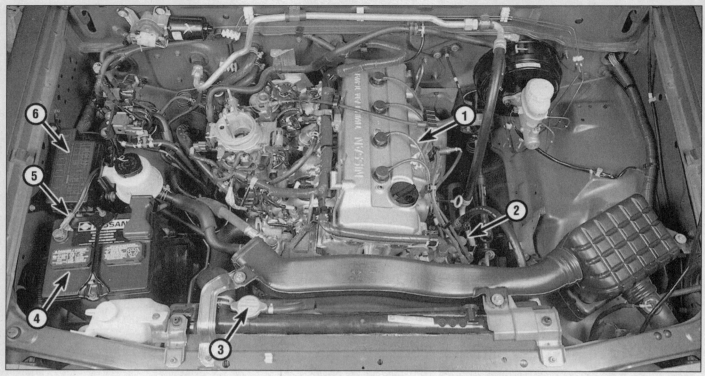

1.1a Engine electrical system components - 2.4L four-cylinder models

| 1 | Spark plug wire | 3 | Alternator (not visible) | 5 | Battery cable (negative) |
| 2 | Distributor | 4 | Battery | 6 | Underhood fuse/relay box |

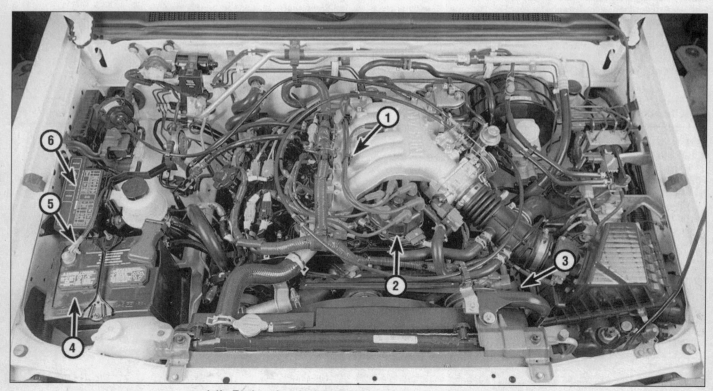

1.1b Engine electrical system components - 3.3L V6 models

| 1 | Spark plug wire | 3 | Alternator (not visible) | 5 | Battery cable (negative) |
| 2 | Distributor | 4 | Battery | 6 | Underhood fuse/relay box |

Chapter 5 Engine electrical systems

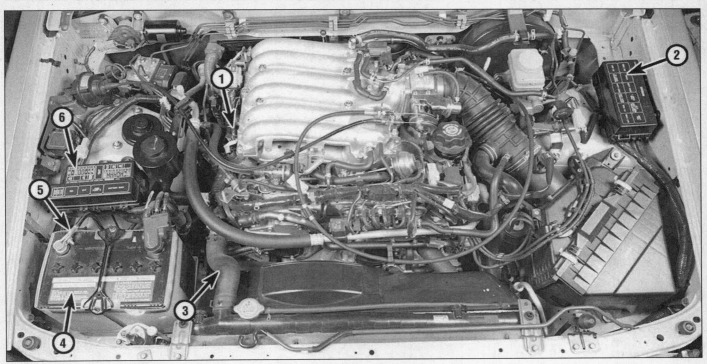

1.1c Engine electrical system components - 3.5L V6 models

1 Ignition coil
2 Auxiliary relay box
3 Alternator (not visible)
4 Battery
5 Battery cable (negative)
6 Underhood fuse/relay box

1 General information

Refer to illustrations 1.1a, 1.1b and 1.1c

The engine electrical systems include all ignition, charging and starting components **(see illustrations)**. Because of their engine-related functions, these components are considered separately from chassis electrical devices like the lights, instruments, etc.

Be very careful when working on the engine electrical components. They are easily damaged if checked, connected or handled improperly. The alternator is driven by an engine drivebelt which could cause serious injury if your hands, hair or clothes become entangled in it with the engine running. Both the starter and alternator are connected directly to the battery and could arc or even cause a fire if mishandled, overloaded or shorted out.

Never leave the ignition switch on for long periods of time with the engine off. Don't disconnect the battery cables while the engine is running. Correct polarity must be maintained when connecting battery cables from another source, such as another vehicle, during jump starting. Always disconnect the negative cable first and hook it up last or the battery may be shorted by the tool being used to loosen the cable clamps.

Additional safety related information on the engine electrical systems can be found in *Safety first* near the front of this manual. It should be referred to before beginning any operation included in this Chapter.

2 Battery - emergency jump starting

Refer to the *Booster battery (jump) starting* procedure at the front of this manual.

3 Battery - check and replacement

Note: *Anytime the battery is disconnected stored operating parameters may be lost from the PCM causing the engine to run rough for a period of time while the PCM relearns the information.*

Check

Refer to illustrations 3.1a, 3.1b, 3.1c and 3.1d

1 A battery cannot be accurately tested until it is at or near a fully charged state. Disconnect the negative battery cable, then the positive cable from the battery and perform the following tests:

a) **Battery state of charge test** - *Visually inspect the indicator eye (if equipped) on the top of the battery. If the indicator eye is dark in color, charge the battery as described in Chapter 1. If the battery is equipped with removable caps, check the battery electrolyte. The electrolyte level should be above the upper edge of the plates. If the level is low, add distilled water. DO NOT OVERFILL. The excess electrolyte may spill over during periods of heavy charging. Test the specific gravity of the electrolyte using a hydrometer* **(see illustration)**. *Remove the caps and extract a sample of the electrolyte and observe the float inside the barrel of the hydrometer. Follow the instructions from the tool manufacturer and determine the specific gravity of the electrolyte for each cell. A fully charged battery will indicate approximately 1.270 (green zone). If the specific gravity of the electrolyte is low (red zone), charge the battery as described in Chapter 1.*

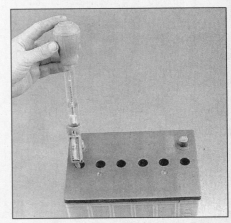

3.1a Use a battery hydrometer to draw electrolyte from the battery cell - this hydrometer is equipped with a thermometer to make temperature corrections

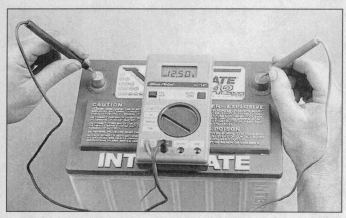

3.1b To test the open circuit voltage of the battery, connect the black probe of the voltmeter to the negative terminal and the red probe to the positive terminal of the battery - a fully charged battery should indicate approximately 12.5 volts depending on the outside air temperature

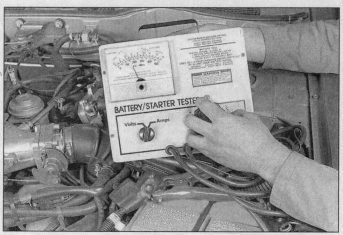

3.1c Some battery load testers are equipped with an ammeter which enables the battery load to be precisely dialed in, as shown – less expensive testers have a load switch and a voltmeter only

b) **Open circuit voltage test** - Using a digital voltmeter, perform an open circuit voltage test **(see illustration)**. *Note: The battery's surface charge must be removed before accurate voltage measurements can be made. Turn On the high beams for ten seconds, then turn them Off and let the vehicle stand for two minutes.* With the engine and all accessories Off, connect the negative probe of the voltmeter to the negative terminal of the battery and the positive probe to the positive terminal of the battery. The battery voltage should be approximately 12.5 volts. If the voltage is less than specified, charge the battery before proceeding to the next test. Do not proceed with the battery load test until the battery is fully charged.

c) **Battery load test** - An accurate check of the battery condition can only be performed with a load tester (available at most auto parts stores). This test evaluates the ability of the battery to operate the starter and other accessories during periods of heavy amperage draw (load). Install a special battery load testing tool onto the battery terminals **(see illustration)**. Load test the battery according to the tool manufacturer's instructions. This tool utilizes a carbon pile to increase the load demand (amperage draw) on the battery. Maintain the load on the battery for 15 seconds and observe that the battery voltage does not drop below 9.6 volts. If the battery condition is weak or defective, the tool will indicate this condition immediately. *Note: Cold temperatures will cause the voltage reading to drop slightly. Follow the chart given in the tool manufacturer's instructions to compensate for cold climates. Minimum load voltage for freezing temperatures (32-degrees F/0-degrees C) should be approximately 9.1 volts.*

3.1d To find out whether there's a drain on the battery, detach the negative cable and connect a test light between the battery post and the cable clamp

d) **Battery drain test** - This test will indicate whether there's a constant drain on the vehicle's electrical system that can cause the battery to discharge. Make sure all accessories are turned Off. If the vehicle has an underhood light, verify it's working properly, then disconnect it. Disconnect the cable from the negative terminal of the battery and attach one lead of a test light to the negative battery cable and the other end to the negative battery terminal **(see illustration)**. The test light should not glow. If the test light glows, it indicates a constant drain on the battery which could cause the battery to discharge. *Note: On vehicles equipped with engine control computers, digital clocks, digital radios, power seats with memory and/or other components which normally cause a key-off battery drain, it's normal for the test light to glow dimly. If you suspect the drain is excessive, install an ammeter in place of the test light. The reading should not exceed 0.05 amps (50 milliamps).*

3.3 Remove the nuts (arrows) from the battery hold-down clamp

Replacement

Refer to illustration 3.3

Warning: *Always disconnect the negative cable first and hook it up last or the battery may be shorted by the tool being used to loosen the cable clamps.*

2 Disconnect the negative battery cable, then the positive cable from the battery.

3 Remove the battery hold-down clamp **(see illustration)**.

4 Lift out the battery. Be careful - it's heavy. *Note: Battery straps and handlers are available at most auto parts stores for a reasonable price. They make it easier to remove and carry the battery.*

5 While the battery is out, inspect the battery tray for corrosion.

6 If corrosion exists on the battery tray, clean the deposits with a mixture of water and baking soda to prevent further corrosion.

7 If you are replacing the battery, make sure you replace it with a battery with the identical dimensions, amperage rating, cold cranking rating, etc.

8 Installation is the reverse of removal.

Chapter 5 Engine electrical systems

Terminal end corrosion or damage.

Insulation cracks.

Chafed insulation or exposed wires.

Burned or melted insulation.

4.2 Typical battery cable problems

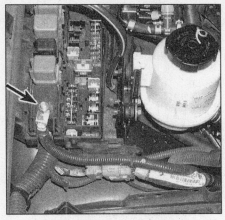

4.4a One branch of the positive cable is fastened to the underhood fuse box

4.4b Detach any battery cable fasteners or retaining clips (arrows)

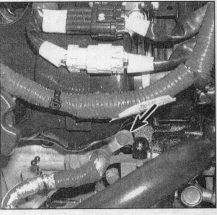

4.4c The negative cable is fastened to the engine block (arrow)

4 Battery cables - check and replacement

Refer to illustrations 4.2, 4.4a, 4.4b and 4.4c

1 Periodically inspect the entire length of each battery cable for damage, cracked or burned insulation and corrosion. Poor battery cable connections can cause starting problems and decreased engine performance.

2 Check the cable-to-terminal connections at the ends of the cables for cracks, loose wire strands and corrosion **(see illustration)**. The presence of white, fluffy deposits under the insulation at the cable terminal connection is a sign that the cable is corroded and should be replaced. Check the terminals for distortion, missing mounting bolts and corrosion.

3 When replacing the cables, always disconnect the negative cable first and hook it up last or the battery may be shorted by the tool used to loosen the cable clamps. Even if only the positive cable is being replaced, be sure to disconnect the negative cable from the battery first.

4 Disconnect and remove the cable **(see illustrations)**. Make sure the replacement cable is the same length and diameter.

5 Clean the threads of the starter solenoid or ground connection with a wire brush to remove rust and corrosion. Apply a light coat of petroleum jelly to the threads to prevent future corrosion.

6 Attach the cable to the starter solenoid or ground connection and tighten the mounting nut/bolt securely.

7 Before connecting the new cable to the battery, make sure that it reaches the battery post without having to be stretched. Clean the battery posts thoroughly and apply a light coat of petroleum jelly to prevent corrosion (see Chapter 1).

8 Connect the positive cable first, followed by the negative cable.

5 Ignition system - general information

1 The ignition system is designed to ignite the fuel/air charge entering each cylinder at just the right moment. It does this by producing a high voltage spark between the electrodes of each spark plug.

Four-cylinder and 3.3L V6 models

2 The four-cylinder and 3.3L V6 models are equipped with an electronic ignition system which consists of the distributor, camshaft position sensor, the power transistor, the ignition coil, an ignition circuit resistor/condenser and the primary and secondary wiring. The camshaft position sensor, power transistor and ignition coil are contained within the distributor and are not serviced separately. In the event any component fails, the entire distributor must be replaced.

3 The camshaft position sensor is the basis of this computer controlled ignition system. It monitors engine speed and piston position and relays this data to the PCM which in turn controls the fuel injection duration (fuel injector on/off time) and ignition timing. The camshaft position sensor consists of a rotor plate, Light Emitting Diodes (LED) and photo diodes, which produce a wave forming circuit. This signal is then sent to the PCM, which produces an ignition signal. The power transistor amplifies the ignition signal from the PCM and intermittently grounds the primary circuit to the ignition coil which generates high voltage in the secondary circuit, thus sending spark from the ignition coil to the distributor, through the spark plug wires and to the spark plugs.

4 The Powertrain Control Module (PCM) receives input signals from various sensors and switches and controls all spark timing advance and retard functions through the power transistor and ignition coil (see Chapter 6 for more information).

5 The ignition system is also integrated with a spark control system which uses a knock sensor in conjunction with the Powertrain Control Module to retard spark timing. The knock sensor system allows the engine to have maximum spark advance without spark knock which also improves driveability and fuel economy.

6 The secondary (spark plug) wires are a carbon-impregnated cord conductor encased in a rubber jacket with an outer silicone jacket. This type of wire will withstand very high temperatures and provides an excellent insulator for the high secondary ignition voltage. Silicone spark plug boots

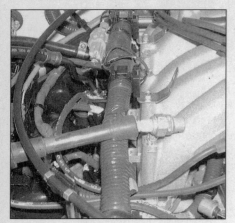 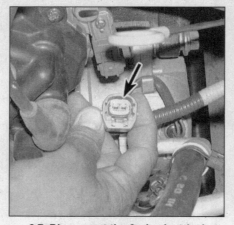 

6.2 To use a calibrated ignition tester on a four-cylinder or 3.3L V6 model, disconnect a spark plug wire, connect the wire to the tester, clip the tester to a convenient ground and operate the starter - if there is enough power to fire the plug, spark will be visible between the electrode tip and the tester body

6.7 Disconnect the 2-pin electrical connector from the distributor and check for battery voltage at the black/white wire terminal with the ignition key On

a) Make sure the battery cable clamps, where they connect to the battery, are clean and tight.
b) Test the condition of the battery (see Section 3). If it does not pass all the tests, replace it with a new battery.
c) Check the external distributor and ignition coil wiring and connections.
d) Check the fusible links (if equipped) inside the engine compartment fuse box (see Chapter 12). If they're burned, determine the cause and repair the circuit.

6.8 Connect the positive lead of a voltmeter to terminal no.1 of the distributor 6-pin connector and check for a voltage signal as the engine is cranked

form a tight seal on the plug. The boot should be twisted 1/2-turn before removing (for more information on the spark plug wires refer to Chapter 1).

3.5L V6 models

7 3.5L V6 models are equipped with a distributorless ignition system. The system consists of six individual ignition coils located above and connected directly to each spark plug. The PCM controls the operation of the ignition coils, firing each coil in sequence. The PCM uses information primarily supplied by the crankshaft position sensor and camshaft position sensor to determine the firing order sequence. In addition to the crankshaft and camshaft sensor signals, the PCM looks at the input from various other sensors to determine the optimum ignition timing.
8 The system is equipped with a knock sensor to detect detonation, or spark knock (usually caused by the use of sub-standard fuel). If a knock signal is received, the PCM will retard the timing until the knock is eliminated.

6 Ignition system - check

Warning: *Because of the high voltage generated by the ignition system, extreme care should be taken whenever an operation is performed involving ignition components. This not only includes the power transistor, coil, distributor and spark plug wires, but related components such as plug connectors, tachometer and other test equipment.*

1 If a malfunction occurs and the vehicle won't start, do not immediately assume that the ignition system is causing the problem. First, check the following items:

Four-cylinder and 3.3L V6 models

Refer to illustrations 6.2, 6.7, 6.8, 6.9a, 6.9b, 6.10a, 6.10b and 6.10c

2 If the engine turns over but won't start, make sure there is sufficient secondary ignition voltage to fire the spark plugs. Disconnect the spark plug wire from any spark plug and attach it to a calibrated ignition system tester (available at most auto parts stores). Connect the clip on the tester to a bolt or metal bracket on the engine **(see illustration)**. Crank the engine and watch the end of the tester to see if a bright blue, well-defined spark occurs (weak spark or intermittent spark is the same as no spark).
3 If spark occurs, sufficient voltage is reaching the plug to fire it (repeat the check at the remaining plug wires to verify that the distributor cap and spark plug wires are OK). However, the plugs themselves may be fouled, so remove and check them as described in Chapter 1.
4 If no spark or intermittent spark occurs, check the cap, rotor and spark plug wires for damage and corrosion as described in Chapter 1.
5 Using an ohmmeter, check each spark plug wire for an open or high resistance. Attach the spark plug wire to the distributor cap terminal and measure the resistance between the terminal inside the distributor cap and the end of the spark plug wire. Com-

pare your measurement with the spark plug wire resistance listed in this Chapter's Specifications. If in doubt about a spark plug wire condition, substitute a known good wire and retest for spark.
6 If moisture is present, dry out the cap and rotor, then reinstall the cap and repeat the spark test.
7 If no spark occurs, check for battery voltage to the ignition coil from the ignition switch with the ignition key On (engine not running). Attach a 12 volt test light to the battery negative (-) terminal or other good ground. Disconnect the distributor 2-pin electrical connector and check for power at the black/white wire terminal (vehicle harness side) **(see illustration)**. Battery voltage should be available. If there is no battery voltage, check the wiring and/or circuit between the coil and ignition switch. Also check the ground circuit for continuity to a good engine ground point. **Note:** *Refer to the wiring diagrams at the end of Chapter 12 for wire color identification for testing and additional information on the circuits.*
8 If battery voltage is available to the ignition coil, reconnect the 2-pin electrical connector. Using a suitable probe, backprobe terminal no.1 of the distributor 6-pin harness connector **(see illustration)**. **Note:** *Refer to Chapter 12 for additional information on how to backprobe an electrical connector.* Connect the positive lead of a voltmeter to the probe. Connect the negative lead to a good engine ground point, then crank the engine. Approximately 0.2-volt (200 millivolts) should be indicated on the meter. This test checks for the trigger signal from the computer. If a trigger signal is present at the coil, the computer and camshaft position sensor are functioning properly. **Note:** *An accurate check of the ignition coil trigger signal must be performed with an oscilloscope. The trigger signal should pulse from zero to 0.2 volts (average) several times per second as the engine is cranked.*
9 If a trigger signal is not present at the ignition coil, check the resistance of the igni-

Chapter 5 Engine electrical systems

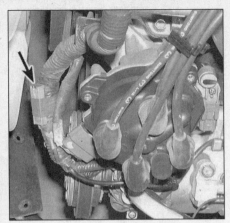

6.9a Remove the resistor from its square case and check the resistance between the two terminals of the resistor - there should be approximately 2.2 K-ohms resistance (four-cylinder shown)

6.9b Ignition resistor location - 3.3L V6

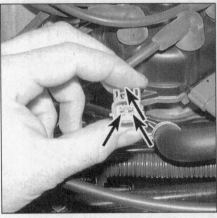

6.10a To check the primary resistance of the ignition coil, measure the resistance across the terminals of the 2-pin distributor connector (distributor side)

tion circuit resistor **(see illustrations)**. Note: *The resistor and connector is taped to the wiring harness near the distributor.* If the resistor is good, refer to Chapter 6 and check the camshaft position sensor.

10 If battery voltage and a trigger signal exist at the ignition coil and there is no spark, check the primary and secondary resistance of the ignition coil **(see illustrations)**. Compare your measurements with the resistance listed in this Chapter's Specifications. Check the power transistor for a short **(see illustration)**. If either component is defective, replace the distributor.

11 If all the components are good and there is no spark, have the PCM checked by a dealer service department or other qualified repair shop.

3.5L V6 models

Refer to illustrations 6.12 and 6.14

12 If the engine turns over but won't start, make sure there is sufficient secondary ignition voltage to fire the spark plug. Remove an ignition coil (see Section 7) and attach a calibrated ignition system tester (available at most auto parts stores) to the spark plug boot (be sure to reconnect the electrical connector to the coil). Connect the clip on the tester to a bolt or metal bracket on the engine **(see illustration)**. Crank the engine and watch the end of the tester to see if a bright blue, well-defined spark occurs (weak spark or intermittent spark is the same as no spark).

13 If spark occurs, sufficient voltage is reaching the plug to fire it (repeat the check at the remaining ignition coils to verify that the ignition coils are good). However, the plugs themselves may be fouled, so remove and check them as described in Chapter 1.

14 If no spark occurs, check the spark plug boot terminal for damage. Check for battery voltage to the ignition coil from the ignition switch with the ignition key On (engine not running). Attach a 12 volt test light to the battery negative (-) terminal or other good ground. Disconnect the coil electrical connector and check for power at terminal no. 3

6.10b To check the secondary resistance of the ignition coil, measure the resistance across terminal no. 2 of the 2-pin distributor connector and the ignition coil high-tension terminal

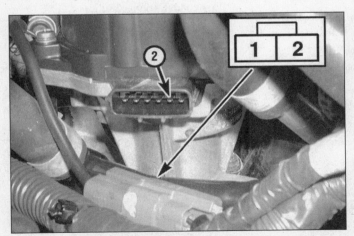

6.10c To check the power transistor, measure the resistance between terminal no. 1 of the 2-pin connector and terminal no. 2 of the 6-pin connector - anything other than a direct short (zero ohms) is good; if the power transistor is shorted, replace the distributor

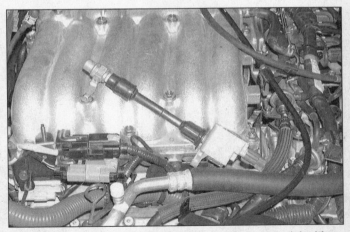

6.12 To check for spark on a 3.5L V6 model, remove an ignition coil and insert a calibrated ignition tester into the spark plug boot, clip the tester to a convenient ground and operate the starter - if there is enough power to fire the plug, spark will be visible between the electrode tip and the tester body

6.14 Disconnect the electrical connector from the ignition coil and check for battery voltage to the coil at terminal no. 3 with the ignition key On, then connect the voltmeter to terminal no. 1 of the coil harness connector and check for a trigger signal when the engine is cranked

7.5 Disconnect the electrical connector, remove the mounting screw (arrow) and pull the coil up with a twisting motion

8.5a Apply an alignment mark on the perimeter of the distributor body in line with the rotor tip (arrows)

(red wire) **(see illustration)**. Battery voltage should be available. If there is no battery voltage, check the wiring and/or circuit between the coil, PCM relay and ignition switch. Also check terminal no. 2 for continuity to a good engine ground point. **Note:** *Refer to the wiring diagrams at the end of Chapter 12 for wire color identification for testing and additional information on the circuits.*

15   If battery voltage is available to the ignition coil, connect the positive probe of a voltmeter to terminal no. 1 of the coil harness connector (vehicle harness side) **(see illustration 6.14)**, connect the negative probe to a good engine ground point then crank the engine. Approximately 0.2-volt (200 millivolts) should be indicated on the meter. This test checks for the trigger signal from the computer. If a trigger signal is present at the coil, the computer and crankshaft position sensors are functioning properly. **Note:** *An accurate check of the ignition coil trigger signal must be performed with an oscilloscope. The trigger signal should pulse from zero to 0.2 volts (average) several times per second as the engine is cranked.*

16   If a trigger signal is not present at the ignition coil, check the ignition circuit condenser. **Note:** *The condenser and connector are taped to the wiring harness at the transmission end of the engine.* Disconnect the condenser from the harness connector and measure the resistance across the two terminals of the condenser; resistance should be very high (1 M-ohm or greater). If the condenser is good, refer to Chapter 6 and check the camshaft position sensor and crankshaft position sensors.

17   If battery voltage, continuity to ground and a trigger signal exist at the ignition coil and there is no spark, replace the coil.

18   If all the components are good and there is no spark, have the PCM checked by a dealer service department or other qualified repair shop.

7   Ignition coil(s) (3.5L V6 models) - removal and installation

Refer to illustration 7.5

1   Disconnect the cable from the negative terminal of the battery.
2   Remove the engine cover.
3   Remove the air intake duct from the throttle body.
4   Disconnect the ignition coil electrical connector.
5   Remove the coil mounting screw and pull the coil up with a twisting motion **(see illustration)**.
6   Installation is the reverse of the removal procedure. Before installing the ignition coil, coat the interior of the boot with silicone dielectric compound.

8   Distributor (four-cylinder and 3.3L V6 models) - removal and installation

Removal

Refer to illustrations 8.5a, 8.5b and 8.6

1   Disconnect the cable from the negative terminal of the battery.
2   Remove the distributor cap cover (if equipped) and note the position of the raised "1" on the distributor cap. This marks the location for the number one cylinder spark plug wire terminal. **Note:** *Some distributor caps may not be marked with the number 1 terminal position.*
3   Disconnect the electrical connectors from the distributor. Follow the wires as they exit the distributor to find the connector, if necessary.
4   Remove the distributor cap (see Chapter 1). Using a socket and breaker bar on the crankshaft pulley bolt, rotate the engine until the rotor is pointing toward the number one spark plug terminal (see the TDC locating procedure in Chapter 2A or 2B).
5   Make a mark on the edge of the distributor base directly below the rotor tip and inline with it **(see illustration)**. Also, mark the distributor base and cylinder head to ensure that the distributor can be reinstalled correctly **(see illustration)**.
6   Remove the distributor hold-down bolt, then pull the distributor straight out to remove it **(see illustration)**. **Caution:** *DO NOT turn the engine while the distributor is removed, or the alignment marks will be useless.*

8.5b Mark the base of the distributor body and the intake manifold, cylinder head or engine block to clearly define the position of the distributor

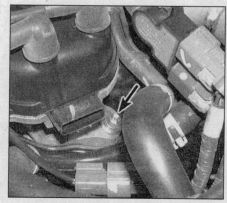

8.6 Remove the distributor hold-down bolt (arrow) and pull the distributor straight out of the engine

Chapter 5   Engine electrical systems

Installation

7   Insert the distributor into the engine in exactly the same relationship to the block that it was in when removed.

8   To mesh the helical gears on the camshaft and the distributor, it may be necessary to turn the rotor slightly. Make sure the distributor is seated completely and the alignment marks made previously are aligned. If not, remove the distributor and reposition it. **Note:** *If the crankshaft has been moved while the distributor is out, locate Top Dead Center (TDC) for the number one piston (see Chapter 2A or 2B) and position the distributor and rotor accordingly.*

9   Loosely install the hold-down bolt.

10   Install the distributor cap and tighten the screws securely.

11   Plug in the electrical connectors.

12   Reattach the spark plug wires to the plugs (if removed).

13   Connect the cable to the negative terminal of the battery.

14   Check and, if necessary, adjust the ignition timing (see Section 9) and tighten the distributor hold-down bolt securely. Reinstall the distributor cap cover (if equipped).

9   Ignition timing - check and adjustment

Refer to illustrations 9.3a and 9.3b

Note: *Timing is adjustable on 2.4L four-cylinder and 3.3L V6 models only.*

1   With the ignition switch off, connect a timing light in accordance with the tool manufacturer's instructions. Install the inductive pick-up onto the number one cylinder spark plug wire.

2   Set the parking brake firmly and block the wheels to prevent the vehicle from rolling. Place the transmission in Park or Neutral.

3   Locate the timing notches on the crankshaft pulley and the pointer on the timing cover **(see illustrations)**. The Top Dead Center (TDC) notch on the crankshaft pulley is marked with yellow paint. The notches on the pulley are spaced in 5 degree increments, clockwise from the TDC notch. Locate the specified notch (three notches clockwise from the yellow TDC mark indicates 15-degrees BTDC, for instance). Clean the marks, if necessary, so they will be easy to see.

4   Start the engine, allow it to warm up to normal operating temperature then shut if off. PCM control of the timing must be disabled before the timing can be checked or adjusted. Refer to Chapter 6, Section 2 and place the system in the self-diagnosis mode (Obtaining diagnostic system trouble codes). If any trouble codes are present, repair the problem before proceeding. If no codes are present the Check Engine light will flash a code 55 (or 0505).

5   With the system in the self-diagnosis mode, start the engine. Rev the engine to approximately 2000 rpm a few times, then let the engine idle. Verify that the engine idle speed is correct (see Chapter 4).

6   Aim the timing light at the timing marks on the front of the engine and check the ignition timing. The specified notch on the pulley will appear stationary and be aligned with the pointer if the timing is correct.

7   If an adjustment is required, loosen the distributor hold-down bolt and rotate the distributor slightly until the timing is correct. Tighten the bolt and recheck the timing.

8   Shut the engine off and disconnect the timing light.

10   Charging system - general information and precautions

The charging system includes the alternator, a voltage regulator (mounted inside the alternator), a charge indicator or warning light, the battery, a large fusible link and the wiring between all the components. The charging system maintains the battery in a fully charged condition and supplies electrical power for the ignition system, the lights, the radio, etc. The alternator is driven by a drivebelt at the front of the engine.

The purpose of the voltage regulator is to limit the alternator's voltage to a preset value. This prevents power surges, circuit overloads, etc., during peak voltage output. All models are equipped with integral type voltage regulator, If a voltage regulator malfunctions, it will be necessary to replace the entire alternator.

The charging system is protected by a large fusible link which is located in the engine compartment fuse box. In the event of charging system problems, check the fusible link for damage or broken contacts.

The charging system doesn't ordinarily require periodic maintenance. However, the drivebelt, battery, wiring and connections should be inspected at the intervals outlined in Chapter 1.

Be very careful when making electrical circuit connections to a vehicle equipped with an alternator and note the following:

a) When reconnecting wires to the alternator from the battery, be sure to note the polarity.

b) Before using arc welding equipment to repair any part of the vehicle, disconnect the battery terminals and the wiring from the alternator.

c) Never start the engine with a battery charger connected.

d) Always disconnect both battery cables before using a battery charger (always disconnect negative cable first, positive cable last).

11   Charging system - check

Refer to illustrations 11.3 and 11.7

1   If a malfunction occurs in the charging circuit, do not immediately assume that the alternator is causing the problem. First, check the following items:

a) The battery cables where they connect to the battery. Make sure the connections are clean and tight.

b) Check the battery state of charge (see Section 3).

c) Check the external alternator wiring and connections.

d) Check the drivebelt condition and tension (see Chapter 1).

e) Check the alternator mounting bolts for tightness.

f) Run the engine and check the alternator for abnormal noise.

g) Check the fusible links in the engine compartment fuse box (see Chapter 12). If they're burned, determine the cause and repair the circuit.

h) Refer to wiring diagrams in Chapter 12 and check all the fuses in series with the charging system. The location and the designations of the fuses may vary by model.

2   Using a voltmeter, check the battery voltage with the engine off. It should be approximately 12-volts.

3   Start the engine and check the battery voltage again. The voltage should be greater than the voltage measured in Step 2, but not

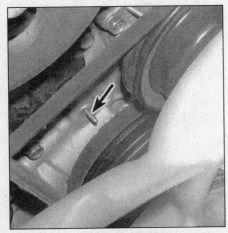

9.3a   Locate the timing pointer (arrow) and notches on the crankshaft pulley - the TDC notch is marked with yellow paint

9.3b   Each notch represents 5-degree increments from TDC

11.3 To measure charging voltage, attach the voltmeter leads to the battery terminals, start the engine and record the voltage reading

11.7 Alternator terminal identification

more than 15-volts **(see illustration)**.
4 Turn the headlights On. The voltage may drop slightly but should be greater than the voltage measured in Step 2, if the charging system is working properly.
5 If the charging voltage is greater than 15-volts, the voltage regulator is defective. Replace the alternator (see Section 12).
6 If the charging voltage is less than the voltage measured in Step 2, check the alternator as follows:
7 Using a voltmeter and working on the backside of the alternator, backprobe the "B+" terminal. There should be 12 volts present with the ignition key Off **(see illustration)**.
8 With the ignition key On (engine not running), backprobe each terminal. There should be 12 volts at the "S" terminal, 1.5 to 2.0 volts at the "L" terminal and 12 volts at the "B+" terminal.
9 **Warning:** *Make sure the meter leads, loose clothing, long hair, etc. are away from the moving parts of the engine (drivebelt, cooling fan, etc.) before starting the engine.* Start the engine, then raise the engine speed to 2000 rpm and backprobe each terminal again. There should be 14.0 to 14.7 volts at the "S" terminal and "B+" terminal and 13.0 to 14.0 volts at the "L" terminal.
10 If the voltages are not as specified, check the wiring harness. If the wiring harness is not defective, replace the alternator.
11 If you suspect that there is a voltage drain on the battery while the vehicle is sitting in the driveway, see Section 3 and perform a battery drain test.
12 If a drain is indicated, carefully remove the fuses one-by-one that govern accessories such as radio, blower motor, trunk lights, etc. until the test light goes out. Trace the short circuit in the particular fused circuit and repair the problem. Recheck the electrical system as described.
13 If all the fuses are pulled out and the test light remains lit, remove the alternator output cable at the rear of the alternator then unplug all the connectors from the backside of the alternator. If the test light goes out, then there is an internal drain in the alternator or voltage regulator. Replace the alternator.

12 Alternator - removal and installation

Four-cylinder and 3.3L V6 models

Refer to illustrations 12.3a, 12.3b and 12.5

1 Disconnect the cable from the negative terminal of the battery.
2 Raise the vehicle and support it securely on jackstands. Remove the lower splash shield (if equipped) from beneath the engine and the side splash shield.
3 Disconnect the electrical connector and the alternator output wire from the alternator. Remove the screw and the ground wire from the alternator **(see illustrations)**.
4 Loosen the adjustment bolt and remove the drivebelt (see Chapter 1).
5 Remove the mounting bolts and separate the alternator from the engine **(see illustration)**.
6 Installation is the reverse of removal.
7 Install the drivebelt and reconnect the

12.3a Disconnect the electrical connector and output wire from the alternator (arrows)

12.3b Remove the ground wire from the alternator (arrow)

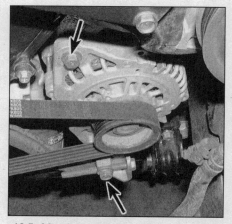

12.5 After loosening the adjustment bolt and removing the drivebelt, remove the mounting bolts (arrows) and remove the alternator from the engine

Chapter 5 Engine electrical systems

12.11 Remove the mounting bolts and remove the alternator - 3.5L V6 engine

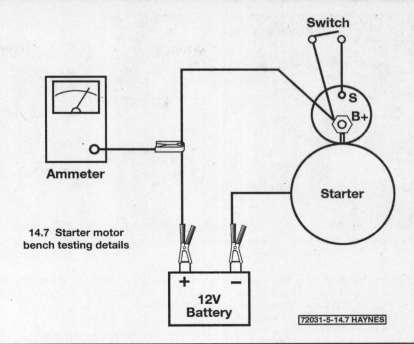

14.7 Starter motor bench testing details

cable to the negative terminal of the battery. Adjust the drivebelt following the procedure in Chapter 1.

3.5L V6 models

Refer to illustration 12.11

8 Disconnect the cable from the negative terminal of the battery.
9 Remove the drivebelt (see Chapter 1).
10 Disconnect the electrical connector, the alternator output wire and the ground wire from the alternator **(see Illustrations 12.3a and 12.3b)**.
11 Remove the alternator mounting bolts and remove the alternator from the engine **(see Illustration)**.
12 Installation is the reverse of removal.
13 Install the drivebelt and reconnect the cable to the negative terminal of the battery.

13 Starting system - general information and precautions

The starting system is composed of the battery, ignition switch, starter inhibitor switch (Park/Neutral Position switch) and inhibitor relay (automatic transmission models) or clutch interlock switch and clutch interlock relay (manual transmission models), starter motor and connecting wiring.

Automatic transmission models are equipped with a Park/Neutral Position switch and inhibitor relay in the starter control circuit, which prevents operation of the starter unless the shift lever is in Neutral or Park. Manual transmission models are equipped with a clutch interlock switch and clutch interlock relay in the starter control circuit, which prevents starter operation unless the clutch pedal is depressed.

Turning the ignition key to the Start position actuates the starter control circuit. If the transmission is in Park or Neutral or the clutch depressed, the inhibitor relay or clutch interlock relay then connects battery power to the starter solenoid. The starter solenoid connects battery power to the starter motor and the starter motor turns. The inhibitor or clutch interlock relay is located in the engine compartment fuse/relay box (see Chapter 12). The starter/solenoid assembly is mounted to the transmission bellhousing.

Never operate the starter motor for more than 15 seconds at a time without pausing to allow it to cool for at least two minutes. Excessive cranking can cause overheating, which can seriously damage the starter.

14 Starter motor and circuit - in-vehicle check

Refer to illustration 14.7

1 If a malfunction occurs in the starting circuit, do not immediately assume that the starter is causing the problem. First, check the following items:

a) Make sure the battery cable clamps, where they connect to the battery, are clean and tight.
b) Check the condition of the battery cables (see Section 4). Replace any defective battery cables with new parts.
c) Test the condition of the battery (see Section 3). If it does not pass all the tests, replace it with a new battery.
d) Check the starter solenoid wiring and connections.
e) Check the starter mounting bolts for tightness.

2 If the starter does not activate when the ignition switch is turned to the start position, check for battery voltage to the solenoid with the ignition switch Off. There should be battery voltage at the positive battery cable on the solenoid if the battery and/or cables are in good working order.
3 Backprobe the S terminal on the starter solenoid and check for voltage as the ignition switch is turned to the start position. This will determine if the solenoid is receiving the correct voltage signal from the ignition switch. If voltage is not available, check the fusible links in the engine compartment fuse box (see Chapter 12). If they're burned, determine the cause and repair the circuit. Also, check the related fuses in the passenger compartment fuse panel (see Chapter 12). If the fuses and fusible links are OK, check the starter inhibitor or clutch interlock relay and circuits for proper operation. Refer to Chapter 12 for the relay locations, wiring diagrams and the relay checking procedure.
4 If the starter circuit is not functioning, check the operation of the Park/Neutral position switch or clutch interlock switch (see Chapter 7 or 8). Make sure the shift lever is in PARK or NEUTRAL or the clutch pedal is fully depressed when attempting to start the engine.
5 If the vehicle is equipped with an anti-theft alarm, check the circuit and the control module for shorts or damaged components.
6 If the starter is receiving voltage but does not activate, most likely the solenoid is defective, but in some rare cases, the engine may be seized. Verify the engine is not seized by rotating the crankshaft pulley (see Chapter 2A, 2B or 2C) before proceeding.
7 If voltage is available at the starter solenoid and there is no movement from the starter motor, remove the starter from the engine (see Section 16) and bench test the starter. Mount the starter/solenoid assembly in a large vise on a sturdy bench. Install one jumper cable from the negative terminal (-) of a fully charged 12-volt automotive battery to the body of the starter **(see illustration)**. Install another jumper cable from the positive terminal (+) of the battery to the battery terminal on the starter. Install a starter switch

15.3 Disconnect the electrical connector and battery cable from the starter motor

15.4a Starter motor mounting bolt (arrow) (lower bolt shown, upper bolt not visible) - 2.4L four-cylinder engine

between the positive terminal of the battery (or the B+ terminal on the starter) and the starter solenoid terminal. Apply battery voltage to the solenoid terminal (for 10 seconds or less) and observe the solenoid plunger, shift lever and overrunning clutch extend and rotate the pinion drive. If the pinion drive extends but does not rotate, the solenoid is operating but the starter motor is defective. If there is no movement but the solenoid clicks, the solenoid and/or the starter motor is defective. If the solenoid plunger extends and rotates the pinion drive at approximately 3,000 rpm, the starter/solenoid assembly is working properly.

15 Starter motor - removal and installation

Refer to illustrations 15.3, 15.4a, 15.4b and 15.4c

1　Disconnect the cable from the negative terminal of the battery.
2　Raise the vehicle and support it securely on jackstands, Remove the splash shield from under the engine (if equipped).
3　Disconnect the battery cable and the solenoid terminal connection from the starter solenoid **(see illustration)**.
4　Remove the starter motor mounting bolts **(see illustrations)** and detach the starter from the engine.
5　Installation is the reverse of removal.

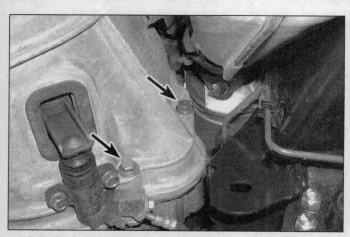

15.4b Starter motor mounting bolts (arrows) - 3.3L V6 engine

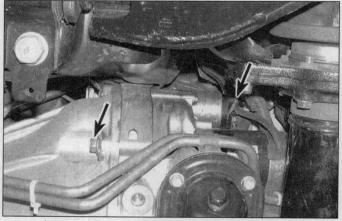

15.4c Starter motor mounting bolts (arrows) - 3.5L V6 engine

Chapter 6
Emissions and engine control systems

Contents

Section		Section	
Camshaft Position (CMP) sensor - check and replacement	10	Manifold Absolute Pressure (MAP) sensor and solenoid valve - check and replacement	6
Catalytic converter	22	Mass Airflow (MAF) sensor - check and replacement	5
Crankshaft Position (CKP) sensor - check and replacement	9	On-Board Diagnosis (OBD) system and trouble codes	2
Engine Coolant Temperature (ECT) sensor - check and replacement	8	Oxygen (O2) sensor - check and replacement	12
Evaporative Emissions Control (EVAP) system	21	Positive Crankcase Ventilation (PCV) system	19
Exhaust Gas Recirculation (EGR) system	20	Power steering pressure switch - check and replacement	11
Fuel temperature sensor - check and replacement	14	Powertrain Control Module (PCM) - removal and installation	3
General information	1	Throttle Position Sensor (TPS) - check, replacement and adjustment	4
Idle Air Control (IAC) system	16	Variable valve timing control system	18
Intake Air Temperature (IAT) sensor - check and replacement	7	Vehicle Speed Sensor (VSS) - check and replacement	15
Intake manifold runner control system	17		
Knock Sensor (KS) - general information	13		

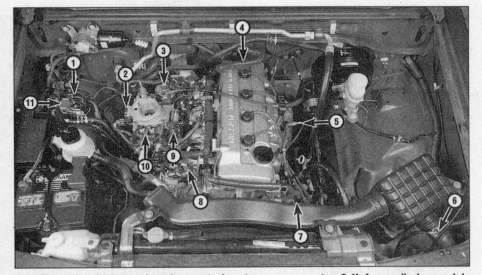

1.1a Typical emission and engine control system components - 2.4L four-cylinder models

1. Manifold Absolute Pressure (MAP) sensor
2. Mass Airflow (MAF) sensor
3. Exhaust Gas Recirculation (EGR) valve
4. Crankshaft Position sensor (on transmission bellhousing)
5. Oxygen sensor
6. Intake Air Temperature (IAT) sensor
7. Camshaft Position sensor (inside distributor)
8. Engine Coolant Temperature (ECT) sensor
9. Idle Air Control (IAC) valve
10. Throttle Position Sensor (TPS)
11. EVAP purge valve

1 General information

Refer to illustrations 1.1a, 1.1b, 1.1c and 1.7

To prevent pollution of the atmosphere from incompletely burned and evaporating gases, and to maintain good driveability and fuel economy, a number of emission control systems are incorporated **(see illustrations)**. They include the:

Electronic engine control system
Evaporative Emission Control (EVAP) system
Positive Crankcase Ventilation (PCV) system
Exhaust Gas Recirculation (EGR) system
Catalytic converter

All of these systems are linked, directly or indirectly, to the emission control system. The Sections in this Chapter include general descriptions, checking procedures within the scope of the home mechanic (when possible) and component replacement procedures for each of the systems listed above.

Before assuming that an emissions control system is malfunctioning, check the fuel and ignition systems carefully. The diagnosis of some emission control devices requires specialized tools, equipment and training. If checking and servicing become too difficult

Chapter 6 Emissions and engine control systems

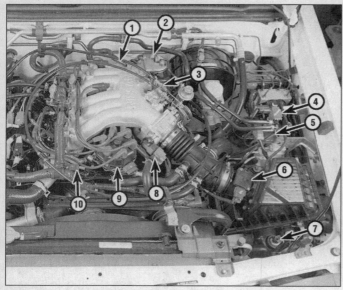

1.1b Typical emission and engine control system components - 3.3L V6 models

1. Crankshaft Position sensor (on transmission bellhousing)
2. Exhaust Gas Recirculation (EGR) backpressure transducer
3. Idle Air Control (IAC) valve (under upper intake manifold)
4. EVAP purge valve
5. Manifold Absolute Pressure (MAP) sensor
6. Mass Airflow (MAF) sensor
7. Intake Air Temperature (IAT) sensor
8. Throttle Position Sensor (TPS)
9. Camshaft Position sensor (inside distributor)
10. Engine Coolant Temperature (ECT) sensor

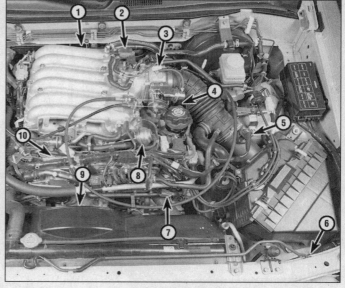

1.1c Typical emission and engine control system components - 3.5L V6 models

1. Engine Coolant Temperature (ECT) sensor (behind upper intake manifold)
2. EVAP purge valve
3. Throttle Position Sensor (TPS)
4. Idle Air Control (IAC) valve (under throttle body)
5. Mass Airflow (MAF) sensor
6. Intake Air Temperature (IAT) sensor (located in air duct leading into air filter housing)
7. Camshaft Position sensor
8. Power valve actuator
9. Crankshaft Position sensor (under crankshaft pulley)
10. Variable valve timing solenoid

or if a procedure is beyond your ability, consult a dealer service department or other qualified repair shop. Remember, the most frequent cause of emissions problems is simply a loose or broken vacuum hose or wire, so always check the hose and wiring connections first.

This doesn't mean, however, that emission control systems are particularly difficult to maintain and repair. You can quickly and easily perform many checks and do most of the regular maintenance at home with common tune-up and hand tools. *Note: Because of a Federally mandated warranty which covers the emission control system components, check with your dealer about warranty coverage before working on any emissions-related systems. Once the warranty has expired, you may wish to perform some of the component checks and/or replacement procedures in this Chapter to save money.*

Pay close attention to any special precautions outlined in this Chapter. It should be noted that the illustrations of the various systems may not exactly match the system installed on the vehicle you're working on because of changes made by the manufacturer during production or from year-to-year.

A Vehicle Emissions Control Information (VECI) label is located in the engine compartment **(see illustration)**. This label contains important emissions specifications and adjustment information, as well as a vacuum hose schematic with emissions components identified. When servicing the engine or emissions systems, the VECI label in your particular vehicle should always be checked for up-to-date information.

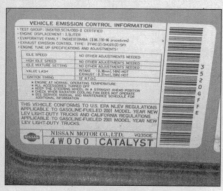

1.7 The Vehicle Emission Control Information (VECI) label is located in the engine compartment and contains information on the emission devices on your vehicle - a vacuum hose routing diagram provides the specific vacuum hose routing information for the vehicle

2.1 Digital multimeters can be used for testing all types of circuits; because of their high impedance, they are much more accurate than analog meters for measuring low-voltage computer circuits

2 On-Board Diagnosis (OBD) system and trouble codes

Diagnostic tool information

Refer to illustrations 2.1, 2.2 and 2.4

1 A digital multimeter is necessary for checking fuel injection and emission related components **(see illustration)**. A digital volt-ohmmeter is preferred over the older style analog multimeter for several reasons. The analog multimeter cannot display the volts-

Chapter 6 Emissions and engine control systems

2.2 Scanners like the Actron Scantool and the AutoXray XP240 are powerful diagnostic aids - programmed with comprehensive diagnostic information, they can tell you just about anything you want to know about your engine management system

2.4 Trouble code readers like the Actron OBD-II diagnostic tester simplify the task of extracting the trouble codes

ohms or amps measurement in hundredths and thousandths increments. When working with electronic circuits which are often very low voltage, this accurate reading is most important. Another good reason for the digital multimeter is the high impedance circuit. The digital multimeter is equipped with a high resistance internal circuitry (10 million ohms). Because a voltmeter is hooked up in parallel with the circuit when testing, it is vital that none of the voltage being measured should be allowed to travel the parallel path set up by the meter itself. This dilemma does not show itself when measuring larger amounts of voltage (9 to 12 volt circuits) but if you are measuring a low voltage circuit such as the oxygen sensor signal voltage, a fraction of a volt may be a significant amount when diagnosing a problem.

2 Hand-held scanners are the most powerful and versatile tools for analyzing engine management systems used on later model vehicles **(see illustration)**. Each brand scan tool must be examined carefully to match the year, make and model of the vehicle you are working on. Often interchangeable cartridges are available to access the particular manufacturer (Ford, GM, Chrysler, etc.). Some manufacturers will specify by continent (Asia, Europe, USA, etc.).

3 With the arrival of the second generation Federally mandated emission control system (OBD-II), a specially designed scanner has been developed. Several tool manufacturers have released OBD-II scan tools for the home mechanic. Ask the parts salesman at a local auto parts store for additional information concerning dates and costs.

4 OBD-II code readers may be available at parts stores **(see illustration)**. These tools simplify the procedure for extracting codes from the engine management computer on some models by simply "plugging in" to the diagnostic connector. Code readers are less expensive than scan tools and are not equipped with the scan tools' diagnostic functions.

On-Board Diagnostic system general description

5 All models are equipped with the second generation OBD-II system. The system consist of an onboard computer, known as the Powertrain Control Module (PCM), and information sensors, which monitor various functions of the engine and send data to the PCM. Based on the data and the information programmed into the computer's memory, the PCM generates output signals to control various engine functions via control relays, solenoids and other output actuators.

6 The PCM is the "brain" of the electronic engine control system. It receives data from a number of sensors and other electronic components (switches, relays, etc.). Based on the information it receives, the PCM generates output signals to control various relays, solenoids and other actuators. The PCM is specifically calibrated to optimize the emissions, fuel economy and driveability of the vehicle.

7 Because of a Federally mandated warranty which covers the emissions system components and because any owner-induced damage to the PCM, the sensors and/or the control devices may void the warranty, it isn't a good idea to attempt diagnosis or replacement of the PCM at home while the vehicle is under warranty. Take the vehicle to a dealer service department if the PCM or a system component malfunctions.

Information sensors

8 **Camshaft Position (CMP) sensor** - The Camshaft Position sensor provides information on camshaft position and the engine speed signal to the PCM. The PCM uses this signal to control ignition timing and fuel injection.

9 **Crankshaft Position (CKP) sensors** - 3.5L V6 models use two Crankshaft Position sensors, while all other models have only one. The Crankshaft Position sensor (REF) is used to detect engine speed. Crankshaft Position sensor (POS) is used to detect TDC for each cylinder. The PCM uses the signals to control ignition timing and fuel injection. They are also used to detect engine misfires.

10 **Engine Coolant Temperature (ECT) sensor** - The Engine Coolant Temperature sensor monitors engine coolant temperature and sends the PCM a voltage signal that affects PCM control of the fuel mixture, ignition timing, and EGR operation.

11 **Exhaust Gas Recirculation (EGR) temperature sensor** - The EGR temperature sensor is used to monitor the rate and flow of exhaust gas recirculation into the intake system.

12 **Fuel temperature sensor** - The fuel temperature sensor provides the PCM with fuel temperature information. The PCM uses this input signal for diagnostic purposes only.

13 **Intake Air Temperature (IAT) sensor** - The Intake Air Temperature sensor provides the PCM with intake air temperature information. The PCM uses this information to control fuel injection, ignition timing, and EGR system operation.

14 **Knock Sensor (KS)** - The Knock Sensor is a piezoelectric element that detects the sound of engine detonation, or "pinging". The PCM uses the input signal from the Knock Sensor to recognize detonation and retard spark advance to avoid engine damage.

15 **Manifold Absolute Pressure (MAP) sensor** - The MAP sensor is used in conjunction with a MAP sensor solenoid valve to monitor intake manifold pressure and ambient barometric pressure. The PCM uses this input signal for diagnostic purposes only.

16 **Mass Airflow (MAF) sensor** - The Mass Airflow sensor measures the molecular mass of the intake airflow entering the engine. The Mass Airflow sensor, along with the Intake Air Temperature sensor, provide mass airflow and air temperature information for the most precise fuel metering.

17 **Oxygen (O2) sensor** - The oxygen sensor generates a voltage signal that varies with the difference between the oxygen content of the exhaust and the oxygen in the surrounding air. The PCM uses this information to determine if the fuel system is running rich or lean.

18 **Power steering pressure switch** - The power steering pressure switch is used to detect excessive line pressure in the power steering system. The PCM uses this input signal to adjust the idle speed under increased engine loads during low-speed vehicle maneuvers.

19 **Throttle Position Sensor (TPS)** - The Throttle Position Sensor senses throttle movement and position, then transmits a voltage signal to the PCM. This signal enables the PCM to determine when the throttle is closed, in a cruise position, or wide open.

20 **Variable valve timing position sensor** - 3.5L V6 models are equipped with a variable valve timing position sensor at each intake camshaft, informing the PCM of the valve timing position.

21 **Vehicle Speed Sensor** - The Vehicle Speed Sensor provides information to the PCM to indicate vehicle speed.
22 **Miscellaneous PCM inputs** - In addition to the various sensors, the PCM monitors various switches, circuits and systems to determine vehicle operating conditions. The switches, circuits and systems include:
 a) Air conditioning system
 b) Antilock brake system
 c) Battery voltage
 d) EVAP system
 e) Ignition switch
 f) Park/neutral position switch
 g) Sensor signal and ground circuits
 h) Transmission control system

Output actuators

23 **Air conditioning clutch relay** - The PCM will de-energize the air conditioning compressor relay during periods of heavy acceleration.
24 **Check Engine light** - The PCM will illuminate the Check Engine light if a malfunction in the electronic engine control system occurs.
25 **Cooling fan control relay** - On models equipped with an electric cooling fan, the PCM controls the operation of the cooling fan according to information received from the engine coolant temperature sensor.
26 **EGR vacuum control solenoids** - The EGR vacuum solenoid is controlled by the PCM to regulate the opening of the vacuum-operated EGR valve.
27 **EVAP canister purge valve** - The evaporative emission canister purge valve is a solenoid valve, operated by the PCM to purge the fuel vapor canister and route fuel vapor to the intake manifold for combustion.
28 **Fuel injectors** - The PCM opens the fuel injectors individually in firing order sequence. The PCM also controls the time the injector is open, called the "pulse width." The pulse width of the injector (measured in milliseconds) determines the amount of fuel delivered. For more information on the fuel delivery system and the fuel injectors, including injector replacement, refer to Chapter 4.
29 **Fuel pump relay** - The fuel pump relay is activated by the PCM with the ignition switch in the Start or Run position. When the ignition switch is turned on, the relay is activated to supply initial line pressure to the system. Refer to Chapter 12 or your owner's manual for more information on relay location. For more information on fuel pump check and replacement, refer to Chapter 4.
30 **Idle Air Control (IAC) valve** - The IAC valve controls the amount of air to bypass the throttle plate when the throttle valve is closed or at idle position. The IAC valve opening and the resulting airflow is controlled by the PCM.
31 **Oxygen sensor heater** - The PCM controls the operation of the oxygen sensor heater. The oxygen sensor heater allows the oxygen sensor to reach operating temperature quickly.
32 **Power transistor** - The power transistor amplifies the ignition signal from the PCM and intermittently grounds the primary circuit

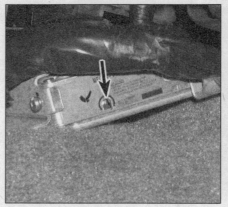

2.38 To read the trouble codes on either the inspection lamps or Check Engine light, use the mode selector on the side of the PCM (2000 and earlier models only)

to the ignition coils which generates high voltage in the secondary circuit, thus sending spark from the ignition coil to the distributor or directly to the spark plug. Refer to Chapter 5 for more information on the power transistor and ignition coil(s).
33 **Intake manifold runner control system** - On 3.5L V6 models, the power valve and the swirl control valve are controlled by the PCM.
34 **Variable valve timing control system** - On 3.5L V6 models, the intake valve timing control solenoid valves are controlled by the PCM.
35 **Transmission Control Module (TCM)** - The TCM receives input signals from various sensors and switches such as the Vehicle Speed Sensor, Park/Neutral position switch, turbine shaft speed sensor, throttle position sensor and the camshaft position sensor to determine shifting points, required line pressure and torque converter lock-up operations of the transmission. The TCM is a separate control module from the PCM although both control modules are used to determine operational characteristics of the transmission.

Obtaining diagnostic system trouble codes

Refer to illustrations 2.38, 2.39a and 2.39b
36 The PCM will illuminate the Check Engine light (also known as the Malfunction Indicator Lamp) on the dash if it recognizes a component fault. It will continue to set the light until the codes are cleared or the PCM does not detect any malfunction for several consecutive drive cycles.
37 On 2000 and earlier models, the diagnostic codes can be extracted from the PCM using two methods, via the Check Engine light or with a scan tool. On 2001 models a scan tool must be used to access the diagnostic codes.
38 The first method, on 2000 and earlier models, requires access to the PCM and the diagnostic mode selector. When the diagnostic mode is entered, the trouble codes are flashed on either the inspection lamps or the

2.39a The Data Link Connector (DLC), typically located under the instrument panel, is used to access the On-Board Diagnostic system with a generic scan tool

Check Engine light. To extract the diagnostic trouble codes using this method, remove the PCM from its mounting bracket (without disconnecting the electrical connectors) and proceed as follows **(see illustration)**:

 a) *Turn the ignition key ON (engine not running). The Check Engine light on the dash should remain ON. This indicates that the PCM is receiving power and the Check Engine light bulb is not defective.* **Note:** *Failure to follow this procedure exactly as described may erase stored trouble codes from the PCM memory.*
 b) *Using a screwdriver, turn the mode selector on the side of the PCM fully clockwise, wait at least two seconds and turn the mode selector fully counterclockwise.*
 c) *Carefully observe the Check Engine light on the instrument panel. The Check Engine light will flash the first two digits of the trouble code with long (approximately 0.6 second) flashes, pause approximately 2.0 seconds and then flash the second two digits with short (approximately 0.3 second) flashes. Record each trouble code number displayed onto paper. For example, code 0403 (throttle position sensor) is indicated by four long flashes, pause, followed by three short flashes. If everything in the self diagnosis system is functioning properly, the Check Engine light will flash a code 0505. Refer to the accompanying trouble code charts for trouble code identification.*
 d) *If the ignition key is turned OFF during the code extraction process and then turned back ON, the self diagnostic system will automatically invalidate the procedure. Restart the procedure to extract the codes.* **Note:** *The self diagnostic system cannot be accessed if the engine is running.*

39 The second method requires the use of a special scan tool that is programmed to interface with the OBD-II system by plugging into the Data Link Connector (DLC) **(see**

Chapter 6 Emissions and engine control systems

illustration). When used, the scan tool has the ability to diagnose in-depth driveability problems and it allows freeze frame data to be retrieved from the PCM stored memory. Freeze frame data is an OBD-II feature that records all related sensor and actuator activity on the PCM data stream whenever an engine control or emissions fault is detected and a DTC is set. This ability to look at the circuit conditions and values when the malfunction occurs provides a valuable tool when trying to diagnose intermittent driveability problems. **Note:** *OBD-II scan tools use different trouble code number designations (referred to as P0 or P1 codes) than the Check Engine light/ MIL lamp codes described above. Refer to the scan tool column of the trouble code chart below for trouble code identification.* If the tool is not available and intermittent driveability problems exist, have the vehicle checked at a dealer service department or other qualified repair facility. **Note:** *Before making any diagnosis or repairs to the engine control system, make sure the PCM ground connections are clean and tight* **(see illustration).** *Many times erroneous sensor data may be generated and false diagnostic trouble codes set due to poor grounding of the PCM and sensors.*

Clearing codes

40 After the system has been repaired, the codes can be cleared from the PCM memory. The codes can be cleared using a scan tool, or on 2000 and earlier models, using the mode selector. On 2000 and earlier models, clear the codes as follows: **Caution:** *Do not disconnect the battery from the vehicle to clear the codes. This will erase stored operating parameters from the memory and cause the engine to run rough for a period of time while the computer relearns the information.* **Note:** *If using an OBD-II scan tool, scroll the menu for the function that describes "CLEARING CODES" and follow the prescribed method for that particular scan tool.*

  a) Obtain the trouble codes as described previously.
  b) Wait at least two seconds, then turn the mode selector clockwise.
  c) Wait at least two seconds, then turn the mode selector counterclockwise.
  d) Turn the ignition key Off.

41 Always clear the codes from the PCM before starting the engine for the first time after installing a new electronic emission control component. The PCM will often store trouble codes during sensor malfunctions. The PCM will also record new trouble codes if a new sensor is allowed to operate before the parameters from the old sensor have been erased. Clearing the codes will allow the computer to relearn the new operating parameters relayed by the new component. During the computer relearning process, the engine may experience a rough idle or slight driveability changes. This period of time, however, should last no longer than 15 to 20 minutes.

2.39b Before making any diagnosis or repairs to the engine control system, make sure the PCM ground connections are clean and tight

Diagnostic trouble code identification

42 The accompanying list of diagnostic trouble codes is a compilation of all the codes that may be encountered. Not all codes pertain to all models and not all codes will illuminate the Check Engine light when set. The codes listed under the "Check Engine Light Flash Code" column are codes that may be displayed by the Check Engine light.

Scan tool trouble code	Check Engine light flash code	Code identification
P0000	0505	No codes identified
P0100	0102	Mass Airflow sensor or circuit fault
P0105	0803	Manifold Absolute Pressure sensor or circuit fault
P0110	0401	Intake Air Temperature sensor or circuit fault
P0115	0103	Engine Coolant Temperature sensor or circuit fault
P0120	0403	Throttle Position Sensor or circuit fault
P0125	0908	Engine Coolant Temperature sensor or circuit fault
P0130	0307	Unable to obtain closed loop operation (right bank)
P0130	0503	Pre-converter oxygen sensor or circuit fault (right bank)
P0131	0411	Pre-converter oxygen sensor lean shift monitor fault (right bank)
P0132	0410	Pre-converter oxygen sensor rich shift monitor fault (right bank)
P0133	0409	Pre-converter oxygen sensor circuit slow response fault (right bank)
P0134	0412	Pre-converter oxygen sensor high voltage fault (right bank)
P0135	0901	Pre-converter oxygen sensor heater fault (right bank)
P0136	0707	Post-converter oxygen sensor or circuit fault
P0137	0511	Post-converter oxygen sensor minimum voltage monitor fault
P0138	0510	Post-converter oxygen sensor maximum voltage monitor fault
P0139	0707	Post-converter oxygen sensor circuit slow response fault

Chapter 6 Emissions and engine control systems

Scan tool trouble code	Check Engine light flash code	Code identification
P0140	0512	Post-converter oxygen sensor high voltage fault
P0141	0902	Post-converter oxygen sensor heater or circuit fault
P0150	0303	Pre-converter oxygen sensor or circuit fault (left bank)
P0150	0308	Unable to obtain closed loop operation (left bank)
P0151	0415	Pre-converter oxygen sensor lean shift monitor fault (left bank)
P0152	0414	Pre-converter oxygen sensor rich shift monitor fault (left bank)
P0153	0413	Pre-converter oxygen sensor circuit slow response fault (left bank)
P0154	0509	Pre-converter oxygen sensor high voltage fault (left bank)
P0155	0101	Pre-converter oxygen sensor heater or circuit fault (left bank)
P0156	0708	Post-converter oxygen sensor or circuit fault (left bank)
P0157	0314	Post-converter oxygen sensor minimum voltage monitor fault
P0158	0313	Post-converter oxygen sensor maximum voltage monitor fault
P0159	0708	Post-converter oxygen sensor circuit slow response fault
P0160	0315	Post-converter oxygen sensor high voltage fault
P0161	1002	Post-converter oxygen sensor heater or circuit fault
P0171	0115	Fuel injection system lean (right bank)
P0172	0114	Fuel injection system rich (right bank)
P0174	0210	Fuel injection system lean (left bank)
P0175	0209	Fuel injection system rich (left bank)
P0180	0402	Fuel tank temperature sensor or circuit fault
P0217	0208	Engine overheating
P0300	0701	Multiple cylinder misfire detected
P0301	0608	Cylinder no. 1 misfire detected
P0302	0607	Cylinder no. 2 misfire detected
P0303	0606	Cylinder no. 3 misfire detected
P0304	0605	Cylinder no. 4 misfire detected
P0305	0604	Cylinder no. 5 misfire detected
P0306	0603	Cylinder no. 6 misfire detected
P0325	0304	Knock sensor or circuit fault
P0335	0802	Crankshaft Position sensor (POS) or circuit fault
P0340	0101	Camshaft Position sensor or circuit fault
P0400	0302	EGR insufficient or excessive flow detected
P0402	0306	EGR backpressure transducer or circuit fault
P0420	0702	Catalyst system fault (right bank)
P0430	0703	Catalyst system fault (left bank)
P0440	0705	EVAP system leak
P0443	0807	EVAP canister purge control valve circuit fault
P0443	1008	EVAP canister purge control valve circuit fault
P0446	0903	EVAP canister vent control valve circuit fault
P0450	0704	EVAP system pressure sensor or circuit fault
P0455	0715	EVAP system gross leak

Chapter 6 Emissions and engine control systems

Scan tool trouble code	Check Engine light flash code	Code identification
P0460	N/A	Fuel level sensor or circuit fault
P0461	N/A	Fuel level sensor or circuit fault
P0464	N/A	Fuel level sensor or circuit fault
P0500	0104	Vehicle Speed Sensor or circuit fault
P0505	0205	Idle Air Control valve or circuit fault
P0510	0203	Closed throttle position switch or circuit fault
P0600	0504	A/T control unit or circuit fault
P0605	0301	PCM or circuit fault
P0705	1003	Park/Neutral position switch or circuit fault
P0705	1101	Park/Neutral position switch or circuit fault
P0710	1208	Transmission temperature sensor or circuit fault
P0720	1102	Vehicle Speed Sensor or circuit fault
P0725	1207	Engine speed signal or circuit fault
P0731	1103	A/T first gear signal fault
P0732	1104	A/T second gear signal fault
P0733	1105	A/T third gear signal fault
P0734	1106	A/T fourth gear signal fault
P0740	1204	TCC solenoid or circuit fault
P0744	1107	A/T TCC solenoid valve fault
P0745	1205	Line pressure solenoid or circuit fault
P0750	1108	Shift solenoid A or circuit fault
P0755	1201	Shift solenoid B or circuit fault
P1105	1302	Manifold Absolute Pressure solenoid valve or circuit fault
P1110	N/A	Variable valve timing control system malfunction (right bank)
P1111	N/A	Variable valve timing control solenoid or circuit fault (right bank)
P1130	N/A	Swirl control valve solenoid or circuit fault
P1135	N/A	Variable valve timing control system malfunction (left bank)
P1136	N/A	Variable valve timing control solenoid or circuit fault (left bank)
P1140	N/A	Variable valve timing position sensor or circuit fault (right bank)
P1145	N/A	Variable valve timing position sensor or circuit fault (left bank)
P1148	0307	Closed loop control fault (right bank)
P1165	N/A	Swirl control valve vacuum check valve fault
P1168	0308	Closed loop control fault (left bank)
P1217	N/A	Engine overheating
P1320	0201	Ignition signal primary circuit fault
P1335	0407	Crankshaft Position sensor (REF) or circuit fault
P1336	0905	Crankshaft Position sensor (POS) fault or signal plate damage
P1400	1005	EGR solenoid valve or circuit fault
P1401	0305	EGR temperature sensor or circuit fault
P1402	0514	EGR high flow detected
P1440	0213	EVAP system small leak

Chapter 6 Emissions and engine control systems

Scan tool trouble code	Check Engine light flash code	Code identification
P1441	0801	Vacuum cut bypass valve or circuit fault
P1444	0214	EVAP canister purge volume control valve or circuit fault
P1445	1008	EVAP canister purge volume control valve or circuit fault
P1446	0215	EVAP canister vent control valve closed
P1447	0111	EVAP purge flow monitor fault
P1448	0309	EVAP canister vent control valve open
P1464	N/A	Fuel level sensor ground circuit fault
P1490	0801	Vacuum cut bypass valve or circuit fault
P1491	0311	Vacuum cut bypass valve or circuit fault
P1605	0804	A/T control unit or circuit fault
P1610	N/A	Vehicle immobilizer system malfunction
P1615	N/A	Vehicle immobilizer system malfunction
P1705	1206	Throttle Position Sensor to A/T signal fault
P1706	1003	Park/Neutral position switch or circuit fault
P1760	1203	Overrun clutch solenoid or circuit fault
P1775	0904	Torque converter clutch solenoid or circuit fault
P1776	0513	Torque converter clutch solenoid or hydraulic control system fault
P1900	0208	Engine cooling fan or circuit fault

3 Powertrain Control Module (PCM) - removal and installation

Refer to illustrations 3.4, 3.5a and 3.5b

Warning: *All models are equipped with a Supplemental Restraint System (SRS) (more commonly known as airbags), always disable the airbag system before working in the vicinity of the airbag system components to avoid the possibility of accidental deployment of the airbag, which could cause personal injury (see Chapter 12).*

Caution: *Static electricity can damage the PCM. Be sure to ground yourself to the vehicle body before touching the PCM or the electrical connector. Store the PCM on an anti-static pad once it is removed.*

Note: *Anytime the battery is disconnected, stored operating parameters may be lost from the PCM causing the engine to run rough for a period of time while the PCM relearns the information.*

1 The Powertrain Control Module (PCM) is located in front of the center console, to the right of the accelerator pedal.

2 Disconnect the cable from the negative terminal of the battery.

3 Remove the driver's side lower trim panel.

4 Disconnect the harness connector from the PCM **(see illustration)**.

5 Remove the retaining bolts from the PCM mounting bracket **(see illustrations)**.

6 Carefully remove the PCM from under the instrument panel.

7 Installation is the reverse of removal. On 3.5L V6 models, perform the idle air volume relearn procedure (see Section 16).

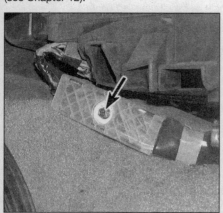

3.4 Remove the bolt (arrow) and carefully detach the electrical connector from the PCM

3.5a On Frontier and Xterra models, remove the PCM bracket mounting screws (arrows)

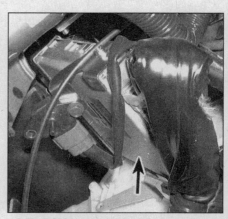

3.5b PCM location - Pathfinder models

Chapter 6 Emissions and engine control systems

4.2a Throttle Position Sensor location - four-cylinder models

4.2b Throttle Position Sensor location - 3.3L V6 models

4.2c Throttle Position Sensor location - 3.5L V6 models

4 Throttle Position Sensor (TPS) - check, replacement and adjustment

1 The Throttle Position Sensor (TPS) is a variable-resistance potentiometer, mounted on the side of the throttle body and connected to the throttle shaft. By monitoring the output voltage from the TPS, the PCM can determine fuel delivery based on throttle valve angle (driver demand). A defective TPS can cause stalling, hesitation, poor acceleration and idle problems. On most models, a wide open/closed throttle switch is incorporated into the throttle position sensor. The wide open/closed throttle switch closes the particular throttle circuit in response to the position of the throttle valve (wide open or closed). A problem with the TPS, wide open/closed throttle switch or related circuits will set a diagnostic trouble code.

Check

Refer to illustrations 4.2a, 4.2b, 4.2c, 4.2d and 4.3

Note: *Performing the following test may set a diagnostic trouble code and illuminate the Check Engine light. Clear the diagnostic trouble code after performing the tests and making the necessary repairs* (see Section 2).

2 Before checking the TPS, check the voltage supply and ground circuits from the PCM to the TPS. Disconnect the TPS electrical connector **(see illustrations)**. Connect the positive probe of a voltmeter to the reference voltage terminal at the harness connector **(see illustration)**. Connect the negative probe to the ground wire terminal. Turn the ignition key ON (engine not running), the voltmeter should indicate approximately 5.0 volts. If the specified voltage is not present, check the circuits from the connector to the PCM. If the circuits are good, have the PCM diagnosed by a dealer service department or other qualified repair facility.

3 To check the operation of the TPS, connect an ohmmeter to the signal terminal (+) and ground terminal (-) on the TPS **(see illustration)**. Check the resistance of TPS with the throttle fully closed, then gradually open the throttle to full throttle - there should be approximately 500 ohms of resistance with the throttle closed and 4,000 ohms at full throttle. The resistance should increase smoothly as the throttle is opened.

4 Disconnect the wide open/closed throttle position switch connector. Connect an ohmmeter to the indicated terminals of the switch **(see illustration 4.3)**. Check for continuity at the switch with the throttle fully closed, then slightly open the throttle. There should be continuity with the throttle completely closed and no continuity when the throttle is opened. If the readings are incorrect, replace the TPS.

Replacement

Refer to illustration 4.7

5 Disconnect the cable from the negative terminal of the battery.

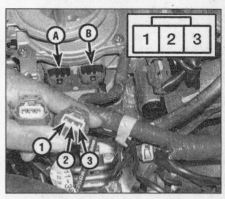

4.2d Throttle Position Sensor wiring harness terminal identification

A Wide open/closed throttle switch
B Throttle Position Sensor
1 Ground (3.3L V6); 5-volt reference (four-cylinder and 3.5L V6)
2 Signal
3 5-volt reference (3.3L V6); ground (four-cylinder and 3.5L V6);

6 Remove the air filter housing (four-cylinder) or air intake duct (V6) (see Chapter 4). Disconnect the TPS and wide open/closed throttle switch electrical connectors.

7 Remove the two retaining screws and separate the TPS from the throttle body **(see illustration)**.

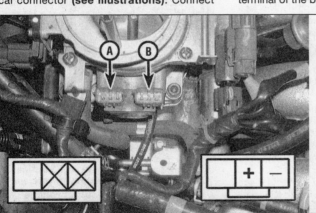

4.3 Wide open/closed throttle switch and TPS terminal identification for testing (four-cylinder model shown)

A Wide open/closed throttle switch (probe the terminals indicated by an "X")
B Throttle Position Sensor

4.7 Remove the TPS retaining screws and separate the TPS from the throttle body

6-10 Chapter 6 Emissions and engine control systems

8 Install the new TPS leaving the mounting screws loose.
9 Adjust the TPS as described and tighten the screws securely.

Adjustment

10 Remove the air filter housing (four-cylinder) or air intake duct (V6) (see Chapter 4).
11 If equipped with a vacuum operated throttle opener, disconnect the vacuum hose and using a hand-held vacuum pump, apply vacuum to the throttle opener until the throttle opener rod moves off the throttle lever (see Section 16).
12 Loosen the TPS mounting screws. Disconnect the wide open/closed throttle switch electrical connector and connect an ohmmeter to the closed throttle switch as described in Step 4. Insert the following feeler gauge, depending on engine type, between the throttle lever and the throttle stop screw - continuity should be indicated on the meter.

  a) Four-cylinder - 0.004-inch (0.1-mm)
  b) 3.3L V6 - 0.012-inch (0.3-mm)
  c) 3.5L V6 - 0.0020-inch (0.05-mm)

13 Remove the feeler gauge and insert the following feeler gauge, depending on engine type, between the throttle lever and the throttle stop screw - no continuity should be indicated.

  a) Four-cylinder - 0.012-inch (0.4-mm)
  b) 3.3L V6 - 0.016-inch (0.4-mm)
  c) 3.5L V6 - 0.0059-inch (0.15-mm)

14 Rotate the TPS until the desired results are achieved and tighten the mounting screws. **Caution:** *Do not alter the position of the throttle stop screw. The throttle stop screw is preset at the factory and does not require adjustment.*

5 Mass Airflow (MAF) sensor - check and replacement

1 On four-cylinder models, the Mass Airflow (MAF) sensor is located on the throttle body. On V6 models, the Mass Airflow (MAF) sensor is installed in the air intake duct. This sensor uses a hot-wire sensing element to measure the molecular mass (or weight) of air entering the engine. The air passing over the hot wire causes it to cool, and the sensor converts this temperature change into a voltage signal to the PCM. The PCM in turn calculates the required fuel injector pulse width to obtain the necessary air/fuel ratio. A defective MAF sensor can cause surging, stalling, rough idle and other driveability problems. The electronic engine control system can detect several different MAF sensor problems and set trouble codes to indicate the specific fault.

Check

Refer to illustration 5.2a, 5.2b, 5.2c, 5.2d and 5.4

Note: *Performing the following test may set a diagnostic trouble code and illuminate the Check Engine light. Clear the diagnostic trouble code after performing the tests and mak-*

5.2a Mass Airflow sensor location - four-cylinder models

5.2b Mass Airflow sensor location - V6 models

5.2c Mass Airflow sensor wiring harness terminal identification - four-cylinder and 3.3L V6 models

1 Signal (+)   3 B+
2 Signal (-)   4 Not used

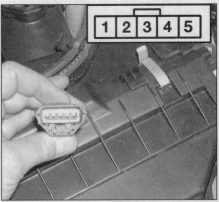

5.2d Mass Airflow sensor wiring harness terminal identification - 3.5L V6 models

1 Signal (+)   3 Signal (-)
2 5-volt   4 B+
   reference   5 Not used

ing the necessary repairs (see Section 2).
2 Before checking the MAF sensor operation, check the power and ground circuits to the MAF sensor. Disconnect the MAF sensor electrical connector **(see illustrations)**. Connect the positive lead of a voltmeter to the B+ wire terminal of the harness connector **(see illustration)**. Connect the negative lead to a good engine ground point. Turn the ignition On but do not start the engine. The meter should indicate approximately battery voltage. If battery voltage is not present, check the ECCS relay and related circuits (see Chapter 12 and the wiring diagrams). Check the ground circuit for continuity between the connector and the PCM.
3 On 3.5L V6 models, connect the positive lead of a voltmeter to the reference wire terminal of the harness connector **(see illustration 5.2d)**. Connect the negative lead to a good engine ground point. Turn the ignition On. The meter should indicate approximately 5.0 volts. If reference voltage is not present, check the wiring harness from the connector to the PCM. If the harness is good, have the PCM diagnosed by a dealership service department or other qualified repair facility.
4 Reconnect the electrical connector to the MAF sensor. Using a suitable probe, backprobe the MAF sensor signal wire terminal **(see illustration)**. See Chapter 12 for additional information on how to backprobe a connector. Connect the positive lead of a voltmeter to the probe and connect the negative lead to a good engine ground point. Turn

5.4 Backprobe the signal wire terminal of the MAF sensor connector (arrow) and check the sensor output voltage (four-cylinder model shown)

Chapter 6 Emissions and engine control systems

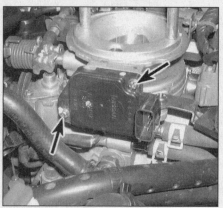

5.9 Remove the retaining screws (arrows) and separate the MAF sensor from the throttle body

5.12 Loosen the retaining clamp, disconnect the air inlet duct, remove the screws (three of four shown) and separate the MAF sensor from the air filter housing

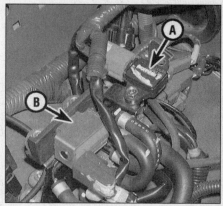

6.2a Manifold Absolute Pressure sensor (A) and solenoid valve location - four-cylinder models (the MAP solenoid valve is located under the EVAP purge solenoid valve (B)

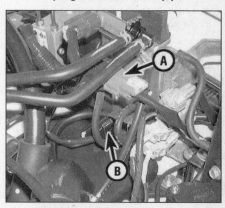

6.2b Manifold Absolute Pressure sensor (A) and solenoid valve (B) location - 3.3L V6 models

the ignition key On. The meter should indicate less than 1.0 volt.
5 Start the engine and check the voltage; it should be 1.0 to 1.8 volts at idle. Increase the engine rpm. The MAF signal voltage should increase from 1.6 to 3.0 volts. It is impossible to simulate driving conditions in the driveway, but it is necessary to watch the voltmeter for an increase in signal voltage as the engine speed is raised. The engine is not under load, but signal voltage should vary slightly.
6 If the voltage readings are correct the MAF sensor is operating properly. Refer to the wiring diagrams and check the wiring harness for open circuits or a damaged harness. If the circuits are good, have the PCM diagnosed by a dealer service department or other qualified repair facility. **Note:** *If MAF related driveability problems continue but these general tests don't indicate a MAF fault, have the sensor tested by a dealer service department or other qualified repair facility. A MAF sensor can develop voltage signal problems that can't be seen on a voltmeter. The PCM can see such signal faults and a driveability problem will result.*

Replacement

Four-cylinder models
Refer to illustration 5.9
7 Remove the air filter housing (see Chapter 4).
8 Disconnect the electrical connector from the MAF sensor.
9 Remove the screws retaining the MAF sensor to the throttle body and remove the sensor **(see illustration)**.
10 Install and connect the new sensor.

V6 models
Refer to illustration 5.12
Note: *The plastic MAF sensor body and the air duct on which it is mounted are an assembly that must be replaced as a unit. Do not try to separate the sensor body from the duct.*
11 Disconnect the electrical connector from the MAF sensor.
12 Remove the clamp that secures the sensor to the intake air duct. Remove the four fasteners securing the sensor to the air filter housing **(see illustration)**.
13 Install and connect the new sensor.

6 Manifold Absolute Pressure (MAP) sensor and solenoid valve/absolute pressure sensor - check and replacement

1 Four-cylinder and 3.3L V6 models use a Manifold Absolute Pressure (MAP) sensor to monitor the intake manifold pressure changes (resulting from changes in engine load and speed) and ambient barometric pressure. 3.5L V6 models monitor ambient barometric pressure only. Four-cylinder and 3.3L V6 models use a MAP sensor solenoid valve in conjunction with the MAP sensor to switch between manifold pressure and ambient barometric pressure. When the PCM applies voltage to the solenoid valve (ON) it switches the vacuum signal and allows the MAP sensor to monitor ambient barometric pressure. When voltage is not supplied to the solenoid valve (OFF) it switches the vacuum signal again and allows the MAP sensor to monitor intake manifold pressure. As the vacuum signal changes the MAP sensor converts this information into a voltage output signal and sends it to the PCM. The voltage will vary from 0.5 volts at closed throttle (high vacuum) and approximately 5.0 volts at wide open throttle (low vacuum). The voltage range values will vary slightly according to changes in altitude. The PCM uses the MAP sensor signal for diagnostic purposes. The PCM can detect several different MAP sensor problems and set trouble codes to indicate the specific fault.

Check
Refer to illustrations 6.2a, 6.2b and 6.2c
Note 1: *The following check procedures apply to four-cylinder and 3.3L V6 models only. A scan tool is required to check the absolute pressure sensor on 3.5L V6 models.*
Note 2: *Performing the following test may set a diagnostic trouble code and illuminate the* Check Engine light. Clear the diagnostic trouble code after performing the tests and making the necessary repairs (see Section 2).
2 On four-cylinder and 3.3L V6 models, disconnect the vacuum hose from the MAP sensor and attach a vacuum gauge to the hose **(see illustrations)**. Start the engine and allow it to idle; after approximately 5 seconds

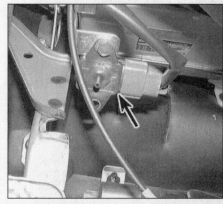

6.2c On 3.5L V6 models, the absolute pressure sensor is located under the instrument panel, near the PCM

Chapter 6 Emissions and engine control systems

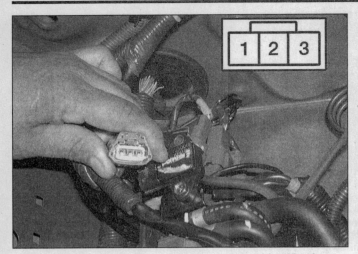

6.3 MAP sensor harness connector terminal identification

1. Ground (four-cylinder and 3.3L V6); 5-volt reference (3.5L V6)
2. Signal
3. 5-volt reference (four-cylinder and 3.3L V6); ground (3.5L V6)

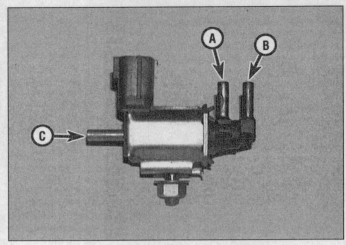

6.9 MAP sensor solenoid valve vacuum port identification

A. To MAP sensor
B. To intake air duct
C. To intake manifold vacuum source

vacuum should be indicated on the gauge. If vacuum is not present, check the vacuum hose from the MAP sensor to the solenoid valve for cracks and clogging. If the hose is good, proceed with the solenoid valve check. If vacuum is present, proceed with the MAP sensor check.

MAP sensor

Refer to illustration 6.3

3 Disconnect the electrical connector from the MAP sensor. Connect a voltmeter to the reference wire terminal (+) and the ground wire terminal (-) **(see illustration)**. With the ignition key ON (engine not running), the meter should indicate approximately 5.0 volts. If the specified voltage is not present, check the circuits from the connector to the PCM. If the circuits are good, have the PCM diagnosed by a dealer service department or other qualified repair facility.

4 Turn the ignition key Off and connect the electrical connector to the MAP sensor. Using a suitable probe, backprobe the signal wire terminal of the connector (see Chapter 12 for additional information on how to backprobe a connector). Connect the positive lead of a voltmeter to the probe and connect the negative lead to a good engine ground point. Turn the ignition key ON (engine not running). There should be approximately 2.2 to 4.8 volts indicated on the meter.

5 Disconnect the vacuum hose from the sensor. Connect a hand-held vacuum pump to the sensor and apply approximately 8 in-Hg of vacuum. The MAP signal voltage should decrease approximately 1.0 to 1.4 volts from the value recorded in Step 4 as vacuum is applied to the sensor. If the MAP sensor does not respond as described, replace the sensor.

Solenoid valve

Refer to illustration 6.9

6 Disconnect the intake manifold vacuum source hose from the solenoid valve and attach a vacuum gauge to the hose. Start the engine and allow it to idle; intake manifold vacuum should be indicated on the gauge. If vacuum is not present, check the hose from the vacuum source to the solenoid valve for cracks and clogging. If vacuum is present, turn the engine off, connect the vacuum hose to the solenoid valve and proceed with the solenoid valve check.

7 Disconnect the vacuum hose from the MAP sensor and attach a vacuum gauge to the hose. Start the engine and allow it to idle; intake manifold vacuum should NOT be indicated on the gauge for approximately five seconds. After five seconds vacuum should be indicated. If the solenoid valve does not operate as described, proceed with the solenoid valve check.

8 Disconnect the electrical connector from the solenoid valve and check for battery voltage between the two terminals of the harness connector with the ignition key ON (engine not running). If voltage is not present, refer to the wiring diagrams and check the wiring harness for an open circuit or a damaged harness from the fuse box to the connector (don't forget to check the fuses first).

9 If battery voltage is present, remove the solenoid valve and test it off the vehicle. Using a pair of fused jumper wires, connect a 12 volt battery source and ground to the two terminals of the solenoid valve. The solenoid should click as voltage is applied. With battery voltage applied, air should pass between ports A and B **(see illustration)**. With no voltage applied, air should pass between ports A and C. If the solenoid does not operate as described, replace the solenoid.

Replacement

10 Disconnect the cable from the negative terminal of the battery.

11 Disconnect the electrical connector from the MAP sensor and/or solenoid valve.

12 Remove the MAP sensor and/or solenoid valve mounting screws. Detach the vacuum hose(s) and remove the MAP sensor and/or solenoid valve.

13 Installation is the reverse of removal.

7 Intake Air Temperature (IAT) sensor - check and replacement

1 The IAT sensor is a thermistor that changes resistance as temperature changes. The sensor is installed in the intake air duct to sense air temperature. As temperature increases, sensor resistance decreases and vice versa. The PCM uses this information to compute the intake temperature and fine tune fuel metering. A problem in the IAT sensor circuit will set a trouble code. The fault may be in the circuit wiring or connections or in the sensor itself.

Check

Refer to illustrations 7.2a, 7.2b, 7.2c, 7.2d and 7.3

Note: *Performing the following test will set a diagnostic trouble code and illuminate the Check Engine light. Clear the diagnostic trouble code after performing the tests and making the necessary repairs (see Section 2).*

2 Before checking the Intake Air Temperature sensor, check the voltage supply and ground circuits from the PCM. Disconnect the electrical connector from the Intake Air Temperature sensor and connect a voltmeter to the two terminals of the harness connector **(see illustrations)**. Turn the ignition key On - the voltage should read approximately 5.0 volts. If the voltage is incorrect, check the wiring from the Intake Air Temperature sensor to the PCM. If the circuits are good, have the PCM checked at a dealer service department or other properly equipped repair facility.

3 With the ignition switch OFF, disconnect the electrical connector from the Intake Air

Chapter 6 Emissions and engine control systems

7.2a Intake Air Temperature sensor location - four-cylinder models

7.2b Intake Air Temperature sensor location (near left headlight) - 3.3L V6 models

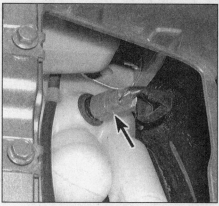

7.2c On 3.5L V6 models, the Intake Air Temperature sensor is located in the air inlet duct (view through the fog lamp opening)

7.2d Disconnect the electrical connector from the Intake Air Temperature sensor and measure the voltage across the two terminals of the harness connector

Temperature sensor. Using an ohmmeter, measure the resistance between the two terminals on the sensor with the engine cool. Reconnect the electrical connector to the sensor, start the engine and warm it up until it reaches operating temperature, disconnect the connector and check the resistance again. Compare your measurements to the resistance chart **(see illustration)**. If the sensor resistance test results are incorrect, replace the Intake Air Temperature sensor.
Note: *A more accurate check may be performed by removing the sensor and suspending the tip of the sensor in a container of water. Heat the water on the stove while you monitor the resistance of the sensor.*

Replacement

4   On 3.5L models, remove the fog lamp assembly.
5   Disconnect the electrical connector, then carefully remove the IAT sensor from the air intake duct. Be careful not to damage any of the plastic parts.
6   Install and connect the new sensor.

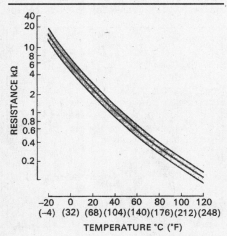

7.3 Intake Air Temperature sensor and Engine Coolant Temperature sensor approximate temperature vs. resistance values

8 Engine Coolant Temperature (ECT) sensor - check and replacement

1   Like the Intake Air Temperature sensor, the Engine Coolant Temperature sensor is a thermistor, which is a variable resistor that changes its resistance as temperature changes. The Engine Coolant Temperature sensor is threaded into an intake manifold coolant passage on four-cylinder and 3.3L V6 models. On 3.5L V6 models the sensor is threaded into the coolant pipe that runs between the rear of each cylinder head. The Engine Coolant Temperature sensor senses coolant temperature. As coolant temperature increases, sensor resistance decreases and vice versa. The PCM uses this information to compute the engine operating temperature. A problem in the Engine Coolant Temperature sensor circuit will set a trouble code. The fault may be in the circuit wiring or connections or in the sensor itself.

Check

Refer to illustration 8.2a, 8.2b, 8.2c, 8.2d and 8.3
Note: *Performing the following test will set a diagnostic trouble code and illuminate the*

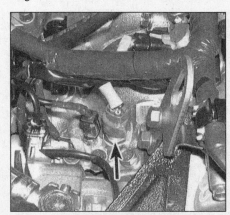

8.2a Engine Coolant Temperature sensor location - four-cylinder models

Check Engine light. Be prepared to clear the diagnostic trouble code after performing the tests and making the necessary repairs (see Section 2).

2   Before checking the Engine Coolant Temperature sensor, check the voltage supply and ground circuits from the PCM. Disconnect the electrical connector from the Engine Coolant Temperature sensor and con-

8.2b Engine Coolant Temperature sensor location - 3.3L V6 models

Chapter 6 Emissions and engine control systems

8.2c Engine Coolant Temperature sensor location (in the coolant pipe near the rear of the right cylinder head) - 3.5L V6 models

8.2d Disconnect the electrical connector from the Engine Coolant Temperature sensor and measure the voltage across the two terminals of the harness connector

8.3 Measure the resistance across the two terminals of the Engine Coolant Temperature sensor

nect a voltmeter to the two terminals of the harness connector **(see illustrations)**. Turn the ignition key On - the voltage should read approximately 5.0 volts. If the voltage is incorrect, check the wiring from the Engine Coolant Temperature sensor to the PCM. If the circuits are good, have the PCM checked at a dealer service department or other properly equipped repair facility.

3 Using an ohmmeter, measure the resistance between the two terminals on the sensor with the engine cool **(see illustration)**. Reconnect the electrical connector to the sensor, start the engine and warm it up until it reaches operating temperature (then turn it off), disconnect the connector and check the resistance again. Compare your measurements to the resistance chart **(see illustration 7.3)**. If the sensor resistance test results are incorrect, replace the Engine Coolant Temperature sensor. **Note:** *A more accurate check may be performed by removing the sensor and suspending the tip of the sensor in a container of water. Heat the water on the stove while you monitor the resistance of the sensor.*

Replacement

Warning: *Wait until the engine is completely cool before performing this procedure.*

4 Before installing the new sensor, wrap the threads with Teflon sealing tape to pre-

vent leakage and thread corrosion.
5 Disconnect the electrical connector and remove the Engine Coolant Temperature sensor from the coolant passage. Install the new sensor as quickly as possible to minimize coolant loss. Tighten the sensor securely and reconnect the electrical connector.
6 Check the coolant level as described in Chapter 1, adding coolant, if necessary. Start the engine and allow it to reach normal operating temperature, then check for coolant leaks. Check the coolant level in the expansion tank after the engine has warmed up and then cooled down again.

9 Crankshaft Position (CKP) sensor - check and replacement

Four-cylinder and 3.3L V6 models

1 The Crankshaft Position sensor is mounted at the rear of the engine, on the transmission bellhousing. It detects gaps in the flywheel/driveplate gear teeth to determine crankshaft speed. The sensor is comprised of a permanent magnet, core and coil. The changing gap causes the magnetic field near the sensor to change, this in turn varies the voltage signal to the PCM. The sensor signal is used by the PCM to detect engine misfire.

Check

Refer to illustration 9.2
2 Remove the sensor from the transmission bellhousing. Using an ohmmeter, measure the resistance across the two Crankshaft Position sensor terminals **(see illustration)**. Resistance should be as follows:
  a) Four cylinder engine - 166 to 204 ohms at 68-degrees F (20-degrees C).
  b) 3.3L V6 engine - 512 to 632 ohms at 68-degrees F (20-degrees C).
3 If the resistance values are incorrect, replace the sensor. If the resistance values

are correct, refer to the wiring diagrams and check the wiring harness for an open circuit to the PCM or a damaged harness. If the sensor and the wiring harness are both good, have the PCM diagnosed by a dealer service department or other qualified repair facility.

Replacement

Refer to illustration 9.6
4 Remove the heat shield (if equipped).
5 Disconnect the electrical connector from the sensor.
6 Remove the Crankshaft Position sensor retaining bolt and remove the sensor **(see illustration)**.
7 Installation is the reverse of removal.

3.5L V6 models

8 Two Crankshaft Position sensors are used on 3.5L V6 models. Crankshaft Position sensor (REF) is mounted at the front of the engine, on the aluminum oil pan section below the crankshaft pulley. It detects gaps in the crankshaft pulley sensor ring corresponding to TDC of each cylinder (120-degree signal). Crankshaft Position sensor

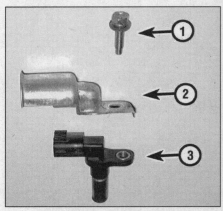

9.6 Crankshaft Position sensor installation details - four-cylinder and 3.3L V6 models

1 Bolt
2 Heat shield
3 Crankshaft Position sensor

9.2 Measure the resistance across the two terminals of the Crankshaft Position sensor

Chapter 6 Emissions and engine control systems

9.9 On 3.5L V6 models, the Crankshaft Position sensor (REF) (arrow) is located on the aluminum oil pan below the crankshaft pulley

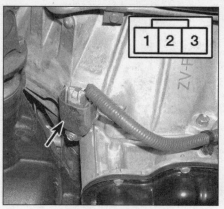

9.10 Crankshaft Position sensor (POS) location and harness terminal identification - 3.5L V6 models

1 B+
2 Signal
3 Ground

10.1 On four-cylinder and 3.3L V6 models, the Camshaft Position sensor (arrow) is an integral part of the distributor

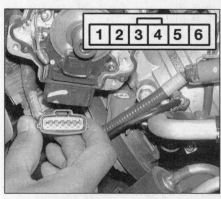

10.2 Camshaft Position sensor harness terminal identification - four-cylinder and 3.3L V6 models

1 Ignition control
2 Ground
3 120-degree signal
4 1-degree signal
5 B+
6 Ground

(POS) is mounted on the aluminum oil pan section at the rear of the engine near the transmission bellhousing. It detects gaps in the flywheel/driveplate gear teeth to determine crankshaft speed (one degree signal). Both sensors use a permanent magnet, core and coil. The changing gap causes the magnetic field near the sensor to change, this in turn varies the voltage signal to the PCM. The sensor signals are used by the PCM for ignition timing, fuel synchronization and detecting engine misfire.

Check

Refer to illustrations 9.9 and 9.10

Crankshaft Position sensor (REF)

9 Disconnect the sensor electrical connector **(see illustration)**. Using an ohmmeter, measure the resistance across the two Crankshaft Position sensor terminals. Resistance should be 470 to 570 ohms at 68-degrees F (20-degrees C). If the resistance values are incorrect, replace the sensor. If the resistance values are correct, refer to the wiring diagrams and check the wiring harness for an open circuit to the PCM or a damaged harness. Check for continuity to ground on the black wire of the harness connector. If the sensor and the wiring harness are both good, have the PCM diagnosed by a dealer service department or other qualified repair facility.

Crankshaft Position sensor (POS)

Note: *Performing the following test will set a diagnostic trouble code and illuminate the Check Engine light. Be prepared to clear the diagnostic trouble code after performing the tests and making the necessary repairs (see Section 2).*

10 Disconnect the sensor electrical connector **(see illustration)**. Turn the ignition key On. Using a voltmeter check for battery voltage at the B+ wire terminal of the harness connector. If voltage is not present, check the circuit from the battery to the ECCS relay (don't forget to check the fuses), check the ECCS relay and the circuit from the relay to the sensor connector (see Chapter 12 and the wiring diagrams). Check for continuity to ground at the black wire terminal of the sensor connector.

11 Remove the sensor. Turn the ignition key Off and connect the electrical connector to the sensor. Using a suitable probe, back-probe the signal wire terminal of the sensor connector (see Chapter 12 for additional information on how to backprobe a connector). Connect the positive lead of a voltmeter to the probe and connect the negative lead to a good engine ground point. Turn the ignition key On. Touch the tip of the sensor with a metal object (such as a screwdriver) and quickly pull it away. The voltmeter should indicate 5.0 volts when contacted and quickly drop to zero volts when the metal object is pulled away. If the sensor does not react as described, replace the sensor. **Note:** *If in the rare case that the wiring, PCM and sensor are all good and the system continues to set a diagnostic trouble code, check the flywheel/driveplate for broken or stripped gear teeth.*

Replacement

12 Disconnect the electrical connector from the sensor.
13 Remove the Crankshaft Position sensor retaining bolt and remove the sensor.
14 Installation is the reverse of removal.

10 Camshaft Position (CMP) sensor - check and replacement

Four-cylinder and 3.3L V6 models

Refer to illustrations 10.1 and 10.2

Note: *Performing the following test will set a diagnostic trouble code and illuminate the Check Engine light. Be prepared to clear the diagnostic trouble code after performing the tests and making the necessary repairs (see Section 2).*

1 The Camshaft Position sensor monitors engine speed and piston position and relays this data to the computer which in turn controls the fuel injection duration (fuel injector on/off time) and ignition timing. The Camshaft Position sensor consists of a rotor plate and a wave forming circuit. The rotor plate has 360 slits for each degree, or one percent signal (engine speed signal) and six slits for the 120-degree camshaft position signal. Light Emitting Diodes (LED) and photo diodes are built into the wave forming circuit. When the rotor plate passes the space between the LED and the photo diode, the slits on the rotor plate continually cut the beam of light sent to the photo diode from the LED. They are then converted into on-off pulses by the wave forming circuit and then sent to the PCM. The Camshaft Position sensor is an integral part of the distributor **(see illustration)**.

2 Disconnect the 6-pin electrical connector from the distributor. Turn the ignition key On. Using a voltmeter, check for battery voltage at terminal no. 5 of the harness connector **(see illustration)**. If voltage is not present,

check the circuit from the battery to the ECCS relay (don't forget to check the fuses), check the ECCS relay and the circuit from the relay to the distributor connector (see Chapter 12 and the wiring diagrams). Using an ohmmeter, check for continuity to ground at terminal no. 6.

3   Turn the ignition OFF and remove the distributor from the engine (see Chapter 5). Reconnect the distributor connector. Using a suitable probe, backprobe terminal no. 3 of the distributor connector (see Chapter 12 for additional information on how to backprobe a connector). Connect the positive lead of a voltmeter to the probe and connect the negative lead to a good engine ground point. With the ignition ON, slowly rotate the distributor shaft and check the voltage; the meter should fluctuate between zero volts and 5.0 volts four or six times per revolution of the distributor shaft. This tests the Camshaft Position sensor 120-degree signal.

4   Turn the ignition OFF and backprobe terminal no. 4 with the voltmeter. With the ignition ON, slowly rotate the distributor shaft; the meter should fluctuate between zero volts and 5.0 volts 360 times per revolution of the distributor shaft. This tests the Camshaft Position sensor 1-degree signal.

5   If the Camshaft Position sensor doesn't operate as described, the distributor must be replaced (see Chapter 5). The Camshaft Position sensor is an integral part of the distributor and not serviced separately.

3.5L V6 models

Refer to illustration 10.6

6   The Camshaft Position sensor is located on the timing chain cover at the front of the engine **(see illustration)**. The sensor uses a permanent magnet, core and coil to detect a gap in the camshaft sprocket. The changing gap causes the magnetic field near the sensor to change, this in turn varies the voltage signal to the PCM. The Camshaft Position sensor provides information on cylinder TDC position to the PCM.

Check

Refer to illustration 10.7

7   Disconnect the electrical connector from the Camshaft Position sensor. Using an ohmmeter, measure the resistance across the two sensor terminals **(see illustration)**. Resistance should be 1,440 to 1,760 ohms at 68-degrees F (20-degrees C) for a Hitachi made sensor or 2,090 to 2,550 ohms at 68-degrees F (20-degrees C) for a Mitsubishi made sensor. If the resistance values are incorrect, replace the sensor.

8   If the resistance values are correct, refer to the wiring diagrams and check the wiring harness for an open circuit to the PCM or a damaged harness. Check for continuity to ground on the black/white wire of the harness connector. If the sensor and the wiring harness are both good, have the PCM diagnosed by a dealer service department or other qualified repair facility.

10.6 3.5L V6 models, the Camshaft Position sensor (arrow) is located on the timing chain cover

Replacement

9   Disconnect the electrical connector from the Camshaft Position sensor.
10 Remove the sensor retaining bolt and remove the sensor.
11 Installation is the reverse of removal.

11 Power steering pressure switch - check and replacement

1   The power steering pressure switch is a normally open switch, mounted in the pressure line between the steering gear and the power steering pump. When steering system pressure reaches a high-pressure setpoint, the power steering pressure switch closes and sends a signal to the PCM that the PCM uses to maintain engine idle speed during parking maneuvers. The PCM can detect switch problems and set trouble codes to indicate specific faults. Check the operation of the power steering pressure switch if the engine stalls during parking or if the engine idles continuously at high rpm.

Check

Refer to illustration 11.2

Note: *Performing the following test will set a diagnostic trouble code and illuminate the Check Engine light. Be prepared to clear the diagnostic trouble code after performing the tests and making the necessary repairs (see Section 2).*

2   Disconnect the power steering pressure switch connector and connect an ohmmeter to the terminals on the switch body or connector **(see illustration)**.
3   Start the engine and let it idle. **Warning:** *Make sure the meter leads, loose clothing, long hair, etc. are away from the moving parts of the engine (drivebelt, cooling fan, etc.) before starting the engine.*
4   Turn the steering wheel to point the front wheels straight ahead and read the ohmmeter. It should indicate an open circuit (infinite resistance).

10.7 Measure the resistance across the two terminals of the Camshaft Position sensor

11.2 Using an ohmmeter, check for continuity across the two terminals of the power steering pressure switch as the steering wheel is rotated

5   Turn the steering wheel to either side and watch the ohmmeter. The power steering pressure switch should close as the wheel nears the steering stop on either side, and the meter should indicate continuity of close to zero ohms.
6   If the switch fails either test, replace it. If the switch is OK, troubleshoot the engine idle control operation if high idle speed or stalling problems continue. Also check the wiring harness for an open circuit to the PCM or a damaged harness and check for continuity to ground on the black wire of the harness connector.

Replacement

7   Disconnect the electrical connector from the switch.
8   Using a back-up wrench on the hex fitting, unscrew the switch from the junction block.
9   Install and connect the new switch. Refer to Chapter 10 and bleed air from the power steering system. Add fluid as required (see Chapter 1).

12 Oxygen (O2) sensor - check and replacement

1 The oxygen in the exhaust reacts with the oxygen sensor to produce a voltage output that varies from 0.1 volt (high oxygen, lean mixture) to 0.9 volt (low oxygen, rich mixture). The pre-converter oxygen sensor in the exhaust system provides a feedback signal to the PCM that indicates the amount of leftover oxygen in the exhaust. The PCM monitors this variable voltage continuously to determine the required fuel injector pulse width and to control the engine air/fuel ratio. A mixture ratio of 14.7 parts air to 1 part fuel is the ideal ratio for minimum exhaust emissions, as well as the best combination of fuel economy and engine performance. Based on oxygen sensor signals, the PCM tries to maintain this air/fuel ratio of 14.7:1 at all times.

2 The post-converter oxygen sensor in the exhaust system has no effect on PCM control of the air/fuel ratio. This sensor is identical to the pre-converter sensor and operates in the same way. The PCM uses the post-converter signal, however, for the catalyst monitor system. A post-converter oxygen sensor will produce a slower fluctuating voltage signal that reflects the lower oxygen content in the post-catalyst exhaust. **Note:** *Four-cylinder models are equipped with one (pre-converter) oxygen sensor and one post-converter oxygen sensor, while V6 models are equipped with two pre-converter (one per cylinder bank) and two post-converter oxygen sensors.*

3 An oxygen sensor produces no voltage when it is below its normal operating temperature of about 600-degrees F (318-degrees C). During this warm-up period, the PCM operates in an open-loop fuel control mode. It does not use the oxygen sensor signal as a feedback indication of residual oxygen in the exhaust. Instead, the PCM controls fuel metering based on the inputs of other sensors and its own programs.

4 Proper operation of an oxygen sensor depends on four conditions:

a) *Electrical - The low voltages generated by the sensor require good, clean connections which should be checked whenever a sensor problem is suspected or indicated.*
b) *Outside air supply - The sensor needs air circulation to the internal portion of the sensor. Whenever the sensor is installed, make sure the air passages are not restricted.*
c) *Proper operating temperature - The PCM will not react to the sensor signal until the sensor reaches approximately 600-degrees F (318-degrees C). This factor must be considered when evaluating the performance of the sensor.*
d) *Unleaded fuel - Unleaded fuel is essential for proper operation of the sensor.*

5 The PCM can detect several different oxygen sensor problems and set diagnostic

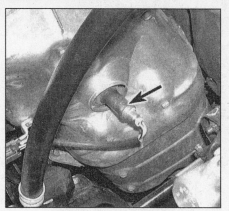

12.6a The pre-converter oxygen sensor is located in exhaust manifold ahead of the catalytic converter

12.6b The post-converter oxygen sensor is located in the exhaust pipe after the catalytic converter

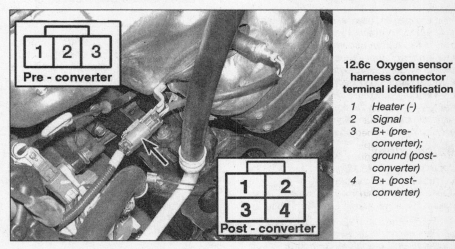

12.6c Oxygen sensor harness connector terminal identification

1 Heater (-)
2 Signal
3 B+ (pre-converter); ground (post-converter)
4 B+ (post-converter)

trouble codes to indicate the specific fault. When an oxygen sensor fault occurs that sets a diagnostic trouble code, the PCM will disregard the oxygen sensor signal voltage and revert to open-loop fuel control as described previously.

Check

Refer to illustrations 12.6a, 12.6b and 12.6c
Note: *Performing the following test will set a diagnostic trouble code and illuminate the Check Engine light. Clear the diagnostic trouble code after performing the tests and making the necessary repairs (see Section 2).*

6 Backprobe the signal wire terminal of the oxygen sensor connector **(see illustrations). Caution:** *The oxygen sensor is very sensitive to excessive circuit loads and circuit damage of any kind. Do not puncture the oxygen sensor wires or try to backprobe the sensor itself. Use only a digital voltmeter to test an oxygen sensor. Refer to Chapter 12 for additional information on how to backprobe an electrical connector.* Connect the positive lead of the voltmeter to the probe and the negative lead to a good engine ground point. Start the engine, warm it up to normal operating temperature and monitor the oxygen sensor signal voltage.

a) *Voltage from a pre-converter sensor should range from 100 to 900 millivolts (0.1 to 0.9 volt) and switch actively between high and low readings.*
b) *Voltage from a post-converter sensor should also read between 100 to 900 millivolts (0.1 to 0.9 volt) but it should not switch actively. The post-converter oxygen sensor voltage may stay toward the center of its range (about 400 millivolts) or stay for relatively longer periods of time at the upper or lower limits of the range.*

7 Check the battery voltage supply and ground circuits to the oxygen sensor heater. Disconnect the electrical connector and connect the positive lead of the voltmeter to the B+ terminal of the sensor connector **(see illustration 12.6c)**. Connect the negative lead to a good chassis ground point. With the ignition ON, the meter should indicate approximately battery voltage. Refer to the wiring diagrams in Chapter 12 for more information on the oxygen sensor circuits.

8 Check the resistance of the oxygen sensor heater with the connector disconnected. Connect an ohmmeter to the two oxygen sensor heater terminals of the connector (oxygen sensor side) **(see illustration 12.6c)**. The oxygen sensor heater resistance should be 2.3 to 4.3 ohms.

Chapter 6 Emissions and engine control systems

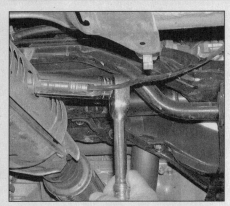

12.13 A special socket (available at many auto parts stores) that allows clearance for the wiring harness may be required for oxygen sensor removal

9 If an open circuit or excessive resistance is indicated, replace the oxygen sensor.
Note: *If the tests indicate that a sensor is good, and not the cause of a driveability problem or diagnostic trouble code, check the wiring harness and connectors between the sensor and the PCM for an open or short circuit. If no problems are found, have the vehicle checked by a dealer service department or other qualified repair facility.*

Replacement

Refer to illustration 12.13

10 The exhaust system must be cool before removing the oxygen sensor. Also observe these guidelines when replacing an oxygen sensor.

a) The sensor has a permanently attached pigtail and electrical connector which should not be removed from the sensor. Damage or removal of the pigtail or electrical connector can harm operation of the sensor.
b) Keep grease, dirt and other contaminants away from the electrical connector and the louvered end of the sensor.
c) Do not use cleaning solvents of any kind on the oxygen sensor.
d) Do not drop or roughly handle the sensor.

11 Raise the vehicle and place it securely on jackstands.
12 Disconnect the electrical connector from the sensor.
13 Using a suitable wrench or special oxygen sensor socket, unscrew the sensor from the exhaust manifold or exhaust pipe **(see illustration)**.
14 Anti-seize compound must be used on the threads of the sensor to aid future removal. The threads of most new sensors will be coated with this compound. If not, be sure to apply anti-seize compound before installing the sensor.
15 Install the sensor and tighten it securely.
16 Lower the vehicle and reconnect the electrical connector for the sensor.

13 Knock Sensor (KS) - check and replacement

1 The Knock Sensor detects abnormal vibration (spark knock or pinging) in the engine. The knock control system is designed to reduce spark knock during periods of heavy detonation. This allows the engine to use maximum spark advance to improve driveability. Knock sensors produce an AC output voltage which increases with the severity of the knock. The signal is fed into the PCM and the timing is retarded to compensate for the severe detonation. On four-cylinder models, the Knock Sensor is located on the side of the engine block below the intake manifold. On V6 models, the Knock Sensor is located in the middle of the engine block underneath the intake manifold.

Check

Refer to illustrations 13.2a, 13.2b and 13.2c

2 Disconnect the electrical connector from the Knock Sensor. Using an ohmmeter, measure the resistance between the Knock Sensor terminal no. 2 (Pathfinder) or terminal no. 1 (Frontier/Xterra) and a good engine ground point **(see illustrations)**. The resistance should be 500 to 620 K-ohms at 77-degrees F (25-degrees C). If the resistance is not as specified, replace the Knock Sensor.

Note: *The Knock Sensor resistance is very high, so use an ohmmeter cable of measuring at least 10 M-ohms.*

Replacement

3 On V6 models, remove the intake manifold (see Chapter 2B or 2C).
4 Disconnect the electrical connector from the Knock Sensor.
5 Remove the sensor retaining bolt and remove the sensor from the engine block.
6 Installation is the reverse of removal. On V6 models, install the intake manifold (see Chapter 2B or 2C).

14 Fuel temperature sensor - check and replacement

1 1997 and later models are equipped with a fuel temperature sensor. The fuel temperature sensor senses the fuel temperature of the fuel inside the tank. The sensor is a thermistor, which is a variable resistor that changes its resistance as temperature changes. As fuel temperature increases, sensor resistance decreases and vice versa. The sensor is located in the fuel tank on the fuel level sending unit. The PCM uses this information for diagnostic purposes. A problem in the fuel temperature sensor circuit will set a trouble code.

Check

Refer to illustrations 14.3 and 14.4
Note: *Performing the following test will set a diagnostic trouble code and illuminate the Check Engine light. Clear the diagnostic trouble code after performing the tests and making the necessary repairs (see Section 2).*

2 On Pathfinder and Xterra models, remove the rear seat and the fuel pump/fuel level sending unit access cover (see Chapter 4). On Frontier models, raise the vehicle, support it securely on jackstands and locate the fuel pump/fuel level sending unit electrical connector.

13.2a On four-cylinder models, the Knock Sensor is located on the side of the engine block

13.2b On V6 models, the Knock Sensor is located under the intake manifold

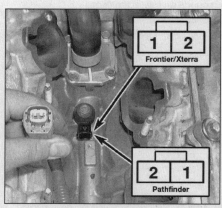

13.2c Measure the resistance of the Knock Sensor between terminal no. 2 (Pathfinder) or terminal no. 1 (Frontier/Xterra) -and a good engine ground point

Chapter 6 Emissions and engine control systems

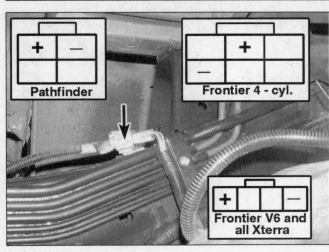

14.3 Disconnect the fuel pump/fuel level sending unit electrical connector and measure the voltage across the fuel temperature sensor reference (+) and ground (-) terminals on the harness connector (vehicle side) (Frontier model shown)

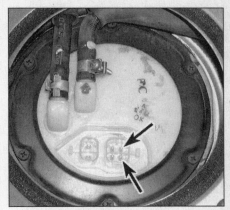

14.4 Disconnect the fuel pump/fuel level sending unit electrical connector and measure the resistance across the two fuel temperature sensor terminals (Pathfinder model shown)

3   Before checking the fuel temperature sensor, check the voltage supply and ground circuits from the PCM. Disconnect the electrical connector from the fuel level sending unit/fuel temperature sensor. Connect a voltmeter to the reference terminal (+) and ground terminal (-) of the harness connector **(see illustration)**. Turn the ignition key On - the voltage should read approximately 5.0 volts. If the voltage is incorrect, check the wiring from the sensor connector to the PCM and from the connector to the engine ground point. If the circuits are good, have the PCM checked at a dealer service department or other properly equipped repair facility.

4   Using an ohmmeter, measure the resistance between the two fuel temperature sensor terminals on the sending unit or harness connector **(see illustration)**. With the system at room temperature (68-degrees F, 20-degrees C), sensor resistance should be 2,300 to 2,700 ohms. A more accurate check may be performed by removing the sensor and suspending the tip of the sensor in a container of water. Heat the water on the stove while you monitor the resistance of the sensor. Compare your measurements to the intake air temperature sensor resistance chart **(see illustration 7.3)**. If the sensor resistance test results are incorrect, replace the fuel temperature sensor.

Replacement

5   Remove the fuel level sending unit (see Chapter 4).
6   Remove the fuel temperature sensor from the fuel level sending unit.
7   Installation is the reverse of removal.

15 Vehicle Speed Sensor (VSS) - check and replacement

1   The Vehicle Speed Sensor is a permanent magnet generator mounted on the transmission case. It produces an AC voltage sine wave, the frequency of which is proportional to vehicle speed. The PCM uses the sensor input signal for several different engine and transmission control functions. The Vehicle Speed Sensor signal also drives the speedometer on the instrument panel. A defective Vehicle Speed Sensor can cause various driveability and transmission problems. The PCM can detect sensor problems and set trouble codes to indicate specific faults.

Check

Refer to illustration 15.3

2   Remove the Vehicle Speed Sensor from the transmission as described below.

3   Connect a voltmeter to the two terminals of the Vehicle Speed Sensor, set the meter on the AC scale and spin the sensor drive gear by hand **(see illustration)**. The sensor should generate approximately 0.5 volts AC.
4   If no AC voltage signal is produced, replace the sensor.
5   If the Vehicle Speed Sensor is good, check for continuity between the sensor connector and the instrument cluster (refer to the wiring diagrams). If the wiring is good, have the instrument cluster and PCM diagnosed by a dealer service department or other qualified repair facility.

Replacement

Refer to illustrations 15.8 and 15.9

6   Raise the vehicle and support it securely on jackstands.
7   Disconnect the electrical connector from the Vehicle Speed Sensor.
8   Remove the hold-down bolt and clamp and remove the Vehicle Speed Sensor from the transmission **(see illustration)**.
9   Inspect the O-ring on the sensor **(see illustration)** and replace it if damaged. If you

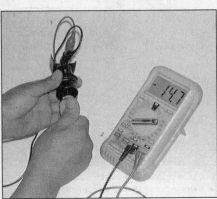

15.3 Remove the Vehicle Speed Sensor and check for a pulsing AC voltage signal as the Vehicle Speed Sensor gear is turned

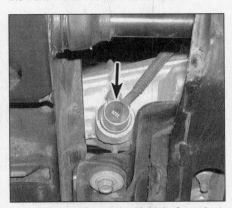

15.8 Location of the Vehicle Speed Sensor (arrow) (two-wheel drive model shown)

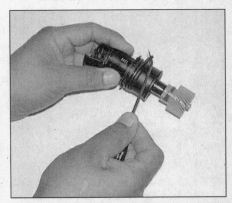

15.9 Inspect the sensor O-ring – install a new one if damaged or if you're installing a new sensor

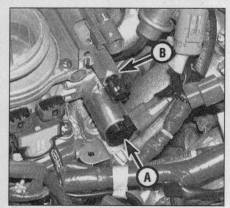

16.1a Idle Air Control valve (A) and fast idle solenoid (B) locations - four-cylinder models

16.1b On 3.3L V6 models, the Idle Air Control valve and fast idle solenoid are located under the rear of the upper intake manifold (arrow)

16.1c Idle Air Control valve location (arrow) - 3.5L V6 models

are installing a new sensor, use a new O-ring.
10 Installation is the reverse of removal.

16 Idle Air Control (IAC) system

Idle Air Control (IAC) valve

Refer to illustrations 16.1a, 16.1b and 16.1c

1 The Idle Air Control (IAC) valve controls the amount of air that bypasses the throttle valve, which controls the engine idle speed. The IAC valve is a stepper motor type actuator controlled by voltage pulses from the PCM. The IAC valve pintle moves in or out, allowing more or less intake air into the system. To increase idle speed, the PCM commands the stepper motor to pull the IAC valve pintle from the seat, allowing more air to bypass the throttle bore. To decrease idle speed, the PCM commands the IAC valve pintle towards the seat, reducing the air flow. On four-cylinder and 3.5L V6 models, the IAC valve is mounted on the throttle body. On 3.3L V6 models, the IAC valve is mounted on the upper intake manifold **(see illustrations)**.

Check

Note: *Performing the following test will set a diagnostic trouble code and illuminate the Check Engine light. Clear the diagnostic trouble code after performing the tests and making the necessary repairs (see Section 2).*

Four-cylinder and 3.3L V6 models

Refer to illustrations 16.2 and 16.3

2 Before checking the IAC valve, check the power supply to the valve. Disconnect the electrical connector from the IAC valve. Turn the ignition key On. Connect the negative probe of a voltmeter to a good engine ground point and probe the B+ wire terminal of the IAC valve electrical connector (harness side) **(see illustration)**. Battery voltage should be present at the B+ wire terminal. If battery voltage is not present, check the circuit from the ignition switch to the IAC valve (don't forget to check the fuses) (see Chapter 12 and the wiring diagrams).

3 To check the IAC valve, use an ohmmeter to measure the resistance across the two terminals on the IAC valve **(see illustration)**. There should be approximately 10 ohms resistance at 77-degrees F (25-degrees C) across the terminals. If the resistance is incorrect, replace the IAC valve. If the IAC valve is good, remove the IAC valve and check the plunger for sticking or complete seizure. If the IAC valve is good have the PCM diagnosed by a dealer service department or other qualified repair facility.

3.5L V6 models

Refer to illustrations 16.4 and 16.5

4 Before checking the IAC valve, check the power supply to the valve. Disconnect the electrical connector from the IAC valve. Turn the ignition key On. Connect the negative probe of a voltmeter to a good engine ground point and probe each of the two B+ wire terminals of the IAC valve electrical connector (harness side) in turn **(see illustration)**. Battery voltage should be present at each of the B+ wire terminals. If battery voltage is not present, check the circuit from the battery to the ECCS relay (don't forget to check the fuses), the ECCS relay and the circuit from the ECCS relay to the IAC valve (see Chapter 12 and the wiring diagrams).

5 To check the IAC valve, use an ohmme-

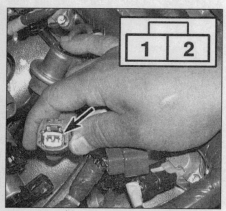

16.2 Idle Air Control valve harness connector terminal identification - four-cylinder and 3.3L V6 models

1 B+   2 IAC control

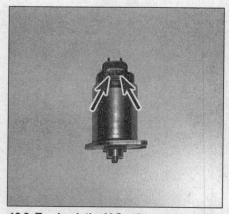

16.3 To check the IAC valve, measure the resistance across the two terminals on the valve - the resistance should be approximately 10 ohms

16.4 Idle Air Control valve harness connector terminal identification - 3.5L V6 models

1 IAC control no. 1   4 IAC control no.2
2 B+   5 B+
3 IAC control no. 3   6 IAC control no.4

Chapter 6 Emissions and engine control systems

16.5 To check the IAC valve, measure the resistance across terminals 1 and 2; 2 and 3; 4 and 5; 5 and 6 - the resistance should be approximately 20 to 24 ohms across each pair

ter to measure the resistance across the indicated terminals on the IAC valve **(see illustration)**. There should be approximately 20 to 24 ohms resistance at 68-degrees F (20-degrees C) across each pair. If the resistance is incorrect, replace the IAC valve. If the IAC valve is good, have the PCM diagnosed by a dealer service department or other qualified repair facility.

Replacement

6 On four-cylinder models, remove the air filter housing (see Chapter 4). On 3.3L V6 models, remove the upper intake manifold (see Chapter 2B). On 3.5L V6 models, remove the air intake duct.
7 Disconnect the electrical connector from the IAC valve.
8 Remove the two IAC valve attaching screws and withdraw the valve from the IAC housing.
9 Clean the sealing surface and the bore of the IAC housing with a shop rag or soft cloth to ensure a good seal. **Caution:** *The IAC valve itself is an electrical component and must not be soaked in any liquid cleaner, as damage may result.*
10 Position a new O-ring on the housing. Lubricate the O-ring with a light film of engine oil.
11 Install the IAC valve and tighten the screws securely.
13 The remainder of the installation is the reverse of the removal. On 3.5L V6 models, perform the idle air volume relearn procedure.

Idle air volume relearn

Note: *Performing the following procedure may set a diagnostic trouble code and illuminate the Check Engine light. If so, refer to Section 2 for the trouble code clearing procedure.*
14 On 3.5L V6 models, anytime the IAC valve is replaced (or the throttle body or PCM) the idle air volume relearn procedure must be performed as follows:
   a) *Place the transmission in Park and set the parking brake. Make sure all the electrical accessories are Off and the front wheels are pointing straight ahead.*
   b) *Start the engine, warm it up to operating temperature then turn it Off.*
   c) *Turn the ignition switch On for one second then turn it Off for at least ten seconds.*
   d) *Start the engine and warm it up to operating temperature then turn it Off.*
   e) *Wait at least ten seconds, then start the engine and allow it to idle for thirty seconds.*
   f) *With the engine idling, disconnect the electrical connector from the Throttle Position Sensor and reconnect it within five seconds.*
   g) *Wait twenty seconds, then rev the engine several times and make sure the idle speed returns to 700 to 800 rpm.*
15 If the idle is not within specifications, make sure the throttle valve is fully closed and not binding. Check the air intake duct for air leaks. Check the Throttle Position Sensor and adjust the closed throttle switch, if necessary.
16 If the idle is still not within specifications, have the system diagnosed by a dealer service department or other qualified repair shop.

Fast idle control solenoid

17 Four-cylinder models and 3.3L V6 models are equipped with a fast idle control solenoid **(see illustration 16.1a)**. The solenoid is used to maintain a constant idle speed due to increased engine loads. The solenoid is energized when the air conditioning is activated. When the fast idle control solenoid is energized it allows additional air to enter the intake manifold thus raising the idle speed. A fault in this system will not be detected by the OBD system.

Check

18 To check the system, start the engine and warm it up to operating temperature. Switch the air conditioning system On, the idle speed should raise slightly. If the idle speed drops, test the solenoid.
19 Disconnect the electrical connector from the solenoid. Connect a voltmeter to the two terminals of the harness connector. Turn the ignition key On and switch the air conditioning system On. Battery voltage should be indicated on the meter. If battery voltage is not present, check the circuit from the ignition switch to the solenoid (don't forget to check the fuses). If the air conditioning clutch is not energized when the system is switched On, check the air conditioning relay as well. Also check the ground circuits for continuity from the connector to ground (see the wiring diagrams).
20 Turn the ignition key OFF. Using a pair of fused jumper wires apply battery voltage and ground to the solenoid terminals. The solenoid should make a distinct clicking sound. If the solenoid does not click, replace the solenoid. Remove the solenoid (proceed to Step 5) and check the plunger for free movement. If the plunger is seized or the detent spring is broken replace the solenoid.

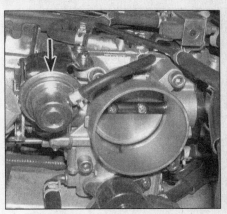

16.26 The throttle opener (arrow) is located on the throttle body

Replacement

21 Disconnect the electrical connector from the fast idle control solenoid.
22 Using a wrench remove the solenoid from the IAC valve housing.
23 Install a new copper washer onto the solenoid and tighten the solenoid securely.
24 The remainder of the installation is the reverse of removal.

Fast idle cam

25 A fast idle cam and thermo element is used on four-cylinder and 3.3L V6 models to increase idle speed during cold operating conditions. The fast idle cam is a mechanical system not controlled by the electronic engine control system. Refer to Chapter 4, Section 16 for information on the fast idle cam.

Throttle opener

Refer to illustration 16.26
26 Some V6 models may be equipped with a throttle opener located on the throttle body **(see illustration)**. The throttle opener opens the throttle slightly to aid in starting the engine. When the engine is started, vacuum is applied to the throttle opener, the plunger retracts and the engine returns to normal idle speed. If the engine idle speed is high and you suspect a problem with the throttle opener, disconnect the vacuum hose from the throttle opener (with the engine off) and connect a hand-held vacuum pump to the fitting on the throttle opener. Apply vacuum while watching the plunger; the plunger should extend against the throttle lever with no vacuum applied and retract with vacuum applied. If the throttle opener does not operate as described, replace the throttle opener.

17 Intake manifold runner control system

1 3.5L V6 models are equipped with an intake manifold runner control system. The system consists of two sub-systems; the power valve system and the swirl control valve system.

17.2 Intake manifold runner control system component locations

1. Power valve actuator
2. Power valve control solenoid
3. Swirl valve control solenoid

17.3 Check the power valve actuator rod (arrow) for movement as the engine is revved

17.9 The swirl control valve actuator is located at the front of the lower intake manifold

Power valve system

Refer to illustrations 17.2 and 17.3

2 The power valve system consists of the power valve (located inside the intake manifold), the power valve actuator, the power valve control solenoid and the PCM **(see illustration)**. The power valve diverts the path of the incoming air through the intake manifold, one path being longer than the other. At low engine speeds vacuum is applied to the power valve actuator, the power valve is closed and air is diverted through the longer path to enhance maximum torque at low speed. At a preset engine speed the PCM de-energizes the solenoid, vacuum is vented from the actuator and the power valve opens. Air is then drawn through the shorter path enhancing high speed power.

3 To check the system, place the transmission in Park or Neutral, apply the parking brake and block the rear wheels. Start the engine and rev the engine to 5,000 rpm several times while watching the power valve actuator. The actuator rod should move when engine speed exceeds 5,000 rpm **(see illustration)**. If the actuator is not responding, stop the engine and continue testing.

4 Check the vacuum hoses from the vacuum source to the actuator for leaks or damage. Check the vacuum tank to make sure it holds vacuum. Check the one-way valve - it should allow air to pass one way but not the other.

5 Disconnect the electrical connector from the power valve control solenoid. Turn the ignition key on and using a voltmeter, check for battery voltage at the red/yellow wire terminal of the harness connector. If battery voltage is not present, check the circuit from the connector to the fuse box. Check for continuity from the connector to the PCM on the yellow/green wire.

6 Remove the vacuum lines from the power valve control solenoid ports. Using a pair of fused jumper wires, connect battery voltage and ground to the two terminals of the power valve control solenoid. The solenoid should click and air should be allowed to pass between the two ports. Remove the jumpers - no air should pass between the two ports. If the solenoid does not operate as described, replace the power valve control solenoid.

7 Remove the vacuum hose from the power valve actuator. Connect a hand-held vacuum pump to the actuator and apply vacuum. The rod should move and the actuator should hold vacuum. If it doesn't, replace the actuator.

8 If all the above tests are good, have the PCM diagnosed by a dealer service department or other qualified repair facility.

Swirl control valve system

Refer to illustration 17.9

9 The swirl control valve system consists of a swirl control valve (located in each intake port of the lower intake manifold), the swirl control valve actuator, the swirl control valve control solenoid and the PCM **(see illustration)**. The swirl control valve increases the velocity of the incoming air through the intake manifold at low engine speeds, producing a swirl effect in the combustion chamber, improving fuel vaporization. At low engine speeds vacuum is applied to the swirl control valve actuator and the swirl control valves are closed. At a preset engine speed the PCM de-energizes the solenoid and the swirl control valves open, eliminating the air restriction through the intake manifold. On automatic transmission models, the swirl control valves open at 3,200 to 3,600 rpm; on manual transmission models, the swirl control valves open at 2,400 to 2,800 rpm. **Note:** *When the engine coolant temperature is below 50-degrees F (10-degrees C) or above 130-degrees F (55-degrees C), the swirl control valves are held open regardless of engine speed.*

10 Check the vacuum hoses from the vacuum source to the actuator for leaks or damage. Check the vacuum tank to make sure it holds vacuum. Check the one-way valve - it should allow air to pass one way but not the other.

11 Disconnect the electrical connector from the swirl control valve solenoid. Turn the ignition key on and using a voltmeter, check for battery voltage at the red/yellow wire terminal of the harness connector. If battery voltage is not present, check the circuit from the connector to the fuse box. Check for continuity from the connector to the PCM on the green wire.

12 Remove the vacuum lines from the swirl valve control solenoid ports. Using a pair of fused jumper wires, connect battery voltage and ground to the two terminals of the swirl valve control solenoid. The solenoid should click and air should be allowed to pass between the two ports. Remove the jumpers - no air should pass between the two ports. If the solenoid does not operate as described, replace the swirl control valve solenoid.

13 Remove the vacuum hose from the swirl valve actuator. Connect a hand-held vacuum pump to the actuator and apply vacuum. The rod should move and the actuator should hold vacuum. If it doesn't, replace the lower intake manifold.

14 If all the above tests are good, have the PCM diagnosed by a dealer service department or other qualified repair facility.

18 Variable valve timing control system

Refer to illustrations 18.1 and 18.6

1 3.5L V6 models are equipped with a variable valve timing control system. The system consists of the actuator pistons (incorporated into each intake camshaft sprocket), two intake valve timing control solenoids, two intake valve timing control position sensors and the PCM **(see illustration)**. The system varies the intake camshaft timing by directing engine oil pressure to the intake camshaft actuator/sprockets, advancing or retarding each intake camshaft timing during certain engine operating conditions. Advancing the intake camshaft timing during high-speed operation increases the mid-to-high range power output, while retarding the intake

Chapter 6 Emissions and engine control systems

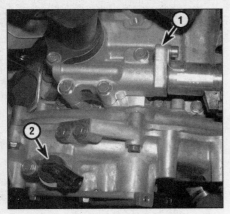

18.1 Variable valve timing control system component locations

1. *Intake valve timing control solenoid*
2. *Intake valve timing control position sensor*

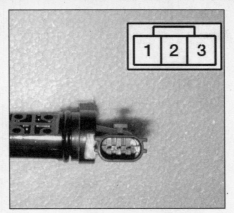

18.6 Intake valve timing control position sensor terminal identification

camshaft during low engine speeds increases engine torque. The intake valve timing control solenoid is controlled by the PCM. The PCM sends an On/Off signal (pulse width) to the intake valve timing solenoid, controlling the amount of oil and the direction of oil flow to the actuator/sprocket. The intake camshaft timing advances with an increase in pulse width and retards with a decrease in pulse width. When the pulse width is equal, the intake valve timing is fixed at the normal position.

2 Disconnect the electrical connector from the intake valve timing control solenoid. Turn the ignition key on and using a voltmeter, check for battery voltage at the red/yellow wire terminal of the harness connector. If battery voltage is not present, check the circuit from the connector to the fuse box. Check for continuity from the connector to the PCM on the orange/blue or pink/blue wire.

3 To check the intake valve timing control solenoid, use an ohmmeter to measure the resistance across the two terminals on the intake valve timing control solenoid. There should be approximately 5.5 to 9.2 ohms resistance at 68-degrees F (20-degrees C) across the terminals. If the resistance is incorrect, replace the intake valve timing control solenoid.

4 Using a pair of fused jumper wires, connect battery voltage and ground to the two terminals of the intake valve timing control solenoid. The solenoid should click as voltage is applied. It may be necessary to place a mechanics stethoscope or other suitable tool against the valve body to hear the valve click. If the solenoid does not operate as described, replace the intake valve timing control solenoid.

5 Disconnect the electrical connector from the intake valve timing control position sensor. Turn the ignition key on and using a voltmeter, check for battery voltage at the black/white wire terminal of the harness connector. If battery voltage is not present, check the ECCS relay and the circuit from the connector to the fuse box. Check for continuity to a good engine ground point on the black wire.

6 To check the intake valve timing control position sensor, remove the sensor from the timing cover. Using an ohmmeter, measure the resistance between terminals no. 3 (+) and 1 (-); no. 2 (+) and 1 (-); then 3 (+) and 2 (-) **(see illustration)**. The meter should indicate a resistance value, if zero ohms or infinity are indicated, replace the sensor. **Note:** *Correct polarity must be observed during meter connection for the test results to be accurate.*

7 If all the above tests are good, have the PCM diagnosed by a dealer service department or other qualified repair facility.

19 Positive Crankcase Ventilation (PCV) system

Refer to illustrations 19.1, 19.2a, 19.2b and 19.2c

1 The Positive Crankcase Ventilation (PCV) system reduces hydrocarbon emissions by scavenging crankcase vapors. It does this by circulating fresh air from the air filter through the crankcase, where it mixes with blow-by gases and is then rerouted through a PCV valve to the intake manifold **(see illustration)**.

2 The PCV system consists of a replaceable PCV valve and the crankcase ventilation hoses **(see illustrations)**.

3 To maintain idle quality, the PCV valve restricts the flow when the intake manifold vacuum is high. If abnormal operating conditions arise, the system is designed to allow excessive amounts of blow-by gases to flow

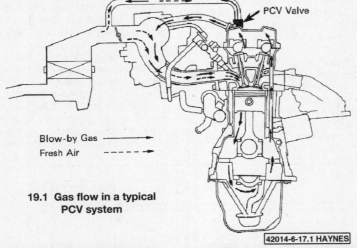

19.1 Gas flow in a typical PCV system

19.2a PCV valve location - four-cylinder models

19.2b PCV valve location - 3.3L V6 models

Chapter 6 Emissions and engine control systems

19.2c PCV valve location - 3.5L V6 models

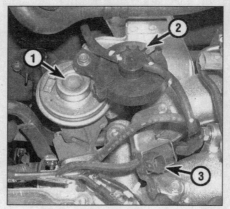

20.2a EGR system component locations - four-cylinder models

1 EGR valve
2 EGR backpressure transducer
3 EGR control solenoid

20.2b EGR system component locations - 3.3L V6 models

1 EGR valve
2 EGR backpressure transducer

back through the crankcase vent tube into the air filter to be consumed by normal combustion.
4 Checking and replacement of the PCV valve is covered in Chapter 1.

20 Exhaust Gas Recirculation (EGR) system

General description

Refer to illustrations 20.2a, 20.2b and 20.2c

1 Four-cylinder and 3.3L V6 models are equipped with an Exhaust Gas Recirculation (EGR) system. The Exhaust Gas Recirculation (EGR) system is used to lower NOx (oxides of nitrogen) emission levels caused by high combustion temperatures. The EGR valve recirculates a small amount of exhaust gases into the intake manifold. The additional mixture lowers the temperature of combustion thereby reducing the formation of NOx compounds.
2 The EGR system is equipped with an EGR valve, an EGR control solenoid valve which receives ported and manifold vacuum, an EGR temperature sensor and a EGR control backpressure transducer valve (see illustrations). The operation of the system is controlled by the PCM which operates the EGR control solenoid. The manifold vacuum system utilizes a vacuum tap in the air intake system positioned after the throttle valve. The ported vacuum control system uses a vacuum tap in the throttle body which is exposed to an increasing percentage of manifold vacuum as the throttle valve is opened during acceleration.
3 The backpressure transducer valve monitors the exhaust backpressure as the engine rpm increases or decreases to aid in controlling the amount of the EGR vacuum signal. The EGR temperature sensor is used to inform the PCM of temperature changes in the EGR passage way. This helps the PCM determine the EGR On/Off time.

System check

Refer to illustration 20.5

4 Check all hoses for cracks, kinks, broken sections and proper connection. Inspect all system connections for damage, cracks and leaks.
5 To check the EGR system operation, bring the engine up to operating temperature and, with the transmission in Neutral (parking brake set and tires blocked to prevent movement), allow it to idle. Open the throttle so the engine speed is between 2,000 and 4,000 rpm and then allow it to close. The EGR valve stem should move if the control system is working properly **(see illustration)**. The test should be repeated several times. Movement of the stem indicates the control system is functioning correctly. If the EGR valve stem does not move, Check the vacuum signal to the EGR valve. Remove the hose from the valve, place your finger over the end of the

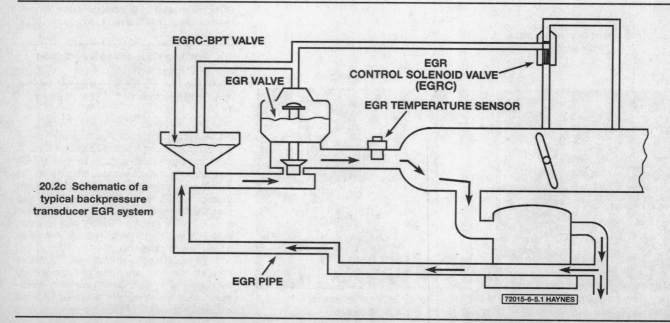

20.2c Schematic of a typical backpressure transducer EGR system

Chapter 6 Emissions and engine control systems

20.5 The EGR valve can get very hot; be careful not to burn your hand when checking for EGR valve diaphragm movement - the diaphragm and rod beneath the top of the valve (arrow) should move when the engine is revved

20.11 EGR control solenoid harness connector terminal identification (3.3L V6 shown)

1 Solenoid control
2 B+

20.15a The EGR temperature sensor (arrow) is located in the EGR passage to the intake manifold (3.3L V6 shown)

20.15b EGR temperature sensor location (arrow) - four-cylinder models

20.18 Loosen the EGR pipe flange nut (not visible), disconnect the EGR pipe and the backpressure transducer line (A) from the EGR valve and remove the two mounting fasteners (B) (one of two visible)

hose and perform the procedure again. If no vacuum is present at the end of the hose check the hose connections to make sure they are not leaking or clogged, then check the EGR control solenoid (see Step 11).

Component checks

EGR Valve

6 With the engine Off, disconnect the vacuum hose and apply ten inches of vacuum with a hand-held vacuum pump to the EGR valve. If the valve opens, measure the valve travel to make sure it is approximately 1/8-inch. If the stem does not move, replace the EGR valve with a new one.

7 Next apply vacuum with the pump and then clamp the hose shut. The valve should stay open for 30 seconds or longer. If it does not, the diaphragm is leaking and the valve should be replaced with a new one.

8 Start the engine and apply vacuum to the valve. The engine should idle roughly when the valve is open. If it doesn't, the passages in the manifold are probably clogged. If there's no difference in idle quality with the valve open or closed the EGR valve is probably not closing all the way. Remove the EGR valve and inspect the poppet and seat area for deposits.

9 If the deposits are more than a thin film of carbon, the valve should be cleaned. To clean the valve, apply solvent and allow it to penetrate and soften the deposits, making sure that none gets on the valve diaphragm, as it could be damaged.

10 Use a vacuum pump to hold the valve open and carefully scrape the deposits from the seat and poppet area with a tool. Inspect the poppet and stem for wear and replace the valve with a new one if wear is found.

EGR control solenoid

Refer to illustration 20.11

11 First check for a vacuum signal to the solenoid with the engine running. If no vacuum is present check the hose to and from the solenoid for cracks and clogging. If the hose is OK, check the throttle body port for clogging. If vacuum is present at the solenoid, disconnect the solenoid valve harness connector and check for battery voltage at the B+ terminal with the ignition key ON (engine not running) **(see illustration)**. Battery voltage should be present. If voltage is not present at the solenoid valve refer to the wiring diagrams and check the wiring harness for open circuits or a damaged harness (don't forget to check the fuses first).

12 To check the operation of this EGR control solenoid, remove the solenoid valve and connect fused battery power and ground to the solenoid terminals. The solenoid should click as power is applied.

13 The solenoid operates identical to the MAP sensor solenoid valve **(see illustration 6.8)**. With battery voltage applied, air should pass between ports A and B. With no voltage applied, air should pass between ports A and C. If the solenoid does not operate as described, replace the solenoid.

EGR control backpressure transducer valve

14 Locate the EGR control backpressure transducer valve **(see illustration 20.2a or 20.2b)**. Remove the vacuum hoses from the valve and plug one of the ports with a finger. Use a hand-held vacuum pump and apply vacuum to the valve. The valve should leak. **Note:** *It should stop leaking when pressure greater than 0.145 psi (1.0 kPa) is applied to the port under the valve (the valve must be removed from the engine to apply pressure at this point)*. If the valve is defective replace it.

EGR temperature sensor

Refer to illustrations 20.15a and 20.15b

15 Remove the EGR temperature sensor **(see illustrations)**. Submerge the tip of the sensor in a container of water. Heat the water on the stove while monitoring the resistance of the sensor. Resistance should decrease as temperature increases and at 212-degrees F (100-degrees C) the sensor resistance should measure 80 to 100 K-ohms.

Component replacement

EGR valve

Refer to illustration 20.18

16 On four-cylinder models, remove the air filter housing (see Chapter 4).

17 Detach the vacuum line from the EGR valve.

18 Disconnect the EGR pipes from the EGR valve and remove the EGR valve mounting fasteners **(see illustration)**.

Chapter 6 Emissions and engine control systems

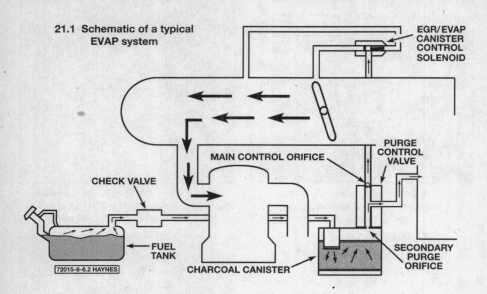

21.1 Schematic of a typical EVAP system

19 Remove the EGR valve and gasket from the manifold. Discard the gasket.
20 With a wire wheel, buff the exhaust deposits from the EGR valve mounting surface on the manifold and, if you plan to use the same valve, the mounting surface of the valve itself. Look for exhaust deposits in the valve outlet. Remove deposit build-up with a screwdriver. **Caution:** *Never wash the valve in solvents or degreaser - both agents will permanently damage the diaphragm. Sandblasting is also not recommended because it will affect the operation of the valve.*
21 If the EGR passage contains an excessive build-up of deposits, clean it out with a wire brush. Make sure that all loose particles are completely removed to prevent them from clogging the EGR valve or from being ingested into the engine.
22 If there are large amounts of deposits within the EGR valve, remove the EGR pipe from the exhaust manifold and clean out the deposits inside the tube. **Note:** *Remove the EGR control backpressure transducer valve and pipe and clean any deposits from the passages.*
23 Installation is the reverse of removal.

EGR control solenoid
24 Unplug the electrical connector from the solenoid.
25 Clearly label and detach the vacuum hoses.
26 Remove the solenoid mounting nut and remove the solenoid.
27 Installation is the reverse of removal.

EGR temperature sensor
28 Disconnect the electrical connector and unscrew the sensor from the manifold **(see illustration 20.15a or 20.15b)**.
29 Apply anti-seize to the threads of the sensor before installing it.
30 Installation is the reverse of removal.

21 Evaporative Emissions Control (EVAP) system

General description
Refer to illustrations 21.1, 21.3a, 21.3b, 21.3c and 21.3d

1 The Evaporative Emissions Control (EVAP) system absorbs fuel vapors and, during engine operation, releases them into the engine intake where they mix with the incoming air-fuel mixture. The EVAP system consists of a charcoal-filled canister and the lines connecting the canister to the fuel tank, ported vacuum and intake manifold vacuum **(see illustration)**. The fuel filler cap is also a component of the EVAP system. If the fuel filler cap is loose, missing or damaged, a diagnostic trouble code may be set.
2 When the engine is not operating, fuel vapors are transferred from the fuel tank, throttle body and intake manifold to the charcoal canister where they are stored. When the engine is running, the fuel vapors are purged from the canister by the purge control valve. The gasses are consumed in the normal combustion process. The electronic purge control valve is controlled by the PCM.
3 1996 models differ from other models. The 1996 system is less complex than later models. 1997 and later models are equipped with several components such as the canister vent valve, canister purge valve and an EVAP control system pressure sensor to provide on-board diagnostic capabilities **(see illustrations)**. The on-board diagnostic system monitors the EVAP system for proper operation and is able to detect a leak in the EVAP system.

Check
4 Poor idle, stalling and poor driveability can be caused by a defective canister purge control valve, a damaged canister, split or cracked hoses or hoses connected to the wrong tubes.
5 Evidence of fuel loss or fuel odor can be caused by the following items: fuel leaking

21.3a Typical EVAP purge valve location (right-hand side of engine compartment)

21.3b Typical EVAP vacuum cut valve location (under rear of vehicle)

Chapter 6 Emissions and engine control systems

6-27

21.3c Typical EVAP pressure sensor location (Frontier model shown, left side frame rail near shock absorber)

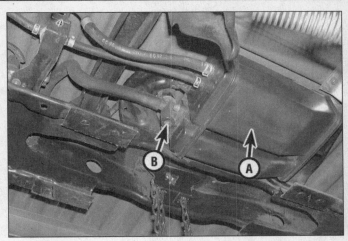

21.3d Typical EVAP canister (A) and vent valve (B) location (Frontier model shown, spare tire removed for clarity)

from fuel lines or a cracked or damaged canister, an inoperative fuel tank check valve, an inoperative purge valve or vent valve, disconnected, misrouted, kinked, deteriorated or damaged vapor or control hoses or an improperly seated air filter or air filter gasket. Inspect each hose attached to the canister for kinks, leaks and breaks along its entire length. Repair or replace as necessary.

6 Inspect the canister. If it is cracked or damaged, replace it. Look for fuel leaking from the bottom of the canister. If fuel is leaking, replace the canister and check the hoses and hose routing.

7 The purge valve and vent valve operation are similar to the EGR control solenoid and may be tested in the same manner (see Section 20). The purge valve is a normally closed valve; no air flows through the ports until the solenoid is energized. The vent valve is a normally open valve; with the solenoid energized, no air flows through the ports.

8 The EVAP system pressure sensor is similar in operation to the MAP sensor and may be tested in the same manner (see Section 6), although the EVAP pressure sensor is much more sensitive. With no vacuum applied to the sensor, the signal voltage is approximately 3.5 volts. With 3.0 in-Hg applied to the sensor, the signal voltage should drop to approximately 0.5 volt. **Caution:** *Do not apply more than 6.0 in-Hg to the sensor or damage may result.*

Canister replacement

1996 Pathfinder models

Note: *On 1996 Pathfinder models, the EVAP canister is located in the engine compartment.*

9 Clearly label, then detach, all vacuum lines from the canister.
10 Remove the canister mounting bolts and remove the canister from the mounting bracket.
11 Installation is the reverse of removal.

All except 1996 Pathfinder models

Refer to illustration 21.14

Note: *On 1997 and later Pathfinder models, the EVAP canister is located on the left frame rail, behind the fuel tank. On Frontier and Xterra models, the EVAP canister is located on a crossmember behind the rear axle assembly.*

12 Raise the vehicle and support it securely on jackstands. On Frontier and Xterra models, remove the spare tire.
13 Clearly label, then detach, all vacuum lines from the canister.
14 Remove the canister mounting bolts and remove the canister **(see illustration)**.
15 Installation is the reverse of removal.

22 Catalytic converter

Note: *Because of a Federally mandated warranty which covers emissions-related components such as the catalytic converter, check with a dealer service department before replacing the converter at your own expense.*

General description

1 The catalytic converter is an emission control device added to the exhaust system to reduce pollutants from the exhaust gas stream. A monolithic three-way (reduction) catalyst design is used. The catalytic coating on the monolith contains palladium, platinum and rhodium, which lowers the levels of oxides of nitrogen (NOx) as well as hydrocarbons (HC) and carbon monoxide (CO).

Check

2 The test equipment for a catalytic converter is expensive and highly sophisticated. If you suspect that the converter on your vehicle is malfunctioning, take it to a dealer or authorized emissions inspection facility for diagnosis and repair.

3 Whenever the vehicle is raised for servicing of underbody components, check the converter for leaks, corrosion, dents and other damage. Check the welds/flange bolts that attach the front and rear ends of the converter to the exhaust system. If damage is discovered, the converter should be replaced.

4 Although catalytic converters don't break too often, they can become plugged. The easiest way to check for a restricted converter is to use a vacuum gauge to diagnose the effect of a blocked exhaust on intake vacuum.

a) Connect a vacuum gauge to an intake manifold vacuum source.
b) Warm the engine to operating temperature, place the transmission in park and apply the parking brake.
c) Note and record the vacuum reading at idle.
d) Open the throttle until the engine speed is about 2000 rpm.
e) Release the throttle quickly and record the vacuum reading.
f) Perform the test three more times, recording the reading after each test.
g) If the reading after the fourth test is more than one in-Hg lower than the reading recorded at idle, the catalytic converter, muffler or exhaust pipes may be plugged or restricted.

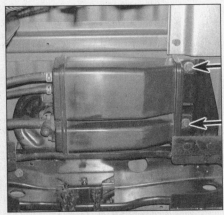

21.14 Typical EVAP canister mounting bolt locations (arrows)

Catalytic converter replacement

Refer to illustration 22.6

5 Remove the flange bolts and disconnect the exhaust pipe from the catalytic converter. If necessary, remove the necessary exhaust system hangers and move the exhaust pipe rearward. Be sure to support the exhaust system.

6 Remove the flange bolts and separate the catalytic converter from the exhaust pipe or exhaust manifold **(see illustration)**.

7 Installation is the reverse of removal.

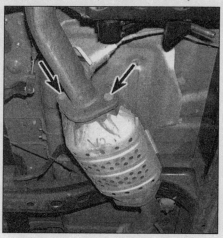

22.6 Remove the flange bolts (arrows) and separate the catalytic converter from the exhaust pipe or exhaust manifold

Chapter 7 Part A
Manual transmission

Contents

	Section
Back-up light switch - check and replacement	3
Extension housing oil seal	See Chapter 7B
General information	1
Manual transmission - removal and installation	6
Manual transmission lubricant change	See Chapter 1
Manual transmission lubricant level check	See Chapter 1

	Section
Manual transmission overhaul - general information	7
Neutral position switch - check and replacement	4
Oil seals - replacement	5
Shift lever - removal and installation	2
Transmission mount - replacement	See Chapter 7B
Vehicle speed sensor	See Chapter 6

Specifications

General
Transmission lubricant type... See Chapter 1

Torque specifications

	Ft-lbs (unless otherwise indicated)	Nm
Transmission-to-engine bolts		
Models with a four-cylinder engine (see illustration 6.26a)		
Bolt A	29 to 36	39 to 49
Bolt B	29 to 36	39 to 49
Bolt C	144 to 192 in-lbs	16 to 22
Bolt D	144 to 192 in-lbs	16 to 22
Models with a 3.3L V6 engine (see illustration 6.26b)		
Bolt A	29 to 36	39 to 49
Bolt B	29 to 36	39 to 49
Bolt C	144 to 192 in-lbs	16 to 22
Gusset-to-engine block bolts	22 to 28	29 to 39
Models with a 3.5L V6 engine (see illustration 6.26c)		
Bolt A	52 to 58	70 to 79
Bolt B	52 to 58	70 to 79
Bolt C	22 to 28	29 to 39
Input shaft bearing retainer bolts	16 to 21	144 to 180 in-lbs

1 General information

The vehicles covered by this manual are equipped with a five-speed manual transmission or a four-speed automatic transmission. Information on the manual transmission is included in this Part of Chapter 7. Information on the automatic transmission can be found in Part B of this Chapter. Information on the transfer case used on 4WD models can be found in Part C of this Chapter.

All four-cylinder Frontier and Xterra models equipped with a manual transmission use an FS5W71C unit; V6 models use an FS5R30A unit. All Pathfinder models with a manual transmission use the FS5R30A unit. Both units are fully synchronized five-speeds with internal shift mechanisms.

Depending on the cost of having a transmission overhauled, it might be a better idea to replace it with a used or rebuilt unit. Your local dealer or transmission shop (and even some auto parts stores) should be able to supply information concerning cost, availability and exchange policy. Regardless of how you decide to remedy a transmission problem, you can still save a lot of money by removing and installing the unit yourself.

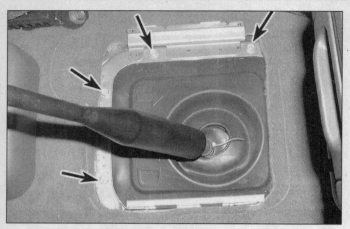

2.2 Remove the bolts (not all are visible here) and pull up the floor pan boot at the bottom of the shifter . . .

2.3 . . . then loosen the clamp, pull the shifter boot up and remove the snap-ring

2 Shift lever - removal and installation

Refer to illustrations 2.2 and 2.3

1 Unscrew the shift knob, then remove the shift lever boot and trim (see Chapter 11).
2 Unscrew the bolts securing the floor pan boot retainer to the floor pan. Loosen the clamps and pull up the boot at the base of the shift lever **(see illustration)**.
3 Remove the snap-ring and withdraw the shift lever from the transmission **(see illustration)**. Note the order of the washers and rubber seat under the shift lever (the proper order, from the top, is: washer, wave washer, washer, rubber seat). **Note:** *Discard the snap-ring; a new one should be used when installing the lever.*
4 Before installing the shift lever, lubricate the rubber seat, the ball on the shift lever and the end of the lever with multi-purpose grease.
5 Installation is the reverse of removal. Be sure to install a new snap-ring, and make sure it seats fully.

3 Back-up light switch - check and replacement

1 The back-up light switch is located on the right side of the transmission case; it's the switch closer to the front of the transmission (the other one is the neutral position switch).

Check

2 Turn the ignition key to the On position and move the shift lever to the Reverse position; the back-up lights should go on.
3 If the lights don't go on, check the back-up light fuse first (see Chapter 12). If the fuse is blown, trace the back-up light circuit for a short-circuit condition.
4 If the fuse is okay, raise the vehicle and support it securely on jackstands. Place the shifter in Reverse.
5 Working under the vehicle, unplug the back-up light switch electrical connector. Using an ohmmeter, check for continuity across the terminals of the switch. Continuity should exist. If not, replace the switch. **Note:** *If the switch has continuity, make sure it doesn't have continuity in any position other than Reverse.*
6 If the switch has continuity, check for voltage at the electrical connector; one of the two terminals should have battery voltage present with the ignition key in the On position. If no voltage is present, trace the circuit between the fuse block and the electrical connector for an open circuit condition.
7 If voltage is present, trace the back-up light circuit between the electrical connector and the back-up light bulbs for an open circuit condition. **Note:** *Although not very likely, the back-up light bulbs could both be burned out; don't rule out this possibility.*

Replacement

8 Raise the vehicle and support it securely on jackstands, if not already done.
9 Unplug the back-up light switch electrical connector.
10 Unscrew the back-up light switch from the transmission case.
11 Apply RTV sealant or Teflon tape to the threads of the new switch to prevent leakage. Install the switch in the transmission case and tighten it securely. Plug in the electrical connector. **Note:** *The replacement switch may come equipped with thread sealant already applied to the threads. Do not apply RTV sealant or Teflon tape to the threads of the new switch if thread sealant has already been applied by the manufacturer.*
12 Lower the vehicle and check the operation of the back-up lights.

4 Neutral position switch - check and replacement

Refer to illustration 4.1

1 The Neutral position switch is located on the right side of the transmission case **(see illustration)**; it's the switch closer to the rear of the transmission (the other one is the back-up light switch).

Check

2 Place the shift lever in the Neutral position. Raise the vehicle and support it securely on jackstands. Unplug the electrical connector from the Neutral position switch and, using an ohmmeter, check for continuity across the terminals of the switch. Continuity should exist. If not, replace the switch. **Note:** *If the switch has continuity, make sure it doesn't have continuity in any position other than Neutral.*

Replacement

3 Raise the vehicle and support it securely on jackstands, if not already done.
4 Unplug the Neutral position switch electrical connector.
5 Unscrew the switch from the transmission case.
6 Apply RTV sealant or Teflon tape to the threads of the new switch to prevent leakage. Install the switch in the transmission case and tighten it securely. Plug in the electrical

4.1 Location of the Neutral position switch - FS5W71C transmission

Chapter 7 Part A Manual transmission

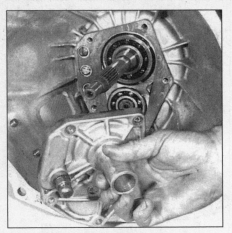

5.3 Unscrew the bolts and detach the input shaft bearing retainer from the transmission

5.4 Carefully pry the seal out of the input shaft bearing retainer

connector. **Note:** *The replacement switch may come equipped with thread sealant already applied to the threads. Do not apply RTV sealant or Teflon tape to the threads of the new switch if thread sealant has already been applied by the manufacturer.*

5 Oil seals - replacement

Front oil seal

Refer to illustrations 5.3 and 5.4

1 Remove the transmission (see Section 6).
2 Remove the release bearing (see Chapter 8).
3 Remove the bolts that retain the transmission input shaft bearing retainer to the case and lift off the retainer **(see illustration)**.
4 Being careful not to nick or damage the retainer, pry out the seal **(see illustration)**.
5 Drive the new seal into position in the retainer using a seal driver or an appropriate sized socket.
6 Apply a light coat of gear oil to the seal lips and the input shaft, then reinstall the bearing retainer.
7 The remainder of installation is the reverse of removal. Tighten the input shaft bearing retainer bolts to the torque listed in this Chapter's Specifications.

Extension housing oil seal

8 See Chapter 7B for the extension housing oil seal procedure.

6 Manual transmission - removal and installation

Removal

1 Disconnect the cable from the negative terminal of the battery.
2 Remove the Crankshaft Position sensor (see Chapter 6).
3 Remove the shift lever (see Section 2). If you're working on a 4WD model, also remove the transfer case shift lever (see Chapter 7C). **Note:** *The ATX14A transfer case is not equipped with a shift lever.*
4 Raise the vehicle sufficiently to provide clearance to easily remove the transmission. Support the vehicle securely on jackstands.
5 Remove the skid plate, if equipped.
6 Disconnect the electrical connector from the back-up light switch, the neutral position switch and the vehicle speed sensor. Disengage the wiring harness from the clips on the transmission. If you're working on a 4WD model, unplug all of the electrical connectors from the transfer case.
7 Drain the transmission lubricant (see Chapter 1). If you're working on a 4WD model, also drain the transfer case lubricant.
8 Remove the driveshaft(s) (see Chapter 8).
9 Remove exhaust system components as necessary for clearance (see Chapter 4).
10 Remove the starter motor (see Chapter 5).
11 Unbolt the clutch release cylinder from the transmission (see Chapter 8). Tie the cylinder out of the way with a piece of wire. **Caution:** *Don't depress the clutch pedal while the release cylinder is removed.*
12 If you're working on a Frontier or Xterra model with 4WD, remove the torsion bars (see Chapter 10).
13 Support the engine from above with an engine hoist or an engine support fixture, or place a jack (with a block of wood as an insulator) under the engine oil pan. The engine must remain supported at all times while the transmission is out of the vehicle.
14 Support the transmission with a jack - preferably a special jack made for this purpose. **Note:** *These jacks can be obtained at most equipment rental yards.* Safety chains will help steady the transmission on the jack.
15 Raise the engine slightly and disconnect the transmission mount from the extension housing and the center crossmember (see Chapter 7B).
16 Raise the transmission slightly and remove the bolts and nuts attaching the crossmember to the frame rails.
17 Lower the jacks supporting the transmission and engine assembly enough to gain access to all of the mounting bolts.
18 Remove the bolts attaching the transmission to the engine, and also unbolt the engine-to-transmission gussets from the engine block. A long extension with a U-joint socket or adapter will be helpful in unscrewing the upper bolts. **Note:** *The bolts securing the transmission to the engine are different lengths. When removing the bolts, mark their positions or lay them out in order so they can be reinstalled in their original locations.*
19 Make a final check for any wiring or hoses connected to the transmission, then move the transmission and jack toward the rear of the vehicle until the transmission input shaft clears the splined hub in the clutch disc. Keep the transmission level as this is done.
20 Once the input shaft is clear, lower the transmission and remove it from under the vehicle.
21 While the transmission is removed, be sure to remove and inspect all clutch components (see Chapter 8). In most cases, new clutch components should be routinely installed if the transmission is removed. Also inspect the front oil seal and replace it if necessary (see Section 5).

Installation

Refer to illustrations 6.26a, 6.26b and 6.26c

22 Insert a small amount of multi-purpose grease into the pilot bushing in the crankshaft and lubricate the inner surface of the bushing. Also apply a light film of grease on the input shaft splines, the area on the front cover where the release bearing rides, and the release lever/bearing contact points (see Chapter 8).
23 Install the clutch components if removed (see Chapter 8).
24 If you're working on a 4WD model, clean the bellhousing-to-engine block mating surfaces, then apply a 1/8-inch bead of RTV sealant to the bellhousing mating surface. **Caution:** *Don't apply sealant within 1-3/4 inches (45 mm) of the grommet at the bottom of the bellhousing.* Also apply a bead of sealant to the area where the engine rear plate meets the engine block, outboard of the rubber seals.
25 With the transmission secured to the jack as on removal, raise the transmission into position behind the engine and then carefully slide it forward, engaging the input shaft with the clutch plate hub. Do not use excessive force to install the transmission - if the input shaft does not slide into place, readjust the angle of the transmission so it is level and/or turn the input shaft so the splines engage properly with the clutch.
26 Install the transmission-to-engine bolts in their proper locations and tighten them to the torque listed in this Chapter's Specifica-

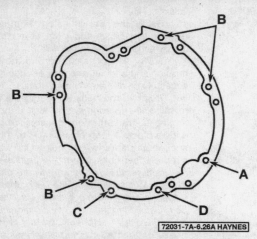

6.26a Transmission mounting bolt location guide - four-cylinder models

A 2-9/16 inches (65 mm)   C 1-inch (25 mm)
B 2-9/32 inches (58 mm)   D 5/8-inch (16 mm)

6.26b Transmission mounting bolt location guide - models with a 3.3L V6 engine

A 2-9/16 inches (65 mm)   C 1-inch (25 mm)
B 2-9/32 inches (58 mm)

tions (see illustrations). Caution: *Don't use the bolts to draw the transmission to the engine. If the transmission doesn't slide forward easily and mate with the engine block, find out why before proceeding.*

27 Raise the transmission into place, install the crossmember and attach it to the frame rails. Install the transmission mount between the extension housing and the crossmember. Carefully lower the transmission extension housing onto the mount and the crossmember. When everything is properly aligned, tighten all nuts and bolts securely.
28 Remove the jacks supporting the transmission and the engine.
29 If you're working on a 4WD Frontier or Xterra, install the torsion bars (see Chapter 10).
30 On 4WD models, install the transfer case and shift linkage (if equipped) (see Chapter 7C).
31 Install the various items removed previously, referring to Chapter 8 for the installation of the driveshaft(s) and clutch release cylinder, Chapter 5 for the starter motor, Chapter 6 for the Crankshaft Position sensor, and Chapter 4 for the exhaust system components.
32 Plug in the electrical connector for the Vehicle Speed Sensor, the back-up light switch and the Neutral position switch. Connect any other wiring attached to the transmission or the transfer case.
33 Remove the jackstands and lower the vehicle.
34 Install the shift lever(s) (see Section 2).
35 Fill the transmission with the specified lubricant to the proper level (see Chapter 1). If you're working on a 4WD model, also fill the transfer case.
36 Connect the cable to the negative terminal of the battery.
37 Road test the vehicle for proper operation and check for leakage.

7 Manual transmission overhaul - general information

Overhauling a manual transmission is a difficult job for the do-it-yourselfer. It involves the disassembly and reassembly of many small parts. Numerous clearances must be precisely measured and, if necessary, changed with select fit spacers and snap-rings. As a result, if transmission problems arise, it can be removed and installed by a competent do-it-yourselfer, but overhaul should be left to a transmission repair shop. Rebuilt transmissions may be available - check with your dealer parts department and auto parts stores. At any rate, the time and money involved in an overhaul is almost sure to exceed the cost of a rebuilt unit.

Nevertheless, it's not impossible for an inexperienced mechanic to rebuild a transmission if the special tools are available and the job is done in a deliberate step-by-step manner so nothing is overlooked.

The tools necessary for an overhaul include internal and external snap-ring pliers, a bearing puller, a slide hammer, a set of pin punches, feeler gauges, a dial indicator and possibly a hydraulic press. In addition, a large, sturdy workbench and a vise or transmission stand will be required.

During disassembly of the transmission, make careful notes of how each piece comes off, where it fits in relation to other pieces and what holds it in place. If you note how each part is installed before removing it, getting the transmission back together again will be much easier.

Before taking the transmission apart for repair, it will help if you have some idea what area of the transmission is malfunctioning. Certain problems can be closely tied to specific areas in the transmission, which can make component examination and replacement easier. Refer to the *Troubleshooting* Section at the front of this manual for information regarding possible sources of trouble.

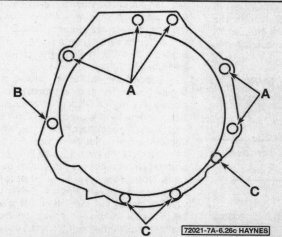

6.26c Transmission mounting bolt location guide - models with a 3.5L V6 engine

A 2-9/16 inches (65 mm)
B 2-11/64 inches (55 mm)
C 1-9/16 inch (40 mm)

Chapter 7 Part B
Automatic transmission

Contents

	Section		Section
Automatic transmission - removal and installation	10	Park/Neutral Position (PNP) switch - check, adjustment and replacement	7
Automatic transmission fluid and filter change	See Chapter 1	Shift cable - check, adjustment and replacement	3
Automatic transmission fluid level check	See Chapter 1	Shift interlock system - description, check and component replacement	5
Diagnosis and trouble codes	2		
Driveplate - removal and installation	See Chapter 2	Shift lever assembly - removal and installation	4
Engine mounts - check and replacement	See Chapter 2	Throttle Valve (TV) cable - removal, installation and adjustment	6
Extension housing oil seal (2WD models) - replacement	8	Transmission Control Module (TCM) - removal and installation	11
General information	1	Transmission mount - check and replacement	9

Specifications

General

Transmission fluid type	See Chapter 1
Throttle Valve cable stroke (four-cylinder Frontier models only)	1.54 to 1.69 inches (39 to 43 mm)

Torque specifications

	Ft-lbs (unless otherwise indicated)	Nm
Transmission fluid pan bolts	See Chapter 1	
Torque converter-to-driveplate bolts	33 to 43	44 to 59
Transmission-to-engine bolts		
With four-cylinder engine **(see illustration 10.25a)**		
Bolt A	29 to 36	39 to 49
Bolt B	26 to 35 in-lbs	3 to 4
Bolt C	144 to 192 in-lbs	16 to 22
With 3.3L V6 engine **(see illustration 10.25b)**		
Bolt A	29 to 36	39 to 49
Bolt B	29 to 36	39 to 49
Bolt C	22 to 29	29 to 39
Gusset-to-engine bolts	22 to 29	29 to 39
With 3.5L V6 engine **(see illustration 10.25c)**		
Bolt A	52 to 59	70 to 80
Bolt B	52 to 59	70 to 80
Bolt C	22 to 29	29 to 39

1 General information

All vehicles covered in this manual come equipped with a five-speed manual transmission or a four-speed automatic transmission. Information on the manual transmission is in Part A of this Chapter. Information on the automatic transmission is included in this Part of Chapter 7. You'll also find certain procedures common to both automatic and manual transmissions - such as oil seal replacement and transmission mount replacement - in this Part of Chapter 7.

Four-cylinder models equipped with an automatic transmission use an RL4R01A automatic transmission; V6 models use an RE4R01A unit. They are both four-speed transmissions - the main difference lies in the shift control system. The RL4R01A transmission uses a Throttle Valve (TV) cable to control shift points, while the RE4R01A transmission uses sensors from the engine control system, the Transmission Control Module (TCM) and solenoids in the transmission valve body to regulate shift points.

These transmissions are equipped with a torque converter clutch (TCC) that provides a direct connection between the engine and the drive wheels for improved efficiency and economy. The TCC consists of a solenoid controlled by the Powertrain Control Module (PCM) that locks the converter when the vehicle is cruising on level ground and the engine is fully warmed up.

Due to the complexity of the automatic transmissions covered in this manual and the need for specialized equipment to perform most service operations, this Chapter contains only general diagnosis, routine maintenance, adjustment and removal and installation procedures.

If the transmission requires major repair work, it should be left to a dealer service department or an automotive or transmission repair shop. You can, however, remove and install the transmission yourself and save the expense, even if the repair work is done by a transmission shop.

2 Diagnosis and trouble codes

Note: *Automatic transmission malfunctions may be caused by five general conditions: poor engine performance, improper adjustments, hydraulic malfunctions, mechanical malfunctions or malfunctions in the computer or its signal network. Diagnosis of these problems should always begin with a check of the easily repaired items: fluid level and condition (see Chapter 1), shift linkage adjustment and throttle linkage adjustment. Next, perform a road test to determine if the problem has been corrected or if more diagnosis is necessary. If the problem persists after the preliminary tests and corrections are completed, additional diagnosis should be done by a dealer service department or transmission repair shop. Refer to the Troubleshooting section at the front of this manual for information on symptoms of transmission problems.*

Preliminary checks

1 Drive the vehicle to warm the transmission to normal operating temperature.
2 Check the fluid level as described in Chapter 1:
 a) *If the fluid level is unusually low, add enough fluid to bring the level within the designated area of the dipstick, then check for external leaks (see below).*
 b) *If the fluid level is abnormally high, drain off the excess, then check the drained fluid for contamination by coolant. The presence of engine coolant in the automatic transmission fluid indicates that a failure has occurred in the internal radiator walls that separate the coolant from the transmission fluid (see Chapter 3).*
 c) *If the fluid is foaming, drain it and refill the transmission, then check for coolant in the fluid, or a high fluid level.*
3 Make sure the ignition timing and engine idle speed is correct (see Chapter 4) (all except Pathfinders with the 3.5L V6 engine).
4 Inspect the shift cable (see Section 3). Make sure that it's properly adjusted and operates smoothly. Also, if you're working on a four-cylinder Frontier, make sure the Throttle Valve cable is properly adjusted (see Section 6).

Fluid leak diagnosis

5 Most fluid leaks are easy to locate visually. Repair usually consists of replacing a seal or gasket. If a leak is difficult to find, the following procedure may help.
6 Identify the fluid. Make sure it's transmission fluid and not engine oil or brake fluid (automatic transmission fluid is a deep red color).
7 Try to pinpoint the source of the leak. Drive the vehicle several miles, then park it over a large sheet of cardboard. After a minute or two, you should be able to locate the leak by determining the source of the fluid dripping onto the cardboard.
8 Make a careful visual inspection of the suspected component and the area immediately around it. Pay particular attention to gasket mating surfaces. A mirror is often helpful for finding leaks in areas that are hard to see.
9 If the leak still cannot be found, clean the suspected area thoroughly with a degreaser or solvent, then dry it.
10 Drive the vehicle for several miles at normal operating temperature and varying speeds. After driving the vehicle, visually inspect the suspected component again.
11 Once the leak has been located, the cause must be determined before it can be properly repaired. If a gasket is replaced but the sealing flange is bent, the new gasket will not stop the leak. The bent flange must be straightened.
12 Before attempting to repair a leak, check to make sure that the following conditions are corrected or they may cause another leak. **Note:** *Some of the following conditions cannot be fixed without highly specialized tools and expertise. Such problems must be referred to a transmission shop or a dealer service department.*

Gasket leaks

13 Check the pan periodically. Make sure the bolts are tight, no bolts are missing, the gasket is in good condition and the pan is flat (dents in the pan may indicate damage to the valve body inside).
14 If the pan gasket is leaking, the fluid level or the fluid pressure may be too high, the vent may be plugged, the pan bolts may be too tight, the pan sealing flange may be warped, the sealing surface of the transmission housing may be damaged, the gasket may be damaged or the transmission casting may be cracked or porous. If sealant instead of gasket material has been used to form a seal between the pan and the transmission housing, it may be the wrong sealant.

Seal leaks

15 If a transmission seal is leaking, the fluid level or pressure may be too high, the vent may be plugged, the seal bore may be damaged, the seal itself may be damaged or improperly installed, the surface of the shaft protruding through the seal may be damaged or a loose bearing may be causing excessive shaft movement.
16 Make sure the dipstick tube seal is in good condition and the tube is properly seated. Periodically check the area around the vehicle speed sensor for leakage. If transmission fluid is evident, check the O-ring for damage.

Case leaks

17 If the case itself appears to be leaking, the casting is porous and will have to be repaired or replaced.
18 Make sure the oil cooler hose fittings are tight and in good condition.

Fluid comes out vent pipe or fill tube

19 If this condition occurs, the transmission is overfilled, there is coolant in the fluid, the case is porous, the dipstick is incorrect, the vent is plugged or the drain-back holes are plugged.

Diagnostic trouble codes (V6 models only)

20 The computer for the automatic transmission has a self-diagnostic capability; it continually monitors important information sensor and output actuator circuits for malfunctions. When a monitored circuit is damaged, shorted or disconnected, a diagnostic trouble code is stored in the computer's memory. At a dealer service department, stored trouble codes are extracted from computer memory with a proprietary diagnostic instrument known as CONSULT.

Chapter 7 Part B Automatic transmission

2.26 Applying vacuum to the throttle opener (3.5L V6 shown)

Codes can also be extracted with some generic scanners. However, the CONSULT tool isn't generally available to anyone outside of a Nissan dealership, and most home mechanics don't have a generic scan tool, so the following procedure is provided to enable you to extract any stored codes by using the O/D OFF indicator light on the instrument cluster. **Note:** *If you do have a scan tool, refer to Chapter 6 for code retrieval information and for the list of trouble codes.*

Make sure the O/D OFF light works

21 Move the shift lever to the P position, if it isn't already there.
22 Start the engine and warm it up to its normal operating temperature.
23 Turn the ignition switch to the OFF position. Wait at least five seconds before proceeding.
24 Turn the ignition switch to the ON position, but don't start the engine.
25 The O/D OFF indicator light should come on for about two seconds.
   a) If the O/D OFF indicator light comes on, proceed to the next Step.
   b) If the O/D OFF indicator light doesn't come on, there's something wrong with the transmission computer or with the indicator light circuit. We don't recommend testing the resistance or voltage of the computer terminals; drawing too much current or putting too much voltage through the terminals can destroy the computer. Have the O/D OFF indicator light circuit tested and repaired by a dealer service department or other qualified repair shop.

Obtaining stored codes using the O/D OFF light

Refer to illustration 2.26

26 Remove the vacuum hose from the throttle opener and connect a hand-held vacuum pump to the throttle opener **(see illustration)**. Apply full vacuum to the opener and make sure it holds.
27 Turn the ignition key to OFF.
28 If you're working on a 2000 or earlier model, turn the ignition switch to the ACC position.
29 Pry off the shift lock release knob cover on the shift indicator cover, then insert a narrow screwdriver into the hole and push it down. Now move the shift lever to the D position.
30 Turn the ignition switch to the ON position, but don't start the engine.
31 Push the O/D switch in and hold it there, then turn the ignition key OFF, then ON again. Release the O/D switch (the light should go out).
32 Wait for at least two seconds, then move the shift lever to the 2 position.
33 Push the O/D switch in then release it (the light should go ON).
34 Push the O/D switch in and hold it depressed, move the shift lever to the 1 position, then release the switch.
35 Push the O/D switch in again - the light should go ON.
36 Push the O/D switch in again - the light should go OFF.
37 Push the overdrive switch in and hold it depressed, push the accelerator pedal all the way to the floor, then release it, then release the O/D switch.
38 The computer is now in its output mode. It will begin displaying any stored trouble code(s) by flashing the O/D OFF indicator light in a sequence of long and short flashes specific to each stored code.
39 If all monitored circuits are operating correctly, you will see one long flash, followed by one long pause, followed by ten short flashes (twelve on 2001 Pathfinder models) of equal duration with short pauses of equal duration between them.
40 Each code is represented by the position of a long flash in the sequence:
   a) *If the first flash of the ten-flash sequence is long, the revolution sensor circuit is shorted or disconnected (code P0720).*
   b) *If the second of the ten flashes is long, the Vehicle Speed Sensor circuit is shorted or disconnected.*
   c) *If the third flash is long, the Throttle Position Sensor circuit is shorted or disconnected (code P1705).*
   d) *If the fourth flash is long, the shift solenoid valve A circuit is shorted or disconnected (code P0750).*
   e) *If the fifth flash is long, the shift solenoid valve B circuit is shorted or disconnected (code P0755).*
   f) *If the sixth flash is long, the overrun clutch solenoid circuit is shorted or disconnected (code P1760).*
   g) *If the seventh flash is long, the torque converter clutch solenoid valve circuit is shorted or disconnected (code P0740).*
   h) *If the eighth flash is long, the fluid temperature sensor is disconnected or the computer power source circuit is damaged.*
   i) *If the ninth flash is long, the engine speed signal circuit is shorted or disconnected (code P0725).*
   j) **All except 2001 Pathfinder:** *If the tenth flash is long, the line pressure solenoid valve circuit is shorted or disconnected.*
   k) **2001 Pathfinder:** *If the tenth flash is long, the turbine revolution sensor circuit is shorted or disconnected.*
   l) **2001 Pathfinder:** *If the eleventh flash is long, the line pressure solenoid valve circuit is shorted or disconnected.*
   m) **2001 Pathfinder:** *If the twelfth flash is long, the PCM-A/T communication line is open or shorted.*
   n) *If the O/D OFF indicator light flashes on and off, alternating back and forth between long flashes and long pauses, all of equal length, either the battery power is low, the battery has been disconnected for a long time, or the battery has been connected incorrectly.*
   o) *If the O/D OFF indicator light comes on but doesn't flash, either the Park/Neutral Position (PNP) switch, the overdrive switch or the wide open/closed throttle switch circuit is disconnected, or the Transmission Control Module (TCM) is damaged.*

41 Except for the Vehicle Speed Sensor and Throttle Position Sensor, both of which are covered in Chapter 6, repairing the rest of the malfunctions listed above is beyond the scope of the home mechanic. If one of more of these codes is displayed, have the transmission repaired by a dealer service department or transmission repair shop.

Erasing a diagnostic trouble code

42 If you have the transmission repaired at a dealer or transmission shop, they will erase the trouble code when they're done making the repair. If you make a repair yourself, here's how to erase the code when you're done:
43 If the ignition switch remains on after a repair, turn it off once, wait at least five seconds, then turn it on again.
44 Perform the self-diagnostic procedure described above.
45 Change the diagnostic test mode from Mode II to Mode I by turning the mode selector on the PCM (see Chapter 6, Section 2).

3 Shift cable - check, adjustment and replacement

Check

1 Move the shift lever from the "P" position to the "1" position. You should be able to feel the detents in each range. If you can't feel the detents, or if the pointer indicating the ranges is incorrectly aligned, adjust the shift cable.

Adjustment
Column shift models

Refer to illustration 3.4

2 Place the shift lever in the "P" position.

Chapter 7 Part B Automatic transmission

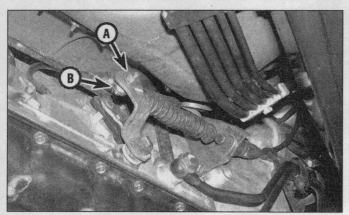

3.4 Loosen the nut that secures the shift cable to the manual lever on the transmission (A), push the cable (B) into its casing, then pull the cable in the direction opposite arrow B approximately 3/64-inch (1.0 mm) and tighten the nut (column shift models)

3.9 Shift cable adjuster details

A   Locknuts   B   Turnbuckle

3.13 On column shift models, the cable end can be pried off the shift lever post (right arrow), and separated from the steering column bracket by removing the clip (left arrow) - instrument panel removed for clarity (you can do this from under the dash, but you'll probably have to work by feel)

3.19 Pull out this clip and detach the cable from the post on the lever

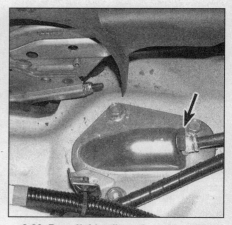

3.20 Pry off this clip to free the cable casing from the floor pan

3   Raise the vehicle and support it securely on jackstands.
4   Loosen the shift cable-to-manual lever locknut **(see illustration)**. place the manual shaft lever in the "P" position.
5   Push the cable into its casing with a force of approximately 4-1/2 pounds (19.6 N), then pull the cable in the opposite direction approximately 3/64-inch (1.0 mm) and tighten the nut.
6   Move the shift lever from "P" to "1" again. Make sure that it moves smoothly and quietly and the detents can be felt.
7   Remove the jackstands and lower the vehicle.

Floor-shift models

Refer to illustration 3.9

8   Remove the center console (see Chapter 11).

9   Loosen the locknuts on either side of the turnbuckle **(see illustration)**, tighten the turnbuckle until all slack is taken out of the inner cable and the shift lever moves back toward the R position, against the P position stop. Now turn the turnbuckle in the opposite direction one turn.
10   Tighten the locknuts securely.
11   Install the center console.

Replacement

Warning: *The models covered by this manual are equipped with airbags. Always disable the airbag system when working in the vicinity of airbag system components* (see Chapter 12).

Column shift models

Refer to illustration 3.13

12   Remove the left side under-dash panel, knee bolster and reinforcement (see Chapter 11).
13   Working under the dash, pry the cable end off the shift lever post, then remove the clip and detach the cable casing from the bracket on the steering column **(see illustration)**.

14   Remove the driver's side kick panel, peel back the carpet and follow the cable to where it goes through the floorpan. Detach the cable casing from the floor pan.
15   Raise the front of the vehicle and support it securely on jackstands.
16   Detach the cable from the manual shift lever on the side of the transmission **(see illustration 3.4)**. Remove the clip securing the cable casing to the bracket on the transmission, then remove the cable from the vehicle.
17   Installation is the reverse of removal. Be sure to adjust the cable when you're done.

Floor shift models

Refer to illustrations 3.19, 3.20 and 3.22

18   Remove the center console (see Chapter 11).
19   Remove the clip and detach the cable end from the post on the lever **(see illustration)**.
20   Pry off the clip from the cable casing where the cable passes through the floor pan **(see illustration)**.
21   Raise the front of the vehicle and sup-

Chapter 7 Part B Automatic transmission

3.22 Remove this clip and detach the cable end from the post on the manual lever

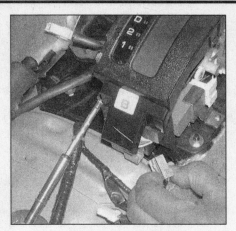

4.3 Unplug the electrical connector at the rear of the shifter assembly

4.4a The shifter assembly is secured by two nuts on the left side . . .

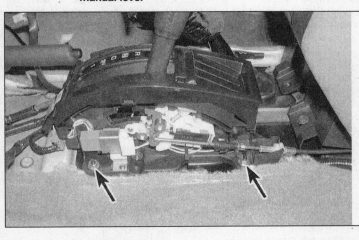

4.4b . . . and two nuts on the right side

Shift lever assembly

Refer to illustrations 4.3, 4.4a and 4.4b

1 Refer to Section 3 and perform Steps 18 through 20.
2 Disconnect the key interlock cable from the shift lock solenoid and from its bracket at the front of the shifter base (see Section 5).
3 Unplug any electrical connectors that may interfere with shifter removal **(see illustration)**.
4 Remove the two nuts from each side of the shifter assembly **(see illustrations)**, then detach the shifter assembly from the floor.
5 Installation is the reverse of removal. Be sure to adjust the shift cable (see Section 3) and the interlock cable (see Section 5).

Shift handle/overdrive switch

Refer to illustrations 4.7a, 4.7b and 4.7c

6 Remove the shift lever assembly (see Steps 1 through 4).
7 Pry the wire retainer clamp on the electrical connector up, pull the retainer clamp back, then remove the wires for the overdrive switch from the electrical connector **(see illustrations)**.
8 Remove the screws securing the shift

port it securely on jackstands.
22 Remove the clip and detach the cable from the manual lever on the side of the transmission **(see illustration)**.
23 Detach the cable from the bracket on the transmission, then pass the cable through the floor pan and remove it.
24 Installation is the reverse of removal. Be sure to adjust the cable when you're done.

4 Shift lever assembly - removal and installation

Warning: *The models covered by this manual are equipped with airbags. Always disable the airbag system when working in the vicinity of airbag system components* (see Chapter 12).
Note: *This procedure applies to floor shift models only.*

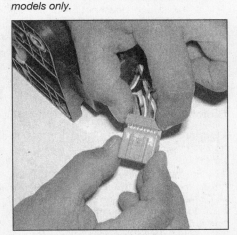

4.7a Pry the wire retainer clamp out of the electrical connector . . .

4.7b . . . pull the clamp back . . .

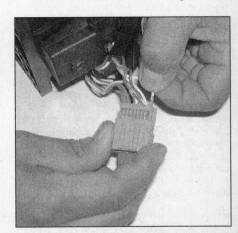

4.7c . . . then detach the wires for the overdrive switch from the connector

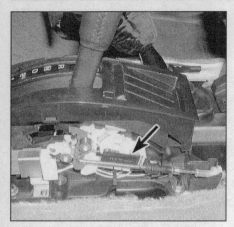

5.3a Location of the shift lock solenoid - floor shift model

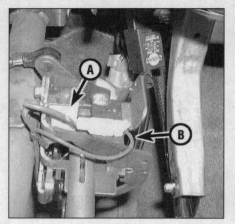

5.3b Location of the shift lock solenoid (A) and the park position switch (B, not visible in photo) - column shift model (shown from above, with the instrument panel removed for clarity)

5.7 The shift lock/cruise control brake switch (arrow) is located on a bracket near the top of the brake pedal (on most models it's the one on the left)

handle to the stalk. Pull the handle up and feed the wires through the stalk.
9 Installation is the reverse of removal.

5 Shift interlock system - description, check and component replacement

Warning: *The models covered by this manual are equipped with airbags. Always disable the airbag system when working in the vicinity of airbag system components (see Chapter 12).*

Description

1 The shift interlock system prevents the shift lever from being shifted out of Park or Neutral until the brake pedal is applied. Other than the following simple component checks, diagnosis of the shift lock system should be left to a dealer service department or other qualified repair shop.

Check

Shift lock solenoid

Refer to illustrations 5.3a and 5.3b

2 If you're working on a floor shift model, remove the center console (see Chapter 11). If you're working on a column shift model, remove the under-dash panel, knee bolster and reinforcement, and the steering column covers (see Chapter 11).
3 Locate the shift lock solenoid **(see illustrations)**. Follow the wiring harness from the shift lock solenoid to the electrical connector, then unplug the connector. Using a pair of jumper wires, momentarily apply battery voltage and ground to the solenoid terminals and verify that there's an audible "click." **Caution:** *Don't apply battery voltage any longer than necessary to perform this check.*
4 If the shift lock solenoid doesn't click when energized, replace it.

Park position switch

5 Follow the wires from the park position switch to the electrical connector. Unplug the electrical connector and, using an ohmmeter, check the continuity of the switch (for switch location **see illustration 5.3b** [column shift] or **illustration 5.23** [floor shift]).
6 The switch should have continuity when the shift lever is in Park. In any other position there should be no continuity. If the switch doesn't work as described, replace it.

Brake switch

Refer to illustration 5.7
Note: *This is the shift lock/cruise control brake switch, not the brake light switch.*
7 Remove the driver's-side under-dash panel and locate the switch; it's screwed into a bracket near the top of the brake pedal **(see illustration)**. On most models it's the switch on the left (the other switch is the brake light switch). In any case, it's the switch *without* any red wires going to it.
8 Unplug the electrical connector from the switch. Using an ohmmeter, check the continuity across the terminals of the switch; when the brake pedal is at rest, there should be continuity. When the pedal is depressed, there should be no continuity.
9 If the switch does not work as described, try adjusting it. It is adjusted using the same procedure as for the brake light switch (see Chapter 9).
10 If the switch still doesn't work right, replace it.

Component replacement

Shift lock solenoid

Floor shift models

11 Remove the center console (see Chapter 11).
12 Follow the wires from the shift lock solenoid to the electrical connector. Remove the wires for the solenoid from the electrical connector **(see illustrations 4.7a, 4.7b and 4.7c)**.
13 Detach the solenoid and its rod from the shifter assembly.
14 Installation is the reverse of removal.

Column shift models

15 Remove the under-dash panel, knee bolster and reinforcement, and the steering column covers (see Chapter 11).
16 Unbolt the steering column and lower it to improve access (see Chapter 10).
17 Detach the shift cable from the post on the lever **(see illustration 3.13)**.
18 Remove the shift lock rod (see Step 38).
19 Unbolt the shift control tube from the steering column.
20 Remove the screws and detach the shift lock solenoid.
21 Installation is the reverse of removal. Check the operation of the shifter and adjust the cable, if necessary.

Park position switch

Floor shift models

Refer to illustration 5.23
22 Remove the shift lever assembly (see Section 4).
23 Unclip the switch from the plastic tangs and remove it **(see illustration)**.

5.23 On floor shift models the park position switch is mounted on the inside of the shift lever housing and is retained by two plastic tangs

Chapter 7 Part B Automatic transmission

5.34a To disconnect the upper end of the key interlock cable from the key lock cylinder, remove the lock plate from the lock cylinder . . .

5.34b . . . and pull the cable out of the key lock cylinder

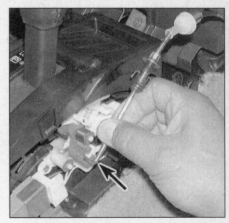

5.35 Detach the cable casing from the shift lever base, then pivot the rod up and detach it from the lever

24 Follow the wires from the switch to the electrical connector. Remove the wires for the switch from the electrical connector.
25 Installation is the reverse of removal.

Column shift models

26 Remove the under-dash panel, knee bolster and reinforcement, and the steering column covers (see Chapter 11).
27 Unbolt the steering column and lower it to improve access (see Chapter 10).
28 Detach the shift cable from the post on the lever (see illustration 3.13).
29 Remove the shift lock rod (see Step 39).
30 Unbolt the shift control tube from the steering column.
31 Remove the screws and detach the park position switch.
32 Installation is the reverse of removal. Check the operation of the shifter and adjust the cable, if necessary.

Key interlock cable (floor shift models)

Refer to illustrations 5.34a, 5.34b and 5.35

33 If the key interlock cable breaks, you'll have to remove the steering column cover and the center console to replace it (see Chapter 11).
34 Up at the key lock cylinder, remove the lock plate from the cylinder and disconnect the upper end of the shift lock cable (see illustrations).
35 Down at the shift lever, detach the key interlock cable casing from the shift lever base, then lift the rod up and disconnect it from the shift lever (see illustration).
36 Unlock the slider from the adjuster holder and remove the key interlock rod.
37 Route the cable into position, making sure it doesn't interfere with any other components and isn't bent too sharply. To insert the key interlock rod into the adjuster holder on the new cable, push it in until it locks into place. Installation is otherwise the reverse of removal. Verify proper operation before driving the vehicle.

Shift lock rod (column shift models)

Refer to illustration 5.39

38 Remove the under-dash panel, knee bolster and reinforcement, and the steering column covers (see Chapter 11).
39 Turn the ignition key to the Accessory position, then squeeze the tabs of the slider and detach the shift lock rod from the key interlock rod at the lock cylinder (see illustration).
40 Detach the other end of the rod from the lever on the steering column.
41 To install the rod, connect the forward end to the lever on the column. Place the shifter in the Park position, turn the ignition key to the Accessory position and insert the rod into the slider.
42 Without holding the shift lock rod, squeeze the tabs of the slider and push the key interlock rod toward the shift lock rod to adjust it.
43 Release the slider and check for proper operation before driving the vehicle.

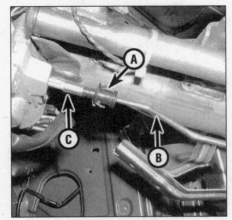

5.39 To detach the shift lock rod from the key interlock rod, squeeze the tabs of the slider, push the slider toward the key lock cylinder and detach the shift lock rod

A Slider
B Shift lock rod
C Key interlock rod

6 Throttle Valve (TV) cable - removal, installation and adjustment

Note: *This procedure applies to four-cylinder Frontier models only.*

1 If the Throttle Valve cable is not adjusted properly, the transmission shift points will be incorrect. If the cable is adjusted too tight, the shift points will occur too early, it the cable is too loose, the shift points will occur late and the transmission may not kick down when the throttle is fully depressed.

Removal and installation

Refer to illustrations 6.3 and 6.7

2 Remove the air filter housing (see Chapter 4).
3 Pull up on the TV cable to create some slack, then pass the cable through the slot in the lever as you slide the cable end out of the lever (see illustration).
4 Detach the cable casing from the bracket.
5 Drain the transmission fluid, then remove the fluid pan (see Chapter 1).

6.3 Using your fingers only, pull up on the Throttle Valve cable and pass the cable through the slot in the lever to detach it

6.7 Pry up the plastic tab (A) and slide the adjuster out of the bracket. B is the adjuster lock plate

7.2 Terminal guide for the Park/Neutral Position (PNP) switch electrical connectors (switch side)

6 Detach the TV cable from its lever, then remove the bolt and detach the cable from the transmission.

7 Detach the cable adjuster from its bracket **(see illustration)** and remove the cable.

8 Installation is the reverse of the removal procedure. Be sure to adjust the cable.

Adjustment

9 Make sure the ignition switch is in the Off position. Raise the vehicle and support it securely on jackstands.

10 Depress the lock plate on the cable adjuster **(see illustration 6.7)** and pull the adjusting tube (the portion of the cable casing above the adjuster) toward the adjuster. Release the lock plate.

11 Have an assistant quickly depress the throttle open to the full throttle position and then release it.

12 Using a piece of tape or a grease pen, mark the cable where it exits the boot above the adjuster. Place a ruler alongside the TV cable and measure the total stroke of the cable as your assistant depresses the throttle fully once again. It should fall within the range listed in this Chapter's Specifications.

13 If the cable stroke does not fall within the specified range, repeat Steps 10 and 11.

7 Park/Neutral Position (PNP) switch - check, adjustment and replacement

Check

Refer to illustration 7.2

1 If the engine will start with the shift lever in any position other than Park or Neutral, check and adjust the Park/Neutral position start switch.

2 Follow the wiring harness from the switch to its electrical connectors. Unplug the switch electrical connectors and check continuity (on the switch side of the connectors) as follows **(see illustration)**:

Four-cylinder Frontier models
a) With the shift lever in Park, there should be continuity between terminals 1 and 2.
b) With the shift lever in Reverse, there should be continuity between terminals 3 and 5.
c) With the shift lever in Neutral, there should be continuity between terminals 1 and 2.

All other models
a) With the shift lever in Park, there should be continuity between terminals 1 and 2, and 3 and 4.
b) With the shift lever in Reverse, there should be continuity between terminals 3 and 5.
c) With the shift lever in Neutral, there should be continuity between terminals 1 and 2, and 3 and 6.
d) With the shift lever in Drive, there should be continuity between terminals 3 and 7.
e) With the shift lever in 2, there should be continuity between terminals 3 and 8.
f) With the shift lever in 1, there should be continuity between terminals 3 and 9.

3 If the switch fails any of these continuity checks, disconnect the shift cable from the manual lever and retest the switch.
a) If the switch passes all the continuity checks this time, reconnect the shift cable and adjust it, then retest the switch.
b) If the switch still fails any of the continuity tests, remove it from the transmission and try testing it again on the bench.
c) If the switch passes all the continuity tests on the bench, install it and adjust it (see below).
d) If the switch still fails any of the continuity tests, replace it (see below).

Adjustment

Refer to illustration 7.4

4 Disconnect the shift cable from the manual lever (see Section 3), loosen the switch retaining screws, set the manual lever at the Neutral position and insert a 0.16-inch pin (or a 5/32-inch drill bit) through the adjustment holes in both the manual shaft lever and the switch **(see illustration)**. Make sure the pin is perpendicular to the switch and the lever. Tighten the switch retaining screws securely and remove the pin.

5 Recheck the switch continuity as described above. If switch continuity is still not as specified, replace the switch. If the switch is working properly, reconnect the shift cable and adjust it (see Section 3).

7.4 Disconnect the shift cable, loosen the switch retaining screws and insert a 5/32-inch drill bit through the adjusting holes in the manual shaft lever and the switch

Replacement

6 Place the ignition key in the Off position. Remove the key from the lock cylinder.

7 Shift the transmission into Neutral.

8 Follow the wiring harness from the switch to its electrical connectors, then unplug the electrical connectors.

9 Remove the switch retaining screws and remove the switch.

10 Installation is the reverse of removal. Don't tighten the retaining screws until you have adjusted the switch as described in Step 4.

11 Reconnect and adjust the shift cable (see Section 3).

8 Extension housing oil seal (2WD models) - replacement

Refer to illustrations 8.4 and 8.5

1 Oil leaks frequently occur due to wear of the extension housing oil seal. Replacement of this seal is relatively easy, since it can be performed without removing the transmission from the vehicle.

2 The extension housing oil seal is located at the extreme rear of the transmission, where the driveshaft is attached. If leakage at the seal is suspected, raise the vehicle and support it securely on jackstands. If the seal is leaking, transmission lubricant will be built up on the front of the driveshaft and may be dripping from the rear of the transmission.

3 Remove the driveshaft (see Chapter 8). On 1999 and later 2WD Pathfinder models with an automatic transmission, remove the

Chapter 7 Part B Automatic transmission   7B-9

8.4 Carefully pry the old seal out of the extension housing - don't damage the splines on the output shaft

8.5 Drive the new seal into place with a hammer and a seal driver or a large socket

Replacement

Refer to illustrations 9.5a and 9.5b

4   Support the transmission with a floor jack. Place a block of wood on the jack head to act as a cushion.

5   Remove the bolts and nuts attaching the mount to the crossmember and transmission **(see illustrations)**. On some models you'll have to also unbolt the plate that supports the exhaust pipe.

6   Raise the transmission slightly with the jack and remove the mount.

7   Installation is the reverse of the removal procedure. Be sure to tighten all nuts and bolts securely.

10 Automatic transmission - removal and installation

Removal

Refer to illustrations 10.10, 10.11a, 10.11b, 10.14a and 10.14b

Caution: *The transmission and torque converter must be removed as a single assembly. If you try to leave the torque converter attached to the driveplate, the converter driveplate, pump bushing and oil seal will be damaged. The driveplate is not designed to support the load, so none of the weight of the transmission should be allowed to rest on the plate during removal.*

1   Disconnect the cable from the negative terminal of the battery.

2   Raise the vehicle and support it securely on jackstands. Remove the skid plate and skid plate crossmember, if equipped.

3   Remove the transmission oil pan drain plug and drain the transmission fluid (see Chapter 1).

4   If you're working on a four-cylinder Frontier model, detach the Throttle Valve cable from the transmission (see Section 6). Temporarily reinstall the transmission fluid pan.

5   Detach the shift cable from the manual lever and from the bracket on the transmission.

9.2 To check the transmission mount, insert a large screwdriver or prybar between the crossmember and the transmission and try to pry the transmission up - it should move very little

companion flange from the transmission output shaft.

4   Using a seal removal tool or a large screwdriver, carefully pry the oil seal out of the rear of the transmission **(see illustration)**. Do not damage the splines on the transmission output shaft.

5   Using a seal driver, a large section of pipe or a very large deep socket as a drift, install the new oil seal **(see illustration)**. Drive it into the bore squarely and make sure it's completely seated.

6   Lubricate the splines of the transmission output shaft and the outside of the driveshaft yoke with lightweight grease, then install the driveshaft (see Chapter 8). Be careful not to damage the lip of the new seal.

9 Transmission mount - check and replacement

Check

Refer to illustration 9.2

1   Raise the vehicle and support it securely on jackstands.

2   Insert a large screwdriver or prybar into the space between the transmission exten-

sion housing and the crossmember and try to pry the transmission up slightly **(see illustration)**.

3   The transmission should not move much at all - if the mount is cracked or torn, replace it.

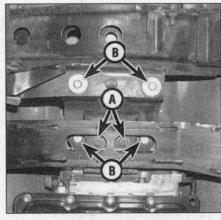

9.5a Remove the nuts that secure the transmission mount to the crossmember (A); on some models you'll also have to remove the bolts that hold the exhaust pipe support plate (B)

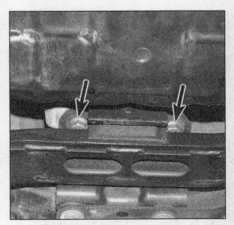

9.5b Transmission mount-to-transmission bolts

7B

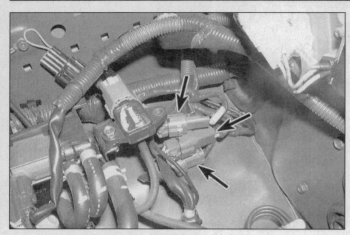

10.10 Some of the connectors for the transmission are located in the right rear corner of the engine compartment (Frontier model shown)

10.11a Remove the bolts and detach the inspection cover

6 Mark the yokes and remove the driveshaft (see Chapter 8). On 4WD models, remove both driveshafts.
7 Remove all exhaust components which would interfere with transmission removal (see Chapter 4).
8 Remove the Crankshaft Position sensor (see Chapter 6).
9 Remove the starter motor (see Chapter 5).
10 Follow the wiring harnesses from the transmission up to their electrical connectors, then unplug the connectors (see illustration). Mark and disconnect any other electrical connectors that would interfere with transmission removal.
11 Remove the inspection cover and mark the relationship of the torque converter to the driveplate so they can be installed in the same position (see illustrations).
12 Remove the torque converter-to-driveplate bolts. Turn the crankshaft for access to each bolt. Turn the crankshaft in a clockwise direction only (as viewed from the front).
13 Remove the fill/dipstick tube bracket bolt and pull the tube out of the transmission. Don't lose the tube seal (it can be reused if it's still in good shape).
14 Remove the fitting bolts and detach the fluid cooler lines from the transmission (see illustrations). Discard the sealing washers that are present on either side of the fittings; new ones must be used when reconnecting the fittings.
15 On 4WD models, remove the transfer case (see Chapter 7C). **Note:** *If you are not planning to replace the transmission, but are removing it in order to gain access to other components such as the torque converter, it isn't really necessary to remove the transfer case. However, the transmission and transfer case are awkward and heavy when removed and installed as a single assembly; they're much easier to maneuver off and on as separate units. If you decide to leave the transfer case attached, disconnect the shift rod (manual shift models only) from the transfer case shift lever. Also disconnect the electrical connectors from the transfer case speed sensors and detach the transfer case vent tube (see Chapter 7C).* **Warning:** *If you decide to leave the transfer case attached to the transmission, be sure to use safety chains to help stabilize the transmission and transfer case assembly and to prevent it from falling off the jack head, which could cause serious dam-*

10.11b Remove one of the torque converter bolts and mark the relationship of the driveplate to the torque converter

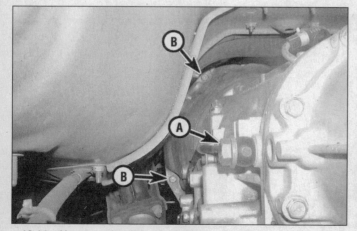

10.14a Unscrew the fitting bolt (A) from the fluid cooler line on the left side of the transmission (B indicates two of the transmission-to-engine bolts)

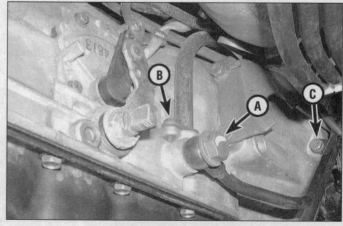

10.14b Also remove the fitting bolt (A) from the fluid cooler line on the right side of the transmission; (B) is the dipstick tube mounting bolt and (C) is another transmission-to-engine bolt

Chapter 7 Part B Automatic transmission

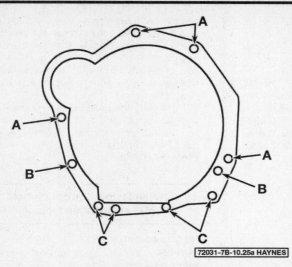

10.25a Transmission mounting bolt location guide - four-cylinder models

A 1-11/16 inches (43 mm)
B 5/8-inch (16 mm)
C 5/8-inch (16 mm)

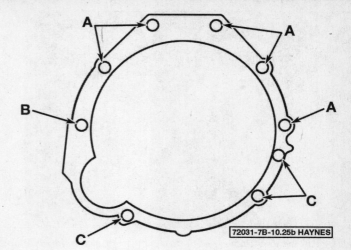

10.25b Transmission mounting bolt location guide - 3.3L V6 models

A 1-7/8 inches (47 mm)
B 2-9/32 inches (58 mm)
C 1-inch (25 mm)

age to the transmission and/or transfer case and serious bodily injury to you.

16   Support the engine with an engine hoist or support fixture from above, or with a jack placed under the oil pan. If you use a floor jack, place a block of wood on the jack head to spread the load.

17   Support the transmission with a jack - preferably a jack made for this purpose (available at most tool rental yards). Safety chains will help steady the transmission on the jack.

18   Remove the bolt securing the transmission mount to the crossmember. Then raise the transmission slightly and remove the crossmember.

19   Lower the engine and transmission slightly and remove the bolts securing the transmission to the engine. A long extension and a U-joint socket will greatly simplify this step. **Note:** *Different length bolts are used - be sure to note the location of each bolt so they can be returned to their original positions when the transmission is installed.*

20   Move the transmission to the rear to disengage it from the engine block dowel pins and make sure the torque converter is detached from the driveplate. Lower the transmission with the jack. Clamp a pair of locking pliers on the bellhousing case. The pliers will prevent the torque converter from falling out while you're removing the transmission.

Installation

Refer to illustrations 10.25a, 10.25b and 10.25c

21   Prior to installation, make sure the torque converter is securely engaged in the pump. If you've removed the converter, apply a small amount of transmission fluid on the torque converter rear hub, where the transmission front seal rides. Install the torque converter onto the front input shaft of the transmission while rotating the converter back and forth. It should engage into the transmission front pump in stages. To make sure the converter is fully engaged, lay a straightedge across the transmission-to-engine mating surface and measure the distance from the straightedge and the converter lugs. If you are working on a model with a four-cylinder engine or 3.3L V6 engine, the converter lugs must be at least 1.024-inches (26 mm) below the straightedge. If you are working on a model with a 3.5L V6 engine, the converter lugs must be at least 0.984-inch (25 mm) below the straightedge.

22   With the transmission secured to the jack, raise it into position.

23   Turn the torque converter to line up the holes with the holes in the driveplate. The marks on the torque converter and driveplate made in Step 11 must line up.

24   Move the transmission forward carefully until the dowel pins engage with the holes in the bellhousing. Make sure the transmission

10.25c Transmission mounting bolt location guide - 3.5L V6 models

A 2-9/16 inches (65 mm)
B 2-11/64 inches (55 mm)
C 1-9/16 inches (40 mm)

mates with the engine with no gap. If there's a gap, make sure there are no wires or other objects pinched between the engine and transmission and also make sure the torque converter is completely engaged in the transmission front pump. Try to rotate the converter - if it doesn't rotate easily, it's probably not fully engaged in the pump. If necessary, lower the transmission and install the converter fully.

25   Install the transmission-to-engine bolts and tighten them to the torque values listed in this Chapter's Specifications **(see illustrations)**. As you're tightening the bolts, make sure that the engine and transmission mate completely at all points. If not, find out why. Never try to force the engine and transmission together with the bolts or you'll break the transmission case!

26   Raise the rear of the transmission and install the transmission crossmember.

27   Remove the jacks supporting the transmission and the engine.

11.3 The Transmission Control Module is located under the left side of the instrument panel

28 Install the torque converter-to-driveplate bolts. Once all the bolts have been installed, tighten them to the torque listed in this Chapter's Specifications.
29 Install the transmission dipstick tube and seal into the transmission housing, then install the bolt and tighten it securely.
30 Install the starter motor (see Chapter 5).
31 Install the torque converter inspection cover.
32 Using new sealing washers, connect the transmission fluid cooler lines to the transmission, tightening the fitting bolts securely.
33 Plug in the transmission electrical connectors.
34 Connect the shift cable (see Section 3).
35 On 4WD models, install the transfer case, if removed (see Chapter 7C).
36 Install the driveshaft(s) (see Chapter 8).
37 Adjust the shift cable (see Section 3).
38 If you're working on a four-cylinder Frontier model, connect the Throttle Valve cable to the transmission, install the fluid pan with a new gasket, then tighten the pan bolts to the torque listed in the Chapter 1 Specifications.
39 Install any exhaust system components that were removed or disconnected (see Chapter 4).
40 Remove the jackstands and lower the vehicle.
41 Fill the transmission with the specified fluid (see Chapter 1), run the engine and check for fluid leaks.

11 Transmission Control Module (TCM) - removal and installation

Refer to illustration 11.3
Note: *This procedure only applies to models with a V6 engine.*
1 Disconnect the cable from the negative terminal of the battery.
2 Remove the driver's side under-dash panel, knee bolster and reinforcement (see Chapter 11).
3 Unscrew the mounting nuts, unplug the electrical connector and remove the TCM **(see illustration)**.
4 Installation is the reverse of removal.

Chapter 7 Part C
Transfer case

Contents

	Section		Section
General information	1	Transfer case lubricant change	See Chapter 1
Oil seals - replacement	4	Transfer case lubricant level check	See Chapter 1
Shift lever - removal and installation	2	Transfer case overhaul - general information	6
Transfer case - removal and installation	5	Transfer case position switches - check and replacement	3

Specifications

General
Transfer case lubricant .... See Chapter 1

Torque specifications
	Ft-lbs (unless otherwise noted)	Nm
Front output shaft flange nut	166 to 239	226 to 324
Transfer case-to-transmission bolts	23 to 30	31 to 41

1 General information

Four-wheel drive (4WD) models are equipped with a transfer case mounted on the rear of the transmission. Drive is transmitted from the engine, through the transmission and the transfer case, to the front and rear axles by driveshafts.

The vehicles covered by this manual are equipped with one of the following transfer cases: Part-time 4WD models are equipped with a TX10A transfer case while Pathfinders with "all-mode" 4WD are equipped with the ATX14A transfer case. The TX10A transfer case has four gear selections: 2WD High, 4WD High, 4WD Low and Neutral. The ATX14A transfer case has five gear selections, 2WD, 4WD High, 4WD Low, and automatic four wheel drive (AUTO). The ATX14A transfer case has its own control module and uses various switches and sensors to control four-wheel drive operation. It is an extremely complex system and can be very difficult to diagnose should a problem occur. It is for this reason that information related to the ATX14A transfer case is not included in this manual. Consult a four-wheel drive specialist, transmission repair shop or dealer service department if a problem develops with this system.

We don't recommend trying to rebuild either of these transfer cases at home. They're difficult to overhaul without special tools, and rebuilt units are available for less than it would cost to rebuild your own. However, there are a number of components that you *can* check, adjust and/or replace - and those are the items covered in this Chapter.

2 Shift lever - removal and installation

Refer to illustrations 2.3 and 2.5

1 Working in the passenger compartment of the vehicle, unscrew the shift lever knob.
2 Remove the trim from around the transfer case shift lever (see Chapter 11).
3 Remove the shift lever boot retaining bolts and remove the boot **(see illustration)**.
4 Raise the front of the vehicle and support it securely on jackstands.
5 Working under the vehicle, remove the nut and detach the shift linkage from the shift lever **(see illustration)**.
6 Remove the shift lever retaining bolts and remove the shift lever.
7 Installation is the reverse of removal.

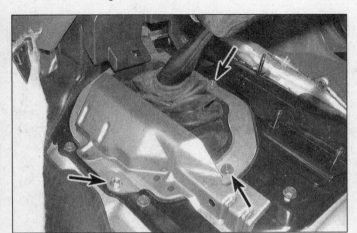

2.3 Remove the bolts securing the transfer case shift lever boot to the floor pan, then pull the boot up and off the lever

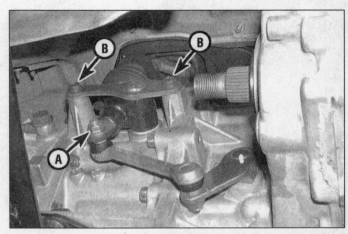

2.5 Unscrew the shift linkage-to-lever nut (A), then remove the lever retaining bolts (B) and detach the lever from the transfer case

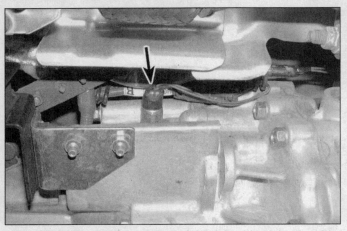

3.2a The 4WD switch is located on the right side of the transfer case - it's the switch that is nearer to the rear of the transfer case

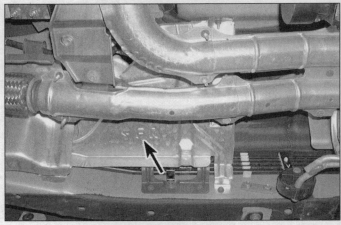

3.2b Remove this heat shield . . .

3 Transfer case position switches - check and replacement

Check

1 Raise the vehicle and support it securely on jackstands.

4WD switch

Refer to illustrations 3.2a, 3.2b and 3.2c

2 Follow the wiring harness from the switch to its electrical connector, then unplug the connector **(see illustrations)**.

3 Working on the switch side of the connector, check the continuity of the switch using an ohmmeter. When the transfer case shift lever is in the 4WD position, there should be continuity. There should be no continuity in any other position. If the switch fails this test, replace it.

Neutral position switch

Refer to illustration 3.4

4 Follow the wiring harness from the switch to its electrical connector, then unplug the connector **(see accompanying illustration and illustrations 3.2b and 3.2c)**.

5 Working on the switch side of the connector, check the continuity of the switch using an ohmmeter. When the transfer case shift lever is in the Neutral position, there should be no continuity. Continuity should be present in any other position. If the switch fails this test, replace it.

Replacement

6 Raise the vehicle and support it securely on jackstands.
7 Follow the wiring harness from the switch to its electrical connector, then unplug the connector.
8 Unscrew the switch from the transfer case.
9 Before installing the new switch, coat its threads with a non-hardening thread locking compound.
10 Installation is the reverse of removal. Tighten the switch securely.

3.2c . . . for access to the transfer case electrical connectors

4 Oil seals - replacement

1 Raise the vehicle and support it securely on jackstands.

3.4 The Neutral position switch is also located on the right side of the transfer case - it's the switch that is nearer to the front of the transfer case

4.3 A chain wrench can be used to prevent the flange from turning while the nut is removed

Chapter 7 Part C Transfer case

4.4 A two-jaw puller will be required to remove the flange if it won't come off by hand

4.5 Use a seal removal tool or a large screwdriver to pry the seal out

Output shaft oil seal(s)

Note: *This procedure applies to both the front and rear output shaft seals.*

2 If you're replacing the front seal, remove the front driveshaft; if you're replacing the rear seal, remove the rear driveshaft (see Chapter 8).

Front seal

Refer to illustrations 4.3 and 4.4

3 A flange holding tool will be required to keep the companion flange from moving while the nut is loosened. A chain wrench will also work **(see illustration)**. Remove the flange nut.

4 Withdraw the flange. It may be necessary to use a two-jaw puller engaged behind the flange to draw it off **(see illustration)**. Do not attempt to pry or hammer behind the flange or hammer on the end of the shaft.

Front or rear seal

Refer to illustrations 4.5 and 4.6

5 Pry out the old seal and discard it **(see illustration)**.

6 Lubricate the lips of the new seal and the seal case with multi-purpose grease, then tap it evenly into position with a seal installation tool or a large socket **(see illustration)**. Make sure it enters the housing squarely and is tapped in to its full depth.

Front seal

7 Install the companion flange. If necessary, tighten the nut to draw the flange into place. Do not try to hammer the flange into position.

8 Apply a bead of RTV sealant to the ends of the splines visible in the center of the flange so oil will be sealed in.

9 Install the nut. Tighten the nut to the torque listed in this Chapter's Specifications.

Front or rear seal

10 Install the driveshaft (see Chapter 8). Check and, if necessary, add the recommended type of lubricant to bring the level up to the bottom of the filler hole (see Chapter 1).

Shift shaft oil seal

Refer to illustration 4.11

11 Mark the relationship of the lever to the shift shaft **(see illustration)**.

12 Detach the linkage from the lever, then loosen the pinch bolt and slide the lever off the shaft.

13 Carefully pry the seal out of the case.

14 Lubricate the new seal with multi-purpose grease, then install it over the shaft and tap it into place with an appropriately sized deep socket.

15 Install the lever onto the shaft and tighten the pinch bolt securely.

16 Reconnect the linkage to the lever, install the bolt and tighten it securely.

17 Check and, if necessary, add the recommended type of lubricant to bring the level up to the bottom of the filler hole (see Chapter 1).

5 Transfer case - removal and installation

Removal

1 Disconnect the cable from the negative terminal of the battery.

2 Raise the vehicle and support it securely on jackstands. Remove the skid plate, if

4.6 Drive the new seal into place with a seal driver or a socket with an outside diameter slightly smaller than that of the seal

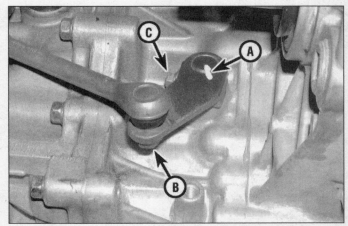

4.11 Mark the lever to the shaft (A), remove the nut (B) and detach the linkage from the lever, then loosen the pinch bolt (C) and slide the lever off the shaft

equipped. If you're working on a Pathfinder, remove the exhaust system (see Chapter 4).
3   Remove the transfer case shift lever (see Section 2).
4   Drain the transfer case lubricant (see Chapter 1).
5   Remove the front and rear driveshafts (see Chapter 8).
6   If you're working on a Frontier or Xterra, remove the torsion bars (see Chapter 10).
7   Unplug all electrical connectors **(see illustrations 3.3b and 3.3c)** and detach the vent hose from the top of the transfer case.
8   Support the transmission with a floor jack. Place a block of wood on the jack head to spread the load.
9   Support the transfer case with a jack - preferably a special jack made for this purpose. Safety chains will help steady the transfer case on the jack.
10   Unbolt the transmission mount from the crossmember, then remove the crossmember.
11   Lower the jacks supporting the transmission and transfer case just enough for access to the upper mounting bolts.
12   Remove the transmission-to-transfer case bolts. **Note:** *On some models there are two different lengths of bolts securing the transfer case to the transmission. Mark their positions or keep them in order when removed so they can be returned to their proper locations.*
13   Make a final check that all wires and hoses have been disconnected from the transfer case, then move the transfer case and jack toward the rear of the vehicle until the transfer case is clear of the transmission. Keep the transfer case level as this is done. Once the input shaft is clear, lower the transfer case and remove it from under the vehicle.

Installation

14   Installation is the reverse of removal, noting the following points:
  a) *Apply anaerobic gasket sealant (not RTV sealant) to the transfer case-to-transmission mating surface.*
  b) *Be sure to install the transmission-to-transfer case bolts in their proper locations and tighten them to the torque listed in this Chapter's Specifications.*
  c) *Refill the transfer case with the proper type and quantity of lubricant* (see Chapter 1).
  d) *If you're working on a Frontier or Xterra, reinstall and adjust the torsion bars* (see Chapter 10).

6   Transfer case overhaul - general information

Overhauling a transfer case is a difficult job for the do-it-yourselfer. It involves the disassembly and reassembly of many small parts. Numerous clearances must be precisely measured and, if necessary, changed with select-fit spacers and snap-rings. As a result, if transfer case problems arise, it can be removed and installed by a competent do-it-yourselfer, but overhaul should be left to a transmission repair shop. Rebuilt transfer cases may be available - check with your dealer parts department and auto parts stores. At any rate, the time and money involved in an overhaul is almost sure to exceed the cost of a rebuilt unit.

Nevertheless, it's not impossible for an inexperienced mechanic to rebuild a transfer case if the special tools are available and the job is done in a deliberate step-by-step manner so nothing is overlooked.

The tools necessary for an overhaul include internal and external snap-ring pliers, a bearing puller, a slide hammer, a set of pin punches, a dial indicator and possibly a hydraulic press. In addition, a large, sturdy workbench and a vise or transmission stand will be required.

During disassembly of the transfer case, make careful notes of how each piece comes off, where it fits in relation to other pieces and what holds it in place. Note how parts are installed when you remove them; this will make it much easier to get the transfer case back together.

Before taking the transfer case apart for repair, it will help if you have some idea what area of the transfer case is malfunctioning. Certain problems can be closely tied to specific areas in the transfer case, which can make component examination and replacement easier. Refer to the *Troubleshooting* section at the front of this manual for information regarding possible sources of trouble.

Chapter 8
Clutch and driveline

Contents

	Section
Axle assembly (front) - removal and installation	23
Axle assembly (rear) - removal and installation	18
Axles - description and check	15
Axleshaft, bearing and oil seals (rear) - removal, bearing/seal replacement and installation	16
Clutch - description and check	2
Clutch components - removal, inspection and installation	6
Clutch fluid level check	See Chapter 1
Clutch hydraulic system - bleeding	5
Clutch master cylinder - removal, installation and reservoir/seal replacement	3
Clutch pedal - adjustment	9
Clutch release bearing - removal, inspection and installation	7
Clutch release cylinder - removal and installation	4
Clutch start switch - check and replacement	10

	Section
Differential lubricant change	See Chapter 1
Differential lubricant level check	See Chapter 1
Driveaxle (4WD models) - removal and installation	21
Driveaxle boot - replacement	22
Driveaxles (4WD models) - general information and inspection	20
Driveshaft and universal joints - general information and inspection	11
Driveshaft center support bearing - replacement	13
Driveshaft(s) - removal and installation	12
Flywheel - removal and installation	See Chapter 2
Free-running hubs (4WD models) - removal and installation	19
General information	1
Pilot bushing - replacement	8
Pinion oil seal - replacement	17
Universal joints - replacement	14

Specifications

General

Clutch pedal height
  Frontier models
    1998 .... 8.70 to 9.09 inches (221 to 231 mm)
    1999 and later
      Four-cylinder engine .... 8.70 to 9.09 inches (221 to 231 mm)
      V6 engine .... 8.94 to 9.33 inches (227 to 237 mm)
  Xterra models
    Four-cylinder engine .... 8.70 to 9.09 inches (221 to 231 mm)
    V6 engine .... 8.94 to 9.33 inches (227 to 237 mm)
  Pathfinder models
    1996 through 1998 .... 7.13 to 7.52 inches (181 to 191 mm)
    1999 and later .... 7.32 to 7.72 inches (186 to 196 mm)
Clutch pedal freeplay .... 0.35 to 0.63 inch (9 to 16 mm)
Clutch start switch - clearance between switch body and pedal (with pedal depressed)
  Frontier and Xterra models .... 0.004 to 0.039 inch (0.1 to 1.0 mm)
  Pathfinder models
    1996 through 1999 .... 0.012 to 0.039 inch (0.3 to 1.0 mm)
    2000 and later .... 0.004 to 0.059 inch (0.1 to 1.5 mm)
Clutch fluid type .... See Chapter 1
Clutch disc lining minimum rivet depth .... 1/32 inch (0.8 mm)
Inner CV joint length .... 3.74 to 3.82 inches (95 to 97 mm)
Outer CV joint boot length .... 3.78 to 3.86 inches (96 to 98 mm)
Driveaxle endplay .... 0.018 inch (0.45 mm) or less
Differential pinion shaft bearing preload (rear, Frontier and Xterra models) .... 10 to 20 in-lbs (1.2 to 2.3 Nm)

Torque specifications

	Ft-lbs (unless otherwise indicated)	Nm
Clutch		
Master cylinder nuts	69 to 95 in-lbs	8 to 11
Release cylinder		
Mounting bolts	23 to 30	31 to 40
Banjo fitting bolt	144 to 168 in-lbs	16 to 19
Pressure plate-to-flywheel bolts		
Frontier and Xterra	16 to 22	22 to 29
Pathfinder		
1996 through 2000	16 to 22	22 to 29
2001	26 to 32	35 to 44

Torque specifications (continued)

	Ft-lbs (unless otherwise indicated)	Nm
Driveshaft		
Flange bolts/nuts		
Front (4WD)		
Frontier and Xterra	29 to 33	39 to 44
Pathfinder		
1996 through 1999	41 to 48	55 to 65
2000 and later	29 to 33	39 to 44
Rear driveshaft-to-differential pinion flange		
Frontier and Xterra	58 to 65	78 to 88
Pathfinder		
1996 through 2000	51 to 58	69 to 78
2001		
2WD	41 to 47	55 to 64
4WD	51 to 58	69 to 78
Rear driveshaft-to-center support bearing flange		
Frontier and Xterra	13 to 18	18 to 25
Pathfinder		
1996 through 2000	58 to 65	78 to 88
2001	29 to 32	40 to 44
Rear driveshaft-to-transmission flange (2WD Pathfinder models with automatic transmission 1999 on)	51 to 58	69 to 78
Center support bearing bolts		
Frontier and Xterra	144 to 172 in-lbs	16 to 22
Pathfinder		
1996	144 to 172 in-lbs	16 to 22
1997 and later	32 to 41	43 to 55
Center support bearing flange nut		
Frontier and Xterra	101 to 123	137 to 167
Pathfinder	181 to 216	246 to 294
Front driveaxle		
CV joint-to-output shaft flange nuts	25 to 33	34 to 44
Free-running hub bolts	18 to 25	25 to 34
Drive flange-to-front hub mounting nuts (Pathfinder)	18 to 26	25 to 35
Front axle/differential mounting bolts/nuts		
Frontier and Xterra	50 to 64	68 to 87
Pathfinder	N/A	
Differential pinion shaft nut		
Front		
Frontier		
Four-cylinder engine	123 to 145	167 to 196
V6 engine	137 to 217	186 to 294
Xterra	137 to 217	186 to 294
Pathfinder	137 to 217	186 to 294
Rear		
Frontier		
1998	94 to 217	127 to 294
1999, 2000		
Four-cylinder engine	94 to 217	127 to 294
V6 engine	123 to 181	167 to 245
2001	94 to 217	127 to 294
Xterra		
2000		
C200 rear axle (differential integral with axle housing)	94 to 217	127 to 294
H233B rear axle (differential assembly bolts to front of housing)	145 to 210	196 to 284
2001	94 to 217	127 to 294
Pathfinder		
1996 through 2000	146 to 210	196 to 284
2001	109 to 144	148 to 196
Rear axleshaft-to-axle housing nuts	40 to 54	54 to 74

Chapter 8 Clutch and driveline

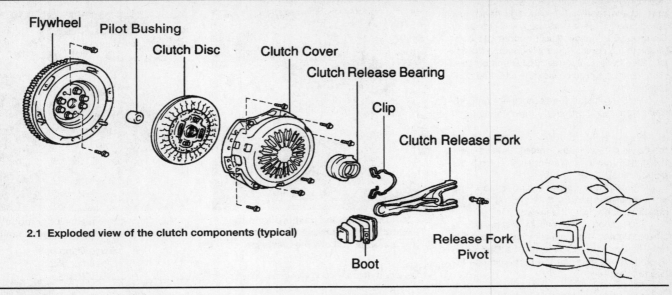

2.1 Exploded view of the clutch components (typical)

1 General information

The Sections in this Chapter deal with the components from the rear of the engine to the rear wheels (except for the transmission and transfer case, which are dealt with in Chapter 7) and forward to the front wheels on four-wheel drive (4WD) models. In this Chapter, the components are grouped into three categories: clutch, driveshaft(s) and axle(s). Separate Sections within this Chapter cover checks and repair procedures for components in each of these three groups.

Since nearly all these procedures involve working under the vehicle, make sure it's safely supported on sturdy jackstands or a hoist where the vehicle can be safely raised and lowered.

2 Clutch - description and check

Refer to illustration 2.1

1   All vehicles with a manual transmission have a single dry plate, diaphragm spring type clutch **(see illustration)**. The clutch disc has a splined hub which allows it to slide along the splines of the transmission input shaft. The clutch and pressure plate are held in contact by spring pressure exerted by the diaphragm spring in the pressure plate.
2   The clutch release system is operated by hydraulic pressure. The hydraulic release system consists of the clutch pedal, a master cylinder and fluid reservoir, the hydraulic line, a release (or slave) cylinder which actuates the clutch release lever and the clutch release (or throwout) bearing.
3   When pressure is applied to the clutch pedal to release the clutch, hydraulic pressure is exerted against the outer end of the clutch release lever. As the lever pivots, the shaft fingers push against the release bearing. The bearing pushes against the fingers of the diaphragm spring of the pressure plate

assembly, which in turn releases the clutch plate.
4   Terminology can be a problem when discussing the clutch components because common names are in some cases different from those used by the manufacturer. For example, the driven plate is also called the clutch plate or disc, the pressure plate assembly is sometimes called the clutch cover, the clutch release bearing is sometimes called a throwout bearing, and the release cylinder is sometimes called the slave cylinder.
5   Other than to replace components with obvious damage, some preliminary checks should be performed to diagnose clutch problems.

a)   *The first check should be of the fluid level in the clutch master cylinder. If the fluid level is low, add fluid as necessary and inspect the hydraulic system for leaks. If the master cylinder reservoir is dry, bleed the system as described in Section 5 and recheck the clutch operation.*

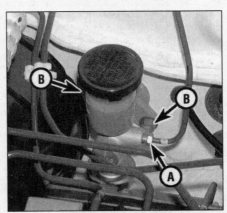

3.2 Clutch master cylinder mounting details

A   *Hydraulic line fitting*
B   *Mounting nuts*

b)   *To check "clutch spin-down time," run the engine at normal idle speed with the transmission in Neutral (clutch pedal up - engaged). Disengage the clutch (pedal down), wait several seconds and shift the transmission into Reverse. No grinding noise should be heard. A grinding noise would most likely indicate a bad pressure plate or clutch disc.*
c)   *To check for complete clutch release, run the engine (with the parking brake applied to prevent vehicle movement) and hold the clutch pedal approximately 1/2-inch from the floor. Shift the transmission between 1st gear and Reverse several times. If the shift is rough, component failure is indicated. Check the release cylinder pushrod travel. With the clutch pedal depressed completely, the release cylinder pushrod should extend substantially (you may have to remove an inspection plug to see the pushrod). If it doesn't, check the fluid level in the clutch master cylinder.*
d)   *Visually inspect the pivot bushing at the top of the clutch pedal to make sure there's no binding or excessive play.*
e)   *Crawl under the vehicle and make sure the clutch release lever is securely attached to the ballstud.*

3   Clutch master cylinder - removal, installation and reservoir/seal replacement

Removal and installation

Refer to illustrations 3.2 and 3.4

1   Remove as much fluid as possible from the reservoir with a suction gun, large syringe or a poultry baster. **Warning:** *If a poultry baster is used, never again use it for the preparation of food.*
2   Disconnect the clutch fluid hydraulic line **(see illustration)**; use a flare-nut wrench to

protect the tube nut. Have rags handy, as some fluid will be lost as the line is removed. **Caution:** *Don't allow fluid to come into contact with the paint since it will damage the finish. Also have a plug ready and immediately plug the line to prevent leakage and fluid contamination.*

3   Working inside the passenger compartment, remove the driver's side under-dash panel.

4   Remove the retaining pin and pull out the clevis pin **(see illustration)** to disconnect the clutch master cylinder pushrod from the clutch pedal.

5   Remove the master cylinder mounting nuts **(see illustration 3.2)** and detach the master cylinder from the firewall.

6   Installation is the reverse of removal. Be sure to tighten the master cylinder mounting nuts to the torque listed in this Chapter's Specifications. Tighten the hydraulic line fitting nut securely.

7   Fill the clutch master cylinder reservoir with the fluid specified in Chapter 1 and bleed the clutch hydraulic system (see Section 5).

Reservoir/seal replacement

Note: *The clutch fluid reservoir can be replaced separately from the master cylinder body if it becomes damaged. If there is leakage between the reservoir and the master cylinder body, the seal in the master cylinder can be replaced.*

8   Remove as much fluid as possible from the reservoir with a suction gun, large syringe or a poultry baster. **Warning:** *If a poultry baster is used, never again use it for the preparation of food.*

9   Place rags under the clutch master cylinder to absorb any fluid that may spill out once the reservoir is detached from the master cylinder. **Caution:** *Brake fluid will damage paint. Cover all body parts and be careful not to spill fluid during this procedure.*

10   Drive out the roll pin, then carefully pry the reservoir straight up.

11   If you are simply replacing the seal, carefully pry the old seal out of the cylinder body and install a new one.

12   Lubricate the reservoir seal with clean brake fluid, then press the reservoir into place on the cylinder body. Install a new roll pin.

13   Refill the reservoir with the recommended brake fluid (see Chapter 1) and check for leaks.

14   Bleed the clutch hydraulic system (see Section 5).

4   Clutch release cylinder - removal and installation

Refer to illustration 4.2

1   Raise the vehicle and support it securely on jackstands.

2   Unscrew the fluid hose banjo fitting bolt **(see illustration)**. **Note:** *There is a sealing washer on either side of the brake hose inlet fitting; be sure to replace these with new ones when reconnecting the hose.* Have rags handy, as some fluid will be lost as the line is removed. **Caution:** *Don't allow fluid to come into contact with the paint - it will damage the finish. Also have a plug ready and immediately plug the line to prevent leakage and fluid contamination.*

3   Unscrew the two release cylinder mounting bolts and detach the release cylinder.

4   Installation is the reverse of removal. Make sure the pushrod dust boot is in good condition and the pushrod is seated correctly in its pocket in the release lever. Tighten the release cylinder mounting bolts to the torque listed in this Chapter's Specifications.

5   Fill the clutch fluid reservoir with the recommended fluid (see Chapter 1).

6   Bleed the clutch hydraulic system (see Section 5).

7   Lower the vehicle and check for proper operation.

5   Clutch hydraulic system - bleeding

Refer to illustration 5.3

1   The hydraulic system should be bled of all air whenever any part of the system has been removed or if the fluid level has been allowed to fall so low that air has been drawn into the master cylinder. The procedure is similar to bleeding a brake system.

2   Fill the master cylinder with new brake fluid conforming to DOT 3 specifications. **Caution:** *Do not re-use any of the fluid coming from the system during the bleeding operation or use fluid which has been inside an open container for an extended period of time.*

3   Locate the bleeder valve on the clutch release cylinder **(see illustration)**. Remove the dust cap which fits over the bleeder valve and push a length of snug-fitting (preferably clear) hose over the valve. Place the other end of the hose into a clear container with about two inches of brake fluid in it. The hose end must be submerged in the fluid.

4   Have an assistant depress the clutch pedal and hold it. Open the bleeder valve on the release cylinder, allowing fluid to flow through the hose. Close the bleeder valve when fluid stops flowing from the hose. Once closed, have your assistant release the pedal.

5   Continue this process until all air is evacuated from the system, indicated by a full, solid stream of fluid being ejected from the bleeder valve each time and no air bubbles in the hose or container. Keep a close watch on the fluid level inside the clutch master cylinder reservoir; if the level drops too low, air will be sucked back into the system and the process will have to be started over again.

6   Install the dust cap on the bleeder valve. Check carefully for proper operation before placing the vehicle in normal service.

6   Clutch components - removal, inspection and installation

Warning: *Dust produced by clutch wear and deposited on clutch components is hazardous to your health. DO NOT blow it out with compressed air and DO NOT inhale it. DO NOT use gasoline or petroleum-based*

3.4 To detach the pushrod from the clutch pedal, remove this retaining pin then pull out the clevis pin

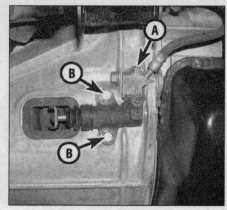

4.2 Clutch release cylinder mounting details (Xterra shown, others similar)
  A   Banjo fitting bolt
  B   Mounting bolts

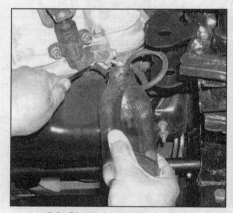

5.3 Clutch hydraulic system bleeding setup

Chapter 8 Clutch and driveline

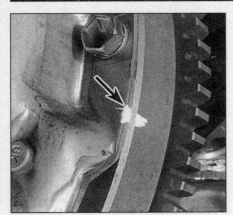

6.5 Be sure to mark the pressure plate and flywheel to insure proper alignment during installation (this won't be necessary if a new pressure plate is to be installed)

6.8 Check the flywheel for cracks, hot spots and other obvious defects (slight imperfections can be removed by a machine shop)

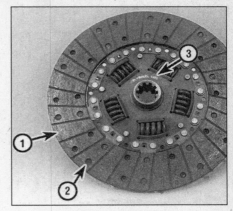

6.10 The clutch plate
1   Lining - This will wear down in use
2   Rivets - These secure the lining and will damage the flywheel or pressure plate if allowed to contact the surfaces
3   Markings - "Flywheel side" or something similar

solvents to remove the dust. Brake system cleaner should be used to flush the dust into a drain pan. After the clutch components are wiped clean with a rag, dispose of the contaminated rags and cleaner in a covered, marked container.

Removal

Refer to illustration 6.5

1   Access to the clutch components is normally accomplished by removing the transmission, leaving the engine in the vehicle. If, of course, the engine is being removed for major overhaul, then check the clutch for wear and replace worn components as necessary. However, the relatively low cost of the clutch components compared to the time and trouble spent gaining access to them warrants their replacement anytime the engine or transmission is removed, unless they are new or in near perfect condition. The following procedures are based on the assumption the engine will stay in place.
2   Referring to Chapter 7 Part A, remove the transmission from the vehicle. Support the engine while the transmission is out. Preferably, an engine hoist or support fixture should be used to support it from above. However, if a jack is used underneath the engine, make sure a piece of wood is positioned between the jack and oil pan to spread the load. **Caution:** *The pickup for the oil pump is very close to the bottom of the oil pan. If the pan is bent or distorted in any way, engine oil starvation could occur.*
3   If you're working on a 2001 Pathfinder, push the wedge collar on the release bearing towards the engine, then pry the release bearing from the diaphragm spring of the pressure plate.
4   To support the clutch disc during removal, install a clutch alignment tool through the clutch disc hub.
5   Carefully inspect the flywheel and pressure plate for indexing marks. The marks are usually an X, an O or a white letter. If they cannot be found, scribe marks yourself so the pressure plate and the flywheel will be in the same alignment during installation **(see illustration)**.
6   Turning each bolt only 1/4-turn at a time, loosen the pressure plate-to-flywheel bolts. Work in a criss-cross pattern until all spring pressure is relieved, then hold the pressure plate securely and completely remove the bolts, followed by the pressure plate and clutch disc.

Inspection

Refer to illustrations 6.8, 6.10, 6.12a and 6.12b

7   Ordinarily, when a problem occurs in the clutch, it can be attributed to wear of the clutch driven plate assembly (clutch disc). However, all components should be inspected at this time.
8   Inspect the flywheel for cracks, heat checking, grooves and other obvious defects **(see illustration)**. If the imperfections are slight, a machine shop can machine the surface flat and smooth, which is highly recommended regardless of the surface appearance. Refer to Chapter 2 for the flywheel removal and installation procedure.
9   Inspect the pilot bushing (see Section 8). It's a good idea to replace it at this time, regardless of its condition.
10 Inspect the lining on the clutch disc. There should be at least 1/32-inch (0.8 mm) of lining above the rivet heads. Check for loose rivets, distortion, cracks, broken springs and other obvious damage **(see illustration)**. As mentioned above, ordinarily the clutch disc is routinely replaced, so if in doubt about the condition, replace it with a new one.
11 The release bearing should also be replaced along with the clutch disc (see Section 7).
12 Check the machined surfaces and the diaphragm spring fingers of the pressure plate **(see illustrations)**. If the surface is grooved or otherwise damaged, replace the

NORMAL FINGER WEAR

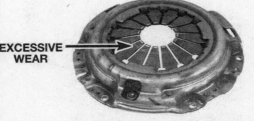
EXCESSIVE FINGER WEAR

BROKEN OR BENT FINGERS

6.12a Replace the pressure plate if excessive wear is noted

6.12b Examine the pressure plate friction surface for score marks, cracks and evidence of overheating

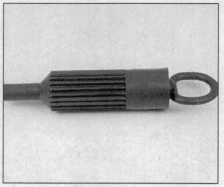

6.14 Center the clutch disc using a clutch alignment tool

pressure plate. Also check for obvious damage, distortion, cracking, etc. Light glazing can be removed with sandpaper or emery cloth. If a new pressure plate is required, new and factory-rebuilt units are available.

Installation

Refer to illustration 6.14

13 Before installation, clean the flywheel and pressure plate machined surfaces with brake system cleaner. It's important that no oil or grease is on these surfaces or the lining of the clutch disc. Handle the parts only with clean hands.
14 Position the clutch disc and pressure plate against the flywheel with the clutch held in place with an alignment tool **(see illustration)**. Make sure it's installed properly (most replacement clutch plates will be marked "flywheel side" or something similar - if not marked, install the clutch disc with the damper springs toward the transmission).
15 Tighten the pressure plate-to-flywheel bolts only finger tight, working around the pressure plate.
16 Center the clutch disc by ensuring the alignment tool extends through the splined hub and into the pilot bushing in the crankshaft. Wiggle the tool up, down or side-to-side as needed to bottom the tool in the pilot bushing. Tighten the pressure plate-to-flywheel bolts a little at a time, working in a criss-cross pattern to prevent distorting the cover. After all of the bolts are snug, tighten them to the torque listed in this Chapter's Specifications. Remove the alignment tool.
17 Using high-temperature grease, lubricate the inner groove of the release bearing. Also place grease on the transmission input shaft bearing retainer.
18 Install the clutch release bearing as described in Section 7.
19 Install the transmission and all components removed previously. Tighten all fasteners to the proper torque specifications.

7 Clutch release bearing - removal, inspection and installation

Warning: *Dust produced by clutch wear and deposited on clutch components is hazardous to your health. DO NOT blow it out with compressed air and DO NOT inhale it. DO NOT use gasoline or petroleum-based solvents to remove the dust. Brake system cleaner should be used to flush the dust into a drain pan. After the clutch components are wiped clean with a rag, dispose of the contaminated rags and cleaner in a covered, marked container.*

Removal

Refer to illustrations 7.3 and 7.4

1 Remove the release cylinder (see Section 4).
2 Remove the transmission (see Chapter 7, Part A).

All models except 2001 Pathfinders

3 Remove the boot from the side of the transmission and disengage the release lever retainer from the ballstud **(see illustration)**.
4 Note how the release lever fingers are engaged by the wire retainer on the bearing, then disengage the bearing from the lever and slide it off the input shaft **(see illustration)**. Remove the release lever. Inspect the release lever boot for cracks or tears. If it's worn or damaged, replace it. Inspect the wire retainer in the lever. If it's damaged or distorted, replace it.

2001 Pathfinder models

5 Push the wedge collar on the release bearing towards the engine, then pry the release bearing from the diaphragm spring of the pressure plate.
6 To remove the release lever from the transmission, remove the retaining pin and washer, pull out the pivot pin, then detach the lever.

Inspection

Refer to illustration 7.8

7 Hold the outer portion of the bearing and rotate the center while applying pressure. If the bearing doesn't turn smoothly or if it's noisy, replace it with a new one. Wipe the bearing with a clean rag and inspect it for damage, wear and cracks. Don't immerse the bearing in solvent - it's sealed for life and to do so would ruin it.
8 If the bearing needs to be replaced, remove it from the hub with a puller **(see illustration)** (all models except 2001 Pathfinders). The new bearing will have to be pressed onto the hub (if you don't have a press, take the hub and bearing to an automotive machine shop).

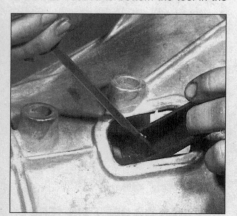

7.3 A screwdriver can be used to disengage the release lever retainer spring from the ballstud

7.4 After detaching the release bearing and hub from the release lever, slide it off the input shaft bearing retainer

7.8 A puller is needed to separate the release bearing from the hub (all except 2001 Pathfinder models)

Chapter 8 Clutch and driveline

7.10 The release bearing retainer must engage the lever like this

8.5 A small slide-hammer puller is handy for removing the pilot bushing

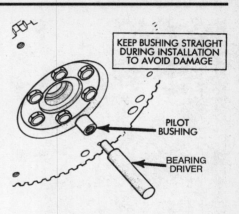

8.6 When installing the pilot bushing in the crankshaft, use a bearing driver

Installation

Refer to illustration 7.10

9 Lightly lubricate the clutch release lever where it contacts the release bearing hub and the ballstud.

10 On all models except 2001 Pathfinders, attach the release bearing to the release lever. Make sure the bearing is properly engaged by the retainer clip (see illustration).

11 On 2001 Pathfinders, attach the release lever to the transmission and install the pivot pin, washer and retaining clip. **Note:** *Be sure to lubricate the pivot pin with high-temperature grease.*

12 On all except 2001 Pathfinders, lubricate the clutch release lever ballstud or pivot pin with high-temperature grease, insert the release lever through the boot, slide the release bearing onto the input shaft bearing retainer and push the lever onto the ballstud until the lever retainer "pops" onto the stud. Make sure that the release lever pivots freely and the release bearing slides freely on the input shaft bearing retainer.

13 Apply a light coat of high-temperature grease to the face of the release bearing, where it contacts the pressure plate diaphragm fingers.

14 The remainder of installation is the reverse of the removal procedure. Tighten all transmission-to-engine bolts to the torque listed in the Chapter 7A specifications.

8 Pilot bushing - replacement

Refer to illustrations 8.5 and 8.6

1 The clutch pilot bushing is pressed into the rear of the crankshaft. It is greased at the factory and does not require additional lubrication. Its primary purpose is to support the front of the transmission input shaft. The pilot bushing should be inspected whenever the clutch components are removed from the engine. Due to its inaccessibility, if you are in doubt as to its condition, replace it with a new one. **Note:** *If the engine has been removed from the vehicle, disregard the following steps which do not apply.*

2 Remove the transmission (refer to Chapter 7, Part A).

3 Remove the clutch components (see Section 6).

4 Inspect for any excessive wear, scoring, lack of grease or obvious damage. If any of these conditions are noted, the bushing should be replaced. A flashlight will be helpful to direct light into the recess.

5 Removal can be accomplished with a slide hammer fitted with a puller attachment (see illustration), which are available at most auto parts stores or equipment rental yards.

6 To install the new bushing, lightly lubricate the outside surface with multi-purpose grease, then drive it into the recess with a hammer and bearing/bushing driver (see illustration). Apply a thin film of grease to the inner surface of the bearing.

7 Install the clutch components, transmission and all other components removed previously, tightening all fasteners properly.

9 Clutch pedal - adjustment

Pedal height

Refer to illustrations 9.1 and 9.3

1 With the clutch pedal fully released, measure the distance from the top of the pad to the floor (see illustration).

2 If the height is not as listed in the Specifications at the beginning of this Chapter it must be adjusted.

3 Loosen the locknut on the pedal stopper or cruise control (ASCD) switch (see illustration).

4 Turn the pedal stopper or cruise control switch until the pedal height is correct.

5 Tighten the locknut and recheck the clutch pedal height.

Pedal freeplay

6 Press down lightly on the clutch pedal and measure the distance that it moves freely before resistance is felt (see illustration 9.1). The freeplay should be within the specified limits listed at the beginning of this Chapter. If it isn't, it must be adjusted.

7 Loosen the locknut on the master cylinder pushrod (see illustration 9.3) and turn the rod clockwise or counterclockwise to adjust the clutch pedal freeplay.

8 Tighten the locknut and recheck the clutch pedal freeplay.

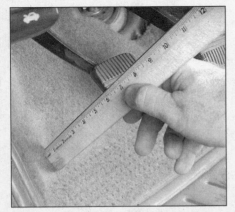

9.1 To check clutch pedal height, measure the distance between the pedal pad and the floor. To check clutch pedal freeplay, measure the distance between the natural resting place of the pedal and the point at which you encounter resistance

9.3 Clutch pedal adjustment details
1 Pushrod locknut
2 Pedal stopper or cruise control switch

Chapter 8 Clutch and driveline

10.1 Loosen the locknut (A) and turn the clutch start switch (B) in its bracket until the gap between the switch body and the depressed pedal is as listed in this Chapter's Specifications

10 Clutch start switch - check and replacement

Refer to illustration 10.1

1 The clutch start switch is located near the top of the clutch pedal, facing the opposite direction of the cruise control switch (or pedal stopper) **(see illustration)**.
2 Verify that the engine will not start when the clutch pedal is released.
3 Verify that the engine will start when the clutch pedal is depressed all the way.
4 If the clutch start switch doesn't perform as described above, loosen the locknut, depress the clutch pedal all the way and turn the switch in its bracket until the gap between the switch body and the pedal is adjusted to the clearance listed in this Chapter's Specifications. Check the operation of the switch again; if it still doesn't work properly, check switch continuity.
5 Verify that there is continuity between the clutch start switch terminals when the pedal is depressed.
6 Verify that no continuity exists between the switch terminals when the pedal is released.
7 If the switch fails either of these continuity tests, replace it: Loosen the nut near the body of the switch, then unscrew the switch from the bracket. Unplug the electrical connector. Installation is the reverse of removal.
8 Adjust the switch as described in Step 4.
9 Verify that the engine doesn't start when the clutch pedal is released.

11 Driveshaft and universal joints - general information and inspection

Refer to illustration 11.1

General information

1 A driveshaft is a tube, or a pair of tubes, that transmits power between the transmission (or transfer case on 4WD models) and the differential. Universal joints are located at either end of the driveshaft and in the center on two-piece driveshafts **(see illustration)**. **Note:** *On 2001 2WD Pathfinders, CV-type joints are used instead of universal joints.*
2 Single-piece rear driveshafts employ a splined yoke at the front, which slips into the extension housing of the transmission. This arrangement allows the driveshaft to slide back-and-forth within the transmission during vehicle operation to compensate for changes in length due to suspension movement. An oil seal prevents leakage of fluid at this point and keeps dirt from entering the transmission. If leakage is evident at the front of the driveshaft, replace the oil seal (see Chapter 7, Part B).
3 Two-piece driveshafts have a center support bearing. The center bearing is a ball-type bearing mounted in a rubber cushion attached to a frame crossmember. The bearing is pre-lubricated and sealed at the factory.
4 On all models, the driveshaft assembly requires very little service. The universal joints are lubricated for life and must be replaced if problems develop. The driveshaft must be removed from the vehicle for this procedure. **Note:** *At the time of writing, the U-joints on models with double-Cardan type joints (sort of a "double U-joint") are not replaceable. Check with your local auto parts store, automotive machine shop or driveline specialist if the joints on one of these types of driveshafts wear out.*
5 Since the driveshaft is a balanced unit, it's important that no undercoating, mud, etc. be allowed to stay on it. When the vehicle is raised for service it's a good idea to clean the driveshaft and inspect it for any obvious damage. Also, make sure the small weights used to originally balance the driveshaft are in place and securely attached. Whenever the driveshaft is removed it must be reinstalled in the same relative position to preserve the balance.
6 Problems with the driveshaft are usually indicated by a noise or vibration while driving the vehicle. A road test should verify if the problem is the driveshaft or another vehicle component. Refer to the *Troubleshooting* section at the front of this manual. If you suspect trouble, inspect the driveline.

Inspection

7 Raise the rear of the vehicle and support it securely on jackstands. Block the front wheels to keep the vehicle from rolling off the stands. Release the parking brake and place the transmission in Neutral.
8 Crawl under the vehicle and visually inspect the driveshaft. Look for any dents or cracks in the tubing. If any are found, the driveshaft must be replaced.
9 Check for oil leakage at the front and rear of the driveshaft. Leakage where the driveshaft enters the transmission or transfer case indicates a defective transmission/transfer case seal (see Chapter 7B). Leakage where the driveshaft enters the differential indicates a defective pinion seal (see Section 17).
10 While under the vehicle, have an assistant rotate a rear wheel so the driveshaft will rotate. As it does, make sure the universal joints are operating properly without binding, noise or looseness. Listen for any noise from the center bearing (if equipped), indicating it's worn or damaged. Also check the rubber portion of the center bearing for cracking or separation, which will necessitate replacement.
11 The universal joint can also be checked with the driveshaft motionless, by gripping your hands on either side of the joint and

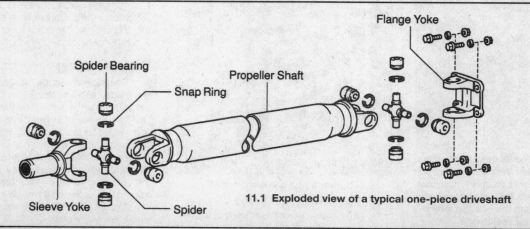

11.1 Exploded view of a typical one-piece driveshaft

Chapter 8 Clutch and driveline

12.2 Mark the relationship of the driveshaft flange to the pinion flange

12.4 On models with a two-piece driveshaft, unbolt the center support bearing from the crossmember

attempting to twist the joint. Any movement at all in the joint is a sign of considerable wear. Lifting up on the shaft will also indicate movement in the universal joints.
12 Finally, check the driveshaft mounting bolts at the ends to make sure they're tight.
13 On 4WD models, the above driveshaft checks should be repeated on the front driveshaft, as well. In addition, check for leakage around the sleeve yoke, indicating failure of the yoke seal.
14 Check for leakage where the driveshafts connect to the transfer case and front differential. Leakage indicates worn oil seals.
15 At the same time, check for looseness in the joints of the front driveaxles. Also check for grease or oil leakage from around the driveaxles by inspecting the rubber boots and both ends of each axle. Oil leakage around the axle flanges indicates a defective axleshaft oil seal. Grease leakage at the CV joint boots means a damaged rubber boot. For servicing of these components, see the appropriate Sections.

12 Driveshaft(s) - removal and installation

Warning: *The manufacturer recommends replacing the universal joint flange nuts and bolts, and the U-joint straps (on models so equipped) whenever they are removed.*

Rear driveshaft

Refer to illustrations 12.2 and 12.4

Removal

1 Raise the vehicle and support it securely on jackstands. Place the transmission in Neutral with the parking brake off.
2 Make reference marks on the driveshaft flange and the pinion flange in line with each other **(see illustration).** If you're working on a 1999 or later 2WD Pathfinder, also mark the front of the driveshaft to the transmission flange. This is to make sure the driveshaft is reinstalled in the same position to preserve the balance.
3 Remove the rear universal joint nuts and bolts. Turn the driveshaft (or wheels) as necessary to bring the bolts into the most accessible position.
4 If the vehicle has a two-piece driveshaft, unbolt the center support bearing from the crossmember **(see illustration).**
5 Lower the rear of the driveshaft. If you're working on a 1999 or later 2WD Pathfinder, unbolt the front of the driveshaft from the transmission flange. On all other models, slide the front of the driveshaft out of the transmission or transfer case.
6 Wrap a plastic bag over the transmission or transfer case housing and hold it in place with a rubber band. This will prevent loss of fluid and protect against contamination while the driveshaft is out.
7 If you intend to separate the rear portion of the driveshaft from the front portion on models with a two-piece shaft, mark the position of the driveshaft to the companion flange (they must be returned to the same relative position when reassembling them).

Installation

8 Remove the plastic bag from the transmission or transfer case and wipe the area clean. Inspect the oil seal carefully. Procedures for replacement of this seal can be found in Chapter 7B.
9 Slide the front of the driveshaft into the transmission or transfer case or, if you're working on a 1999 or later 2WD Pathfinder, line up the matchmarks, install the bolts and tighten them to the torque listed in this Chapter's Specifications.
10 On models with a two-piece driveshaft, position the center support bearing on the crossmember, install the fasteners and tighten them to the torque listed in this Chapter's Specifications.
11 Raise the rear of the driveshaft into position, checking to be sure the marks are in alignment. If not, turn the rear wheels to match the pinion flange and the driveshaft.
12 Install the bolts and nuts, tightening them to the torque listed in this Chapter's Specifications.

Front driveshaft (4WD models)

Removal

13 Raise the front of the vehicle and place it securely on jackstands. Remove the skid plate, if equipped.
14 Mark the relationship of the driveshaft to the front differential companion flange and to the transfer case companion flange.
15 Remove the bolts and nuts from the flanges, then lower the shaft from the vehicle.

Installation

16 Attach the ends of the shaft to the differential and transfer case companion flanges (be sure to line up the marks), install the bolts and nuts and tighten them to the torque listed in this Chapter's Specifications.
17 Install the skid plate (if equipped).

13 Driveshaft center support bearing - replacement

Refer to illustrations 13.5 and 13.6
Warning: *The manufacturer recommends replacing the universal joint flange nuts and bolts, and the U-joint straps (on models so equipped) whenever they are removed.*

1 Raise the vehicle and support it securely on jackstands. Remove the driveshaft (see Section 12).
2 Turn the bearing and verify that it operates freely and smoothly. If it's stiff or noisy, replace it.
3 Mark the relationship of the driveshaft to the front companion flange, then unbolt the center U-joint from the flange.
4 Unstake the flange retaining nut and remove it. A strap wrench or flange holding tool can be used to prevent the driveshaft from turning.
5 Mark the relationship of the front shaft to the flange **(see illustration).**

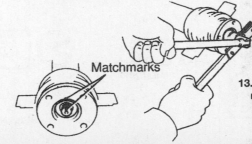

13.5 Remove the nut, then mark the relationship of the flange to the front portion of the driveshaft

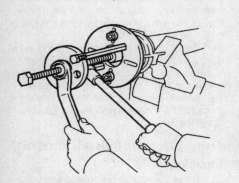

13.6 Remove the flange from the front portion of the shaft with a puller

14.3 Use a small pair of pliers to remove the snap-rings from the ends of the universal joint yokes

14.4 To remove the U-joint from the driveshaft, use a vise as a press - the small socket will push the cross and bearing cap into the large socket

6 Remove the flange from the intermediate shaft **(see illustration)**.

7 Remove the center bearing from the intermediate shaft. **Note:** *A hydraulic press may be required to push the shaft out of the center support bearing. If you can't get the bearing off and don't have a press, take the assembly (and the new center support bearing) to an automotive machine shop to have the old bearing removed and the new one installed.*

8 Installation is the reverse of removal. When installing the bearing on the front portion of the shaft, make sure the "F mark on the bearing faces the front of the vehicle. Be sure to tighten the flange retaining nut to the torque listed in this Chapter's Specifications, then stake the collar of the nut into the groove in the shaft.

14 Universal joints - replacement

Note 1: *Always purchase a universal joint service kit for your model vehicle before beginning this procedure. Also, read through the entire procedure before beginning work.*
Note 2: *On all models, select-fit snap-rings are available to adjust the universal joint endplay, which should be 0.0008-inch (0.02 mm) or less. Check with your local auto parts store or dealer service department if the endplay is greater than specified after the joint is assembled (or if the joint is extremely tight and won't free-up).*

1 Remove the driveshaft (see Section 12).

Outer snap-ring type

Refer to illustrations 14.3, 14.4, 14.5 and 14.6

2 Place the driveshaft on a bench equipped with a vise.
3 Remove the snap-rings with a small pair of pliers **(see illustration)**.
4 Support the cross (also called a spider) on a short piece of pipe or a large socket and use another socket to press out the cross by closing the vise **(see illustration)**.
5 Press the cross through as far as possible, then grip the bearing cap with pliers and remove it **(see illustration)**.
6 A universal joint repair kit will contain a new cross, seals, bearings, caps and snap-rings **(see illustration)**.
7 Inspect the bearing cap bores in the yokes for wear and damage.
8 If the bearing cap bores in the yoke are so worn that the caps are a loose fit, the driveshaft will have to be replaced with a new one.
9 Make sure the dust seals are properly located on the cross.
10 Using a vise, press one bearing cap into the yoke approximately 1/4-inch.
11 Use chassis grease to hold the needle rollers in place in the caps.
12 Insert the cross into the partially installed bearing cap, taking care not to dislodge the needle rollers.
13 Hold the cross in correct alignment and press both caps into place by slowly and carefully closing the jaws of the vise.
14 Use a socket slightly smaller in diameter than the caps to press them into the yoke. Press in one side, install the snap-ring, then press the other side to shift the cross assembly tight against the installed snap-ring and install the other snap-ring.
15 Repeat the operations for the remaining two bearing caps.

Inner snap-ring type

Refer to illustrations 14.16, 14.19 and 14.20

16 Remove the snap-rings located on the inner part of each bearing cap **(see illustration)**.
17 Press out the bearing caps as described in Steps 4 and 5.
18 Remove the cross from the yoke.
19 Reassembly is the same as for the outer snap-ring joint described in Steps 9 through 15, except the snap-rings are on the inner part of each bearing cap **(see illustration)**.

14.5 Locking pliers can be used to remove the bearing caps from the yoke

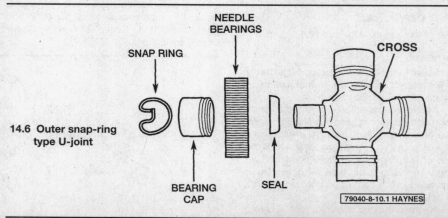

14.6 Outer snap-ring type U-joint

Chapter 8 Clutch and driveline

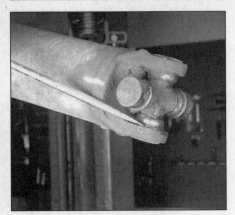

14.16 Remove the inner snap-rings from the U-joint by tapping them off with a screwdriver and hammer

20 When installing the bearing cap, press it in until the snap-ring can be installed **(see illustration)**.

All models

Refer to illustration 12.21

21 If the joint is stiff after assembly, strike the yoke sharply with a hammer **(see illustration)**. This will spring the yoke ears slightly and free up the joint.

15 Axles - description and check

Description

1 The rear axle assembly is a hypoid (the centerline of the pinion gear is below the centerline of the ring gear), semi-floating type. When the vehicle goes around a corner, the differential allows the outer rear wheel to turn at a higher speed than the inner tire. The axleshafts are splined to the differential side gears, so when the vehicle goes around a corner, the inner wheel, which turns more slowly than the outer wheel, turns its side gear more slowly than the outer wheel turns its side gear. The differential pinion (or spider) gears roll around the slower side gear, driving the outer side gear - and tire - more quickly.

2 On 4WD models, a fully independent front axle assembly is used. This consists of a differential and a pair of driveaxles. Each driveaxle has an inner and outer constant velocity (CV) joint.

3 An optional locking limited-slip rear axle is also available. This differential allows for normal operation until one wheel loses traction. A limited-slip unit is similar in design to a conventional differential, except for the addition of a pair of multi-disc clutch packs which slow the rotation of the differential case when one wheel is on a firm surface and the other is on a slippery one. The difference in wheel rotational speed produced by this condition applies additional force to the pinion gears and through the cone, which is splined to the axleshafts, equalizes the rotation speed of the axleshaft driving the wheel with traction.

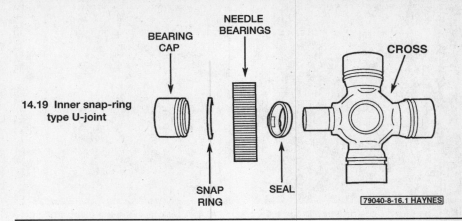

14.19 Inner snap-ring type U-joint

Check

4 Often, a suspected "axle" problem lies elsewhere. Do a thorough check of other possible causes before assuming the axle is the problem.

5 The following noises are those commonly associated with axle diagnosis procedures:

a) Road noise is often mistaken for mechanical faults. Driving the vehicle on different surfaces will show whether or not the road surface is the cause of the noise. Road noise will remain the same if the vehicle is under power or coasting.

b) Tire noise is sometimes mistaken for mechanical problems. Tires which are worn or low on pressure are particularly susceptible to emitting vibrations and noises. Tire noise will remain about the same during varying driving situations, where axle noise will change during coasting, acceleration, etc.

c) Engine and transmission noise can be deceiving because it will travel along the driveline. To isolate engine and transmission noises, make a note of the engine speed at which the noise is most pronounced. Stop the vehicle and place the transmission in Neutral and run the engine to the same speed. If the noise is the same, the axle is not at fault.

6 Because of the special tools needed, overhauling the differential isn't cost effective for a do-it-yourselfer. The procedures included in this Chapter describe axleshaft removal and installation, axleshaft oil seal replacement, axleshaft bearing replacement and removal of the entire unit for repair or replacement. Any further work should be left to a qualified repair shop.

16 Axleshaft, bearing and oil seals (rear) - removal, bearing/seal replacement and installation

Refer to illustrations 16.3, 16.4, 16.6a and 16.6b

Warning: *The manufacturer recommends replacing the axle retaining plate nuts with new ones whenever they are removed.*

Removal

1 Loosen the wheel lug nuts, raise the rear of the vehicle and support it securely on jackstands. Chock the front wheels to prevent the vehicle from rolling. Remove the wheel.

2 Remove the brake drum and brake shoes. Disconnect the parking brake cable from the lever on the trailing shoe and unbolt the cable casing from the backing plate. Also disconnect the hydraulic line fitting from the wheel cylinder (see Chapter 9).

14.20 Installing a snap-ring on an inner snap-ring type U-joint

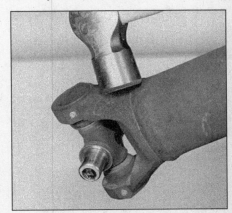

14.21 Strike the yoke sharply with a hammer to "spring" the yoke ears, which will free-up the joint

8-12 Chapter 8 Clutch and driveline

16.3 The rear axle is retained in the axle housing by these four nuts

16.4 After removing the axle retaining nuts, pull the axle from the housing with a slide hammer and adapter connected to the axle flange

16.6a Prying out the axleshaft oil seal with a seal removal tool

3   Remove the nuts securing the axle retaining plate to the axle housing (see illustration).

4   Connect a slide hammer and adapter to the axle flange and pull the axle from the housing (see illustration). Note: *The rear axle on four-cylinder, 2WD Frontier and Xterra models use shims to adjust axleshaft endplay. Be sure to retrieve any shims from between the axle housing and retainer gasket.*

5   If the bearing or bearing grease seal must be replaced, take the assembly to an automotive machine shop to have the old bearing and seal removed and new ones installed.

6   The oil seal in the axle tube can be pried our with a large screwdriver or seal removal tool (see illustration). The new seal should be driven in with a seal installer (see illustration). If you don't have a seal installer, a large socket with an outside diameter just slightly smaller than that of the seal can be used as a driver.

Installation

7   Wipe the bore in the axle housing clean. Apply a thin coat of grease to the outer surface of the bearing.

8   If you're working on a four-cylinder, 2WD Frontier or Xterra, install any shims that were present during removal. Note: *The standard thickness of the shims and gasket is 0.059-inch (1.5 mm).*

9   Guide the axleshaft straight into the axle housing. Rotate the shaft slightly to engage the splines on the shaft with the splines in the differential side gear.

10 Install new axle retaining plate nuts and tighten them to the torque listed in this Chapter's Specifications.

11 Check the axleshaft endplay using a dial indicator.

a) *On four-cylinder, 2WD Frontier and Xterra models the endplay should be 0.0008 to 0.0059-inch (0.02 to 0.15 mm). If not, add or subtract shims as necessary to obtain the desired endplay.*

b) *On all other models the endplay should be 0.0-inch (0.0 mm). If there is play, the axle bearing is either worn out (used bearing) or was not adjusted properly when installed (new bearing).*

12 The remainder of installation is the reverse of removal. Check the differential lubricant level and add some, if necessary. Bleed the brakes (see Chapter 9).

13 Install the wheel and lug nuts. Lower the vehicle and tighten the lug nuts to the torque listed in the Chapter 1 Specifications.

17 Pinion oil seal - replacement

1   Loosen the wheel lug nuts. Raise the front (for front differential) or rear (for rear differential) of the vehicle and support it securely on jackstands. Block the opposite set of wheels to keep the vehicle from rolling off the stands. Remove the wheels.

2   Disconnect the driveshaft from the differential companion flange and fasten it out of the way (see Section 12).

Rear pinion oil seal - Frontier and Xterra models

Refer to illustrations 17.3, 17.4, 17.5, 17.7 and 17.9

3   Rotate the pinion a few times by hand. Use a beam-type or dial-type inch-pound torque wrench to check the torque required to rotate the pinion (see illustration). Record it for use later.

4   Mark the relationship of the pinion flange to the shaft (see illustration), then count and write down the number of exposed threads on the shaft.

5   A flange holding tool will be required to keep the companion flange from moving while the self-locking pinion nut is loosened.

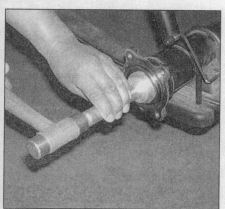

16.6b Driving in the axleshaft oil seal with a seal installer

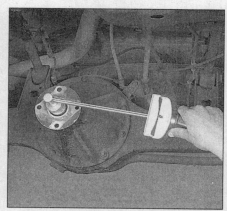

17.3 Use an inch-pound torque wrench to check the torque necessary to rotate the pinion shaft

17.4 Mark the position of the flange to the shaft and count the number of exposed threads above the nut

Chapter 8 Clutch and driveline   8-13

17.5 A chain wrench is being used here to prevent the pinion flange from turning while the nut is loosened

A chain wrench will also work **(see illustration)**.
6 Remove the pinion nut.
7 Withdraw the flange. It may be necessary to use a two-jaw puller engaged behind the flange to draw it off **(see illustration)**. Do not attempt to pry or hammer behind the flange or hammer on the end of the pinion shaft.
8 Pry out the old seal and discard it.
9 Lubricate the lips of the new seal and fill the space between the seal lips with wheel bearing grease, then tap it evenly into position with a seal installation tool or a large socket **(see illustration)**. Make sure it enters the housing squarely and is tapped in to its full depth.
10 Install the pinion flange, lining up the marks made in Step 4. If necessary, tighten the pinion nut to draw the flange into place. Do not try to hammer the flange into position.
11 Apply a bead of RTV sealant to the ends of the splines visible in the center of the flange so oil will be sealed in.
12 Install the pinion nut. Tighten the nut until the number of threads exposed in Step 4 are showing.
13 Measure the torque required to rotate the pinion and tighten the nut in small increments (no more than 5 ft-lbs) until it matches

17.9 Lubricate the lips of the new seal and seat it squarely in the bore, then drive it into the carrier with a seal driver or a large socket

17.7 If you can't pull the pinion flange off by hand, remove it with a puller

the figure recorded in Step 3. To compensate for the drag of the new oil seal, the nut should be tightened a little more until the rotational torque of the pinion exceeds the earlier recording by 5 in-lbs. The preload should fall within the range listed in this Chapter's Specifications. **Caution:** *If the maximum nut torque is reached before the desired preload is obtained, the differential must be disassembled and a new collapsible spacer must be installed.*
14 Reinstall all components removed previously by reversing the removal Steps, tightening all fasteners to their specified torque values.

Rear pinion oil seal (Pathfinder models) and front pinion oil seal (all 4WD models)

15 A flange holding tool will be required to keep the companion flange from moving while the self-locking pinion nut is loosened. A chain wrench will also work **(see illustration 17.5)**.
16 Remove the pinion nut.
17 Withdraw the flange. It may be necessary to use a two-jaw puller engaged behind the flange to draw it off **(see illustration 17.7)**. Do not attempt to pry or hammer behind the flange or hammer on the end of the pinion shaft.
18 Pry out the old seal and discard it.
19 Lubricate the lips of the new seal and fill the space between the seal lips with wheel bearing grease, then tap it evenly into position with a seal installation tool or a large socket **(see illustration 17.9)**. Make sure it enters the housing squarely and is tapped in to its full depth.
20 Install the pinion flange. If necessary, tighten the pinion nut to draw the flange into place. Do not try to hammer the flange into position.
21 Apply a bead of RTV sealant to the ends of the splines visible in the center of the flange so oil will be sealed in.
22 Install the pinion nut. Tighten the nut to the torque listed in this Chapter's Specifications.
23 Reconnect the driveshaft to the pinion flange (see Section 12).

18 Axle assembly (rear) - removal and installation

Removal
1 Loosen the rear wheel lug nuts, raise the rear of the vehicle and support it securely on jackstands placed under the frame rails. Block the front wheels to keep the vehicle from rolling off the stands. Remove the rear wheels.
2 Position a jack under the rear axle differential housing. If you have two floor jacks, position one under each axle tube.
3 Disconnect the driveshaft from the rear axle pinion flange (see Section 12). Fasten the driveshaft out of the way with a piece of wire from the underbody.
4 Disconnect the shock absorbers at their lower mounts.
5 Disconnect the vent hose from the fitting on the axle housing and fasten it out of the way.
6 Unscrew the brake line fittings from the brake hose junction block, then unbolt the brake hose junction block from the axle housing (see Chapter 9). Plug the hose to prevent fluid leakage. Also disconnect the ABS wheel speed sensor electrical connector(s).
7 If you're working on a Pathfinder or a 2WD Frontier model, remove the brake drums and brake shoes, disconnect the parking brake cables from the actuating levers, then unbolt the cable casings from the brake backing plates (see Chapter 9). On all other models, disconnect the parking brake cables from the crank levers on the backing plates.
8 On models with leaf springs, unscrew the nuts from the spring U-bolts. Remove the spring plates.
9 On Pathfinder and Xterra models, unbolt the stabilizer bar brackets from the axle. Remove the coil springs, then unbolt the suspension arms and Panhard rod from the axle housing (see Chapter 10).
10 Lower the jack under the differential, then remove the rear axle assembly from under the vehicle.

Installation
11 Installation is the reverse of removal. Tighten the U-joint strap bolts to the torque listed in this Chapter's Specifications. Tighten all suspension fasteners to the torque values listed in the Chapter 10 Specifications.
12 Bleed the brakes (see Chapter 9).

19 Free-running hubs (4WD models) - removal and installation

Manual-lock free-running hubs
Refer to illustration 19.3
1 Turn the knob to the Free position.
2 Remove the hub mounting bolts and pull the hub assembly off the driveaxle splines. Remove the snap-ring from the end of the driveaxle and remove the spindle washer and thrust washer.

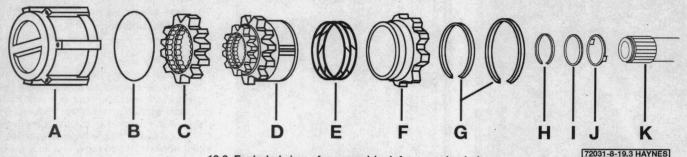

19.3 Exploded view of a manual-lock free-running hub

- A Hub body
- B O-ring
- C Clutch ring
- D Drive gear
- E Wave spring
- F Bronze bearing
- G Snap-rings (in hub assembly)
- H Snap-ring (on driveaxle)
- I Spindle washer
- J Thrust washer
- K Driveaxle

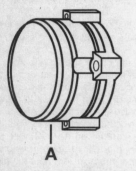  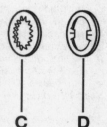 

19.5a Exploded view of an auto-lock free-running hub - 1998 through 2000 Frontier models

- A Auto-lock hub assembly
- B Snap-ring
- C Spindle washer
- D Thrust washer
- E Driveaxle

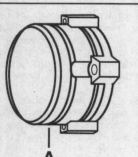  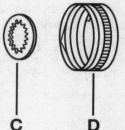 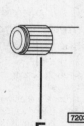

19.5b Exploded view of an auto-lock free-running hub - 2000 and later Xterra models, 2001 Frontier models

- A Auto-lock hub assembly
- B Snap-ring
- C Spindle washer
- D Fixed cam assembly
- E Driveaxle

3 The hub clutch assemblies can become clogged with dirt and water which can make them inoperable. If further disassembly and inspection of the clutch assembly is required pay very close attention to the way the parts fit together when disassembling **(see illustration)**. Lay all the parts out in the order in which they were removed. Clean the parts one at a time and lay them back out in the same sequence. Lubricate the parts with a light coat of multi-purpose grease, then reassemble the clutch assembly in the reverse order.

4 Installation of the hub assembly onto the driveaxle is the reverse of the removal, with the following points:

a) Replace the snap-ring with a new one.
b) Apply multi-purpose grease to the inner hub splines, the driveaxle splines and the snap-ring.
c) The control handle should be set to the Free position when installing the assembly.
d) Tighten the mounting bolts to the torque listed in this Chapter's Specifications.

Auto-lock free-running hubs

Refer to illustrations 19.5a and 19.5b

5 Remove the hub mounting bolts and pull the hub assembly off the driveaxle splines. Remove the snap-ring from the end of the driveaxle and remove the spindle washer and thrust washer (1998 through 2000 Frontier models) or fixed cam assembly (2000 and later Xterra models and 2001 Frontier models) from the driveaxle **(see illustrations)**.

6 Clean the washer(s) and fixed cam assembly (if equipped) and inspect them for signs of wear, replacing parts as necessary.

7 Installation is the reverse of the removal sequence, with the following points:

a) Replace the snap-ring with a new one.
b) Apply multi-purpose grease to the inner hub splines, driveaxle splines, spindle washer, thrust washer or fixed cam assembly, and the snap-ring.
c) To seat the snap-ring, use a deep socket that fits over the driveaxle, then tap on it to seat the snap-ring in the inner locking groove.
d) Tighten the mounting bolts to the torque listed in this Chapter's Specifications.

20 Driveaxles (4WD models) - general information and inspection

1 Power is transmitted from the front differential/axle to the front wheels through a pair of driveaxles. The inner end of each driveaxle is bolted to an axleshaft connected to the differential side gears; the outer end of each driveaxle has a stub shaft that is splined to the front hub and bearing assembly.

2 The inner ends of the driveaxles are equipped with sliding constant velocity (CV)

Chapter 8 Clutch and driveline

joints, which are capable of both angular and axial motion. Each inner CV joint assembly consists of a tripot-type bearing and a housing in which the joint is free to slide in-and-out as the driveaxle moves up-and-down with the wheel.

3   The outer ends of the driveaxles are equipped with "ball-and-cage" type CV joints, which are capable of angular but not axial movement. Each outer CV joint consists of six caged ball bearings running between an inner race and the housing.

4   The boots should be inspected periodically for damage and leaking lubricant. Torn CV joint boots must be replaced immediately or the joints will be damaged. If either boot of a driveaxle is damaged, that driveaxle must be removed in order to replace the boot (see Section 21).

5   Should a boot be damaged, the CV joint can be disassembled and cleaned (see Section 22), but if any parts are damaged, the entire driveaxle assembly must be replaced as a unit.

6   The most common symptom of worn or damaged CV joints, besides lubricant leaks, is a clicking noise in turns, a clunk when accelerating after coasting and vibration at highway speeds. To check for wear in the CV joints and driveaxle shafts, grasp each axle (one at a time) and rotate it in both directions while holding the CV joint housings, feeling for play indicating worn splines or sloppy CV joints. Also check the driveaxle shafts for cracks, dents and distortion.

21 Driveaxle (4WD models) - removal and installation

Frontier and Xterra models

Refer to illustrations 21.4 and 21.9

Removal

1   Loosen the wheel lug nuts, raise the front of the vehicle and support it securely on jackstands. Remove the wheel.

2   Remove the free-running hub (see Section 19), then remove the snap-ring from the driveaxle splines. **Note:** *The manufacturer recommends replacing the snap-ring with a new one whenever it is removed, but save the old one for now; it will be used for endplay measurement purposes when the driveaxle is installed.*

3   Remove the skid plate, if equipped.

4   Remove the driveaxle-to-axleshaft flange bolts **(see illustration)**. Have an assistant apply the brake as you loosen the bolts to prevent the driveaxle from turning. Separate the driveaxle from the axleshaft flange.

5   Place a floor jack under the outer end of the lower control arm. Raise the jack slightly. **Warning:** *The jack must remain in this position throughout the entire procedure.*

6   Separate the tie-rod end from the steering knuckle (see Chapter 10).

7   Remove the cotter pin and loosen, but don't remove, the nut from the upper control

21.4 Remove the driveaxle-to-axleshaft flange bolts (four out of six are shown here)

arm balljoint (see Chapter 10, if necessary). Break the steering knuckle loose from the balljoint by striking the steering knuckle boss where the ballstud passes through.

8   Remove the nut and separate the upper control arm from the steering knuckle. Pull out on the knuckle and push the driveaxle through it, then maneuver the driveaxle out. **Caution:** *Be careful not to stretch the brake hose. Don't lose the thrust washer on the outer end of the driveaxle.*

Installation

9   Inspect the needle bearing in the knuckle. If it shows signs of wear, replace it with a new one. **Note:** *If you install a new bearing, make sure it faces in the proper direction* **(see illustration)**.

10   Install the thrust washer on the driveaxle, if removed. The chamfer on the washer must face the center of the driveaxle.

11   Lubricate the needle bearing in the steering knuckle with multi-purpose grease. Also lubricate the driveaxle splines.

12   Guide the driveaxle into position, pass the outer end of the driveaxle through the steering knuckle, then reconnect the upper balljoint to the steering knuckle. Tighten the nut to the torque listed in the Chapter 10 Specifications and install a new cotter pin.

13   Connect the inner end of the driveaxle to the axleshaft flange and install the bolts, tightening them to the torque listed in this Chapter's Specifications.

14   Connect the tie-rod end to the steering knuckle and tighten the nut to the torque listed in the Chapter 10 Specifications. Install a new cotter pin.

15   Install the snap-ring (which was removed in Step 2) in the inner groove on the driveaxle. Using a dial indicator, measure the endplay of the driveaxle in the hub **(see illustration 21.32)**. The endplay should fall within the range listed in this Chapter's Specifications. If the endplay is OK, remove the snap-ring, measure its thickness with a micrometer and install a new one of the same thickness (the manufacturer recommends replacing the snap-ring with a new one whenever it is removed).

16   If the endplay is not within the range listed in this Chapter's Specifications, install

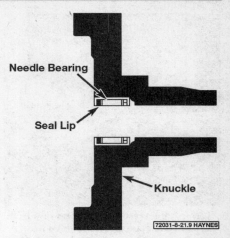

21.9 When installing a new needle bearing in the steering knuckle, make sure the seal faces the inside of the knuckle

a snap-ring of the proper thickness. Select-fit snap-rings are available in seven thicknesses, ranging from 0.043-inch (1.1 mm) to 0.083-inch (2.1 mm), in 0.008-inch (0.2 mm) increments.

17   Install the free-running hub assembly (see Section 19).

18   Install the wheel and lug nuts. Lower the vehicle and tighten the lug nuts to the torque listed in the Chapter 1 Specifications.

Pathfinder models

Refer to illustrations 21.20, 21.21, 21.25, 21.31, 21.32 and 21.34

Removal

19   Loosen the wheel lug nuts, raise the front of the vehicle and support it securely on jackstands. Remove the wheel.

20   Pull off the hub cap **(see illustration)**. **Note:** *Before using pliers to pull the cap off, move the cap off its seat by driving a chisel between the cap and the drive flange, a little at a time, working around the circumference of the cap. (The manufacturer recommends replacing the cap with a new one whenever it is removed, but if you are very careful and don't distort it during removal, it can be reused.)*

21.20 If you're working on a Pathfinder, remove the hub cap . . .

8-16   Chapter 8 Clutch and driveline

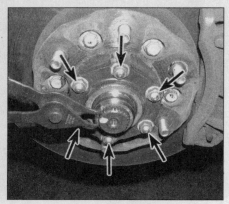

21.21 ... then remove the snap-ring from the driveaxle, unscrew the drive flange nuts and remove the flange from the hub

21.25 Pull out on the steering knuckle and pull the driveaxle out of the hub

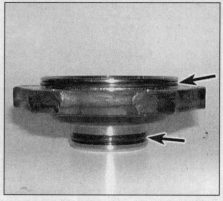

21.31 Replace the drive flange O-rings if they aren't in perfect condition

21 Remove the snap-ring from the end of the driveaxle, then unscrew the nuts and remove the drive flange from the hub **(see illustration)**. See the **Note** at the end of Step 2.
22 Remove the skid plate, if equipped.
23 Remove the driveaxle-to-axleshaft flange bolts **(see illustration 21.9)**. Have an assistant apply the brake as you loosen the bolts to prevent the driveaxle from turning. Separate the driveaxle from the axleshaft flange.
24 Remove the three nuts and detach the balljoint from the control arm (see Chapter 10). **Warning:** *The manufacturer recommends replacing the balljoint nuts with new ones whenever they are removed.*
25 Pull out on the knuckle and push the driveaxle through it, then maneuver the driveaxle out **(see illustration)**. **Caution:** *Be careful not to stretch the brake hose, and don't lose the thrust washer on the outer end of the driveaxle.*

Installation

26 Inspect the needle bearing in the knuckle. If it shows signs of wear, replace it with a new one. **Note:** *If you install a new bearing, make sure it faces in the proper*

direction **(see illustration 21.9)**.
27 Install the thrust washer on the driveaxle, if removed. The chamfer on the washer must face the center of the driveaxle.
28 Lubricate the needle bearing in the steering knuckle with multi-purpose grease. Also lubricate the driveaxle splines.
29 Guide the driveaxle into position, pass the outer end of the driveaxle through the steering knuckle, the connect the inner end to the axleshaft flange. Install the bolts, tightening them to the torque listed in this Chapter's Specifications.
30 Connect the balljoint to the control arm. Install new nuts and tighten them to the torque listed in the Chapter 10 Specifications.
31 Check the condition of the O-rings on the drive flange. If they are cracked, hardened or show any signs of deterioration, replace them **(see illustration)**. Install the drive flange, tightening the nuts to the torque listed in this Chapter's Specifications.
32 Install the snap-ring (which was removed in Step 21) in the groove on the driveaxle. Using a dial indicator, measure the endplay of the driveaxle in the hub **(see illustration)**. The endplay should fall within the range listed in this Chapter's Specifications. If the endplay is OK, remove the snap-ring,

measure its thickness with a micrometer and install a new one of the same thickness (the manufacturer recommends replacing the snap-ring with a new one whenever it is removed).
33 If the endplay is not within the range listed in this Chapter's Specifications, install a snap-ring of the proper thickness. Select-fit snap-rings are available in seven thicknesses, ranging from 0.043-inch (1.1 mm) to 0.083-inch (2.1 mm), in 0.008-inch (0.2 mm) increments.
34 Install the hub cap **(see illustration)**.
35 Install the wheel and lug nuts. Lower the vehicle and tighten the lug nuts to the torque listed in the Chapter 1 Specifications.

22 Driveaxle boot - replacement

Note: *Complete rebuilt driveaxles are available on an exchange basis, which eliminates much time and work.*

1 Remove the driveaxle (see Section 21).

Inner CV joint

Disassembly

Refer to illustrations 22.2a, 22.2b, 22.4a, 22.4b and 22.5

2 Remove both boot clamps and discard them, then slide the boot down the shaft and out of the way **(see illustrations)**.

21.32 Connect a dial indicator like this, grasp the outer CV joint and move it in-and-out, noting the total endplay on the dial

21.34 When installing the hub cap, use a large socket or a piece of pipe that contacts the flange of the cap - don't hit the center of the cap to drive it on

22.2a Pry up the retaining tabs on the boot clamps . . .

Chapter 8 Clutch and driveline

22.2b . . . then open the clamps and remove them from the boot

22.4a Remove the snap-ring from the groove in the end of the axleshaft

22.4b Use a center punch to place marks on the spider and shaft to ensure that they're properly reassembled

3 Mark the relationship of the inner joint housing to the shaft. Mount the driveaxle shaft in a vise with padded jaws, then tap the inner joint housing towards the center of the shaft to dislodge the cover on the end of the joint. Remove the cover and slide the joint housing down the shaft and out of the way.

4 Remove the snap-ring from the end of the shaft, then mark the relationship of the shaft and spider so they can be reassembled

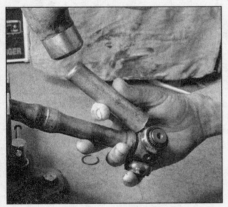

22.5 Drive the spider from the shaft with a hammer and brass punch

in the original position **(see illustrations)**.

5 Using a hammer and a brass punch, drive the spider assembly off the shaft **(see illustration)**.

6 Slide the housing and boot off the axleshaft.

Inspection

7 Clean the components with solvent to remove all traces of grease. Inspect all of the components for pitting, score marks, cracks and other signs of wear and damage. Shiny, polished spots are normal and will not adversely affect CV joint performance. Also check the splines on the shaft and in the spider for wear. If any undesirable conditions exist, the entire joint (or driveaxle assembly) will have to be replaced.

Reassembly

Refer to illustrations 22.8, 22.10, 22.15a, 22.15b, 22.16a and 22.16b

8 Wrap the axleshaft splines with tape to avoid damaging the boot. Slide the small boot clamp and boot onto the axleshaft **(see illustration)**, then remove the tape. Slide the large boot clamp over the boot.

9 Slide the housing onto the shaft.

10 Install the spider assembly onto the shaft, lining up the marks you made in Step 4.

The side with the chamfered splines must face toward the center of the shaft **(see illustration)**.

11 Install a new snap-ring, making sure it seats completely in its groove.

12 Lubricate the spider assembly and housing with CV joint grease, then move the housing up onto the spider, aligning the mark on the housing (made in Step 3) with the mark on the shaft. Apply a bead of RTV sealant to a new cover, then tap the cover into place.

13 Fill the housing with approximately 3-1/2 ounces (100 g) of CV joint grease (normally included with the new boot kit). **Caution:** *Use CV joint grease only.*

14 Wipe any excess grease from the boot grooves on the shaft and the housing. Seat the small diameter of the boot in the recessed area on the axleshaft. Push the other end of the boot onto the housing and move the race in or out until there's no deformation (distortion or dents) in the boot.

15 Adjust the CV joint to the length listed in this Chapter's Specifications (measured from one end of the boot to the other) **(see illustration)**. Equalize the pressure in the boot by inserting a dull screwdriver between the boot and the outer race **(see illustration)**. Don't damage the boot with the tool.

22.8 Wrap the splined area of the axleshaft with tape to prevent damage to the boots when installing them

22.10 Install the spider with the chamfered ends of the splines facing toward the center of the axleshaft

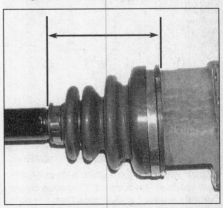

22.15a Adjust the length of the CV joint to the dimension listed in this Chapter's Specifications . . .

8-18 Chapter 8 Clutch and driveline

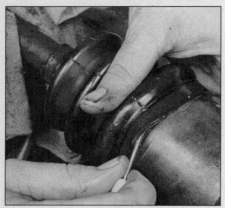

22.15b . . . then equalize the pressure inside the boot by inserting a dull screwdriver between the boot and the housing

22.16a To install the new clamps, bend the tang down . . .

22.16b . . . and tap the tabs down to hold it in place

22.19 To remove the outer CV joint, clamp the axleshaft in a bench vise (with padded jaws) and tap the joint off with a brass hammer

22.21 After the old grease has been rinsed away, move the inner race through its full range of motion and inspect the bearing surfaces for wear or damage - if any of the balls, the race or the cage look damaged, replace the outer joint assembly

22.24 Apply CV joint grease through the splined hole, then insert a wooden dowel (slightly smaller in diameter than the hole) into the hole and push down - the dowel will force the grease into the joint

16 Install the boot clamps **(see illustrations)**.
17 Install the driveaxle (see Section 10).

Outer CV joint

Disassembly

Refer to illustration 22.19

18 Remove the boot clamps and separate the boot from the outer CV joint **(see illustrations 22.2a and 22.2b)**.
19 Clamp the axleshaft in a bench vise (equipped with protective jaws) and drive off the outer CV joint with a brass hammer **(see illustration)**, then remove the retainer ring for the inner race. Slide off the old boot.
20 Thoroughly wash the outer CV joint in clean solvent then rinse it out with brake system cleaner to remove all traces of solvent.

Inspection

Refer to illustration 22.21

21 Rotate the outer CV joint housing at an angle to the driveaxle to expose the bearings, inner race and cage **(see illustration)**. Inspect the bearing surfaces for signs of wear. If the CV joint is worn, replace it.

Reassembly

Refer to illustrations 22.24, 22.25 and 22.26

22 Slide the new outer boot and small clamp onto the driveaxle. It's a good idea to wrap vinyl tape around the shaft splines to prevent damage to the boot **(see illustration 22.8)**. Remove the tape.
23 Install a new inner race retainer ring on the shaft.
24 Pack the CV joint with approximately 5-ounces (140 g) of CV joint grease through the splined hole in the inner race. **Caution:** *Use CV joint grease only.* Force the grease into the joint by inserting a wooden dowel through the splined hole and pushing it to the bottom of the joint. Repeat this procedure until the bearing is completely packed **(see illustration)**. **Note:** *If the specified quantity of grease can't be packed into the joint, place the remainder in the CV joint boot.*
25 Tap the outer CV joint into place with a hammer and a wood block **(see illustration)**.
26 Slide the boot into position. When the boot is in position add the remainder of the grease in the boot replacement kit to the CV

Chapter 8 Clutch and driveline

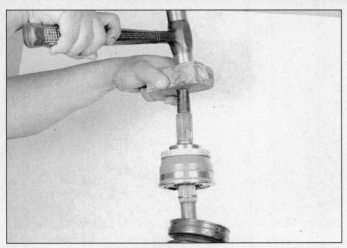

22.25 To install the outer CV joint, put the axleshaft in a bench vise (equipped with protective jaws) and tap on the CV joint with a hammer and a block of wood; drive the joint onto the axleshaft splines until the retainer ring on the shaft seats in the groove in the inner race of the joint

22.26 Adjust the boot to the length indicated in this Chapter's Specifications, then tighten the clamps

joint boot. Slide the boot onto the joint and adjust the length **(see illustration)**, equalize the pressure inside the boot **(see illustration 22.15b)** and install the new clamps **(see illustrations 22.16a and 22.16b)**.

27 Install the driveaxle (see Section 21).

23 Axle assembly (front) - removal and installation

1 Raise the vehicle and support it securely on jackstands. Remove the under-vehicle splash shield, then detach the vent hose from the differential housing.
2 Mark the relationship of the driveshaft to the differential companion flange, then unbolt the driveshaft from the flange (see Section 12). Support the driveshaft with a piece of wire from the underbody.
3 Unbolt the inner ends of the driveaxles (see Section 21) and support them with pieces of wire from the underbody.

Frontier and Xterra models

Refer to illustration 23.6

4 Support the engine from above with an engine hoist or an engine support fixture. Unbolt the engine mounts (see Chapter 2A or 2B) and raise the engine for clearance.
5 Support the front axle assembly with a floor jack placed under the differential.
6 Unbolt the differential and axle tube mounts from the front suspension crossmember **(see illustration)**.
7 Remove the bolt from each end of the differential mounting member, then maneuver the axle/differential assembly out from under the vehicle.
8 Installation is the reverse of the removal procedure, noting the following points:

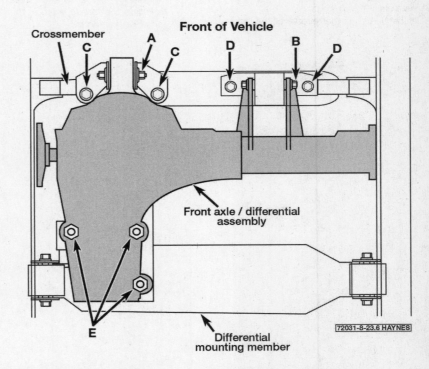

23.6 Front axle/differential mounting details (Frontier and Xterra models) - when installing the unit, temporarily tighten nut A, then nut B; now tighten bolts C and D, nuts A, B and E, in that order, to the torque values listed in this Chapter's Specifications

a) Tighten the mounting fasteners to the torque listed in this Chapter's Specifications, in the order shown in **illustration 23.6**.
b) Tighten the engine mount fasteners to the torque values listed in the Chapter 2A or 2B Specifications.
c) Tighten the driveshaft and driveaxle fasteners to the torque listed in this Chapter's Specifications.
d) Check the differential lubricant level and add some, if necessary, to bring it to the appropriate level (see Chapter 1).

Pathfinder models

Refer to illustrations 23.10a and 23.10b

9 Support the front axle assembly with a floor jack placed under the differential.
10 Remove the mounting bolts from the

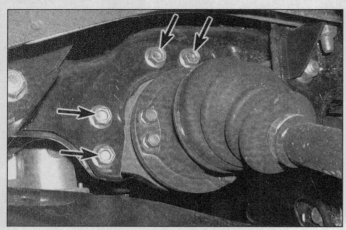

23.10a Front axle/differential assembly mounting bolts (Pathfinder) - right side

23.10b Front axle/differential assembly mounting bolts (Pathfinder) - left side

support brackets at either end of the axle/differential assembly **(see illustrations)**. Slowly lower the assembly and remove it from under the vehicle.

11 Installation is the reverse of the removal procedure, noting the following points:
a) *Tighten the mounting fasteners securely.*
b) *Tighten the driveshaft and driveaxle fasteners to the torque listed in this Chapter's Specifications.*
c) *Check the differential lubricant level and add some, if necessary, to bring it to the appropriate level (see Chapter 1).*

Chapter 9 Brakes

Contents

	Section
Anti-lock Brake System (ABS) - general information and trouble codes	2
Brake caliper - removal and installation	4
Brake check	See Chapter 1
Brake disc - inspection, removal and installation	5
Brake fluid level check	See Chapter 1
Brake hoses and lines - inspection and replacement	9
Brake hydraulic system - bleeding	11
Brake light switch - check, adjustment and replacement	15
Brake pedal adjustments	See Chapter 1
Disc brake pads - replacement	3
Drum brake shoes - replacement	6
General information	1
Load Sensing Valve (LSV) - description and replacement	10
Master cylinder - removal, installation and reservoir/seal replacement	8
Parking brake - adjustment	13
Parking brake cables - replacement	14
Power brake booster - check, replacement and adjustment	12
Wheel cylinder - removal and installation	7

Specifications

General
Brake fluid type .... See Chapter 1

Disc brakes
Minimum pad lining thickness .... See Chapter 1
Brake disc minimum thickness .... Cast into disc
Maximum disc runout
   Frontier, Xterra .... 0.0028 inch (0.07 mm)
   Pathfinder .... 0.004 inch (0.1 mm)
Maximum disc thickness variation
   Frontier, Xterra .... 0.0008 inch (0.02 mm)
   Pathfinder .... 0.0006 inch (0.015 mm)

Drum brakes
Shoe friction material minimum thickness .... See Chapter 1
Maximum inside diameter .... Cast into drum
Maximum out-of-round .... 0.0012 inch (0.03 mm)

Power brake booster
Output rod length .... 0.4045 to 0.4144 inch (10.275 to 10.525 mm)
Booster-to-center of clevis hole dimension
  Frontier
    Four-cylinder models
      2WD .... 6.30 inches (160 mm)
      4WD .... 6.50 inches (165 mm)
    V6 models
      1999, 2000 .... 5.31 inches (135 mm)
      2001 .... 6.50 inches (165 mm)
  Xterra .... 5.45 inches (138.5 mm)
  Pathfinder
    1996 through 2000 .... 5.12 inches (130 mm)
    2001 .... 5.31 inches (135 mm)

Brake light switch
Plunger-to-pedal stopper clearance .... 0.012 to 0.039 inch (0.3 to 1.0 mm)

Load Sensing Valve (LSV) spring installed length
Frontier (2WD models only) .... 7.99 inches (203 mm)
Pathfinder (4WD models only)
   1996 through 1998 .... 7.64 inches (194 mm)
   1999 and later .... 8.56 inches (217 mm)

Torque specifications

	Ft-lbs (unless otherwise indicated)	Nm
ABS wheel speed sensor mounting bolt		
Front		
Frontier and Xterra	13 to 17	18 to 23
Pathfinder	96 to 132 in-lbs	11 to 14
Rear		
2WD Frontier and Xterra		
H190A rear axle	72 to 96 in-lbs	8 to 11
C200 rear axle	13 to 17	18 to 24
4WD Frontier		
2000 and earlier	96 to 132 in-lbs	11 to 14
2001	16 to 19	21 to 26
4WD Xterra	16 to 19	21 to 26
Pathfinder	13 to 17	18 to 23
Brake booster-to-body mounting nuts	108 to 144 in-lbs	13 to 16
Caliper mounting bolts		
Frontier and Xterra models	16 to 23	22 to 31
Pathfinder models	24 to 31	32 to 42
Caliper mounting bracket bolts		
Frontier		
1998 through 2000	53 to 72	72 to 97
2001	101 to 130	137 to 177
Xterra	101 to 130	137 to 177
Pathfinder		
1996 through 1998	53 to 72	72 to 97
1999 and later	127 to 134	172 to 181
Hub-to-disc bolts	36 to 51	48 to 69
Brake hose-to-caliper bolt	144 to 168 in-lbs	17 to 20
Caliper and wheel cylinder bleeder screws	61 to 78 in-lbs	7 to 9
LSV mounting bolts/nuts		
Frontier and Xterra	13 to 17	18 to 23
Pathfinder	25 to 29	35 to 39
LSV bracket bolts	13 to 17	18 to 23
LSV spring-to-bracket bolt/nut	26 to 39 in-lbs	3 to 4.4
Master cylinder-to-brake booster retaining nuts	108 to 132	12 to 15
Wheel cylinder retaining bolts	52 to 95 in-lbs	6 to 11
Wheel lug nuts	See Chapter 1	

1 General information

General

The vehicles covered by this manual are equipped with hydraulically operated front and rear brake systems. The front brakes are disc type and the rear brakes are drum type. Both the front and rear brakes are self adjusting.

Hydraulic system

The hydraulic system consists of two separate circuits, split front-to-rear. The master cylinder has separate reservoirs for the two circuits, and, in the event of a leak or failure in one hydraulic circuit, the other circuit will remain operative and a warning indicator will light up on the instrument panel when a substantial amount of brake fluid is lost, showing that a failure has occurred.

Power brake booster

The power brake booster uses engine manifold vacuum to provide assistance to the brakes. It is mounted on the firewall in the engine compartment, directly behind the master cylinder.

Parking brake

The parking brake operates the rear brakes only, through cable actuation. On Pathfinder models it's activated by a lever mounted in the center console, while on Frontier and Xterra models it's activated by a pull handle mounted under the instrument panel, to the right of the steering column.

Service

After completing any operation involving disassembly of any part of the brake system, always test drive the vehicle to check for proper braking performance before resuming normal driving. When testing the brakes, perform the tests on a clean, dry, flat surface. Conditions other than these can lead to inaccurate test results.

Test the brakes at various speeds with both light and heavy pedal pressure. The vehicle should stop evenly without pulling to one side or the other.

Tires, vehicle load and wheel alignment are factors which also affect braking performance.

2 Anti-lock Brake System (ABS) - general information and trouble codes

Rear-wheel Anti-lock Brake System (2WD Frontier models)

1   The rear-wheel anti-lock brake system is

Chapter 9 Brakes

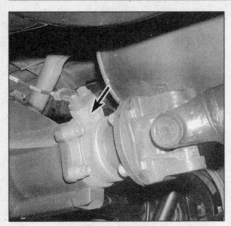

2.2 The speed sensor for the rear-wheel anti-lock brake system is mounted to the front of the differential

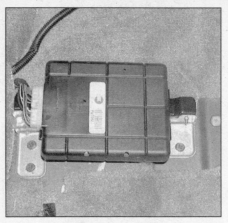

2.3 The control unit for the rear-wheel anti-lock brake system is located under the driver's seat

standard equipment on 2WD Frontier models. It is designed to maintain vehicle maneuverability, directional stability and optimum deceleration under severe braking conditions on most road surfaces. It does so by monitoring the rotational speed of the rear wheels and controlling the brake line pressure to the rear wheels while braking. This prevents the rear wheels from locking up prematurely during hard braking, regardless of the payload.

Components
Refer to illustrations 2.2, 2.3 and 2.4

Rear wheel speed sensor
2   A wheel speed sensor mounted on the front of the rear differential housing transmits speed and rate-of-deceleration inputs, in analog form (variable voltage signal), to the electronic control module **(see illustration)**. The sensor is actuated by an "exciter ring" pressed onto the differential pinion shaft. The exciter ring is the trigger mechanism: as it rotates, the teeth on the ring interrupt the magnetic field around the sensor pole (the sensor is similar in construction and operation to a pick-up coil in an electronic distributor). The rate of interruption produces a variable voltage signal which is transmitted to the control module.

ABS control unit
3   The control unit **(see illustration)** is mounted in the passenger compartment, under the driver's seat. The control unit inputs variable voltage signals from the rear wheel speed sensor, converts this analog signal into digital data, processes this data and controls the ABS actuator, which modulates hydraulic line pressure to the rear wheel cylinders to avoid wheel lock-up. The control unit also has self-diagnostic capabilities which enable it to continuously monitor the system for malfunctions during vehicle operation.

ABS actuator
4   The ABS actuator is located under the vehicle on the right-side frame rail **(see illustration)**. It contains a *dump valve* and an *isolation valve*. If the control unit senses the rear wheels decelerating too rapidly (locking up), it modulates fluid pressure to the rear wheel cylinders to prevent lock-up. The system switches back to normal operation as soon as the rear wheels are no longer locking up.

Warning light
5   If a problem develops within the system, the ABS warning light will glow on the dashboard. A diagnostic code will also be stored, which, when retrieved by a service technician, will indicate the problem area or component.

Diagnosis and repair
Refer to illustration 2.7

6   If the ABS warning light on the dashboard comes on and stays on, make sure the parking brake is not applied and there's no problem with the standard brake hydraulic system. If neither of these is the cause, the ABS system is probably malfunctioning. Check the following:
   a) *Make sure the brake shoes, pads, calipers and wheel cylinders are in good condition.*
   b) *Check the electrical connectors at the control module assembly.*
   c) *Check the fuses.*
   d) *Follow the wiring harness to the speed sensor and valve and make sure all connections are secure and the wiring isn't damaged.*

7   If the above preliminary checks don't identify the problem, you can check for the presence of any trouble codes stored in the control unit. To do this, first find the check connector under the left side of the instrument panel **(see illustration)**.

8   Start the engine and, using a jumper wire, ground terminal 3 of the check connector. Look at the ABS light on the instrument panel to see if it is flashing; if it is, count the number of flashes - these correlate to specific trouble codes that will indicate which part of the system or circuit is malfunctioning.

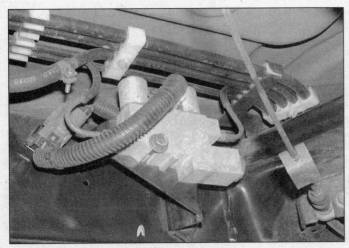

2.4 The ABS actuator for the rear-wheel anti-lock brake system is located under the vehicle, bolted to the right-side frame rail

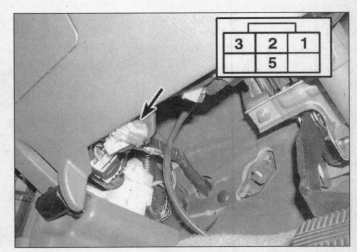

2.7 When checking for ABS trouble codes on a 2WD Frontier model, ground terminal 3 of the check connector

Rear-wheel ABS trouble codes

Code (number of flashes)	Malfunctioning component or circuit
2	Actuator isolation valve solenoid - open circuit
3	Actuator dump solenoid valve - open circuit
4	Actuator isolation valve solenoid - blocked
5	Actuator or rear drum brake problem
6	Rear wheel speed sensor - erratic
7	Actuator isolation valve solenoid - short circuit
8	Actuator dump solenoid valve - short circuit
9	Rear wheel speed sensor - open circuit
10	Rear wheel speed sensor - short circuit
13	Control unit or circuit, or rear sensor
14	Control unit or circuit, or rear sensor
15	Control unit or circuit, or rear sensor

2.12 ABS actuator terminal guide (rear wheel ABS)

9   After the problem has been repaired, erase the trouble code(s) by disconnecting the cable from the negative terminal of the battery for at least ten seconds.

Component resistance checks

10   If a trouble code indicates a problem with the speed sensor or actuator, you can check the resistance of the component to see if it is the source of the problem. If the component checks out OK, there is a problem with the circuit leading to the component, or possibly the ABS control unit.

Rear wheel speed sensor

11   Unplug the electrical connector from the speed sensor and, using an ohmmeter, check the resistance between the terminals of the sensor. The resistance should be 1.22 to 1.48 ohms; if not, replace the sensor.

ABS actuator

Refer to illustration 2.12

12   Unplug the electrical connector to the actuator and measure the resistance between the indicated terminals (on the actuator side of the connector) **(see illustration)**. If the resistance is not as specified, replace the actuator.

Terminals	Resistance
2 and body ground	Continuity
2 and 3	4 ohms
4 and body ground	No continuity
1 and 2	1.5 ohms

Four-wheel Anti-Lock Brake System (all models except 2WD Frontier)

Description

13   The four-wheel Anti-lock Brake System (ABS) prevents wheel lock-up under heavy braking conditions on virtually any road surface. Preventing the wheels from locking up maintains vehicle maneuverability, preserves directional stability, and allows optimal deceleration. It does this by monitoring the rotation speed of the wheels and controlling the brake line pressure to the calipers/wheel cylinders at each wheel during braking. The front wheels are controlled individually, while the rear wheels are paired on the same hydraulic circuit.

Components

Wheel speed sensors

Refer to illustrations 2.14a and 2.14b

14   A wheel speed sensor is mounted at each front wheel **(see illustration)**. A sensor ring, or "tone wheel," is located on the back of the front hubs. As the tone wheel turns, its teeth interrupt the magnetic field around the speed sensor (which is similar to a pick-up coil in an electronic distributor), producing a voltage signal that varies in proportion to the speed of rotation of each wheel. The analog outputs from the two front wheel speed sensors are transmitted to the ABS control module. On 4WD Frontier models, 4WD Xterra models and all Pathfinders, a speed sensor is also present at each rear wheel, with the sensor rings pressed onto the axleshafts **(see illustration)**. On 2WD Xterra models, only one rear wheel speed sensor is used, and is identical in design and operation to the rear wheel speed sensor previously described for rear-wheel ABS systems.

2.14a Front wheel speed sensor (four-wheel ABS)

2.14b Rear wheel speed sensor (four-wheel ABS on Pathfinders and 4WD Xterra and Frontier models)

2.15a Location of the ABS actuator - Pathfinder models

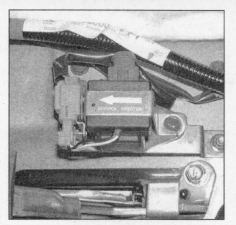

2.17 The G sensor (4WD models only) is located under the center console (Pathfinder shown)

2.19a When checking for ABS trouble codes on a 2000 or earlier Pathfinder or a 4WD Frontier, ground terminal 4 of the data link connector

2.19b When checking for ABS trouble codes on a 2001 Pathfinder or Xterra, ground terminal 9 of the data link connector; if you're checking for ABS codes on a 2000 Xterra, ground terminal 8

2.15b Location of the ABS actuator - Xterra and 4WD Frontier models

ABS actuator

Refer to illustrations 2.15a and 2.15b

15 The ABS actuator is located in the right-rear corner of the engine compartment (Pathfinder models) or the left rear corner of the engine compartment (Xterra and 4WD Frontier models) **(see illustrations)**. The HCU contains the anti-lock valves and the pump/motor assembly. The anti-lock valves consist of a solenoid valve body with individual valves to control pressure to each front brake, and one valve to control pressure to the rear brakes.

ABS control unit

16 The ABS control unit, which is located forward of the center console under the heater unit (1996 through 1998 Pathfinder models) or combined with the ABS actuator (all other models), monitors wheel speeds and controls the hydraulic control unit to prevent wheel lockup. The control unit receives analog voltage signals from each wheel speed sensor, converts these signals into digital data, processes this data and adjusts the solenoids inside the hydraulic control unit accordingly.

G sensor

Refer to illustration 2.17

17 4WD models also have a G sensor, mounted under the center console, which monitors the deceleration rate of the vehicle **(see illustration)**. The control unit uses this data, along with the signals sent from the wheel speed sensors, to determine whether the vehicle is being driven on a road with good traction or on one that is slippery.

Diagnosis and repair

Refer to illustrations 2.19a, 2.19b and 2.23

18 If the ABS warning light on the dashboard comes on and stays on, make sure the parking brake is not applied and there's no problem with the standard brake hydraulic system. If neither of these is the cause, the ABS system is probably malfunctioning. Check the following:

a) Make sure the brake pads, shoes, calipers and wheel cylinders are in good condition.

b) Check the electrical connectors at the control module assembly.
c) Check the fuses.
d) Follow the wiring harness to the speed sensor and valve and make sure all connections are secure and the wiring isn't damaged.

19 If the above preliminary checks don't identify the problem, you can check for the presence of any trouble codes stored in the control unit. To do this, first find the data link connector. On Pathfinder and Xterra models it's mounted just behind the lower edge of the lower dash panel **(see illustration)**. On Frontier models it's attached to the bottom of the fuse panel **(see illustration)**.

20 Drive the vehicle over 20 mph for at least one minute, then park the vehicle and turn the engine off.

21 Ground the proper terminal of the data link connector **(see illustrations 2.19a and 2.19b)**. On 4WD Frontier models and 1996 through 2000 Pathfinder models, ground terminal 4. On 2000 Xterra models ground terminal 8. On 2001 Pathfinders and Xterras, ground terminal 9.

22 Turn the ignition switch to the On position. **Note:** *Do not depress the brake pedal.*

23 Look at the ABS light on the instrument panel to see if it is flashing; if it is, count the number of flashes - the code is determined by counting the number of flashes **(see illustration)**. The sequence always begins with a 3.0-second "off" period, followed by a flash,

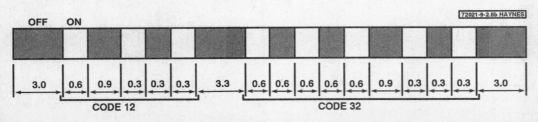

2.23 The ABS light on the dash indicates a problem in a particular circuit by the number of flashes. For example, three 0.6-second flashes, followed by a 0.9-second interval, followed by two 0.3-second flashes indicates a code 32, which means there's a short circuit in the rear right sensor circuit

then a 0.9-second off period, then two flashes. This "Code 12" (the "start" code, not a trouble code) is followed by a 3.3-second off period, then the trouble codes are displayed, in the order in which they were stored, starting with the latest stored code. All codes are two-digit codes, so the first flash(es) indicate the tens place, followed by a pause, followed by the ones-place flash(es). For example, a sequence of four flashes, then a pause, followed by a sequence of five flashes, would indicate a Code 45 (actuator front left outlet solenoid valve).

24 After noting the number of flashes of the ABS light, refer to the accompanying table.

Four-wheel ABS trouble codes

Note: *Not all codes apply to all models.*

Code	Defective part or circuit
12	No malfunctions detected
17	G sensor or circuit
18	Sensor rotor (or abnormal tire size)
21	Front right sensor (open circuit)
22	Front right sensor (short circuit)
25	Front left sensor (open circuit)
26	Front left sensor (short circuit)
31	Rear right sensor (open circuit)
32	Rear right sensor (short circuit)
35	Rear left sensor (open circuit)
36	Rear left sensor (short circuit)
41	Actuator front right outlet solenoid valve
42	Actuator front right inlet solenoid valve
45	Actuator front left outlet solenoid valve
46	Actuator front left inlet solenoid valve
55	Actuator rear outlet solenoid valve
56	Actuator rear inlet solenoid valve
57	Power supply (low voltage)
61	Actuator motor or motor relay
63	Solenoid valve relay or circuit
71	Control unit
Warning light stays on when ignition switch is turned on (1995 - 1998)	Control unit power supply circuit / Warning light bulb circuit / Control unit power supply circuit / Control unit or control unit connector / Solenoid valve relay stuck / Power supply for solenoid valve relay coil
Warning light does not come on when ignition switch is turned on	Fuse, warning light bulb or warning light circuit / Control unit
Vehicle vibrates excessively during ABS operation	Control unit-to-Transmission Control Module (TCM) circuit problem

25 After the problem has been repaired, erase the trouble code(s) as follows: Disconnect the jumper wire at the data link connector from ground; the ABS warning light should remain on. Ground the jumper wire three successive times within 12.5 seconds; each ground must last more than one second. The ABS light should now go out.

a) If the light stays on, take the vehicle to a dealer service department or other qualified repair shop and have the ABS system repaired.

b) If the light doesn't stay on, drive the vehicle above 20 mph for at least one minute and verify that the warning light on the dash doesn't come on again. If it does, take the vehicle to a dealer service department or other qualified repair shop and have the ABS system repaired.

3 Disc brake pads - replacement

Refer to illustrations 3.5a through 3.5n

Warning: *Disc brake pads must be replaced on both front or both rear wheels at the same time - never replace the pads on only one wheel. Also, the dust created by the brake system is harmful to your health. Never blow it out with compressed air and don't inhale any of it. An approved filtering mask should be worn when working on the brakes. Do not, under any circumstances, use petroleum-based solvents to clean brake parts. Use brake system cleaner only!*

1 Remove the cap from the brake fluid reservoir. Remove about two-thirds of the fluid from the reservoir.

2 Loosen the front wheel lug nuts, raise the front of the vehicle and support it

3.5a Before disassembling the brake, wash it thoroughly with brake system cleaner and allow it to dry - position a drain pan under the brake to catch the residue - DO NOT use compressed air to blow off brake dust!

3.5b Use a C-clamp to depress the pistons into the caliper before removing the caliper and pads

Chapter 9 Brakes

3.5c Unscrew the lower caliper mounting bolt . . .

3.5d . . . and pivot the caliper up for access to the brake pads

3.5e Remove the inner brake pad . . .

securely on jackstands. Block the wheels at the opposite end.

3 Remove the wheels. Work on one brake assembly at a time, using the assembled brake for reference if necessary.

4 Inspect the brake disc carefully as outlined in Section 5. If machining is necessary, follow the information in that Section to remove the disc, at which time the pads can be removed as well.

5 Follow the accompanying photo sequence **(see illustrations 3.5a through 3.5n)**. Be sure to stay in order and read the caption under each illustration.

6 When reinstalling the caliper, be sure to tighten the mounting bolt to the torque listed in this Chapter's Specifications.

7 After the job has been completed, firmly depress the brake pedal a few times to bring the pads into contact with the disc. Check the level of the brake fluid, adding some if necessary. Check the operation of the brakes carefully before placing the vehicle into normal service.

3.5f . . . and the outer brake pad

3.5g Remove the lower slide pin from the caliper mounting bracket

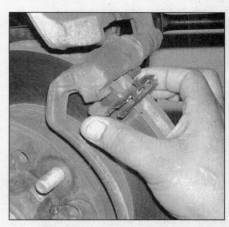

3.5h Remove the upper and lower pad retainers from the caliper mounting bracket

3.5i Apply anti-squeal compound to the back of both pads . . .

3.5j . . . then remove the shims from the old pads and install them on the new ones

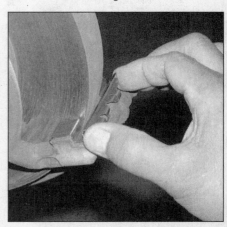

3.5k Install the upper and lower pad retainers

3.5l Install the inner brake pad . . .

3.5m . . . and the outer brake pad

3.5n Clean the slide pin and inspect it for scoring and corrosion; coat the pin with high-temperature grease and install it in the mounting bracket. Swing the caliper down over the pads, install the mounting bolt and tighten it to the torque listed in this Chapter's Specifications

4 Brake caliper - removal and installation

Refer to illustrations 4.2a and 4.2b

Warning: *The dust created by the brake system is harmful to your health. Never blow it out with compressed air and don't inhale any of it. An approved filtering mask should be worn when working on the brakes. Do not, under any circumstances, use petroleum-based solvents to clean brake parts. Use brake system cleaner only!*

Removal

1 Loosen the wheel lug nuts, raise the vehicle and place it securely on jackstands. Block the wheels at the opposite end. Remove the wheel.

2 Remove the inlet fitting bolt and disconnect the brake hose from the caliper. Discard the old sealing washers **(see illustrations)**. Plug the brake hose immediately to keep contaminants and air out of the brake system and to prevent losing any more brake fluid than is necessary. **Note:** *If you are simply removing the caliper for access to other components, leave the brake hose connected and suspend* the caliper with a length of wire - don't let it hang by the hose **(see illustration 5.2)**.

3 Remove the caliper mounting bolts and detach the caliper from the mounting bracket.

Installation

4 Installation is the reverse of removal. Don't forget to use new sealing washers on each side of the brake hose inlet fitting and be sure to tighten the fitting bolt and the caliper mounting bolts to the torque values listed in this Chapter's Specifications.

5 Bleed the brake system (see Section 11). **Note:** *If the brake hose was not disconnected, bleeding won't be required.* Make sure there are no leaks from the hose connections. Test the brakes carefully before returning the vehicle to normal service.

5 Brake disc - inspection, removal and installation

Inspection

Refer to illustrations 5.2, 5.3, 5.4a, 5.4b, 5.5a and 5.5b

1 Loosen the wheel lug nuts, raise the front of the vehicle and support it securely on jackstands. Apply the parking brake. Remove the wheel.

2 Remove the brake caliper (but don't disconnect the brake line or hose from the caliper) and suspend it out of the way with a piece of wire **(see illustration)**. **Caution:** *Don't let the caliper hang by the brake hose.* Remove the brake pads (see Section 3).

3 Visually inspect the disc surface for score marks and other damage **(see illustration)**. Light scratches and shallow grooves are normal after use and won't affect brake operation. Deep grooves - over 0.015-inch (0.38 mm) deep - require disc removal and refinishing by an automotive machine shop. Be sure to check both sides of the disc.

4 To check disc runout, place a dial indicator at a point about 1/2-inch from the outer edge of the disc **(see illustration)**. Set the indicator to zero and turn the disc. The indicator reading should not exceed the allowable runout listed in this Chapter's Specifications. If it does, the disc should be refinished

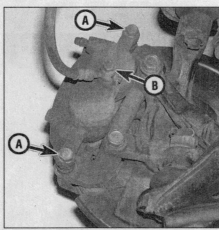

4.2a Caliper mounting details

    A   Mounting bolts
    B   Inlet fitting bolt

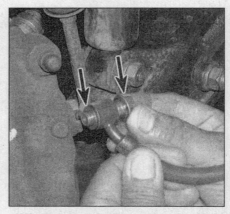

4.2b There is a sealing washer on either side of the brake hose inlet fitting; be sure to replace these with new ones when reconnecting the hose

5.2 Hang the caliper out of the way with a piece of wire - don't let it hang by the brake hose

5.3 The brake pads on this vehicle were obviously neglected, as they wore down completely and cut deep grooves into the disc - wear this severe means the disc must be replaced

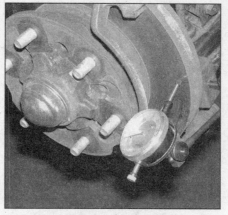

5.4a To check disc runout, mount a dial indicator as shown and rotate the disc

5.4b Using a swirling motion, remove the glaze from the disc with sandpaper or emery cloth

by an automotive machine shop. **Note:** *To produce a smooth finish and ensure a perfectly smooth surface - thereby eliminating brake pedal pulsation or any other undesirable symptoms - the discs should be resurfaced regardless of the dial indicator reading. If you elect not to have the discs resurfaced, deglaze them with sandpaper or emery cloth* **(see illustration)**.

5   The disc must not be machined to a thickness less than the specified minimum thickness, which is cast into the disc **(see illustration)**. Measure disc thickness with a micrometer **(see illustration)**.

Removal and installation

Refer to illustrations 5.6 and 5.8

6   Remove the brake caliper (see Section 4), brake pads (see Section 3) and the caliper mounting bracket **(see illustration)**; it isn't necessary to disconnect the brake hose from the caliper.

7   The disc is bolted to the backside of the hub. Disc/hub removal and installation is part of the front wheel bearing repack and adjustment procedure (see Chapter 1).

8   Once the disc/hub assembly has been removed, unscrew the disc-to-hub bolts and separate the two components **(see illustration)**.

9   While the disc/hub assembly is removed, clean, inspect and repack the wheel bearings (see Chapter 1). When bolting the disc to the hub, tighten the bolts to the torque listed in this Chapter's Specifications.

10   Install the disc/hub assembly and adjust the wheel bearings (see Chapter 1).

11   Install the caliper mounting bracket and tighten the bolts to the torque listed in this Chapter's Specifications. Install the brake pads and the caliper (see Section 3).

12   Install the wheels and lug nuts. Lower the vehicle and tighten the lug nuts to the torque listed in the Chapter 1 Specifications.

13   Before driving the vehicle, pump the brake pedal several times to bring the pads into contact with the disc.

6   Drum brake shoes - replacement

Refer to illustrations 6.2a, 6.2b, 6.4, 6.5a through 6.5w, 6.6a through 6.6aa, 6.7a through 6.7y and 6.8

Warning: *Drum brake shoes must be*

5.5a The minimum thickness is cast into the edge of the disc (typical)

replaced on both wheels at the same time - never replace the shoes on only one wheel. Also, the dust created by the brake system is harmful to your health. Never blow it out with compressed air and don't inhale any of it. An approved filtering mask should be worn when working on the brakes. Do not, under any circumstances, use petroleum-based solvents to clean brake parts. Use brake system cleaner only!

5.5b Use a micrometer to measure disc thickness at several points

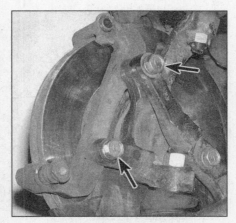

5.6 Caliper mounting bracket bolts (arrows)

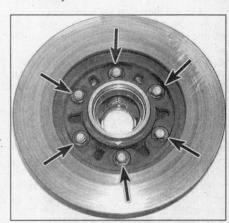

5.8 Remove the bolts and separate the disc from the hub

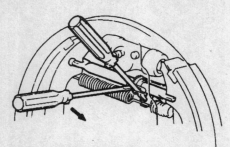

6.2a If the drum binds on the shoes, lift the adjuster lever off the adjuster wheel and rotate the adjuster wheel until the shoes are retracted sufficiently to allow drum removal

2WD Frontier models

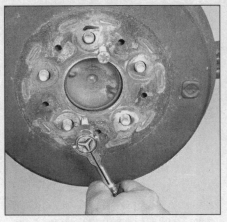

6.2b If the brake drum is rusted to the axle flange, thread two bolts (M8 x 1.25) into the holes provided in the drum and tighten them until they push the drum off

6.4 Wash the brake assembly with brake cleaner - DO NOT blow the dust out with compressed air

6.5a Unhook the lower return spring from the leading shoe

Caution: *Whenever the brake shoes are replaced, the return and hold-down springs should also be replaced. Due to the continuous heating/cooling cycle the springs are subjected to, they lose tension over a period of time and may allow the shoes to drag on the drum and wear at a much faster rate than normal.*

1   Loosen the wheel lug nuts, raise the rear of the vehicle and support it securely on jackstands. Block the front wheels to keep the vehicle from rolling. Remove the wheels.

2   Release the parking brake and remove the brake drums. If the shoes have worn into the drum, preventing drum removal, remove the access plug from the backing plate, insert a small screwdriver through the hole, lift the adjuster lever off the adjusting wheel and turn the wheel with another screwdriver to back off the brake shoes **(see illustration)**. If the drum is rusted to the hub, install two bolts of the proper size and thread pitch into the threaded holes provided **(see illustration)** and turn them in. As the bolts are tightened, they will contact the surface of the hub flange and push the drum off.

3   **Note:** *All four rear brake shoes must be replaced at the same time, but to avoid mix-*

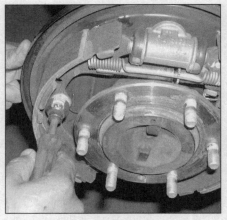

6.5b Push in on the hold-down spring retainer, twist it 1/4-turn to align the slot in the retainer with the blade on the pin, then remove the retainer and spring (the tool shown here is available at auto parts stores and simplifies this step) . . .

ing up parts, work on only one brake assembly at a time.

4   Before disassembling anything, clean off the brake assembly with brake system

6.5d Pull the bottom of the leading shoe out and remove the adjuster assembly

6.5c . . . also remove the spring seat

cleaner **(see illustration)**.

5   If you're working on a 2WD Frontier model, follow the accompanying illustrations **(6.5a through 6.5w)**. Be sure to stay in order and read the caption under each illustration.

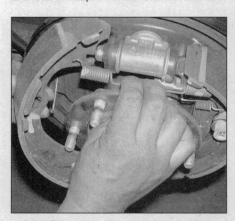

6.5e Disengage the top of the leading shoe from the wheel cylinder piston, then twist the shoe outward and unhook the upper return spring from the hole in the shoe

Chapter 9 Brakes

6.5f Remove the hold-down spring from the trailing shoe

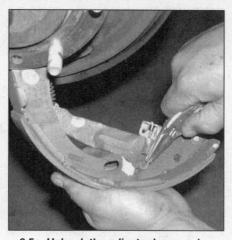

6.5g Unhook the adjuster lever spring from the shoe

6.5h Clean, then lubricate the contact areas on the brake backing plate with high-temperature brake grease

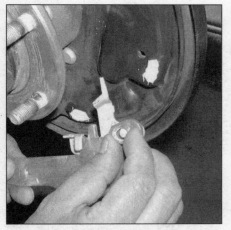

6.5i Pry the c-clip off the retaining pin and separate the parking brake lever from the trailing shoe - don't lose this wave washer that goes between the parking brake lever and the shoe

6.5j Assemble the adjuster lever, parking brake lever and wave washer to the trailing shoe, insert the retaining pin and install a new c-clip . . .

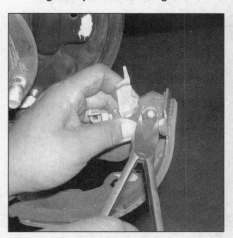

6.5k . . . then bend the ends of the c-clip around the retaining pin, making sure it seats in the groove

6.5l Install the adjuster lever spring and hook it into the hole in the shoe

6.5m Position the trailing shoe assembly against the backing plate, install the pin, spring seat and hold-down spring, then secure the spring with the retainer; make sure the top of the shoe is engaged with the wheel cylinder piston

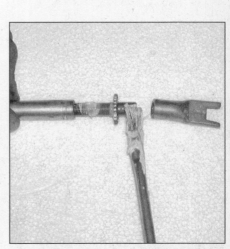

6.5n Clean the adjuster screw assembly, then lubricate the threads and socket end with high-temperature grease

6.5o Install the adjuster assembly - make sure the long end of the adjuster points toward the front of the vehicle

6.5p Hook the upper return spring to the trailing shoe . . .

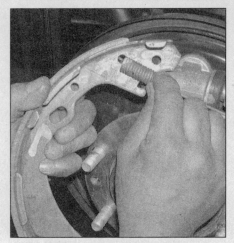

6.5q . . . and to the hole in the leading shoe

6.5r Engage the top of the leading shoe with the slot in the wheel cylinder piston and the adjuster screw

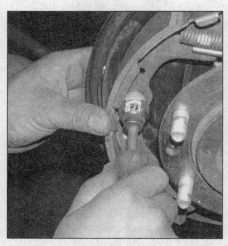

6.5s Position the leading shoe against the backing plate, install the pin, spring seat and spring, then secure the spring with the retainer

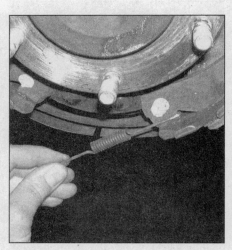

6.5t Connect the lower return spring to the trailing shoe . . .

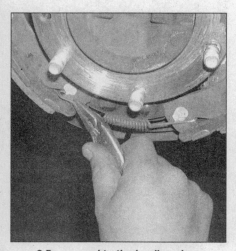

6.5u . . . and to the leading shoe

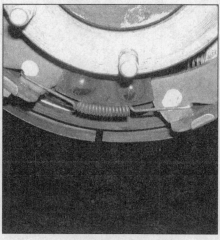

6.5v Make sure the spring fits under the anchor plate

6.5w The brake assembly should look like this when you're done (2WD Frontier models) - left side shown

Xterra and 4WD Frontier models

6.6a Unhook and remove the lower return spring from the shoes

6.6b Unhook the adjuster spring from the adjuster lever, then remove it from the trailing shoe

6.6c Unhook the upper return spring; the tool shown here is available at most auto parts stores and simplifies this step

6.6d Push in on the hold-down spring retainer, twist it 1/4-turn to align the slot in the retainer with the blade on the pin, then remove the retainer, spring and spring seat (the tool shown here is available at auto parts stores and simplifies this step)

6.6e Remove the trailing shoe

6.6f Remove the adjuster screw assembly

6 If you're working on an Xterra or a 4WD Frontier model, follow the accompanying illustrations (6.6a through 6.6aa). Be sure to stay in order and read the caption under each illustration.

7 If you're working on a Pathfinder model, follow the accompanying illustrations (6.7a through 6.7y). Be sure to stay in order and read the caption under each illustration.

8 Before reinstalling the drum, it should be

6.6g Detach the adjuster lever from the pin on the leading shoe

6.6h Remove the hold-down spring from the leading shoe

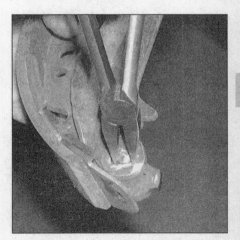

6.6i Remove the c-clip from the parking brake lever retaining pin

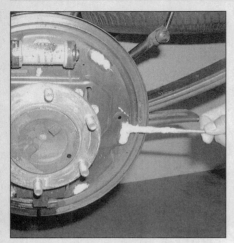

6.6j Clean, then lubricate the contact areas on the brake backing plate with high-temperature brake grease

6.6k Also lubricate the bellcrank post (A) (you'll have to remove the snap-ring and slide the bellcrank off the post to do this) and the ends of the parking brake interconnecting cables

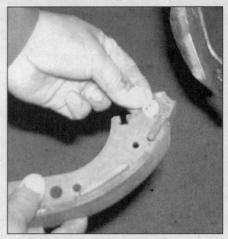

6.6l Install the wave washer on the leading shoe pin . . .

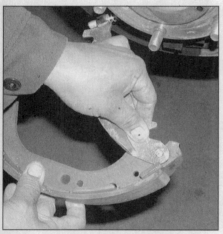

6.6m . . . attach the parking brake lever to the pin . . .

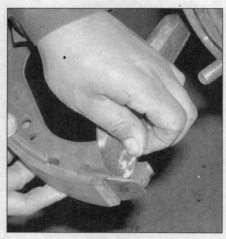

6.6n . . . and secure it with a new c-clip (bend the ends of the clip around the pin)

6.6o Install the leading shoe on the backing plate . . .

6.6p . . . and install the pin, spring seat and hold-down spring, then secure the spring with the retainer

6.6q Connect the lower return spring to the bottom of each shoe . . .

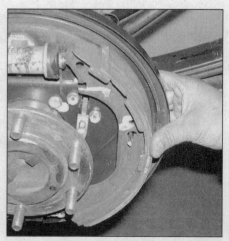

6.6r . . . position the trailing shoe on the backing plate . . .

Chapter 9 Brakes

6.6s ... and install the pin, spring seat, spring and retainer

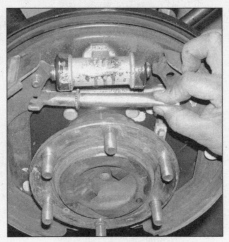

6.6t After cleaning and lubricating the adjuster assembly (see illustration 6.5n), connect the front of the adjuster to the leading shoe ...

6.6u ... and to the trailing shoe (the long end of the adjuster must point to the rear of the vehicle on these models)

6.6v Attach the adjuster lever to the pin on the leading shoe

6.6w Connect the adjuster spring to the hole in the trailing shoe ...

6.6x ... and hook the other end into the notch in the adjuster lever

6.6y Connect the upper return spring to the hole in the trailing shoe ...

6.6z ... and hook the other end into the hole in the leading shoe

6.6aa The brake assembly should look like this when you're done (Xterra and 4WD Frontier models)

Pathfinder models

6.7a Unhook and remove the lower return spring from the shoes

6.7b Unhook and remove the upper return spring; the tool shown here is available at most auto parts stores and simplifies this Step

6.7c Unhook the adjuster spring from the adjuster lever, then remove it from the leading shoe

6.7d Push in on the hold-down spring retainer, twist it 1/4-turn to align the slot in the retainer with the blade on the pin, then remove the retainer, spring and spring seat (the tool shown here is available at auto parts stores and simplifies this step)

6.7e Remove the adjuster assembly

6.7f Detach the adjuster lever from the pin on the trailing shoe

6.7g Remove the hold-down spring from the trailing shoe

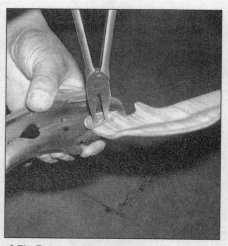

6.7h Remove the c-clip from the parking brake lever retaining pin

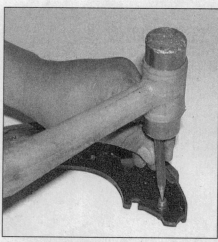

6.7i If the new trailing shoe doesn't come with a retaining pin, drive the pin out of the old shoe . . .

Chapter 9 Brakes 9-17

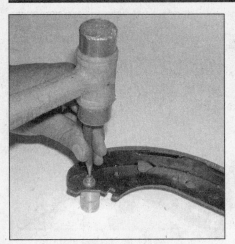

6.7j ...and transfer it to the new one

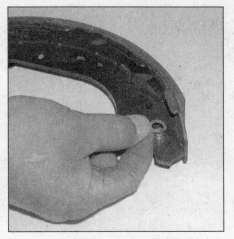

6.7k This wave washer fits on the retaining pin between the parking brake lever and the trailing shoe

6.7l Clean, then lubricate the contact areas on the brake backing plate with high-temperature brake grease

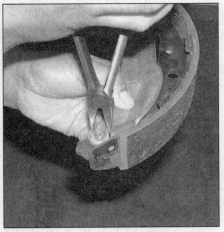

6.7m Secure the parking brake lever to the trailing shoe with a new c-clip - make sure the clip fits into the groove

6.7n Position the trailing shoe assembly on the backing plate, making sure the top of the shoe engages with the slot in the wheel cylinder piston

6.7o Install the pin, spring seat and hold-down spring, then secure the spring with the retainer

6.7p Position the leading shoe against the backing plate, install the pin, spring seat and spring, then secure the spring with the retainer

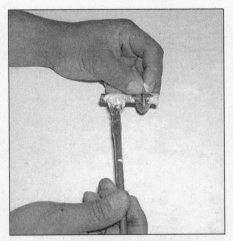

6.7q Clean the adjuster screw assembly, then lubricate the threads and socket end with high-temperature grease

6.7r Install the adjuster assembly - make sure the long end of the adjuster points toward the front of the vehicle on these models

6.7s Hook the adjuster lever to the pin at the top of the trailing shoe - make sure the lever properly engages the adjuster

6.7t Connect the adjuster spring to the leading shoe . . .

6.7u . . . and to the adjuster lever

6.7v Connect the upper return spring to the leading shoe . . .

6.7w . . . and to the trailing shoe

checked for cracks, score marks, deep scratches and hard spots, which will appear as small discolored areas. If the hard spots cannot be removed with fine emery cloth or if any of the other conditions listed above exist, the drum must be taken to an automotive machine shop to have it resurfaced. **Note:** *Professionals recommend resurfacing the drums each time a brake job is done. Resurfacing will eliminate the possibility of out-of-round drums.* If the drums are worn so much that they can't be resurfaced without exceeding the maximum allowable diameter, which is stamped into the drum **(see illustration)**, then new ones will be required. At the very least, if you elect not to have the drums resurfaced, remove the glaze from the surface with emery cloth using a swirling motion.

9 Turn the adjuster screw wheel so the drum just slips over the shoes. Now, working through the backing plate, turn the adjuster

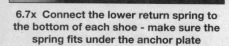

6.7x Connect the lower return spring to the bottom of each shoe - make sure the spring fits under the anchor plate

6.7y The brake assembly should look like this when you're done (Pathfinder models) - left side shown

6.8 The maximum allowable diameter is cast into the drum (typical)

Chapter 9 Brakes

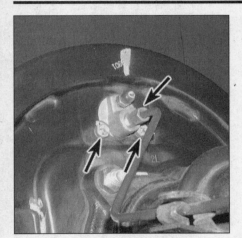

7.4 To remove the wheel cylinder, disconnect the brake line fitting, then remove the two wheel cylinder bolts

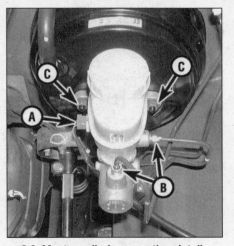

8.2 Master cylinder mounting details
A   Electrical connector for fluid level switch
B   Brake line fittings
C   Mounting nuts

8.8 The best way to bleed air from the master cylinder before installing it on the vehicle is with a pair of bleeder tubes that direct brake fluid into the reservoir during bleeding

screw wheel until the shoes drag on the drum when the drum is turned. Finally, back off the adjuster screw wheel so the shoes don't drag. Depress the brake pedal firmly several times, then rotate the drum to ensure that the brakes are not dragging. If they are, back off the star wheel a little more.
10   Mount the wheel, install the lug nuts, then lower the vehicle. Tighten the wheel lug nuts to the torque listed in the Chapter 1 Specifications.
11   Make a number of forward and reverse stops and operate the parking brake to adjust the brakes until satisfactory pedal action is obtained.
12   Check the operation of the brakes carefully before driving the vehicle.

7   Wheel cylinder - removal and installation

Removal

Refer to illustration 7.4

1   Raise the rear of the vehicle and support it securely on jackstands. Block the front wheels to keep the vehicle from rolling.
2   Remove the brake shoe assembly (see Section 6).
3   Remove all dirt and foreign material from around the wheel cylinder.
4   Disconnect the brake line **(see illustration)** with a flare-nut wrench, if available. Don't pull the brake line away from the wheel cylinder.
5   Remove the wheel cylinder mounting bolts.
6   Detach the wheel cylinder from the brake backing plate and place it on a clean workbench. Immediately plug the brake line to prevent fluid loss and contamination.

Installation

7   Place the wheel cylinder in position and install the bolts finger tight. Connect the brake line to the cylinder, being careful not to cross-thread the fitting. Tighten the wheel cylinder bolts to the torque listed in this Chapter's Specifications.
8   Tighten the brake line securely and install the brake shoes (see Section 6).
9   Bleed the brakes (see Section 11).
10   Check the operation of the brakes carefully before driving the vehicle.

8   Master cylinder - removal, installation and reservoir/seal replacement

Removal

Refer to illustration 8.2

1   Disconnect the cable from the negative battery terminal.
2   Unplug the electrical connector for the fluid level warning switch **(see illustration)**.
3   Remove as much fluid as possible from the reservoir with a suction gun, large syringe or a poultry baster. **Warning:** *If a poultry baster is used, never again use it for the preparation of food.*
4   Place rags under the fittings and prepare caps or plastic bags to cover the ends of the lines once they're disconnected. **Caution:** *Brake fluid will damage paint. Cover all body parts and be careful not to spill fluid during this procedure.* Loosen the fittings at the ends of the brake lines where they enter the master cylinder. To prevent rounding off the flats, use a flare-nut wrench, which wraps around the fitting hex.
5   Pull the brake lines away from the master cylinder and plug the ends to prevent contamination.
6   Remove the nuts attaching the master cylinder to the power booster **(see illustration 8.2)**. Pull the master cylinder off the studs to remove it. Again, be careful not to spill the fluid as this is done.

Installation

Refer to illustrations 8.8 and 8.16

7   Bench bleed the new master cylinder before installing it. Mount the master cylinder in a vise, with the jaws of the vise clamping on the mounting flange.
8   Attach a pair of master cylinder bleeder tubes to the outlet ports of the master cylinder **(see illustration)**.
9   Fill the reservoir with brake fluid of the recommended type (see Chapter 1).
10   Slowly push the pistons into the master cylinder (a large Phillips screwdriver can be used for this) - air will be expelled from the pressure chambers and into the reservoir. Because the tubes are submerged in fluid, air can't be drawn back into the master cylinder when you release the pistons.
11   Repeat the procedure until no more air bubbles are present.
12   Remove the bleed tubes, one at a time, and install plugs in the open ports to prevent fluid leakage and air from entering. Install the reservoir cap.
13   Install the master cylinder over the studs on the power brake booster and tighten the attaching nuts only finger tight at this time. Don't forget to use a new gasket.
14   Thread the brake line fittings into the master cylinder. Since the master cylinder is still a bit loose, it can be moved slightly so the fittings thread in easily. Don't strip the threads as the fittings are tightened.
15   Tighten the mounting nuts to the torque listed in this Chapter's Specifications. Tighten the brake line fittings securely.
16   Fill the master cylinder reservoir with fluid, then bleed the lines at the master cylinder, followed by bleeding the remainder of the brake system (see Section 11). To bleed the lines at the master cylinder, have an assistant depress the brake pedal and hold it down. Loosen the fitting to allow air and fluid

Chapter 9 Brakes

8.16 Have an assistant depress the brake pedal and hold it down, then loosen the fitting nut, allowing air and fluid to escape; repeat this procedure on both fittings until the fluid is clear of air bubbles

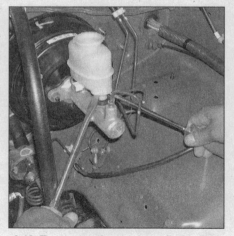

8.19 To remove the master cylinder fluid reservoir, pry it straight up out of the seals using two screwdrivers

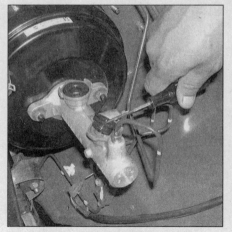

8.20 The reservoir seals can be replaced if they are leaking

to escape **(see illustration)**. Tighten the fitting, then allow your assistant to return the pedal to its rest position. Repeat this procedure on both fittings until the fluid is free of air bubbles, then bleed the rest of the system. Check the operation of the brake system carefully before driving the vehicle. **Warning:** *If you do not have a firm brake pedal at the end of the bleeding procedure, or have any doubts as to the effectiveness of the brake system, DO NOT drive the vehicle. Have it towed to a dealer service department or other qualified repair shop for diagnosis.*

Reservoir/seal replacement

Refer to illustrations 8.19 and 8.20
Note: *The brake fluid reservoir can be replaced separately from the master cylinder body if it becomes damaged. If there is leakage between the reservoir and the master cylinder body, the seals in the master cylinder can be replaced.*

17   Remove as much fluid as possible from the reservoir with a suction gun, large syringe or a poultry baster. **Warning:** *If a poultry baster is used, never again use it for the preparation of food.*
18   Place rags under the master cylinder to absorb any fluid that may spill out once the reservoir is detached from the master cylinder. **Caution:** *Brake fluid will damage paint. Cover all body parts and be careful not to spill fluid during this procedure.*
19   Carefully pry the reservoir straight up **(see illustration)**.
20   If you are simply replacing the seals, carefully pry the old seals out of the master cylinder and install new ones **(see illustration)**.
21   Lubricate the reservoir seals with clean brake fluid, then press the reservoir into place on the master cylinder body.
22   Refill the reservoir with the recommended brake fluid (see Chapter 1) and check for leaks.
23   Bleed the master cylinder **(see illustration 8.16)**.

9   Brake hoses and lines - inspection and replacement

Inspection

1   About every six months, with the vehicle raised and supported securely on jackstands, the rubber hoses which connect the steel brake lines with the front and rear brake assemblies should be inspected for cracks, chafing of the outer cover, leaks, blisters and other damage. These are important and vulnerable parts of the brake system and inspection should be complete. A light and mirror will be helpful for a thorough check. If a hose exhibits any of the above conditions, replace it with a new one.

Replacement

Flexible brake hose

Front

Refer to illustrations 9.3 and 9.4
2   Loosen the wheel lug nuts, raise the vehicle and support it securely on jackstands. Remove the wheel.

9.3 Using a flare-nut wrench, unscrew the threaded fitting on the brake line . . .

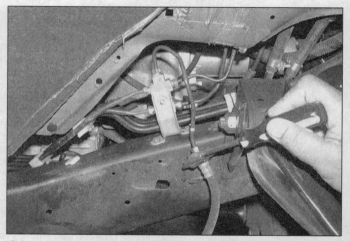

9.4 . . . then pry the U-clip off the hose and detach the hose from the bracket

Chapter 9 Brakes

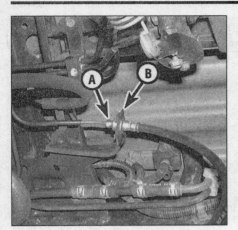

9.12 At the upper end of the chassis-to-rear axle brake hose, unscrew the line fitting (A) with a flare-nut wrench, then remove the U-clip (B)

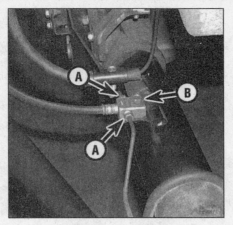

9.14 At the lower end of the chassis-to-rear axle brake hose, unscrew the brake line fittings (A) with a flare-nut wrench, then remove the bolt securing the junction block (B)

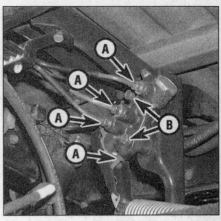

10.5 Load Sensing Valve (LSV) mounting details

A Brake line fittings
B Mounting bolts

3   At the bracket, unscrew the brake line fitting from the hose **(see illustration)**. Use a flare-nut wrench to prevent rounding off the corners of the fitting nut, and hold the hose end with a wrench to prevent twisting the frame bracket.
4   Remove the U-clip from the female fitting at the bracket, then pass the hose through the bracket **(see illustration)**. On Pathfinder models the hose must also be disconnected from the strut bracket.
5   At the caliper end of the hose, remove the inlet fitting bolt, then separate the hose from the caliper. Note that there are two copper sealing washers on either side of the inlet fitting **(see illustration 4.2b)** - they should be replaced with new ones during installation.
6   If you're working on a Pathfinder, route the hose through the bracket on the strut.
7   Connect the fitting to the caliper with the inlet fitting bolt and new sealing washers. Tighten the inlet fitting bolt to the torque listed in this Chapter's Specifications.
8   Route the hose into the frame bracket, making sure it isn't twisted. Connect the brake line fitting, starting the threads by hand. Install the U-clip, then tighten the fitting securely.
9   Bleed the caliper (see Section 8).
10 Install the wheel and lug nuts, lower the vehicle and tighten the lug nuts to the torque listed in the Chapter 1 Specifications.

Rear

Chassis-to-rear axle

Refer to illustrations 9.12 and 9.14

11 Raise the rear of the vehicle and support it securely on jackstands. Block the front wheels to prevent the vehicle from rolling.
12 At the chassis bracket, unscrew the brake line fitting from the hose **(see illustration)**. Use a flare-nut wrench to prevent rounding off the corners of the fitting nut.
13 Remove the U-clip from the female fitting at the bracket, then pass the hose through the bracket.
14 At the axle end of the hose, unscrew the two brake line fittings with a flare-nut wrench, unscrew the bolt securing the fitting block to the axle housing, then separate the lines from the fitting block and remove the hose **(see illustration)**.
15 To install the hose, reverse the removal procedure, then bleed both rear brakes (see Section 11).

Metal brake lines

16 When replacing brake lines, be sure to use the correct parts. Don't use copper tubing for any brake system components. Purchase steel brake lines from a dealer or auto parts store.
17 Prefabricated brake line, with the tube ends already flared and fittings installed, is available at auto parts stores and dealer parts departments. These lines must be bent to the proper shapes using a tubing bender.
18 When installing the new line, make sure it's securely supported in the brackets and has plenty of clearance between moving or hot components.
19 After installation, check the master cylinder fluid level and add fluid as necessary. Bleed the brake system (see Section 11) and test the brakes carefully before driving the vehicle in traffic.

10 Load Sensing Valve (LSV) - description and replacement

Refer to illustrations 10.5 and 10.11
Note: *This Section applies to 2WD Frontier and 4WD Pathfinder models only.*

Description

1   The Load Sensing Valve (LSV) regulates hydraulic pressure to the rear brakes in accordance with the amount of weight present in the rear of the vehicle. When the load is light, pressure to the rear brakes is decreased to avoid locking up the wheels. When the load is heavy, the valve senses the lower ride height and directs more pressure to the rear brakes.
2   Because of the special gauge set needed to check the operation of the LSV, diagnosis of the unit must be left to a dealer service department or other qualified repair shop. However, if the LSV is known to be malfunctioning or is leaking, you can replace it yourself.

Replacement

Warning: *Do not disassemble the LSV.*
3   Raise the rear of the vehicle and support it securely on jackstands. Block the front wheels to prevent the vehicle from rolling.
4   Unscrew the nuts, remove the bolt and disconnect the height sensing spring from the bracket on the axle housing.
5   Loosen the tube nuts that connect the brake lines to the LSV **(see illustration)**. Use a flare-nut wrench to prevent rounding off the flats on these nuts. Pull the brake lines away from the LSV slightly and plug the ends of the lines to prevent contamination.
6   Remove the LSV mounting bolts and detach the valve from the bracket.
7   Unbolt the LSV from the bracket. Disconnect the height sensing spring from the valve.
8   Installation is the reverse of removal. Be sure to tighten the mounting bolts to the torque listed in this Chapter's Specifications. Lubricate the ends of the spring and the stopper bolt.
9   Bleed the brake hydraulic system (see Section 14).
10 Lower the vehicle to the ground. Roll it back and forth a couple of times, then have one person sit in the driver's seat. Have another person stand on the rear bumper or sit on the tailgate, then slowly get off (this will stabilize the suspension).
11 Crawl under the vehicle and push the LSV lever against the stopper bolt (away from the spring), then measure the distance from

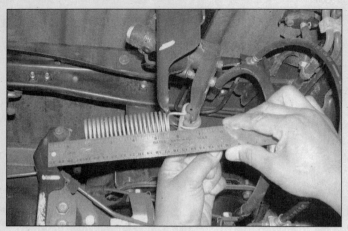

10.11 Push the LSV lever against its stop, then measure the length of the spring

11.8 On models with a Load Sensing Valve (LSV), bleed the LSV first

one end of the spring to the other **(see illustration)**. If the distance is not as listed in this Chapter's Specifications, loosen the bolts connecting the spring bracket to the axle, then reposition the bracket as necessary.

12 Check the operation of the brakes in an isolated are before driving the vehicle in traffic.

11 Brake hydraulic system - bleeding

Refer to illustrations 11.8 and 11.10

Warning: *Wear eye protection when bleeding the brake system. If the fluid comes in contact with your eyes, immediately rinse them with water and seek medical attention.*

Note: *Bleeding the hydraulic system is necessary to remove any air that manages to find its way into the system when it's been opened during removal and installation of a hose, line, caliper or master cylinder.*

1 You'll probably have to bleed the system at all four brakes if air has entered it due to low fluid level, or if the brake lines have been disconnected at the master cylinder.

2 If a brake line was disconnected only at a wheel, then only that caliper must be bled. If a brake line is disconnected at a fitting located between the master cylinder and any of the brakes, that part of the system served by the disconnected line must be bled.

3 Remove any residual vacuum from the brake power booster by applying the brake several times with the engine off.

4 Disconnect the cable from the negative terminal of the battery (this is to disable the ABS system).

5 Remove the master cylinder reservoir cap and fill the reservoir with brake fluid. Reinstall the cap. **Note:** *Check the fluid level often during the bleeding operation and add fluid as necessary to prevent the fluid level from falling low enough to allow air bubbles into the master cylinder.*

6 If air has entered the master cylinder, bleed the master cylinder as described in Section 8.

7 Have an assistant on hand, as well as a supply of new brake fluid, a clear container partially filled with clean brake fluid, a length of clear tubing to fit over the bleeder valve and a wrench to open and close the bleeder valve.

2WD Frontier and 4WD Pathfinder models

8 If you're working on a 2WD Frontier or a 4WD Pathfinder, bleed the LSV first.

a) Loosen the bleeder valve slightly, then tighten it to a point where it's snug but can still be loosened quickly and easily.
b) Place one end of the tubing over the bleeder valve and submerge the other end in brake fluid in the container **(see illustration)**.
c) Have the assistant depress the brake pedal slowly and hold it in the depressed position.
d) While the pedal is held down, open the bleeder valve just enough to allow a flow of fluid to leave the valve. Watch for air bubbles to exit the submerged end of the tube. When the fluid flow slows after a couple of seconds, close the valve and have your assistant release the pedal.
e) Repeat Steps c) and d) until no more air is seen leaving the tube, then tighten the bleeder valve and proceed to the next Step.

All models

9 Working at the left rear wheel, loosen the bleeder valve slightly, then tighten it to a point where it's snug but can still be loosened quickly and easily.

10 Place one end of the tubing over the bleeder valve and submerge the other end in brake fluid in the container **(see illustration)**.

11 Have the assistant depress the brake pedal slowly and hold it in the depressed position.

12 While the pedal is held down, open the bleeder valve just enough to allow a flow of fluid to leave the valve. Watch for air bubbles to exit the submerged end of the tube. When the fluid flow slows after a couple of seconds, close the valve and have your assistant release the pedal.

13 Repeat Steps 11 and 12 until no more air is seen leaving the tube, then tighten the bleeder valve and proceed to the right rear wheel, the left front wheel and the right front wheel, in that order, and perform the same procedure. Be sure to check the fluid in the master cylinder reservoir frequently.

2WD Frontier models

14 Bleed the ABS actuator. Have your assistant depress the pedal and hold it there, then loosen the inlet line at the actuator with a flare-nut wrench. When the flow of fluid slows, tighten the fitting and have your assistant release the brake pedal. Repeat this step on the outlet line.

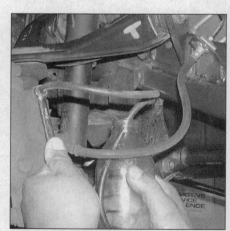

11.10 When bleeding the brakes, a hose is connected to the bleed screw at the component to be bled and then submerged in brake fluid - air will be seen as bubbles in the tube and container (all air must be expelled before moving to the next component)

Chapter 9 Brakes

12.5a Measure the length of the booster output rod from the top of the rod to the master cylinder mounting surface and compare your measurement to the specified length listed in this Chapter's Specifications

Xterra models and 4WD Frontier models

15 If you're working on a 4WD model, perform Step 14 on all five fittings threaded into the ABS actuator (inlet lines first, then the outlet lines).

All models

16 Never use old brake fluid. It contains moisture which can cause the fluid to boil, rendering the brake system inoperative.
17 Refill the master cylinder with fluid at the end of the operation.
18 Check the operation of the brakes. The pedal should feel solid when depressed, with no sponginess. If necessary, repeat the entire process. **Warning:** *Do not operate the vehicle if you're in doubt about the effectiveness of the brake system.*

12 Power brake booster - check, replacement and adjustment

Check

Operating check

1 Depress the brake pedal several times with the engine off and make sure there's no change in the pedal reserve distance.
2 Depress the pedal and start the engine. If the pedal goes down slightly, operation is normal.

Airtightness check

3 Start the engine and turn it off after one or two minutes. Depress the brake pedal slowly several times. If the pedal depresses less each time, the booster is airtight.
4 Depress the brake pedal while the engine is running, then stop the engine with the pedal depressed. If there's no change in the pedal reserve travel after holding the pedal for 30 seconds, the booster is airtight.

12.5b To adjust the length of the output rod, hold the serrated portion of the rod with a pair of pliers and turn the adjusting screw in or out to achieve the desired setting

Output rod length check

Refer to illustrations 12.5a and 12.5b

5 Remove the master cylinder (see Section 8). It isn't necessary to disconnect the lines from the master cylinder, as long as you can move it forward far enough to provide clearance for the following measurement. Just make sure you don't kink the metal lines. Apply about 20 in-Hg of vacuum to the brake booster with a hand-operated vacuum pump, push in on the output rod with approximately 4-1/2 pounds (2 kg) of force, then measure the length of the output rod **(see illustration)** and compare your measurement to the dimensions listed in this Chapter's Specifications. If the rod length is outside specifications, turn the adjusting screw on the end of the output rod until the length is correct **(see illustration)**.

Replacement

Refer to illustration 12.8

Note: *Power brake booster units shouldn't be disassembled; if a problem with the booster develops, replace it with a new or rebuilt one.*

6 Remove the brake master cylinder, if you haven't already done so (see Section 8).
7 Disconnect the vacuum hose leading from the engine to the booster. Be careful not to damage the hose when removing it from the booster fitting.
8 Remove the under-dash panel. Locate the pushrod clevis connecting the booster to the brake pedal **(see illustration)**.
9 Remove the clevis pin retaining clip with pliers and pull out the clevis pin.
10 Remove the four nuts holding the brake booster to the firewall **(see illustration 12.8)**.
11 Slide the booster straight out from the firewall until the studs clear the holes.
12 Installation is the reverse of removal. But be sure to measure the following dimension before installing the power brake booster assembly.

Adjustment

Refer to illustration 12.13

13 Measure the distance between the power brake booster and the hole in the cle-

12.8 Remove the retaining clip from the clevis (right arrow), pull out the clevis pin and detach the pushrod from the brake pedal; two of the power brake booster mounting nuts (left arrows) are visible in this photo

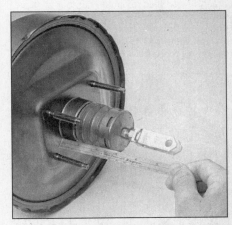

12.13 Measure the distance between the power brake booster and the hole in the clevis and compare your measurement to the dimension listed in this Chapter's Specifications; if necessary, adjust the clevis before installing the booster

vis **(see illustration)** and compare it to the booster-to-clevis dimension listed in this Chapter's Specifications. If it isn't the same, loosen the adjusting nut and turn the clevis in or out to the specified length, then tighten the nut.

13 Parking brake - adjustment

Refer to illustrations 13.6a, 13.6b and 13.6c
Warning: *The models covered by this manual are equipped with airbags. Always disable the airbag system when working in the vicinity of airbag system components (see Chapter 12).*

1 The parking brake lever or handle, when properly adjusted, should travel 10 to 12 clicks (Frontier and Xterra) or 7 or 8 clicks (Pathfinder) when a pulling force of 44 lbs (20 kg) is applied.

Chapter 9 Brakes

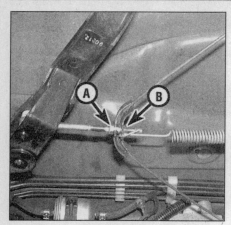

13.6a To adjust the parking brake on a 2WD Frontier model, loosen the locknut (A) and turn the adjusting nut (B) at the cable equalizer

13.6b To adjust the parking brake on a 4WD Frontier or an Xterra model, loosen the locknut (A) and turn the adjusting nut (B)

13.6c To adjust the parking brake on a Pathfinder model, use a ratchet, extension and socket inserted through the slot in the center console to turn the adjusting nut (remove the small cover on the center console for access)

2   If the parking brake lever travels less than the specified minimum number of clicks, it might not be releasing completely and the shoes or pads could even be dragging against the drum or disc. If the lever can be pulled up more than the specified maximum number of clicks, the parking brake may not hold adequately on an incline, allowing the car to roll.
3   Before adjusting the parking brake, make sure the clearance between the brake shoes and brake drums is correct. Release the parking brake lever or handle and loosen the adjusting nut (and locknut, on Xterra and 4WD Frontier models) (see illustration 13.6a, 13.6b or 13.6c), then firmly depress the brake pedal ten times with the engine running.
4   Raise the rear of the vehicle and support it securely on jackstands. Block the front wheels to prevent the vehicle from rolling.
5   Adjust the brake shoe-to-drum clearance on each rear brake as described in Section 6.
6   Pull the handle or lever up four or five clicks, then tighten the adjusting nut until it is fairly tight (see illustrations).
7   Release the parking brake lever then pull it up again with a force of 44 lbs (20 kg). The lever should now travel the number of clicks applied in Step 1. If not, turn the adjusting nut as required.
8   Fully release the parking brake lever. Turn each rear wheel, making sure the brake shoes do not drag on the drum. If necessary, adjust the shoe-to-drum clearance (see Section 6).

14 Parking brake cables - replacement

Warning: *The models covered by this manual are equipped with airbags. Always disable the airbag system when working in the vicinity of airbag system components (see Chapter 12).*

Frontier and Xterra models
Front cable
Refer to illustrations 14.3a, 14.3b, 14.4 and 14.6

1   Disconnect the cable from the negative terminal of the battery.
2   Remove the left-side under-dash panel, the knee bolster and its reinforcement (see Chapter 11).
3   Working under the dash, remove the roll-pin from the parking brake handle shaft (see illustration). Turn the handle and shaft,

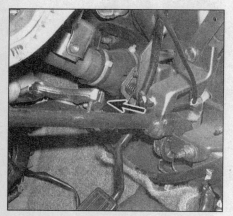

14.3a Using a pair of locking pliers, pull the roll-pin out of the parking brake handle shaft (shown from above, with the instrument panel removed for clarity)

14.3b Line up the slot in the shaft with the slot in the support tube, then pass the cable through the slots

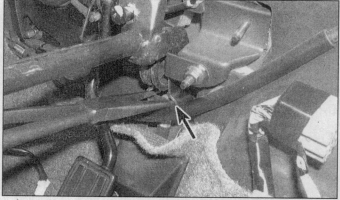

14.4 The upper portion of the cable casing is secured to a bracket with a C-clip

Chapter 9 Brakes

14.6 With the carpet peeled back, unbolt the cable casing from the floor pan

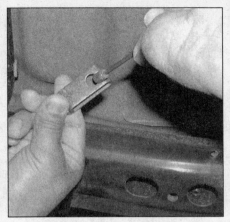

14.12 Detaching the left rear parking brake cable from the coupler on the right rear cable (2WD Frontier)

14.14 Each rear parking brake cable is secured by a number of brackets

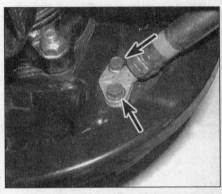

14.15 On 2WD Frontier models and all Pathfinders, the rear parking brake cables are attached to the brake backing plates with two bolts

lining up the slot in the shaft and the slot in the support tube, then pass the cable through the slots **(see illustration)**.

4 Remove the clip securing the cable casing to the bracket **(see illustration)**, then pull the cable and casing down through the bracket.

5 Unscrew the bolt and detach the accelerator pedal stop from the floor.

6 Remove the left-side kick panel and the sill plates, then peel back the driver's side carpet. Unbolt the cable casing from the floor pan **(see illustration)**.

7 Raise the vehicle and support it securely on jackstands.

8 Unscrew the adjuster nut at the cable equalizer (2WD Frontier, **see illustration 13.6a**) or lever (Xterra or 4WD Frontier, **see illustration 13.6b**).

9 Detach the cable casing from any other brackets that may be present, then pull it through the floor pan into the interior of the vehicle.

10 Installation is the reverse of removal. Be sure to adjust the cable as described in Section 13.

Rear cable(s)

Refer to illustrations 14.12, 14.14 and 14.15

11 If you're working on a 2WD Frontier, loosen the rear wheel lug nuts.

12 Raise the vehicle and support it securely on jackstands. If you're working on a 2WD Frontier, loosen the adjusting nut at the equalizer to provide slack in the cables, then detach the right rear cable from the left rear cable at the coupler **(see illustration)**.

13 If you're working on an Xterra or a 4WD Frontier, unscrew the adjusting nut and detach the rear cable from the lever **(see illustration 13.6)**.

14 Unscrew any bracket securing bolts and detach the cable from the frame rails **(see illustration)**.

15 If you're working on a 2WD Frontier, remove the rear brake shoes (see Section 6) and detach the cable from the lever on the trailing brake shoe. Unbolt the cable casing from the backing plate **(see illustration)** and remove the cable and casing assembly.

16 If you're working on an Xterra or 4WD Frontier, detach the rear cable(s) from the crank lever(s) on the brake backing plate(s).

17 Installation is the reverse of removal. Apply a light coat of grease to the portion of the cable end that engages with the levers on the brake shoes. Be sure to adjust the cable as described in Section 13.

Pathfinder models

Front cable

Refer to illustrations 14.20, 14.21 and 14.23

18 Remove the center console (see Chapter 11).

19 Unplug the electrical connector for the parking brake warning light switch.

20 Remove the parking brake cable adjusting nut **(see illustration)**.

21 Remove the two bolts from the parking

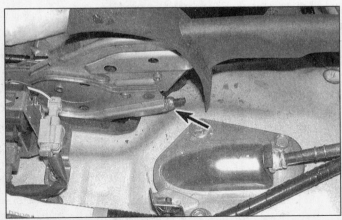

14.20 Remove the parking brake cable adjusting nut . . .

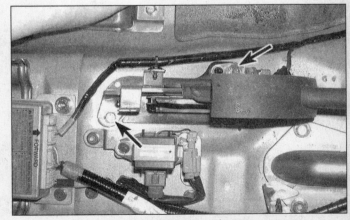

14.21 . . . and the parking brake lever base bolts

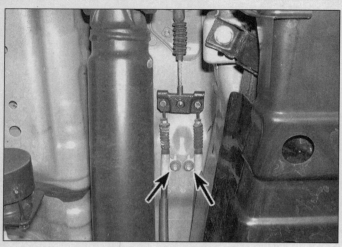

14.23 To disconnect the rear parking brake cables from the equalizer, remove these bracket nuts (arrows), slide the cables forward slightly and disengage the cable ends from the equalizer

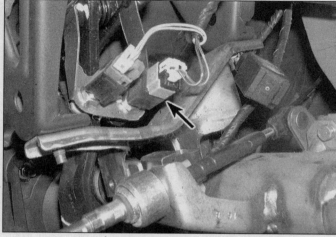

15.1 The brake light switch (arrow) is located on a bracket near the top of the brake pedal (on most models it's the one on the right)

brake lever base **(see illustration)**.

22 Raise the vehicle and support it securely on jackstands. Block the front wheels.

23 Remove the two cable housing nuts just behind the equalizer and disengage the rear cables from the equalizer **(see illustration)**.

24 Working inside the vehicle, pull the cable and equalizer through the floorpan.

25 Installation is the reverse of removal. Be sure to adjust the cable as described in Section 13.

Rear cables

26 Make sure the parking brake is completely released.

27 Loosen the rear wheel lug nuts, raise the rear of the vehicle and support it securely on jackstands. Block the front wheels. Remove the wheel.

28 Remove the brake drum and brake shoes, and disconnect the cable from the parking brake levers on the trailing shoes (see Section 6). Unbolt the cable retainer from the backing plate **(see illustration 14.15)** and pull the cable through the backing plate.

29 Unbolt any cable brackets from the underbody.

30 Remove the cable housing nuts just behind the equalizer and disengage the cables from the equalizer **(see illustration 14.23)**.

31 Installation is the reverse of removal. Apply a light coat of grease to the portion of the cable end that engages with the equalizer, and where the rear ends of the cables engage with the levers on the trailing brake shoes.

32 Adjust the parking brake assembly when you're done (see Section 13).

15 Brake light switch - check, adjustment and replacement

Refer to illustration 15.1

Check and adjustment

1 The brake light switch **(see illustration)** is located on a bracket at the top of the brake pedal. On most models it's the switch on the right (the other switch is the cruise control cut-off/shift lock switch). In any case, it's the switch *with* at least one red wire going to it. The switch activates the brake lights at the rear of the vehicle when the pedal is depressed.

2 To check the brake light switch, simply note whether the brake lights come on when the pedal is depressed and go off when the pedal is released.

3 If the brake lights don't come on when the brake pedal is depressed, make sure the brake pedal is correctly adjusted (see Chapter 1). Then try adjusting the switch as follows.

4 A locknut secures the switch to the bracket. Loosen the locknut (on the firewall side of the bracket), then screw the switch in or out to provide a 0.012 to 0.039-inch (0.3 to 1.0 mm) clearance between the switch plunger and the pedal stopper, with no pressure on the plunger, and tighten the locknut. Recheck the clearance to verify that it didn't change when you tightened the locknut. The switch should now function properly.

5 If the switch still doesn't work properly, either it isn't getting voltage, or the switch itself is defective. Use a voltmeter or test light to verify that there's voltage at the switch connector. With the pedal at rest, voltage should be present at one of the terminals of the switch. With the pedal depressed, voltage should be present at both terminals. If voltage isn't present at both terminals when the pedal is depressed, replace the switch.

Replacement

6 Unplug the electrical connector from the brake light switch.

7 Loosen the locknut and unscrew the switch from the bracket.

8 Installation is the reverse of removal.

9 Make sure the brake pedal is adjusted properly (see Chapter 1), then adjust the switch (see above).

Chapter 10
Suspension and steering systems

Contents

	Section		Section
Balljoints - check and replacement	10	Steering knuckle - removal and installation	11
Coil spring, rear (Pathfinder models) - removal and installation	15	Steering linkage (Frontier and Xterra models) - removal and installation	23
Front end alignment - general information	29	Steering wheel - removal and installation	17
General information	1	Strut or coil spring (Pathfinder models) - replacement	6
Intermediate shaft - removal and installation	19	Strut/coil spring assembly (Pathfinder models) - removal, inspection and installation	5
Leaf spring (Frontier and Xterra models) - removal and installation	14	Suspension arms, rear (Pathfinder models) - removal and installation	16
Lower control arm - removal and installation	8	Suspension, steering and driveaxle boot check	See Chapter 1
Power steering fluid level check	See Chapter 1	Tension rod (2WD Frontier models) - removal and installation	9
Power steering pump - removal and installation	25	Tie-rod ends - removal and installation	21
Power steering system - bleeding	26	Tire and tire pressure checks	See Chapter 1
Shock absorber, rear - removal and installation	12	Tire rotation	See Chapter 1
Shock absorber, front (Frontier and Xterra models) - removal and installation	2	Torsion bar (Frontier and Xterra models) - removal and installation	4
Stabilizer bar and bushings (front) - removal and installation	3	Upper control arm (Frontier and Xterra models) - removal and installation	7
Stabilizer bar and bushings (rear) - removal and installation	13	Wheel studs - replacement	27
Steering column - removal and installation	18	Wheels and tires - general information	28
Steering column transfer gear (Pathfinder models) - removal and installation	20		
Steering gear - removal and installation	24		
Steering gear boots (Pathfinder models) - replacement	22		

Specifications

General

Power steering fluid type .... See Chapter 1
Front suspension ride height (models with torsion bars) **(see illustrations 4.13a and 4.13b)**
  Frontier models
    2WD
      With four-cylinder engine .... 4.37 to 4.53 inches (111 to 115 mm)
      With V6 engine .... 1.484 to 1.642 inches (38 to 42 mm)
    4WD .... 1.791 to 1.949 inches (45.5 to 49.5 mm)
  Xterra models
    2WD .... 1.484 to 1.642 inches (38 to 42 mm)
    4WD .... 1.791 to 1.949 inches (45.5 to 49.5 mm)
Anchor arm adjusting bolt protrusion - 2WD Frontier **(see illustration 4.9a)** .... 0.24 to 0.71 inch (6 to 18 mm)
Anchor arm protrusion below crossmember - Xterra, 4WD Frontier **(see illustration 4.9b)** .... 0.98 to 1.54 inches (25 to 39 mm)

Torque specifications

	Ft-lbs (unless otherwise indicated)	Nm
Front suspension		
Shock absorber		
Upper mounting nut	144 to 192 in-lbs	16 to 22
Lower mounting bolts	87 to 108	118 to 147
Strut/coil spring assembly (Pathfinder)		
Strut-to-steering knuckle bolts/nuts	111 to 122	151 to 165
Strut upper mounting nuts	29 to 40	39 to 54
Strut piston rod nut		
1996 and 1997	30 to 39	41 to 53
1998 and later	43 to 58	59 to 78
Upper control arm pivot bolt nuts	72 to 87	98 to 118

Chapter 10 Suspension and steering systems

Torque specifications (continued)   **Ft-lbs** (unless otherwise indicated)   **Nm**

Front suspension (continued)

Item	Ft-lbs	Nm
Lower control arm pivot bolt nut(s)		
Frontier and Xterra	80 to 105	109 to 142
Pathfinder	69 to 96	94 to 130
Lower control arm bushing bracket bolts (Pathfinder)	87 to 108	118 to 147
Tension rod-to-control arm bolts/nuts	36 to 47	49 to 64
Tension rod-to-frame nut	84 to 108	114 to 147
Balljoint-to-steering knuckle nut*		
Upper	58 to 108	78 to 147
Lower		
Frontier and Xterra	87 to 141	118 to 191
Pathfinder	87 to 123	118 to 167
Balljoint-to-lower control arm nuts (Pathfinder)	76 to 94	103 to 127
Stabilizer bar		
Link nuts		
Frontier and Xterra	12 to 16	16 to 22
Pathfinder	61 to 76	83 to 103
Clamp bolts		
Frontier and Xterra	12 to 16	16 to 22
Pathfinder	46 to 65	63 to 88
Anchor arm adjusting bolt locknut	22 to 30	30 to 40
Torque arm-to-lower control arm bolts/nuts		
2WD Frontier	37 to 50	50 to 68
Xterra and 4WD Frontier		
Small inner nuts	33 to 44	45 to 60
Large outer nut	66 to 88	89 to 108

* Tighten nut to lower torque figure, then if necessary, tighten a little more to allow cotter pin insertion (but do not exceed the upper torque figure).

Rear suspension

Item	Ft-lbs	Nm
Shock absorber		
Frontier and Xterra		
Upper and lower mounting nuts	30 to 37	40 to 50
Pathfinder		
Upper mounting nut/bolt		
1996 and 1997	36 to 49	49 to 67
1998	43 to 58	59 to 78
1999 and later	49 to 65	67 to 88
Lower mounting nut/bolt		
1996 through 2000	49 to 65	67 to 88
2001	44 to 57	59 to 78
Stabilizer bar		
Xterra		
Link nuts	30 to 35	41 to 47
Clamp bolts	32 to 41	43 to 55
Pathfinder		
Link nuts		
Upper	19 to 23	26 to 32
Lower	31 to 34	42 to 47
Clamp bolts (bar and link clamps)	19 to 23	26 to 32
Leaf spring		
U-bolt nuts	72 to 80	98 to 108
Front spring pin nut		
Frontier		
1998 through 2000 2WD models	58 to 72	78 to 98
2001 2WD models	87 to 108	118 to 147
4WD models	87 to 108	118 to 147
Spring pin-to-frame bolt (2000 and earlier 2WD Frontier)	144 to 180 in-lbs	16 to 21
Rear shackle nuts	58 to 72	78 to 98
Trailing arm bolts/nuts		
Upper	103 to 116	140 to 157
Lower		
To body	85 to 98	115 to 133
To rear axle	103 to 116	140 to 157
Panhard rod bolts/nuts		
1996 and 1997	103 to 116	140 to 157
1998 and later		
To body	80 to 94	108 to 127
To rear axle	103 to 116	140 to 157

Chapter 10 Suspension and steering systems

Torque specifications (continued)

	Ft-lbs (unless otherwise indicated)	Nm
Steering		
Airbag mounting bolts	132 to 168 in-lbs	15 to 20
Steering gear-to-frame bolts (Frontier and Xterra)	62 to 71	84 to 96
Steering gear-to-frame bolts/nuts (Pathfinder)	87 to 101	118 to 137
Intermediate shaft pinch bolt	17 to 22	24 to 29
Steering transfer gear-to-frame bolts (Pathfinder)	14 to 22	20 to 29
Coupler shaft pinch bolts (Pathfinder)	17 to 22	24 to 29
Steering linkage (Frontier/Xterra)		
Pitman shaft nut		
Frontier		
1998	174 to 195	235 to 265
1999 and 2000		
Four-cylinder engine		
Manual steering	174 to 195	235 to 265
Power steering	102 to 130	138 to 176
V6 engine	174 to 195	235 to 265
2001		
Four-cylinder engine	102 to 130	138 to 176
V6 engine	174 to 195	235 to 265
Pitman arm-to-relay rod nut*		
1998 and 1999	40 to 72	54 to 98
2000 Frontier		
2WD	40 to 72	54 to 98
4WD	43 to 54	59 to 75
2000 Xterra	43 to 54	59 to 75
2001 Frontier and Xterra	40 to 72	54 to 98
Idler arm-to-frame bolts/nuts	44 to 54	54 to 69
Idler arm ballstud nut*		
1998	40 to 72	54 to 98
1999 and 2000		
Four-cylinder engine		
2WD	40 to 72	54 to 98
4WD	43 to 54	59 to 75
V6 engine	40 to 72	54 to 98
2001	40 to 72	54 to 98
Tie-rod-to-relay rod nut*		
1998 through 2000	40 to 72	54 to 98
2001	44 to 54	59 to 74
Tie-rod adjuster sleeve bolt/nut (2000 and earlier 2WD Frontier)	120 to 168 in-lbs	14 to 20
Tie-rod end-to-steering knuckle nut*		
1998 through 2000	40 to 72	54 to 98
2001	44 to 54	59 to 74

* Tighten nut to lower torque figure, then if necessary, tighten a little more to allow cotter pin insertion (but do not exceed the upper torque figure).

	Ft-lbs (unless otherwise indicated)	Nm
Steering column		
Steering wheel nut	22 to 29	29 to 39
Steering column mounting nuts/bolts	108 to 144 in-lbs	13 to 16
Steering column-to-firewall nuts		
Frontier and Xterra	108 to 144 in-lbs	13 to 16
Pathfinder	78 to 104 in-lbs	9 to 12
Shaft coupler bolt	17 to 22	24 to 29
Power steering		
Pump mounting bolts		
Frontier and Xterra		
Front bolts	144 to 180 in-lbs	16 to 21
Rear bolt	20 to 27	27 to 37
Mounting bracket-to-engine block bolts	16 to 22	22 to 29
Pathfinder		
1996 through 2000		
Pivot bolt	33 to 43	45 to 59
Adjuster locknut	18 to 23	25 to 31
Mounting bracket-to-engine block bolts	32 to 43	43 to 58
2001		
Upper mounting bolt	24 to 31	32 to 42
Lower mounting bolt	33 to 38	45 to 51
Power steering pressure line banjo bolt		
To power steering pump	37 to 50	49 to 68
To power steering gear (Pathfinder)	22 to 26	30 to 35

Chapter 10 Suspension and steering systems

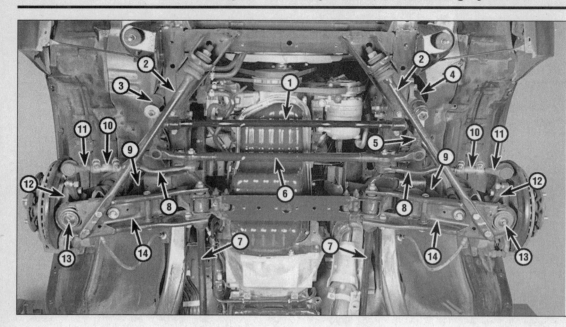

1.1a Front suspension and steering components - 2WD Frontier

1. Stabilizer bar
2. Tension rod
3. Idler arm
4. Steering gear
5. Pitman arm
6. Relay rod
7. Torsion bar
8. Tie-rod
9. Shock absorber
10. Adjuster sleeve
11. Tie-rod end
12. Steering knuckle
13. Lower balljoint
14. Lower control arm

1 General information

Refer to illustrations 1.1a, 1.1b, 1.1c, 1.2 and 1.3

Front suspension

The front suspension (see illustrations) is fully independent. Frontier and Xterra models use upper and lower control arms, tension rods (2WD Frontier models), torsion bars and shock absorbers. Pathfinder models have MacPherson struts instead. On all models a stabilizer bar connected to the frame and to the two lower control arms (Frontier and Xterra) or struts (Pathfinder) reduces body roll during cornering.

Rear suspension

The rear suspension on Frontier and Xterra models consists of a pair of multi-leaf springs and two shock absorbers (see illustration). The rear axle assembly is attached to the leaf springs by U-bolts. The front ends of the springs are attached to the frame at the front hangers, through rubber bushings. The rear ends of the springs are attached to the frame by shackles which allow the springs to alter their length as they compress and rebound.

The rear suspension on Pathfinder models is a five-link design, using coil springs, upper and lower control arms, a Panhard rod, two shock absorbers and a stabilizer bar (see illustration).

Steering system

The steering system on Pathfinder models consists of a rack-and-pinion steering gear and two adjustable tie-rods. Power assist is standard.

The steering system on Frontier and Xterra models consists of a recirculating-ball steering gearbox, Pitman arm, idler arm, relay rod and two adjustable tie-rod assemblies. When the steering wheel is turned, the gear rotates the Pitman arm which forces the relay rod to one side. The tie-rods, which are connected to the relay rod, transfer steering force to the wheels. The tie-rods are adjustable and are used for toe-in adjustments. The relay rod is supported by the Pitman arm and idler arm. The idler arm pivots on a support attached to the opposite frame rail.

Precautions

Frequently, when working on the suspension or steering system components, you may come across fasteners which seem impossible to loosen. These fasteners on the underside of the vehicle are continually subjected to water, road grime, mud, etc., and can become rusted or "frozen," making them extremely difficult to remove. In order to unscrew these stubborn fasteners without damaging them (or other components), be

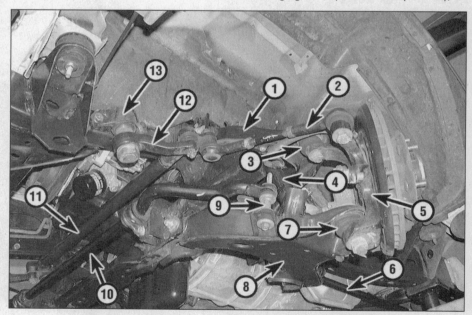

1.1b Front suspension and steering components - 2WD Xterra (4WD Xterra and 4WD Frontier similar)

1. Upper control arm
2. Tie-rod end
3. Upper balljoint
4. Shock absorber
5. Steering knuckle
6. Torsion bar
7. Lower balljoint
8. Lower control arm
9. Stabilizer bar link
10. Stabilizer bar
11. Relay rod
12. Pitman arm
13. Steering gear

Chapter 10 Suspension and steering systems

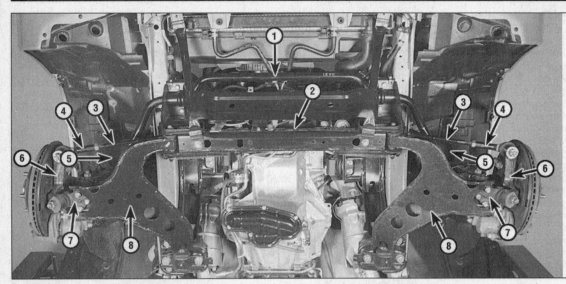

1.1c Front suspension and steering components - 2WD Pathfinder (4WD similar)

1. Stabilizer bar
2. Steering gear (above crossmember)
3. Tie-rod
4. Tie-rod end
5. Strut/coil spring assembly
6. Steering knuckle
7. Balljoint
8. Control arm

1.2 Rear suspension components - Frontier (Xterra similar)

1. Leaf spring shackle
2. Shock absorber
3. Leaf spring
4. Spring plate
5. Rear axle

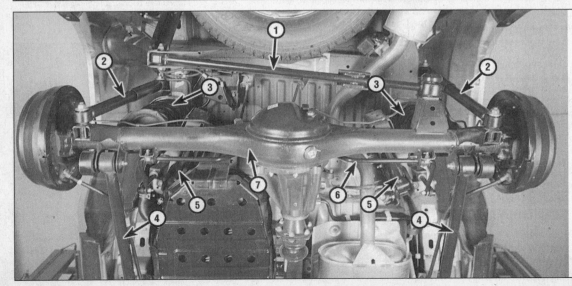

1.3 Rear suspension components - Pathfinder

1. Panhard rod
2. Shock absorber
3. Coil spring
4. Lower control arm
5. Upper control arm
6. Stabilizer bar
7. Rear axle

sure to use lots of penetrating oil and allow it to soak in for a while. Using a wire brush to clean exposed threads will also ease removal of the nut or bolt and prevent damage to the threads. Sometimes a sharp blow with a hammer and punch is effective in breaking the bond between a nut and bolt threads, but care must be taken to prevent the punch from slipping off the fastener and ruining the threads. Heating the stuck fastener and surrounding area with a torch sometimes helps too, but isn't recommended because of the obvious dangers associated with fire. Long breaker bars and extension, or "cheater," pipes will increase leverage, but never use an extension pipe on a ratchet - the ratcheting mechanism could be damaged. Sometimes, turning the nut or bolt in the tightening (clockwise) direction first will help to break it loose. Fasteners that require drastic measures to unscrew should always be replaced with new ones.

Since most of the procedures that are dealt with in this Chapter involve jacking up the vehicle and working underneath it, a good pair of jackstands will be needed. A hydraulic floor jack is the preferred type of jack to lift the vehicle, and it can also be used to support certain components during various operations. **Warning:** *Never, under any circumstances, rely on a jack to support the vehicle while working on it.* Also, whenever any of the suspension or steering fasteners are loosened or removed they must be inspected and, if necessary, replaced with new ones of the same part number or of original equipment quality and design. Torque specifications must be followed for proper reassembly and component retention. Never attempt to heat or straighten suspension or steering components. Instead, replace bent or damaged parts with new ones.

2 Shock absorber, front (Frontier and Xterra models) - removal and installation

Refer to illustrations 2.3 and 2.5

Warning: *The manufacturer recommends replacing the shock absorber mounting nut and bolt with new ones whenever they are removed.*

1 Loosen the front wheel lug nuts. Raise the front of the vehicle and support it securely on jackstands, then remove the wheel.
2 Support the outer end of the lower control arm with a floor jack. **Warning:** *The jack must remain in this position throughout the entire procedure.*
3 Using a back-up wrench on the stem, remove the shock absorber upper mounting nut **(see illustration)**.
4 Remove the metal washers and rubber bushing. Note the order in which they are installed.
5 Remove the bolt that attaches the lower end of the shock absorber to the lower control arm **(see illustration)** and remove the shock absorber.

2.3 Hold the shock absorber stem (A) with a wrench to prevent it from turning when the upper mounting nut (B) is loosened

3.2a Stabilizer bar link bolt nut - Frontier and Xterra

6 Remove the bushing and washer from the stem.
7 Installation is the reverse of removal. Be sure to tighten the upper mounting nut and the lower mounting bolt to the torque values listed in this Chapter's Specifications.
8 Install the wheels and lug nuts, lower the vehicle and tighten the lug nuts to the torque listed in the Chapter 1 Specifications.

3 Stabilizer bar and bushings (front) - removal and installation

Refer to illustrations 3.2a, 3.2b, 3.3a, 3.3b and 3.3c

Warning: *The manufacturer recommends replacing the stabilizer bar link nuts with new ones whenever they are removed.*

1 Raise the vehicle and support it securely on jackstands.
2 Remove the nuts from the links and detach the links from the bar **(see illustrations)**. **Note:** *Be sure to keep the parts for the left and right sides separate.*
3 Remove the stabilizer bar bracket bolts **(see illustrations)**. If you're working on a Pathfinder, you'll have to remove the brace that bolts to the front crossmember and the

2.5 The shock absorber is attached to the lower control arm with one bolt (the nut is welded to the bracket on the other side - you don't have to hold it with a wrench)

3.2b Stabilizer bar link details - Pathfinder
(**Note:** *If the ballstud spins as the nut is loosened, insert an Allen wrench into the ballstud hex*)

A Link-to-bar nut
B Link-to-strut nut

lower radiator support **(see illustration)**.
4 Remove the stabilizer bar.
5 Remove the rubber bushings.
6 Inspect all parts for wear and damage, replacing them as necessary.

3.3a Stabilizer bar bracket bolts - 2WD Frontier shown, Xterra and 4WD Frontier similar

Chapter 10 Suspension and steering systems

3.3b If you're working on a Pathfinder, you'll have to remove this brace before removing the stabilizer bar bracket bolts

3.3c Stabilizer bar bracket bolts - Pathfinder

4.2a Mark the relationship of the torsion bar to the anchor arm . . .

4.2b . . . and to the torque arm

4.3a Measure the length of the adjusting bolt (2WD Frontier shown)

4.3b Location of the torsion bar adjusting bolt nut - Xterra and 4WD Frontier models

7 Installation is the reverse of removal. Be sure to tighten all fasteners to the torque listed in this Chapter's Specifications.

4 Torsion bar (Frontier and Xterra models) - removal and installation

Warning: *The manufacturer recommends replacing the adjuster bolt locknut with a new one whenever it is removed.*

Removal

Refer to illustrations 4.2a, 4.2b, 4.3a, 4.3b, 4.6 and 4.7

1 Loosen the wheel lug nuts, raise the vehicle and support it securely on jackstands. Remove the wheel.
2 Slide the rubber boot forward on the torsion bar, off the anchor arm. Apply match marks from the torsion bar to the anchor arm and the torque arm **(see illustrations)**. **Caution:** *Use paint or a grease pen only. Do not use a scribe to mark the components.*
3 Measure how far the adjusting bolt protrudes (from the washer to the end of the bolt) **(see illustrations)**. Record this figure for use on installation.
4 Remove the anchor arm adjusting bolt locknut while holding the adjusting nut with a wrench to prevent it from turning.
5 Lubricate the threads of the adjusting bolt with grease, then unscrew the adjusting bolt; as the nut nears the end of the bolt, all twisting force should be relieved from the torsion bar. Remove the adjusting nut and bolt.
6 If you're working on a 2WD Frontier, slide the torsion bar forward and remove the anchor arm **(see illustration)**. Now slide the torsion bar out of the torque arm, towards the rear of the vehicle, lower the front of the bar and remove it towards the front of the vehicle. **Note:** *Each anchor arm has a tab on it that must be aligned with the opening in the crossmember to allow removal. If you have difficulty removing the anchor arm, raise the lower control arm with a floor jack far enough to allow anchor arm removal.*
7 If you're working on an Xterra or a 4WD Frontier, unbolt the torque arm from the lower control arm **(see illustration)**. Slide the torsion bar forward far enough to clear the crossmember, then lower the rear end of the bar and remove the torsion bar and torque arm.

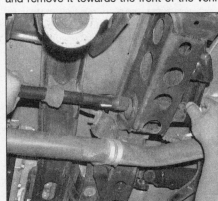

4.6 Slide the torsion bar forward and remove the anchor arm from the crossmember (2WD Frontier models)

4.7 On 4WD models the torque arm is secured to the lower control arm with three nuts and bolts

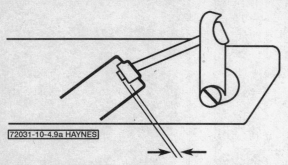
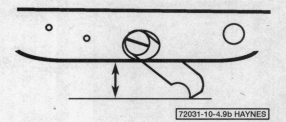

4.9a When installing a new torsion bar on a 2WD Frontier, the anchor arm adjusting bolt must protrude the specified amount

4.9b When installing a new torsion bar on an Xterra or a 4WD Frontier, the anchor arm must hang down past the crossmember the specified amount

Installation and adjustment

Refer to illustrations 4.9a, 4.9b, 4.13a and 4.13b

Note: *There are left and right side identification marks on the rear ends of the torsion bars - if both bars have been removed, be careful not to get them mixed up.*

8 Apply a thin coat of moly-based grease to the splines on each end of the torsion bar.
9 Install the torsion bar by reversing the removal procedure. It may be necessary on some models to raise the lower control arm to allow anchor arm removal (see the Note in Step 6). Be sure to line up the match marks on the bar with the marks on the anchor arm and torque arm. If you're installing a new torsion bar on a 2WD Frontier, position the anchor arm so that the adjusting bolt, when installed through the bracket (unloaded), protrudes the specified amount **(see illustration)**. If you're installing a new torsion bar on an Xterra or a 4WD Frontier, install the anchor arm so that it hangs down the specified amount from the crossmember **(see illustration)**.
10 Install the adjusting bolt, pivot and adjuster nut (don't install the locknut yet). Lubricate the threads of the bolt and tighten the nut until the previously recorded bolt protrusion figure has been attained.
11 Lower the vehicle.
12 Bounce the front end of the vehicle several times, then roll the vehicle back and forth to settle the suspension.
13 Measure the distance from the lower control arm pivot bolt to the ground - this will be called distance "A." Now measure the distance from the rearmost tension-rod-to-lower control arm bolt and the ground (2WD Frontier models) or from the bottom of the steering stopper bracket and ground (Xterra and 4WD Frontier models) - this will be called distance "B" **(see illustrations)**. Subtract distance "B" from distance "A" to calculate the ride height, or distance "C" in the illustrations. If the ride height is not as listed in this Chapter's Specifications, adjust the torsion bar as necessary, then re-check the ride height. **Note:** *If the ride height is too high, don't just back off the adjuster nut to lower the vehicle. Instead, loosen the nut more than is necessary to achieve the desired height,*

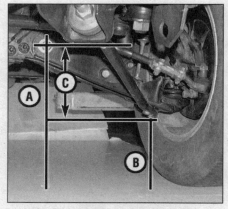

4.13a Ride height measuring points - 2WD Frontier models

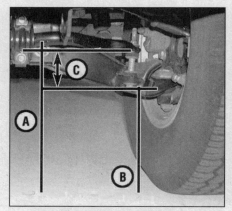

4.13b Ride height measuring points - Xterra and 4WD Frontier models

then tighten it to the point where the proper height is obtained.
14 If adjustment is necessary, be sure to bounce the vehicle several times and roll it back and forth to settle the suspension.
15 Once the proper ride height is attained, install a new locknut on the adjuster bolt and tighten it to the torque listed in this Chapter's Specifications.

5 Strut/coil spring assembly (Pathfinder models) - removal, inspection and installation

Warning: *The manufacturer recommends replacing the strut-to-knuckle nuts and the strut upper mounting nuts with new ones whenever they are removed.*

Removal

Refer to illustrations 5.2, 5.4 and 5.6

1 Loosen the front wheel lug nuts, raise the front of the vehicle and support it securely on jackstands. Remove the wheels.
2 Unclip the brake hose from the strut bracket and detach it from the bracket. Also remove the bolt and unclip the ABS speed sensor harness from the strut **(see illustration)**.
3 Remove the nut and detach the stabilizer bar link from the strut (see Section 3).
4 Remove the strut-to-knuckle nuts **(see**

5.2 Detach the ABS speed sensor harness bracket (A), remove the clip securing the brake hose (B) and detach the stabilizer bar link from the strut (C)

illustration) and knock the bolts out with a hammer and punch. Separate the strut from the steering knuckle. Be careful not to overextend the inner CV joint and don't let the knuckle fall outward, as this could damage the brake hose.
5 If equipped with adjustable shock absorbers, unplug the electrical connector from the shock absorber actuator and unbolt the actuator from the top of the strut tower.
6 Remove the three strut upper mounting nuts **(see illustration)**. **Note:** *Support the*

Chapter 10 Suspension and steering systems

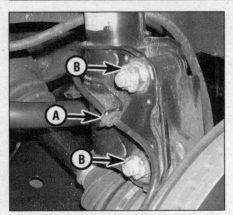

5.4 Detach the ABS speed sensor harness from the bracket (A), remove the two nuts (B), then drive out the strut-to-knuckle bolts with a hammer and punch

5.6 To detach the upper end of the strut from the body, remove the upper mounting nuts. Warning: *Don't unscrew the center nut (under the cover)!*

6.3 Install the spring compressor according to the tool manufacturer's instructions and compress the spring until all pressure is relieved from the upper spring seat

strut with one hand before removing the last nut. Remove the assembly out from the fenderwell.

Inspection

7   Check the strut body for leaking fluid, dents, cracks and other obvious damage which would warrant repair or replacement.
8   Check the coil spring for chips or cracks in the spring coating (this will cause premature spring failure due to corrosion). Inspect the spring seat for cuts, hardness and general deterioration.
9   If any undesirable conditions exist, proceed to the strut disassembly procedure (see Section 6).

Installation

10   Guide the strut assembly up into the fenderwell and insert the upper mounting studs through the holes in the shock tower. Once the studs protrude from the shock tower, install the nuts (new ones) so the strut won't fall back through. This is most easily accomplished with the help of an assistant, as the strut is quite heavy and awkward.
11   Slide the steering knuckle into the strut flange and insert the bolts. Install *new* nuts and tighten them to the torque listed in this Chapter's Specifications.
12   Guide the brake hose through its bracket in the strut and install the retaining clip. Clip the ABS speed sensor grommet into its bracket, then install the harness bolt and tighten it securely.
13   Install the wheel and lug nuts, then lower the vehicle and tighten the lug nuts to the torque listed in the Chapter 1 Specifications.
14   Tighten the upper mounting nuts to the torque listed in this Chapter's Specifications.
15   If equipped with adjustable shock absorbers, install the shock absorber actuator, ensuring that the cutout in the actuator aligns with the control rod in the center of the piston rod. Tighten the bolts securely.

6   Strut or coil spring (Pathfinder models) - replacement

Warning: *The manufacturer recommends replacing the piston rod nut with a new one whenever it is removed.*

1   If the struts or coil springs exhibit the telltale signs of wear (leaking fluid, loss of damping capability, chipped, sagging or cracked coil springs) explore all options before beginning any work. The strut/shock absorber assemblies are not serviceable and must be replaced if a problem develops. However, strut assemblies complete with springs may be available on an exchange basis, which eliminates much time and work. Whichever route you choose to take, check on the cost and availability of parts before disassembling your vehicle. **Warning:** *Disassembling a strut is potentially dangerous and utmost attention must be directed to the job, or serious injury may result. Use only a high-quality spring compressor and carefully follow the manufacturer's instructions furnished with the tool. After removing the coil spring from the strut assembly, set it aside in a safe, isolated area.*

Disassembly

Refer to illustrations 6.3, 6.4, 6.5, 6.6 and 6.7

2   Remove the strut and spring assembly following the procedure described in the previous Section. Mount the strut assembly in a vise. Line the vise jaws with wood or rags to prevent damage to the unit and don't tighten the vise excessively.
3   Following the tool manufacturer's instructions, install the spring compressor (which can be obtained at most auto parts stores or equipment yards on a daily rental basis) on the spring and compress it sufficiently to relieve all pressure from the upper spring seat **(see illustration)**. This can be verified by wiggling the spring.
4   Remove the piston rod nut **(see illustration)**.
5   Remove the upper mount **(see illustration)**. Inspect the bearing in the mount for smooth operation. If it does not turn smoothly, replace it. Check the rubber portion of the mount for cracking and general deterioration. If there is any separation of the rubber, replace it.
6   Lift the spring seat and upper insulator from the piston rod **(see illustration)**. Check

6.4 Remove the piston rod nut

6.5 Lift the upper mount off the piston rod

Chapter 10 Suspension and steering systems

6.6 Remove the spring seat from the piston rod

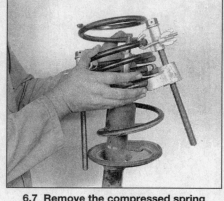

6.7 Remove the compressed spring assembly - keep the ends of the spring pointed away from your body

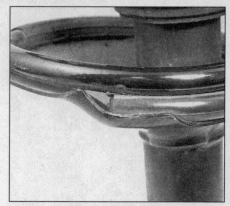

6.11 When installing the spring, make sure the end fits into the recessed portion of the lower seat

6.12a Make sure this cutout (arrow) in the upper seat . . .

6.12b . . . is facing out (toward the strut-to-knuckle flanges)

7.3 Mark the relationship of the adjusting cams to the frame

the rubber spring seat for cracking and hardness, replacing it if necessary.
7 Carefully lift the compressed spring from the assembly **(see illustration)** and set it in a safe place. **Warning:** *Carry the spring carefully and never place any part of your body near the end of the spring!*
8 Slide the dust boot off the piston rod.
9 Check the lower insulator (if equipped) for wear, cracking and hardness and replace it if necessary.

Reassembly

Refer to illustrations 6.11, 6.12a and 6.12b

10 If the lower insulator is being replaced, set it into position with the dropped portion seated in the lowest part of the seat. Extend the damper rod to its full length and install the dust boot.
11 Carefully place the coil spring onto the lower insulator, with the end of the spring resting in the lowest part of the insulator **(see illustration)**.
12 Install the upper insulator and the spring seat. Make sure the cutout on the spring seat is facing out (away from the vehicle), in line with the strut-to-knuckle attachment points **(see illustrations)**.
13 Install the dust seal and upper mount to the piston rod.

14 Install a *new* piston rod nut and tighten it to the torque listed in this Chapter's Specifications.
15 Install the strut assembly (see Section 5).

7   Upper control arm (Frontier and Xterra models) - removal and installation

Refer to illustrations 7.3 and 7.4

Removal

1 Loosen the wheel lug nuts, raise the front of the vehicle and support it securely on jackstands. Remove the wheel. Position a floor jack under the lower control arm in the area underneath the balljoint. Raise the jack slightly to take the spring pressure off the upper control arm. **Warning:** *The jack must remain in this position throughout the entire procedure.*
2 Remove the shock absorber (see Section 2).
3 Mark the relationship of the adjusting cams to the brackets on the frame **(see illustration)**.
4 To disconnect the upper control arm from the steering knuckle, remove the cotter pin, loosen the upper balljoint nut a few turns

(don't remove it), install a balljoint separator and break the balljoint loose from the knuckle. Now remove the nut. **Note:** *If you don't have a press-type balljoint removal tool, a "picklefork" type balljoint separator can be used, but keep in mind that this type of tool will probably destroy the balljoint boot* **(see illustration)**.
5 Remove the upper control arm pivot bolts and nuts, noting which way the bolts are installed. Remove the control arm.

7.4 Separating the upper control arm balljoint from the steering knuckle using a "picklefork" type separator

8.6 Separating the lower control arm balljoint with a "picklefork" type balljoint separator (2WD Frontier model)

8.7a Lower control arm pivot bolt/nut (2WD Frontier)

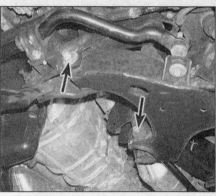

8.7b Lower control arm pivot bolts/nuts (2WD Xterra shown, 4WD Frontier and 4WD Xterra similar)

Installation

6 Position the arm in the frame brackets and install the bolts and nuts, but don't tighten them yet.

7 Attach the balljoint to the steering knuckle and tighten the ballstud nut to the torque listed in this Chapter's Specifications. Install a new cotter pin. **Note:** *If necessary, tighten the nut a little more to align the slots in the nut with the hole in the ballstud; don't loosen the nut to allow cotter pin insertion.*

8 Place a floor jack under the outer end of the lower control arm and raise it to simulate normal ride height. Make sure the marks you made on the adjustment cams prior to disassembly are aligned, then tighten the nuts to the torque listed in this Chapter's Specifications.

9 The remainder of installation is the reverse of removal. Tighten the wheel lug nuts to the torque listed in the Chapter 1 Specifications.

10 Have the front end alignment checked and, if necessary, adjusted.

8 Lower control arm - removal and installation

2WD Frontier and 2WD Xterra models

Refer to illustrations 8.6, 8.7a and 8.7b

Warning: *The manufacturer recommends replacing the pivot bolt nut(s) with new ones whenever they are removed.*

Removal

1 Loosen the wheel lug nuts, raise the vehicle and support it securely on jackstands placed under the frame rails. Remove the wheel.

2 Remove the torsion bar and torque arm (see Section 4).

3 Detach the stabilizer bar link from the lower control arm (see Section 3).

4 Remove the shock absorber lower mounting bolt (see Section 2).

5 If you're working on a 2WD Frontier model, unbolt the tension rod from the lower control arm (see Section 9).

6 To disconnect the lower control arm from the steering knuckle, loosen the balljoint nut a few turns (don't remove it), install a balljoint separator and break the balljoint loose from the knuckle **(see accompanying illustration**, or, if you're working on an Xterra, **see illustration 8.18)**. Now remove the nut. **Note:** *If you don't have a press-type balljoint removal tool, a "picklefork" type balljoint separator can be used, but keep in mind that this type of tool will probably destroy the balljoint boot* **(see illustration)**.

7 Remove the lower control arm pivot bolt(s) and nut(s), noting which way the bolts are installed **(see illustrations)**. Remove the lower control arm.

8 The control arm bushings are replaceable, but special tools are necessary to do the job. Carefully inspect the bushings for hardening, excessive wear and cracks. If they appear to be worn or deteriorated, take the control arm to an automotive machine shop or other repair facility for replacement.

Installation

9 Installation is the reverse of removal. **Note:** *If necessary, tighten the balljoint-to-steering knuckle nut a little more to align the slots in the nut with the hole in the ballstud; don't loosen the nut to allow cotter pin insertion. Also, be sure to use a new cotter pin.* Don't tighten the pivot bolt nut(s) to the torque listed in this Chapter's Specifications until the vehicle has been lowered and is resting at normal ride height (this can also be simulated by raising the lower control arm with a floor jack). Be sure to tighten the wheel lug nuts to the torque listed in the Chapter 1 Specifications. Adjust the ride height following the procedure described in Section 4.

10 Have the front wheel alignment checked and, if necessary, adjusted.

4WD Frontier and 4WD Xterra models

Refer to illustration 8.18

Warning: *The manufacturer recommends replacing the pivot bolt nuts with new ones whenever they are removed.*

Removal

11 Loosen the wheel lug nuts, raise the vehicle and support it securely on jackstands placed under the frame rails. Remove the wheel.

12 Remove the locking hub assembly, then remove the snap-ring from the end of the driveaxle (see Chapter 8).

13 Remove the torsion bar and torque arm (see Section 4).

14 Detach the stabilizer bar link from the lower control arm (see Section 3).

15 Remove the shock absorber lower mounting bolt (see Section 2).

16 Unbolt the driveaxle inner CV joint from the differential drive flange (see Chapter 8).

17 Remove the lower control arm pivot bolts/nuts **(see illustration 8.7b)**. Separate the lower control arm from the frame brackets. Support the driveaxle so it doesn't hang by the outer CV joint. Using a soft face hammer, tap on the end of the driveaxle to free it from the hub splines, then remove the driveaxle.

18 Remove the cotter pin and loosen, but do not remove, the nut from the lower balljoint. Using a two-jaw puller, break loose the ballstud from the steering knuckle **(see illustration)**. **Note:** *A "picklefork" type balljoint sepa-*

8.18 A two-jaw puller can be used to separate the balljoint from the steering knuckle on Xterra and 4WD Frontier models (as well as all Pathfinder models)

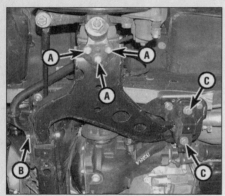

8.25 On Pathfinder models the balljoint can be unbolted from the control arm by removing the three nuts (A); to detach the control arm from the subframe, remove the pivot bolt/nut (B) and the rear bushing bracket bolts (C)

9.3 Remove the nut from the front of the tension rod, then remove the washers and bushing

9.4 When removing these tension rod-to-control arm bolts, hold the nuts on the upper side of the arm with a wrench to prevent them from turning

rator can be used, but keep in mind that this type of tool will probably destroy the balljoint boot **(see illustration 8.6)**.

19 Remove the nut and detach the lower control arm from the steering knuckle.

20 Carefully inspect the bushings for hardening, excessive wear and cracks. The forward control arm bushing is replaceable, but special tools are necessary to do the job; if is appears to be worn or deteriorated, take the control arm to an automotive machine shop or other repair facility for replacement. The rear bushing in the frame boss is also replaceable, with the use of a drawbolt-type bushing extractor (this tool may be available at some auto parts stores or equipment rental yards).

Installation

21 Installation is the reverse of removal. **Note:** *If necessary, tighten the balljoint-to-steering knuckle nut a little more to align the slots in the nut with the hole in the ballstud; don't loosen the nut to allow cotter pin insertion. Also, be sure to use a new cotter pin.* Don't tighten the pivot bolt nut(s) to the torque listed in this Chapter's Specifications until the vehicle has been lowered and is resting at normal ride height (this can also be simulated by raising the lower control arm with a floor jack). Be sure to tighten the wheel lug nuts to the torque listed in the Chapter 1 Specifications. Tighten the driveaxle inner flange bolts and the locking hub bolts to the torque values listed in the Chapter 8 Specifications.

22 Adjust the ride height following the procedure described in Section 4.

23 Have the front wheel alignment checked and, if necessary, adjusted.

Pathfinder models

Refer to illustration 8.25
Warning: *The manufacturer recommends replacing the balljoint-to-control arm nuts and the control arm pivot bolt nut whenever they are removed.*

Removal

24 Loosen the wheel lug nuts, raise the vehicle and support it securely on jackstands placed under the subframe rails. Remove the wheel.
25 Remove the balljoint-to-lower control arm nuts **(see illustration)**. Separate the balljoint from the control arm.
26 Remove the front pivot bolt/nut from the lower arm, and the rear bushing bracket bolts. Detach the control arm from the frame.
27 If you're working on a 4WD model and have to remove the balljoint from the steering knuckle, the driveaxle will have to be removed first (see Chapter 8).
28 Carefully inspect the bushings for hardening, excessive wear and cracks. The forward control arm bushing is replaceable, but special tools are necessary to do the job; if is appears to be worn or deteriorated, take the control arm to an automotive machine shop or other repair facility for replacement. The rear bushing is also replaceable, and can be slid off the pin of the control arm.

Installation

29 Installation is the reverse of removal. Don't tighten the pivot bolt nut or the bushing bracket bolts to the torque values listed in this Chapter's Specifications until the vehicle has been lowered and is resting at normal ride height (this can also be simulated by raising the lower control arm with a floor jack). Be sure to tighten the wheel lug nuts to the torque listed in the Chapter 1 Specifications.

9 Tension rod (2WD Frontier models) - removal and installation

Refer to illustrations 9.3 and 9.4
Warning: *The manufacturer recommends replacing the large nut at the front of the tension rod with a new one whenever it is removed.*

1 Loosen the wheel lug nuts, raise the front of the vehicle and support it securely on jackstands. Remove the wheel.
2 Support the outer end of the lower control arm with a floor jack. Raise the jack slightly.
3 Remove the nut from the front end of the tension rod **(see illustration)**. Also remove the washers and bushing, noting how they are positioned.
4 Unscrew the tension rod-to-lower control arm nuts and bolts and remove the rod **(see illustration)**.
5 Slide the sleeve, bushing and washers off the rod. Check the bushings for cracking, hardness and deterioration, replacing them if necessary.
6 Installation is the reverse of removal. Make sure the concave side of the washers face the bushings. Tighten the fasteners to the torque values listed in this Chapter's Specifications.
7 Install the wheel and lug nuts. Lower the vehicle and tighten the lug nuts to the torque listed in the Chapter 1 Specifications.

10 Balljoints - check and replacement

Check

1 Inspect the control arm balljoint(s) for looseness anytime a balljoint is separated from the steering knuckle. See if you can turn the ballstud in its socket with your fingers. If the balljoint is loose, or if the ballstud can be turned, replace the balljoint. You can also check the balljoints with the suspension assembled as follows.

Upper balljoint (Frontier and Xterra models only)

2 Raise the front of the vehicle and support it securely on jackstands placed under the frame rails. Place a floor jack under the lower control arm and raise it far enough to lift the upper control arm off its stop.
3 Attempt to move the control arm up and down; a prybar may be helpful. If any play is felt, replace the upper control arm and balljoint as an assembly (the balljoint is not replaceable separately).
4 Also try to move the steering knuckle in-

Chapter 10 Suspension and steering systems

and-out. If any play is felt, replace the upper control arm/balljoint assembly.

5   Check the balljoint boot for cracks and tears. If any are present, but the balljoint is still in good condition, separate the upper balljoint from the steering knuckle (see Section 7), pry off the old boot and install a new boot and clamp.

Lower balljoint

6   Raise the front of the vehicle and support it securely on jackstands placed under the frame rails.

Frontier and Xterra models

7   Place a floor jack under the lower control arm and raise it slightly. Attempt to move the steering knuckle up and down; a large prybar underneath the tire, or a prybar placed between the end of the control arm and the steering knuckle1 will be helpful. If any play is felt, replace the control arm and balljoint as an assembly (the balljoint is not replaceable separately).

8   Also try to move the steering knuckle in-and-out. If any play is felt, replace the control arm/balljoint assembly.

9   Check the balljoint boot for cracks and tears. If any are present, but the balljoint is still in good condition, separate the lower balljoint from the steering knuckle (see Section 8), pry off the old boot and install a new boot and clamp.

Pathfinder models

10   Visually inspect the rubber dust boot for damage, deterioration and leaking grease. If the boot is damaged, deteriorated or leaking, replace the balljoint.

11   Place a large prybar under the balljoint and resting on the wheel, then try to pry the balljoint up while feeling for movement between the balljoint and steering knuckle. Now, pry between the control arm and the steering knuckle and try to lever the control arm down while feeling for movement between the balljoint and steering knuckle. If any movement is evident in either check, the balljoint is worn out.

12   Have an assistant grasp the tire at the top and bottom and move the top of the tire in-and-out. Touch the balljoint stud nut. If any looseness is felt, suspect a worn-out balljoint stud or a widened hole in the steering knuckle boss. If the latter problem exists, the steering knuckle should be replaced as well as the balljoint.

Replacement

Frontier and Xterra models

13   As stated previously in this Section, the balljoints on these models are an integral part of the control arm and are not available separately. The entire control arm must be replaced.

Pathfinder models

Warning: *The manufacturer recommends replacing the balljoint-to-control arm nuts whenever they are removed.*

14   Loosen the wheel lug nuts, raise the front of the vehicle and support it securely on jackstands. Remove the wheel.

15   Remove the balljoint-to-control arm nuts **(see illustration 8.25)** and detach the balljoint from the control arm.

16   If you're working on a 4WD model, remove the driveaxle (see Chapter 8).

17   Remove the cotter pin and loosen, but do not remove, the ballstud nut. Using a two-jaw puller, break the balljoint loose from the steering knuckle. **Note:** *A "picklefork" type balljoint separator can be used, but keep in mind that this type of tool will probably destroy the balljoint boot* **(see illustration 8.6).**

18   Remove the nut and detach the balljoint from the steering knuckle.

19   Installation is the reverse of removal. Tighten the balljoint-to-control arm nuts and the balljoint-to steering knuckle nut to the torque values listed in this Chapter's Specifications. **Note:** *If necessary, tighten the balljoint-to-steering knuckle nut a little more to align the slots in the nut with the hole in the ballstud; don't loosen the nut to allow cotter pin insertion. Also, be sure to use a new cotter pin.*

20   Install the wheel and lug nuts. Tighten the lug nuts to the torque listed in the Chapter 1 Specifications.

21   Have the front end alignment checked and, if necessary, adjusted.

11 Steering knuckle - removal and installation

Warning: *The manufacturer recommends replacing the balljoint-to-control arm nuts whenever they are removed (Pathfinder models only).*

1   Loosen the wheel lug nuts, raise the vehicle and support it securely on jackstands placed underneath the frame (or subframe) rails. Remove the wheel.

2   If you're working on a 4WD model, remove the driveaxle (see Chapter 8).

3   If you're working on a 2WD Frontier or Xterra model, support the lower control arm with a floor jack. Raise the jack slightly.
Warning: *The jack must remain in this position throughout the entire procedure.*

4   Remove the brake caliper and brake disc (see Chapter 9). Hang the caliper out of the way on a piece of wire (don't disconnect the brake hose).

5   Remove the disc splash shield from the steering knuckle.

6   If you're working on a Pathfinder model, detach the balljoint from the control arm (see Section 8).

7   Disconnect the tie-rod end from the steering knuckle (see Section 21).

8   If you're working on a Frontier or an Xterra model, disconnect the balljoints from the steering knuckle (see Sections 7 and 8) and remove the steering knuckle.

9   If you're working on a Pathfinder model, remove the nuts and bolts and detach the steering knuckle from the strut (see Section 5), then remove the steering knuckle.

10   Installation is the reverse of removal. Be sure to tighten the balljoint, strut (Pathfinder models) and tie-rod end fasteners to the torque values listed in this Chapter's Specifications. **Note:** *If necessary, tighten the nut a little more to align the slots in the nut with the hole in the ballstud; don't loosen the nut to allow cotter pin insertion. Also, be sure to use a new cotter pin.* Tighten the caliper mounting bolts to the torque values listed in the Chapter 9 Specifications. Tighten the locking hub bolts to the torque listed in the Chapter 8 Specifications (4WD models). Tighten the wheel lug nuts to the torque listed in the Chapter 1 Specifications.

12 Shock absorber, rear - removal and installation

Refer to illustrations 12.3a and 12.3b
Warning: *The manufacturer recommends replacing the upper and lower shock absorber mounting nuts with new ones whenever they are removed.*

1   Raise the rear of the vehicle and support it securely on jackstands placed underneath the frame rails. Block the front wheels so the vehicle doesn't roll off the stands. **Note:** *It isn't necessary to remove the rear wheels, but doing so will improve access to the shock absorbers.*

2   Support the rear axle with a floor jack placed under the axle tube closest to the shock absorber being removed.

3   Remove the shock absorber upper and lower mounting fasteners **(see illustrations)**.

12.3a Shock absorber lower mounting nut

12.3b Shock absorber upper mounting nut

10-14 Chapter 10 Suspension and steering systems

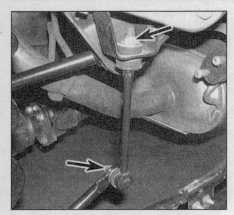

13.2 Remove the stabilizer bar link nuts and detach the links from the bar

13.3 2000 and earlier Pathfinders have one bolt attaching each stabilizer bar clamp; 2001 Pathfinders and all Xterras have two bolts per clamp

14.3 Remove the shock absorber lower mounting bolt and the four U-bolt nuts, then remove the spring plate and the U-bolts

4 Remove the shock absorber.
5 Installation is the reverse of removal. Tighten all fasteners to the torque values listed in this Chapter's Specifications.

13 Stabilizer bar and bushings (rear) - removal and installation

Refer to illustrations 13.2 and 13.3
Warning: *The manufacturer recommends replacing the stabilizer bar link nuts whenever they are removed.*

1 Loosen the rear wheel lug nuts, raise the rear of the vehicle and support it securely on jackstands. Block the front wheels to keep the vehicle from rolling off the stands. Remove the rear wheels.
2 Remove the nuts from the lower ends of the stabilizer bar links, then separate the links from the bar **(see illustration)**.
3 Remove the stabilizer bar clamp bolts **(see illustration)** and remove the stabilizer bar.
4 Inspect the stabilizer bar bushings and link bushings for cracks, tears and other signs of deterioration. Replace as necessary. Also check the ballstuds at the ends of the links for looseness, replacing the links if necessary. The links are attached to the body with two bolts.
5 Installation is the reverse of removal. Be sure to tighten all fasteners to the torque values listed in this Chapter's Specifications.

14 Leaf spring (Frontier and Xterra models) - removal and installation

Refer to illustrations 14.3, 14.4 and 14.5
Warning: *The manufacturer recommends replacing the nut on the front spring hanger bolt whenever it is removed.*

Removal

1 Loosen the rear wheel lug nuts, raise the rear of the vehicle and support it securely on

14.4 At the front end of the leaf spring, remove the nut from the spring pin (left arrow); on 2WD Frontier models, also remove the bolt that secures the pin to the frame (right arrow)

jackstands placed underneath the frame rails. Block the front wheels to keep the vehicle from rolling off the stands. Remove the rear wheels.
2 Support the axle with a floor jack placed under the axle tube and raise it slightly to take the weight of the axle. Remove the shock absorber lower mounting bolt.
3 Remove the four U-bolt nuts and washers **(see illustration)**, the spring plate and the two U-bolts.
4 At the front end of the spring, remove the nut from the spring pin **(see illustration)**. On 2WD Frontier models also remove the bolt securing the pin to the frame bracket. Remove the pin.
5 At the rear end of the spring, remove the spring shackle nuts **(see illustration)**. Remove the shackle plate and shackle.
6 Remove the spring assembly.
7 If the bushings at the ends of the spring are worn or deteriorated, an automotive machine shop or dealer service department can press the old ones out and press new ones in. The upper shackle bushings can be replaced without the use of special tools.

14.5 Remove the nuts from the rear spring shackle, lift off the shackle plate, then push the shackle pins through the bushings and remove it

Installation

8 Place the spring in position and install the front mounting pin and nut, but don't tighten the nut yet. If you're working on a 2WD Frontier, install the pin-to-frame bolt, tightening it to the torque listed in this Chapter's Specifications.
9 Raise the rear of the spring into position and install the shackle, shackle plate and nuts. Don't tighten the nuts yet.
10 Raise or lower the axle on the jack until it mates properly with the spring. Install the spring plate and U-bolts, then install the nuts and washers. Tighten the U-bolt nuts, in a criss-cross pattern, to the torque listed in this Chapter's Specifications.
11 Connect the lower end of the shock absorber to the spring plate. Install the washer and nut, tightening the nut to the torque listed in this Chapter's Specifications.
12 Install the wheel, lower the vehicle to the ground and bounce it a few times, then tighten the front mounting pin nut and the shackle nuts to the torque values listed in this Chapter's Specifications. Tighten the lug nuts to the torque listed in the Chapter 1 Specifications.

Chapter 10 Suspension and steering systems

16.2a When removing an upper trailing arm, unbolt the stabilizer bar link bracket from the frame (A), then remove the pivot nuts and bolts (B) from each end of the arm

16.2b Lower trailing arm mounting details

16.6a Panhard rod-to-frame bracket nut/bolt

16.6b Panhard rod-to-rear axle housing nut

15 Coil spring, rear (Pathfinder models) - removal and installation

1 Raise the rear of the vehicle and support it securely on jackstands. Block the front wheels to prevent the vehicle from rolling.
2 Support the rear axle with a floor jack placed under the differential. If you have two floor jacks, place one under each axle tube (this is the preferable method).
3 Unbolt the rear brake hose junction block from the rear axle, then free the brake line from the securing clips along the rear axle housing. It isn't necessary to disconnect the brake line fittings.
4 Unbolt the Panhard rod from the axle (see Section 16).
5 Unbolt the stabilizer bar links from the bar (see Section 13).
6 Disconnect the lower ends of the shock absorbers from the axle (see Section 12).
7 Slowly lower the rear axle housing until the coil springs are fully extended, but no farther. Keep an eye on the brake hose and lines, making sure they don't get stretched or bent.
8 Remove the coil springs.
9 Check the upper spring seats and bump stops for cracks, hardening and general deterioration, replacing them as necessary.
10 Installation is the reverse of the removal procedure. When installing the springs, make sure the direction marks (usually a paint stripe) is facing the rear of the vehicle.

16 Suspension arms, rear (Pathfinder models) - removal and installation

Warning: *The manufacturer recommends replacing the trailing arm and Panhard rod pivot bolt nuts with new ones whenever they are removed.*

Trailing arms

Refer to illustrations 16.2a and 16.2b
Warning: *Remove and install only one arm at a time. This will prevent the axle housing from shifting on the jack.*
1 Loosen the rear wheel lug nuts. Raise the rear of the vehicle and support it securely on jackstands placed underneath the frame rails. Block the front wheels to prevent the vehicle from rolling. Remove the wheel(s).
2 Support the rear axle with a floor jack, then remove the nuts, washers and bolts from each end of the trailing arm **(see illustrations)**. **Note:** *If you're removing an upper trailing arm, the stabilizer bar link bracket must first be unbolted from the frame.*
3 Remove the arm. Check the bushings in the arm for cracking, hardness or other signs of deterioration. If the bushings are in need of replacement, take the arm to an automotive machine shop or other qualified repair facility to have the old ones pressed out and new ones pressed in.
4 Installation is the reverse of removal. Before tightening the bolts/nuts, either lower the vehicle to the ground or raise the rear axle with a floor jack to simulate normal ride height, then tighten the fasteners to the torque listed in this Chapter's Specifications. Tighten the wheel lug nuts to the torque listed in the Chapter 1 Specifications.

Panhard rod

Refer to illustration 16.6a and 16.6b
5 Raise the rear of the vehicle and support it securely on jackstands placed underneath the subframe rails. Block the front wheels to prevent the vehicle from rolling.
6 Remove the nuts/bolts from each end of the rod **(see illustrations)**.
7 Remove the rod. Check the bushings in the rod for cracking, hardness or other signs of deterioration. If the bushings are in need of replacement, check with your local auto parts store or dealer parts department regarding the availability of replacement bushings. If the bushings are in need of replacement, take the rod to an automotive machine shop or other qualified repair facility to have the old ones pressed out and new ones pressed in.
8 Installation is the reverse of removal. Before tightening the bolts/nuts, either lower the vehicle to the ground or raise the rear axle with a floor jack to simulate normal ride height, then tighten the fasteners to the torque listed in this Chapter's Specifications.

Chapter 10 Suspension and steering systems

17.2 Pry off the access cover from the underside of the steering wheel and unplug the airbag module connector

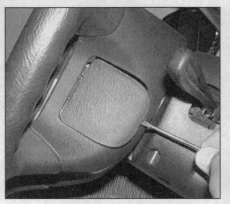

17.3a Pry off the two small side covers . . .

17.3b . . . and remove the Torx bolt (arrow); the Torx bolts are coated with a special bonding agent, so they must be discarded - be sure to replace them with new ones during reassembly

17 Steering wheel - removal and installation

Warning 1: *These models are equipped with airbags. Always disable the airbag system before working in the vicinity of any airbag system component to avoid the possibility of accidental deployment of the airbag, which could cause personal injury (see Chapter 12).*
Warning 2: *The manufacturer recommends replacing the airbag module bolts with new ones whenever they are removed.*

Removal

Refer to illustrations 17.2, 17.3a, 17.3b, 17.5, 17.7 and 17.8

1   Disconnect the cable from the negative battery terminal, then the positive battery terminal and wait at least ten minutes before proceeding.
2   Remove the small cover from the underside of the steering wheel and unplug the airbag module connector **(see illustration)**.
3   Remove the side covers and remove the Torx bolts behind them **(see illustrations)**. Some models may be equipped with a cruise control switch or a stereo control switch, secured by a screw, instead of a cover. **Note:** *Some models are equipped with special tamper-resistant Torx bolts, which require a special driver. The Torx bolts are coated with a special bonding agent, so they must be discarded; be sure to replace them with new ones during reassembly. Lift the airbag module off the steering wheel.* **Warning:** *Handle the airbag module with care, carry the module with the trim cover side facing away from your body and store it in a safe location with the trim side facing up. See the precautions in Chapter 12.*
4   Turn the steering wheel so that the front wheels are pointing straight ahead.
5   Remove the steering wheel retaining nut, then mark the relationship of the steering wheel to the steering shaft **(see illustration)**.
6   Disconnect the electrical connector for the cruise control wiring harness.
7   Use a steering wheel puller to separate the steering wheel from the steering shaft **(see illustration)**. When removing the wheel, make sure the electrical leads for the airbag module and the cruise control system don't snag on the wheel. **Warning:** *Do not turn the steering shaft while the steering wheel is removed.*
8   Remove the clockspring only if the steering column switches must be checked or replaced **(see illustration)**. **Caution:** *Regardless if the clockspring will be removed or not, tape the center hub of the clockspring to the outer ring so the center hub cannot rotate. This will retain the clockspring in the centered position. Also, don't allow the steering shaft to rotate with the steering wheel removed.* If necessary, remove the screws, unlock and disconnect the electrical connector, then separate the clockspring from the steering column.

Installation

Refer to illustrations 17.10a and 17.10b

9   Verify that the front wheels are pointing straight ahead. If the clockspring was removed, make sure the turn signal cancel tab is aligned with the notch in the combination switch.
10 Turn the clockspring for the airbag clockwise by hand until it stops (don't apply too much force), then rotate the clockspring counterclockwise about two and one-half turns (models with power steering) or four turns (models without power steering) until the arrow on the bottom of the clockspring is aligned with the pin on the clockspring hub

17.5 Remove the steering wheel retaining nut, then mark the relationship of the steering wheel to the shaft before removing the wheel

17.7 Use a steering wheel puller to separate the steering wheel from the steering shaft - DO NOT hammer on the shaft in an attempt to remove the wheel!

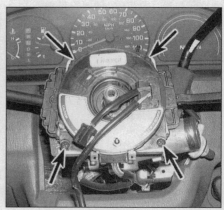

17.8 If it is necessary to remove the clockspring, remove these screws, disconnect the electrical connector and detach it from the steering column

Chapter 10 Suspension and steering systems

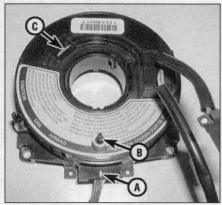

17.10a When the clockspring is properly centered on Frontier and Xterra models, the arrow (A) will be aligned with the pin (B), and the yellow mark will appear in the window (C)

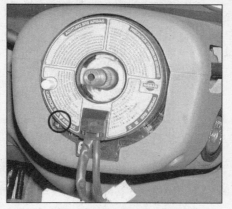

17.10b When the clockspring is properly centered on Pathfinder models the two arrows will be in alignment

18.7 Steering column-to-intermediate shaft pinch bolt

(Frontier and Xterra models) or the two arrows are aligned (Pathfinder models) **(see illustrations)**.

11 Pull the electrical leads for the airbag module and the cruise control system through the steering wheel and install the wheel. Make sure the clockspring pin guides are properly engaged with their corresponding holes in the back of the steering wheel.

12 Install the steering wheel retaining nut and tighten it to the torque listed in this Chapter's Specifications.

13 Install the airbag module and secure it with new Torx bolts. Do not reuse the old bolts. Install cruise control switch and stereo control switch, if equipped, or the side covers.

14 Plug in the airbag module connector. Install the lower cover.

15 Start the engine and turn the steering wheel all the way to the left and all the way to the right to make sure that the clockspring is properly centered.

16 Verify that the airbag circuit is operational by turning the ignition key to the On or Start position. The "AIR BAG" warning light should illuminate for a few seconds, then turn off.

18 Steering column - removal and installation

Refer to illustrations 18.7 and 18.8

Warning: *These models are equipped with airbags. Always disable the airbag system before working in the vicinity of any airbag system component to avoid the possibility of accidental deployment of the airbag, which could cause personal injury* (see Chapter 12).

Removal

1 Park the vehicle with the wheels pointing straight ahead. Disconnect the cable from the negative battery terminal, then the positive battery terminal. Disable the airbag system (see Chapter 12).

2 Remove the steering wheel (see Section 17), then turn the ignition key to the LOCK position to prevent the steering shaft from turning. **Caution:** *If this is not done, the airbag clockspring could be damaged.*

3 Remove the knee bolster and the reinforcement behind it (see Chapter 11). On models with a column-mounted shifter, detach the shift cable from the shift lever on the column (see Chapter 7B).

4 Remove the steering column covers (see Chapter 11).

5 Disconnect the electrical connectors for the steering column harness.

6 On 2WD Frontier models equipped with an automatic transmission, disconnect the shift cable from the lever on the column (see Chapter 7B). Also detach the shift interlock cable (see Chapter 7B).

7 Remove the shaft coupler nut and remove the bolt securing the steering shaft to the intermediate shaft **(see illustration)**. Mark the relationship of the intermediate shaft to the steering column shaft.

8 Remove the steering column mounting fasteners **(see illustration)**, lower the column and pull it to the rear, making sure nothing is still connected. Separate the intermediate shaft from the steering shaft and remove the column.

Installation

9 Guide the steering column into position, connect the intermediate shaft, then install the mounting fasteners, but don't tighten them yet.

10 Install the coupler bolt, tightening it to the torque listed in this Chapter's Specifications.

11 Tighten the column mounting fasteners to the torque listed in this Chapter's Specifications.

12 The remainder of installation is the reverse of removal. On automatic transmission-equipped models, adjust the shift cable and interlock cable following the procedures described in Chapter 7B.

19 Intermediate shaft - removal and installation

Refer to illustrations 19.3a and 19.3b

1 Park the vehicle with the wheels pointing straight ahead. Disconnect the cable from the negative battery terminal, then the positive battery terminal. Disable the airbag system (see Chapter 12).

2 Turn the ignition key to the LOCK position to prevent the steering shaft from turn-

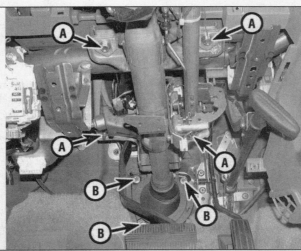

18.8 Typical steering column mounting details (2WD Frontier shown, other models similar)

A Mounting bolts/nuts
B Column-to-firewall nuts

10-18　Chapter 10　Suspension and steering systems

19.3a Intermediate shaft-to-steering gear pinch bolt (Frontier and Xterra models)

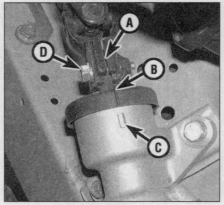

19.3b When the steering is centered, the gap on the intermediate shaft (A) is aligned with the mark on the end cap (B) and the projection on the transfer gear (C). (D) is the pinch bolt (Pathfinder models)

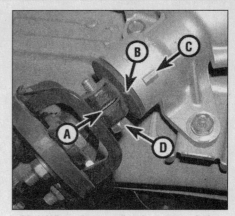

20.4 When the steering is centered, the gap on the coupler shaft (A) is aligned with the mark on the end cap (B) and the projection on the transfer gear (C). (D) is the pinch bolt

ing. **Caution:** *If this is not done, the airbag clockspring could be damaged.*

3　Working under the hood, mark the relationship of the intermediate shaft to the steering gear input shaft (Frontier and Xterra models) **(see illustration)** or the transfer gear assembly (Pathfinder models) **(see illustration)**. Remove the coupler bolt.

4　Mark the relationship of the intermediate shaft to the steering column shaft, then remove the pinch bolt **(see illustration 18.7)**.

5　Slide the intermediate shaft up and off the steering gear input shaft (Frontier and Xterra models) or the transfer gear assembly (Pathfinder models), then pull the shaft down off the steering gear input shaft.

6　Installation is the reverse of removal. Be sure to align the matchmarks and tighten the coupler bolts to the torque values listed in this Chapter's Specifications.

20　Steering column transfer gear (Pathfinder models) - removal and installation

Refer to illustrations 20.4 and 20.5

1　Park the vehicle with the wheels pointing straight ahead. Disconnect the cable from the negative battery terminal, then the positive battery terminal. Disable the airbag system (see Chapter 12).

2　Turn the ignition key to the LOCK position to prevent the steering shaft from turning. **Caution:** *If this is not done, the airbag clockspring could be damaged.*

3　Working under the hood, remove the air filter housing (see Chapter 4) and note the relationship of the intermediate shaft to the transfer gear assembly **(see illustration 19.3b)**. Remove the coupler bolt.

4　Note the relationship of the transfer gear coupler shaft to the transfer gear **(see illustration)**, then remove the pinch bolt.

5　Remove the transfer gear mounting bolts **(see illustration)**, separate the shafts from the transfer gear and remove it.

6　Installation is the reverse of the removal procedure, noting the following:

a) *Make sure the gaps in the intermediate shaft and the coupler shaft align with the marks on the cover caps and the projections on the gear housing **(see illustrations 19.3b and 20.4)**.*

b) *Tighten the mounting bolts to the torque listed in this Chapter's Specifications.*

c) *Tighten the pinch bolts to the torque values listed in this Chapter's Specifications.*

20.5 Transfer gear mounting bolts (Pathfinder)

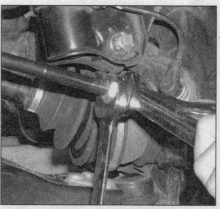

21.2 Hold the tie-rod end with a wrench while loosening the jam nut

21　Tie-rod ends - removal and installation

1　Loosen the wheel lug nuts, raise the front of the vehicle and support it securely on jackstands. Apply the parking brake and block the rear wheels to keep the vehicle from rolling off the jackstands. Remove the wheel.

Pathfinder, Xterra and 4WD Frontier models

Refer to illustrations 21.2 and 21.3

Removal

2　Break loose the tie-rod end jam nut **(see illustration)**. Don't back the nut off; once it has just been loosened, it will serve as the point to which the tie-rod end will be threaded. If you are removing the tie-rod end to replace the steering gear boot, mark the threads of the tie-rod on the *inner* side of the nut.

3　Remove the cotter pin and loosen (but don't remove) the nut on the tie-rod end ballstud. Break loose the tie-rod end from the steering knuckle arm with a puller **(see illustration)**. Remove the nut and disconnect the tie-rod end from the steering knuckle.

4　Unscrew the tie-rod end from the tie-rod.

Installation

5　If the jam nut was removed, thread it onto the tie-rod until it meets the mark applied in Step 2. Thread the tie-rod end onto the tie-rod until it contacts the jam nut, then connect the tie-rod end to the steering arm. Install the ballstud nut and tighten it to the torque listed in this Chapter's Specifications. Install a new cotter pin. **Note:** *If necessary, tighten the nut a little more to align the slots in the nut with the hole in the ballstud; don't loosen the nut to allow cotter pin insertion.*

6　Tighten the jam nut securely and install

Chapter 10 Suspension and steering systems

21.3 Back off the ballstud nut a few turns, then separate the tie-rod end from the steering knuckle with a puller (leaving the nut on the ballstud will prevent the tie-rod end from separating violently)

21.8 On 2WD Frontier models, count the number of threads showing before removing the tie-rod end

the wheel. Lower the vehicle and tighten the lug nuts to the torque listed in the Chapter 1 Specifications.

7 Have the front end alignment checked and, if necessary, adjusted.

2WD Frontier models

Refer to illustration 21.8

Removal

8 Count the number of threads protruding from the adjuster sleeve (write down the number), then loosen the adjuster sleeve clamp bolt nut **(see illustration)**.

9 Remove the cotter pin and loosen (but don't remove) the nut on the tie-rod end ballstud. Break loose the tie-rod end from the steering knuckle arm with a puller **(see illustration 21.3)**. Remove the nut and disconnect the tie-rod end from the steering knuckle.

10 Unscrew the tie-rod end from the tie-rod.

Installation

11 Thread the new tie-rod end into the adjuster sleeve until the previously recorded number of threads are showing. Tighten the adjuster sleeve clamp bolt/nut to the torque listed in this Chapter's Specifications.

12 Connect the tie-rod end to the steering arm, install the ballstud nut and tighten it to the torque listed in this Chapter's Specifications. Install a new cotter pin. **Note:** *If necessary, tighten the nut a little more to align the slots in the nut with the hole in the ballstud; don't loosen the nut to allow cotter pin insertion.*

13 Install the wheel and lug nuts, lower the vehicle and tighten the lug nuts to the torque listed in the Chapter 1 Specifications.

14 Have the front end alignment checked and, if necessary, adjusted.

22 Steering gear boots (Pathfinder models) - replacement

Refer to illustration 22.4

1 Loosen the wheel lug nuts, raise the front of the vehicle and support it securely on jackstands, then remove the wheels. Remove the under-vehicle splash shield.

2 Remove the tie-rod end from the tie-rod (see Section 21).

3 Remove the tie-rod end jam nut.

4 Remove the outer boot clamp with a pair of pliers, then cut off the inner boot clamp and discard it **(see illustration)**.

5 Remove the boot.

6 Install a new clamp on the inner end of the boot.

7 Apply multi-purpose grease to groove on the tie-rod (where the outer end of the boot will ride).

8 Slide the new boot over the tie-rod and onto the steering gear housing.

9 Make sure the boot isn't twisted, then tighten the inner clamp.

10 Install the outer clamp and tie-rod end jam nut.

11 Install the tie-rod ends (see Section 21).

12 Install the steering gear assembly.

13 Have the front end alignment checked and, if necessary, adjusted.

23 Steering linkage (Frontier and Xterra models) - removal and installation

Tie-rod end

1 Refer to Section 21 for the tie-rod end replacement procedure.

Tie-rod

Refer to illustration 23.3

2 Separate the tie-rod end from the steering knuckle (see Section 21).

3 Remove the cotter pin and loosen the inner tie-rod-to-relay rod nut. Break loose the ballstud from the relay rod with a puller or a "picklefork" type balljoint separator **(see illustration)**. **Note:** *Keep in mind that the use of a picklefork-type separator will most likely damage the balljoint boot.*

4 Remove the nut and detach the tie-rod from the relay rod.

5 Installation is the reverse of removal. If the ballstud spins when attempting to tighten the nut, force it into the tapered hole with a large pair of pliers. Be sure to tighten the nut to the torque listed in this Chapter's Specifications, and install a new cotter pin. **Note:** *If necessary, tighten the nut a little more to align the slots in the nut with the hole in the ballstud; don't loosen the nut to allow cotter pin insertion.* Have the front end alignment checked and, if necessary, adjusted.

Relay rod

6 Raise the vehicle and support it securely on jackstands. Apply the parking brake.

7 Detach the inner ends of the tie-rods from the relay rod **(see illustration 23.3)**.

8 Separate the relay rod from the Pitman arm.

9 Separate the relay rod from the idler arm.

10 Installation is the reverse of the removal procedure. If the ballstuds spin when attempting to tighten the nuts, force them into the tapered holes with a large pair of pliers.

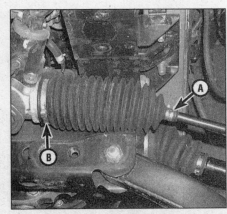

22.4 The outer clamp (A) on the steering gear boot can be squeezed and removed with a pair of pliers; the inner clamp (B) must be cut off

23.3 Separating the inner tie-rod balljoint from the relay rod with a picklefork-type tool

Chapter 10 Suspension and steering systems

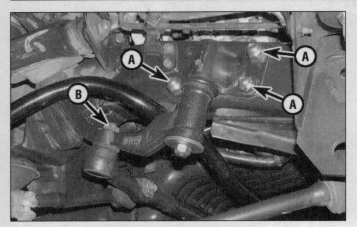

23.14 Idler arm mounting details
- A Idler arm mounting nuts/bolts
- B Idler arm-to-relay rod nut

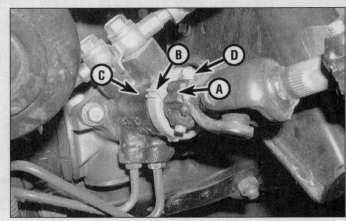

24.4 When the steering gear is centered, the gap in the coupler shaft (A) and the line (and arrow) on the cover cap (B) will be aligned with the protrusion on the housing (C). (D) is the pinch bolt

Be sure to tighten all of the nuts to the torque listed in this Chapter's Specifications, and install new cotter pins. **Note:** *If necessary, tighten the nut a little more to align the slots in the nut with the hole in the ballstud; don't loosen the nut to allow cotter pin insertion.*

Idler arm

Refer to illustration 23.14

11 Loosen the right front wheel lug nuts. Raise the vehicle and support it securely on jackstands, then remove the wheel. Apply the parking brake.
12 Loosen but do not remove the idler arm-to-relay rod nut.
13 Separate the idler arm from the relay rod with a two jaw puller or a picklefork type balljoint separator **(see illustration 23.3)**. Remove the nut.
14 Remove the idler arm-to-frame nuts and bolts **(see illustration)**.
15 To install the idler arm, position it on the frame and install the bolts and nuts, tightening them to the torque listed in this Chapter's Specifications.
16 Connect the relay rod to the idler arm ballstud and install the nut. Tighten the nut to the torque listed in this Chapter's Specifications, and install a new cotter pin. If the ballstud spins when attempting to tighten the nut, force it into the tapered hole with a large pair of pliers. **Note:** *If necessary, tighten the nut a little more to align the slots in the nut with the hole in the ballstud; don't loosen the nut to allow cotter pin insertion.*

Pitman arm

17 Raise the vehicle and support it securely on jackstands.
18 Remove the cotter pin, loosen the nut and detach the relay rod from the Pitman arm; use the technique shown in **illustration 23.3**.
19 Remove the Pitman arm nut and washer (see Section 24). Mark the Pitman arm and the steering gear shaft to ensure proper alignment at reassembly time (only if the same Pitman arm is going to be used).

20 Remove the Pitman arm with a Pitman arm puller or a two-jaw puller.
21 Installation is the reverse of removal. Make sure the marks you made on the Pitman arm and steering gear shaft are aligned. Tighten the nut to the torque listed in this Chapter's Specifications, and install a new cotter pin. If the ballstud spins when attempting to tighten the nut, force it into the tapered hole with a large pair of pliers. **Note:** *If necessary, tighten the nut a little more to align the slots in the nut with the hole in the ballstud; don't loosen the nut to allow cotter pin insertion.*

24 Steering gear - removal and installation

Warning: *DO NOT allow the steering column shaft to rotate with the steering gear removed or damage to the airbag system could occur. As a method of preventing the shaft from turning, pass the seat belt through the rim of the steering wheel and buckle the belt in place.*

Pathfinder models

Refer to illustrations 24.4 and 24.7

1 Loosen the front wheel lug nuts, raise the front of the vehicle and support it securely on jackstands. Apply the parking brake. Remove the wheels.
2 Remove the under-vehicle splash shield.
3 Remove the brace from the front crossmember **(see illustration 3.3b)**.
4 Note the relationship of the transfer gear coupler shaft to the steering gear input shaft and remove the pinch bolt **(see illustration)**. The protrusion on the rear cover cap should be aligned with the match mark on the housing, and the gap in the coupler shaft must also be aligned with the protrusion.
5 Detach the tie-rod ends from the steering knuckles (see Section 21).
6 Remove the stabilizer bar brackets (see Section 3).
7 Position a drain pan under the steering gear. Remove the banjo bolts securing the power steering pressure and return lines, then disconnect the lines from the steering gear **(see illustration)**. Cap the lines to prevent leakage. Discard the sealing washers that are present on each side of the fittings - new ones must be used during installation.
8 Unbolt the power steering pressure line brackets from the front crossmember.
9 Unscrew the mounting bolts and remove the brackets. Detach the coupler shaft as you lower the steering gear from the vehicle.

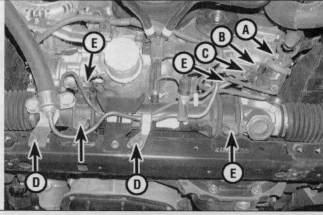

24.7 Rack-and-pinion steering gear mounting details (Pathfinder models)

- A Coupler shaft pinch bolt
- B Fluid return line
- C Fluid pressure line
- D Pressure line support brackets
- E Steering gear mounting bolts

Chapter 10 Suspension and steering systems

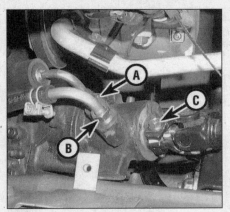

24.11 Unscrew the pressure line (A), return line (B) and intermediate shaft pinch bolt (C)

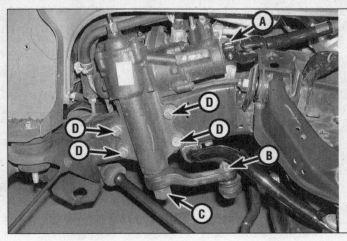

24.16 Steering gear details - Frontier and Xterra models

- A Intermediate shaft pinch bolt
- B Relay rod-to-Pitman arm nut
- C Pitman arm nut
- D Steering gear mounting bolts

10 Installation is the reverse of removal, noting the following:
 a) Make sure the rack is centered in the steering gear, and the protrusion on the cover cap is aligned with the match mark on the housing **(see illustration 24.4)**; the gap in the coupler shaft must also be aligned with the match mark.
 b) Be sure to tighten all fasteners to the torque values listed in this Chapter's Specifications.
 c) Tighten the wheel lug nuts to the torque listed in the Chapter 1 Specifications.
 d) Check the power steering fluid level and add some, if necessary (see Chapter 1), then bleed the system as described in Section 26.

Frontier and Xterra models

Refer to illustrations 24.11 and 24.16

11 If the vehicle is equipped with power steering, place a drain pan under the steering gear. Open the hood and unscrew the power steering pressure and return lines from the steering gear **(see illustration)**. **Caution:** *Use a flare-nut wrench, if available, to avoid rounding off the corners of the fittings. Plug the lines to prevent excessive fluid loss and contamination.*

12 Mark the relationship of the intermediate shaft to the steering gear input shaft, then remove the intermediate shaft-to-steering gear pinch bolt **(see illustration 24.11)**.

13 Loosen the left front wheel lug nuts, raise the front of the vehicle and support it securely on jackstands. Apply the parking brake and remove the wheel.

14 Remove the inner fender splash shield.

15 Separate the relay rod from the Pitman arm (see Section 23).

16 Remove the steering gear mounting bolts and nuts from the frame rail, then detach the steering gear from the frame **(see illustration)**.

17 If you're installing a new steering gear, remove the Pitman arm from the steering gear sector shaft.

18 Installation is the reverse of removal. Be sure to tighten all fasteners to the torque values listed in this Chapter's Specifications. Tighten the wheel lug nuts to the torque listed in the Chapter 1 Specifications.

19 Check the power steering fluid level and add some, if necessary (see Chapter 1), then bleed the system as described in Section 26.

25 Power steering pump - removal and installation

Refer to illustrations 25.5a, 25.5b and 25.5c

Removal

1 Disconnect the cable from the negative terminal of the battery. If you're working on a Pathfinder with a 3.5L V6, remove the battery for better access to the pump (see Chapter 5).

2 If you're working on a Pathfinder with a 3.3L V6 engine, raise the front of the vehicle and support it securely on jackstands. Remove the under-vehicle splash shield.

3 If you're working on a Frontier or Xterra with a four-cylinder engine, remove the air filter housing and the air intake duct from above the radiator (see Chapter 4).

4 Remove the drivebelt (see Chapter 1).

5 Detach the pressure line and return hose from the pump **(see illustrations)**. Discard

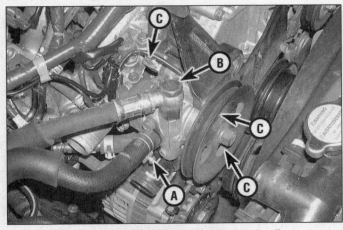

25.5a Power steering pump mounting details - four-cylinder engine

- A Return hose clamp
- B Pressure line banjo bolt
- C Pump mounting bolts

25.5b Power steering pump mounting details - V6 Frontier and Xterra models

- A Pressure line banjo bolt
- B Pump mounting bolts

25.5c Power steering pump mounting details - 3.5L V6 Pathfinder

A   Return hose
B   Pressure line banjo bolt
C   Upper mounting bolt
D   Lower mounting bolt

27.3 Use a small press tool such as this to push the stud out of the flange

the sealing washers on either side of the pressure line banjo fitting - new ones must be used upon installation. Plug the hoses to prevent contaminants from entering.
6   Unscrew the mounting fasteners and remove the pump from the vehicle, taking care not to spill fluid on the painted surfaces.
Note: *On models with holes in the pulley, the mounting bolts at the front of the pump are accessed through the holes in the pulley.*

Installation

7   Position the pump in the mounting bracket and install the mounting fasteners. Tighten the fasteners to the torque listed in this Chapter's Specifications.
8   Connect the pressure line and return hose to the pump. Be sure to use new sealing washers on the pressure line fitting and tighten the banjo bolt to the torque listed in this Chapter's Specifications.
9   Install and adjust the drivebelt (see Chapter 1).
10   The remainder of installation is the reverse of removal.
11   Fill the power steering reservoir with the recommended fluid (see Chapter 1) and bleed the system following the procedure described in the next Section.

26 Power steering system - bleeding

1   Following any operation in which the power steering fluid lines have been disconnected, the power steering system must be bled to remove all air and obtain proper steering performance.
2   With the front wheels in the straight ahead position, check the power steering fluid level and, if low, add fluid until it reaches the Cold mark on the dipstick or reservoir.
3   Start the engine and allow it to run at fast idle. Recheck the fluid level and add more if necessary to reach the Cold mark on the dipstick or reservoir.
4   Bleed the system by turning the wheels from side-to-side, without hitting the stops. This will work the air out of the system. Keep the reservoir full of fluid as this is done.
5   When the air is worked out of the system, return the wheels to the straight ahead position and leave the vehicle running for several more minutes before shutting it off. Recheck the fluid level.
6   Road test the vehicle to be sure the steering system is functioning normally and noise free.
7   Recheck the fluid level to be sure it's up to the Hot mark on the dipstick or reservoir while the engine is at normal operating temperature. Add fluid if necessary (see Chapter 1).

27 Wheel studs - replacement

Refer to illustrations 27.3 and 27.4
Note: *This procedure applies to both the front and rear wheel studs.*
1   Loosen the wheel lug nuts, raise the vehicle and support it securely on jackstands. Remove the wheel.
2   Remove the brake disc or drum (see

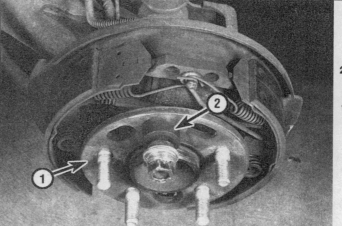

27.4 Install a spacer or washers and a lug nut on the stud, then tighten the nut to draw the stud into place

1   Axle flange
2   Spacer

Chapter 9).
3   Push the stud out of the hub flange or axle flange with a press tool **(see illustration)**.
4   Insert the new stud into the hub flange or axle flange from the back side and install some flat washers and a lug nut on the stud **(see illustration)**.
5   Tighten the lug nut until the stud is seated in the flange.
6   Reinstall the disc and caliper or brake drum (see Chapter 9).
7   Install the wheel and lug nuts. Lower the vehicle and tighten the lug nuts to the torque listed in the Chapter 1 Specifications.

28 Wheels and tires - general information

Refer to illustration 28.1
Most vehicles covered by this manual are equipped with metric-size fiberglass or steel belted radial tires **(see illustration)**, or inch-pattern light truck tires. Use of other size or type of tires may affect the ride and handling of the vehicle. Don't mix different types

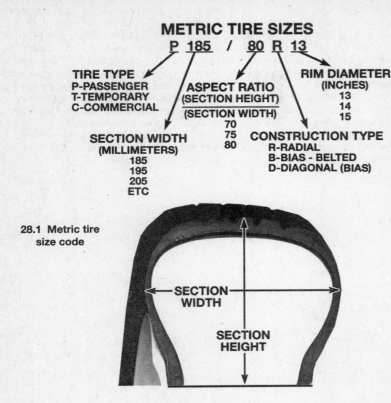

28.1 Metric tire size code

of tires, such as radials and bias belted, on the same vehicle as handling may be seriously affected. It's recommended that tires be replaced in pairs on the same axle, but if only one tire is being replaced, be sure it's the same size, structure and tread design as the other.

Because tire pressure has a substantial effect on handling and wear, the pressure on all tires should be checked at least once a month or before any extended trips (see Chapter 1).

Wheels must be replaced if they're bent, dented, leak air, have elongated bolt holes, are heavily rusted, out of vertical symmetry or if the lug nuts won't stay tight. Wheel repairs that use welding or peening are not recommended.

Tire and wheel balance is important to the overall handling, braking and performance of the vehicle. Unbalanced wheels can adversely affect handling and ride characteristics as well as tire life. Whenever a tire is installed on a wheel, the tire and wheel should be balanced by a shop with the proper equipment.

29 Front end alignment - general information

Refer to illustration 29.1

A front end alignment **(see illustration)** refers to the adjustments made to the front wheels so they're in proper angular relationship to the suspension and the ground. Front wheels that are out of proper alignment not only affect steering control, but also increase tire wear.

Getting the proper front wheel alignment is a very exacting process, one in which complicated and expensive machines are necessary to perform the job properly. Because of this, you should have a technician with the proper equipment perform these tasks. We will, however, use this space to give you a basic idea of what is involved with front end alignment so you can better understand the process and deal intelligently with the shop that does the work.

Toe-in is the turning in of the front wheels. The purpose of a toe specification is to ensure parallel rolling of the front wheels. In a vehicle with zero toe-in, the distance between the front edges of the wheels will be

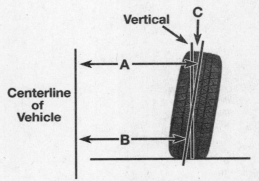

CAMBER ANGLE (FRONT VIEW)

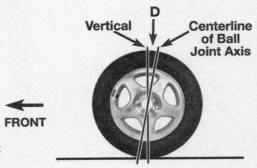

29.1 Front end alignment details

A minus B = C
 (degrees camber)
D = degrees camber
E minus F = toe-in
 (measured in inches)
G = toe-in (expressed
 in degrees)

CASTER ANGLE (SIDE VIEW)

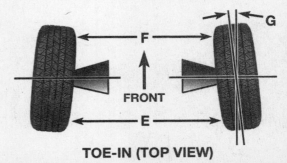

TOE-IN (TOP VIEW)

the same as the distance between the rear edges of the wheels. The actual amount of toe-in is normally only a fraction of an inch. Toe-in is adjusted on Frontier and Xterra models by turning the adjustment sleeves on the tie-rods; on Pathfinder models it's adjusted by turning the tie-rod in the tie-rod end to lengthen or shorten the tie-rod. Incorrect toe-in will cause the tires to wear improperly by making them scrub against the road surface.

Camber is the tilting of the front wheels from vertical when viewed from the front of the vehicle. When the wheels tilt out at the top, the camber is said to be positive (+). When the wheels tilt in at the top the camber is negative (-). The amount of tilt is measured in degrees from the vertical and this measurement is called the camber angle. This angle affects the amount of tire tread which contacts the road and compensates for changes in the suspension geometry when the vehicle is cornering or traveling over an undulating surface. On Frontier and Xterra models, camber is adjusted by turning the upper control arm pivot bolts, one way or the other, in equal amounts. Camber isn't adjustable on Pathfinder models.

Caster is the tilting of the top of the front steering axis from vertical. A tilt toward the rear is positive caster and a tilt toward the front is negative caster. On Frontier and Xterra models, caster is adjusted by turning the upper control arm pivot bolts, one way or the other, in opposite directions. Caster isn't adjustable on Pathfinder models.

When making adjustments to the front end alignment, the caster is set first, then the camber, then the toe-in.

Chapter 11 Body

Contents

Section		Section
Body - maintenance	2	Hood latch and release cable - removal and installation	11
Body repair - major damage	6	Instrument panel - removal and installation	28
Body repair - minor damage	5	Liftgate and liftgate glass (SUV models) - removal and installation	23
Bumpers - removal and installation	12	
Center console - removal and installation	25	Liftgate latch, lock cylinder, handle and support struts (SUV models) – removal, installation and adjustment	24
Cowl cover - removal and installation	14	
Dashboard trim panels - removal and installation	26	Mirrors - removal and installation	20
Door - removal, installation and adjustment	16	Radiator grille - removal and installation	9
Door latch, lock cylinder and handles - removal and installation	17	Seats - removal and installation	29
		Steering column covers - removal and installation	27
Door trim panels - removal and installation	15	Tailgate (pick-up models) – removal, installation and adjustment	21
Door window glass - removal and installation	18	
Door window glass regulator - removal and installation	19	Tailgate latch and handle (pick-up models) - removal and installation	22
Front fender - removal and installation	13	
General information	1	Upholstery and carpets - maintenance	4
Hinges and locks - maintenance	7	Vinyl trim - maintenance	3
Hood - removal, installation and adjustment	10	Windshield and fixed glass - replacement	8

1 General information

Warning: *The models covered by this manual are equipped with Supplemental Restraint Systems (SRS), more commonly known as airbags. Always disable the airbag system before working in the vicinity of any airbag system components to avoid the possibility of accidental deployment of the airbags, which could cause personal injury (see Chapter 12).*

Frontier and Xterra models are built with body-on-frame construction. The frame is a, ladder-type, with boxed frame rails that run the entire length of the vehicle and welded-in crossmembers. The Frontier body is in two separate sections, the cab and the bed, while the Xterra body incorporates the cab, backseat area and cargo compartment in one unitized structure.

Pathfinder models feature a "unibody" layout, using a floor pan with front and rear frame side rails which supports (and is welded to) the main body structure.

Certain components are particularly vulnerable to accident damage and can be unbolted and repaired or replaced. Among these parts are the hood, doors, tailgate, liftgate, bumpers and front fenders.

Only general body maintenance practices and body panel repair procedures within the scope of the do-it-yourselfer are included in this Chapter.

2 Body - maintenance

1 The condition of your vehicle's body is very important, because the resale value depends a great deal on it. It's much more difficult to repair a neglected or damaged body than it is to repair mechanical components. The hidden areas of the body, such as the wheel wells, the frame and the engine compartment, are equally important, although they don't require as frequent attention as the rest of the body.
2 Once a year, or every 12,000 miles, it's a good idea to have the underside of the body steam cleaned. All traces of dirt and oil will be removed and the area can then be inspected carefully for rust, damaged brake lines, frayed electrical wires, damaged cables and other problems. The front suspension components should be greased after completion of this job.
3 At the same time, clean the engine and the engine compartment with a steam cleaner or water-soluble degreaser.
4 The wheel wells should be given close attention, since undercoating can peel away and stones and dirt thrown up by the tires can cause the paint to chip and flake, allowing rust to set in. If rust is found, clean down to the bare metal and apply an anti-rust paint.
5 The body should be washed about once a week. Wet the vehicle thoroughly to soften the dirt, then wash it down with a soft sponge and plenty of clean, soapy water. If the surplus dirt is not washed off very carefully, it can wear down the paint.
6 Spots of tar or asphalt thrown up from the road should be removed with a cloth soaked in tar remover or kerosene lamp oil.
7 Once every six months, wax the body and chrome trim. If a chrome cleaner is used to remove rust from any of the vehicle's plated parts, remember that the cleaner also removes part of the chrome, so use it sparingly.

3 Vinyl trim - maintenance

Don't clean vinyl trim with detergents, caustic soap or petroleum-based cleaners. Plain soap and water works just fine, with a soft brush to clean dirt that may be ingrained. Wash the vinyl as frequently as the rest of the vehicle. After cleaning, application of a high-quality rubber and vinyl protectant will help prevent oxidation and cracks. The protectant can also be applied to weather-stripping, vacuum lines and rubber hoses, which often fail as a result of chemical degradation, and to the tires.

4 Upholstery and carpets - maintenance

1 Every three months remove the floormats and clean the interior of the vehicle (more frequently if necessary). Use a stiff whiskbroom to brush the carpeting and loosen dirt and dust, then vacuum the upholstery and carpets thoroughly, especially along seams and crevices.
2 Dirt and stains can be removed from carpeting with basic household or automotive carpet shampoos available in spray cans. Follow the directions and vacuum again, then use a stiff brush to bring back the "nap" of the carpet.
3 Most interiors have cloth or vinyl upholstery, either of which can be cleaned and maintained with a number of material-specific cleaners or shampoos available in auto supply stores. Follow the directions on the product for usage, and always spot-test any

upholstery cleaner on an inconspicuous area (bottom edge of a backseat cushion) to ensure that it doesn't cause a color shift in the material.

4   After cleaning, vinyl upholstery should be treated with a protectant. **Note:** *Make sure the protectant container indicates the product can be used on seats - some products may make a seat too slippery.* **Caution:** *Do not use protectant on vinyl-covered steering wheels.*

5   Leather upholstery requires special care. It should be cleaned regularly with saddlesoap or leather cleaner. Never use alcohol, gasoline, nail polish remover or thinner to clean leather upholstery.

6   After cleaning, regularly treat leather upholstery with a leather conditioner, rubbed in with a soft cotton cloth. Never use car wax on leather upholstery.

7   In areas where the interior of the vehicle is subject to bright sunlight, cover leather seating areas of the seats with a sheet if the vehicle is to be left out for any length of time.

5   Body repair - minor damage

Plastic body panels

The following repair procedures are for minor scratches and gouges. Repair of more serious damage should be left to a dealer service department or qualified auto body shop. Below is a list of the equipment and materials necessary to perform the following repair procedures on plastic body panels.

   Wax, grease and silicone removing solvent
   Cloth-backed body tape
   Sanding discs
   Drill motor with three-inch disc holder
   Hand sanding block
   Rubber squeegees
   Sandpaper
   Non-porous mixing palette
   Wood paddle or putty knife
   Curved-tooth body file
   Flexible parts repair material

Flexible panels (bumper trim)

1   Remove the damaged panel, if necessary or desirable. In most cases, repairs can be carried out with the panel installed.
2   Clean the area(s) to be repaired with a wax, grease and silicone removing solvent applied with a water-dampened cloth.
3   If the damage is structural, that is, if it extends through the panel, clean the backside of the panel area to be repaired as well. Wipe dry.
4   Sand the rear surface about 1-1/2 inches beyond the break.
5   Cut two pieces of fiberglass cloth large enough to overlap the break by about 1-1/2 inches. Cut only to the required length.
6   Mix the adhesive from the repair kit according to the instructions included with the kit, and apply a layer of the mixture approximately 1/8-inch thick on the backside

of the panel. Overlap the break by at least 1-1/2 inches.
7   Apply one piece of fiberglass cloth to the adhesive and cover the cloth with additional adhesive. Apply a second piece of fiberglass cloth to the adhesive and immediately cover the cloth with additional adhesive in sufficient quantity to fill the weave.
8   Allow the repair to cure for 20 to 30 minutes at 60-degrees to 80-degrees F.
9   If necessary, trim the excess repair material at the edge.
10   Remove all of the paint film over and around the area(s) to be repaired. The repair material should not overlap the painted surface.
11   With a drill motor and a sanding disc (or a rotary file), cut a "V" along the break line approximately 1/2-inch wide. Remove all dust and loose particles from the repair area.
12   Mix and apply the repair material. Apply a light coat first over the damaged area; then continue applying material until it reaches a level slightly higher than the surrounding finish.
13   Cure the mixture for 20 to 30 minutes at 60-degrees to 80-degrees F.
14   Roughly establish the contour of the area being repaired with a body file. If low areas or pits remain, mix and apply additional adhesive.
15   Block sand the damaged area with sandpaper to establish the actual contour of the surrounding surface.
16   If desired, the repaired area can be temporarily protected with several light coats of primer. Because of the special paints and techniques required for flexible body panels, it is recommended that the vehicle be taken to a paint shop for completion of the body repair.

Steel body panels

See photo sequence

Repair of minor scratches

17   If the scratch is superficial and does not penetrate to the metal of the body, repair is very simple. Lightly rub the scratched area with a fine rubbing compound to remove loose paint and built up wax. Rinse the area with clean water.
18   Apply touch-up paint to the scratch, using a small brush. Continue to apply thin layers of paint until the surface of the paint in the scratch is level with the surrounding paint. Allow the new paint at least two weeks to harden, then blend it into the surrounding paint by rubbing with a very fine rubbing compound. Finally, apply a coat of wax to the scratch area.
19   If the scratch has penetrated the paint and exposed the metal of the body, causing the metal to rust, a different repair technique is required. Remove all loose rust from the bottom of the scratch with a pocketknife, then apply rust inhibiting paint to prevent the formation of rust in the future. Using a rubber or nylon applicator, coat the scratched area with glaze-type filler. If required, the filler can

be mixed with thinner to provide a very thin paste, which is ideal for filling narrow scratches. Before the glaze filler in the scratch hardens, wrap a piece of smooth cotton cloth around the tip of a finger. Dip the cloth in thinner and then quickly wipe it along the surface of the scratch. This will ensure that the surface of the filler is slightly hollow. The scratch can now be painted over as described earlier in this Section.

Repair of dents

20   When repairing dents, the first job is to pull the dent out until the affected area is as close as possible to its original shape. There is no point in trying to restore the original shape completely as the metal in the damaged area will have stretched on impact and cannot be restored to its original contours. It is better to bring the level of the dent up to a point that is about 1/8-inch below the level of the surrounding metal. In cases where the dent is very shallow, it is not worth trying to pull it out at all.
21   If the backside of the dent is accessible, it can be hammered out gently from behind using a soft-face hammer. While doing this, hold a block of wood firmly against the opposite side of the metal to absorb the hammer blows and prevent the metal from being stretched.
22   If the dent is in a section of the body which has double layers, or some other factor makes it inaccessible from behind, a different technique is required. Drill several small holes through the metal inside the damaged area, particularly in the deeper sections. Screw long, self-tapping screws into the holes just enough for them to get a good grip in the metal. Now pulling on the protruding heads of the screws with locking pliers can pull out the dent.
23   The next stage of repair is the removal of paint from the damaged area and from an inch or so of the surrounding metal. This is easily done with a wire brush or sanding disk in a drill motor, although it can be done just as effectively by hand with sandpaper. To complete the preparation for filling, score the surface of the bare metal with a screwdriver or the tang of a file or drill small holes in the affected area. This will provide a good grip for the filler material. To complete the repair, see the Section on filling and painting.

Repair of rust holes or gashes

24   Remove all paint from the affected area and from an inch or so of the surrounding metal using a sanding disk or wire brush mounted in a drill motor. If these are not available, a few sheets of sandpaper will do the job just as effectively.
25   With the paint removed, you will be able to determine the severity of the corrosion and decide whether to replace the whole panel, if possible, or repair the affected area. New body panels are not as expensive as most people think and it is often quicker to install a new panel than to repair large areas of rust.
26   Remove all trim pieces from the affected

area except those which will act as a guide to the original shape of the damaged body, such as headlight shells, etc. Using metal snips or a hacksaw blade, remove all loose metal and any other metal that is badly affected by rust. Hammer the edges of the hole in to create a slight depression for the filler material.

27 Wire-brush the affected area to remove the powdery rust from the surface of the metal. If the back of the rusted area is accessible, treat it with rust inhibiting paint.

28 Before filling is done, block the hole in some way. This can be done with sheet metal riveted or screwed into place, or by stuffing the hole with wire mesh.

29 Once the hole is blocked off, the affected area can be filled and painted. See the following subsection on filling and painting.

Filling and painting

30 Many types of body fillers are available, but generally speaking, body repair kits which contain filler paste and a tube of resin hardener are best for this type of repair work. A wide, flexible plastic or nylon applicator will be necessary for imparting a smooth and contoured finish to the surface of the filler material. Mix up a small amount of filler on a clean piece of wood or cardboard (use the hardener sparingly). Follow the manufacturer's instructions on the package, otherwise the filler will set incorrectly.

31 Using the applicator, apply the filler paste to the prepared area. Draw the applicator across the surface of the filler to achieve the desired contour and to level the filler surface. As soon as a contour that approximates the original one is achieved, stop working the paste. If you continue, the paste will begin to stick to the applicator. Continue to add thin layers of paste at 20-minute intervals until the level of the filler is just above the surrounding metal.

32 Once the filler has hardened, the excess can be removed with a body file. From then on, progressively finer grades of sandpaper should be used, starting with a 180-grit paper and finishing with 600-grit wet-or-dry paper. Always wrap the sandpaper around a flat rubber or wooden block, otherwise the surface of the filler will not be completely flat. During the sanding of the filler surface, the wet-or-dry paper should be periodically rinsed in water. This will ensure that a very smooth finish is produced in the final stage.

33 At this point, the repair area should be surrounded by a ring of bare metal, which in turn should be encircled by the finely feathered edge of good paint. Rinse the repair area with clean water until all of the dust produced by the sanding operation is gone.

34 Spray the entire area with a light coat of primer. This will reveal any imperfections in the surface of the filler. Repair the imperfections with fresh filler paste or glaze filler and once more smooth the surface with sandpaper. Repeat this spray-and-repair procedure until you are satisfied that the surface of the filler and the feathered edge of the paint are perfect. Rinse the area with clean water and allow it to dry completely.

35 The repair area is now ready for painting. Spray painting must be carried out in a warm, dry, windless and dust free atmosphere. These conditions can be created if you have access to a large indoor work area, but if you are forced to work in the open, you will have to pick the day very carefully. If you are working indoors, dousing the floor in the work area with water will help settle the dust that would otherwise be in the air. If the repair area is confined to one body panel, mask off the surrounding panels. This will help minimize the effects of a slight mismatch in paint color. Trim pieces such as chrome strips, door handles, etc., will also need to be masked off or removed. Use masking tape and several thickness of newspaper for the masking operations.

36 Before spraying, shake the paint can thoroughly, then spray a test area until the spray painting technique is mastered. Cover the repair area with a thick coat of primer. The thickness should be built up using several thin layers of primer rather than one thick one. Using 600-grit wet-or-dry sandpaper, rub down the surface of the primer until it is very smooth. While doing this, the work area should be thoroughly rinsed with water and the wet-or-dry sandpaper periodically rinsed as well. Allow the primer to dry before spraying additional coats.

37 Spray on the top coat, again building up the thickness by using several thin layers of paint. Begin spraying in the center of the repair area and then, using a circular motion, work out until the whole repair area and about two inches of the surrounding original paint is covered. Remove all masking material 10 to 15 minutes after spraying on the final coat of paint. Allow the new paint at least two weeks to harden, then use a very fine rubbing compound to blend the edges of the new paint into the existing paint. Finally, apply a coat of wax.

6 Body repair - major damage

1 Major damage must be repaired by an auto body shop specifically equipped to perform body and frame repairs. These shops have the specialized equipment required to do the job properly.

2 If the damage is extensive, the frame (Frontier and Xterra) or body (Pathfinder) must be checked for proper alignment or the vehicle's handling characteristics may be adversely affected and other components may wear at an accelerated rate.

3 Due to the fact that all of the major body components (hood, fenders, etc.) are separate and replaceable units, any seriously damaged components should be replaced rather than repaired. Sometimes the components can be found in a wrecking yard that specializes in used vehicle components, often at considerable savings over the cost of new parts.

7 Hinges and locks - maintenance

Once every 3000 miles, or every three months, the hinges and latch assemblies on the doors, hood and trunk should be given a few drops of light oil or lock lubricant. The door latch strikers should also be lubricated with a thin coat of grease to reduce wear and ensure free movement. Lubricate the door and trunk locks with spray-on graphite lubricant.

8 Windshield and fixed glass - replacement

Replacement of the windshield and fixed glass requires the use of special fast-setting adhesive/caulk materials and some specialized tools and techniques. These operations should be left to a dealer service department or a shop specializing in glass-work. On SUV models, the fixed glass also includes the quarter windows.

9 Radiator grille - removal and installation

Refer to illustrations 9.3 and 9.5

1 Open the hood and remove the front side marker/turn signal light housing(s) (see Chapter 12).

2 On 1998 and earlier Pathfinder models, remove the screws securing the grille in the front side marker/turn signal light openings.

3 Using a screwdriver, rotate the grille retaining clips 45 degrees in either direction **(see illustration)**. There are typically four retaining clips on most models. **Note:** *On 2001 Frontier models, the grille is an integral part of the front bumper fascia (see Section 12).*

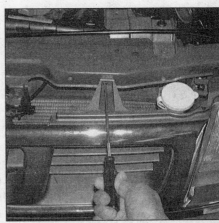

9.3 Pull outward on the grille while using a screwdriver to release the grille retaining clips

These photos illustrate a method of repairing simple dents. They are intended to supplement *Body repair - minor damage* in this Chapter and should not be used as the sole instructions for body repair on these vehicles.

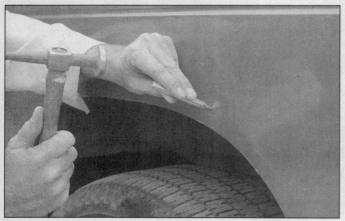

1   If you can't access the backside of the body panel to hammer out the dent, pull it out with a slide-hammer-type dent puller. In the deepest portion of the dent or along the crease line, drill or punch hole(s) at least one inch apart . . .

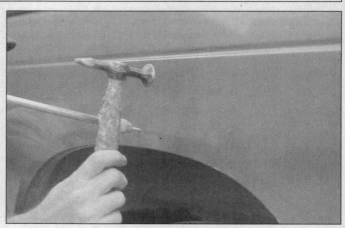

2   . . . then screw the slide-hammer into the hole and operate it. Tap with a hammer near the edge of the dent to help 'pop' the metal back to its original shape. When you're finished, the dent area should be close to its original contour and about 1/8-inch below the surface of the surrounding metal

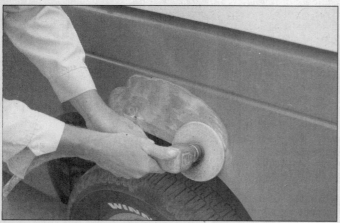

3   Using coarse-grit sandpaper, remove the paint down to the bare metal. Hand sanding works fine, but the disc sander shown here makes the job faster. Use finer (about 320-grit) sandpaper to feather-edge the paint at least one inch around the dent area

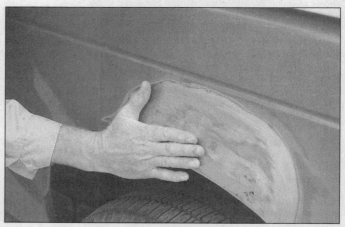

4   When the paint is removed, touch will probably be more helpful than sight for telling if the metal is straight. Hammer down the high spots or raise the low spots as necessary. Clean the repair area with wax/silicone remover

5   Following label instructions, mix up a batch of plastic filler and hardener. The ratio of filler to hardener is critical, and, if you mix it incorrectly, it will either not cure properly or cure too quickly (you won't have time to file and sand it into shape)

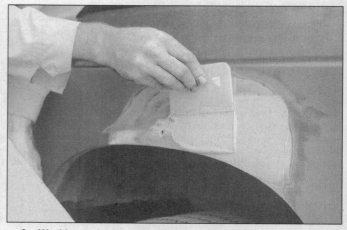

6   Working quickly so the filler doesn't harden, use a plastic applicator to press the body filler firmly into the metal, assuring it bonds completely. Work the filler until it matches the original contour and is slightly above the surrounding metal

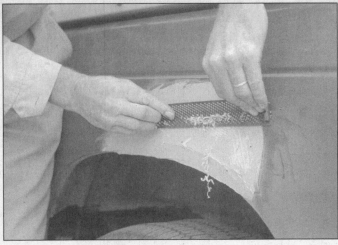

7 Let the filler harden until you can just dent it with your fingernail. Use a body file or Surform tool (shown here) to rough-shape the filler

8 Use coarse-grit sandpaper and a sanding board or block to work the filler down until it's smooth and even. Work down to finer grits of sandpaper - always using a board or block - ending up with 360 or 400 grit

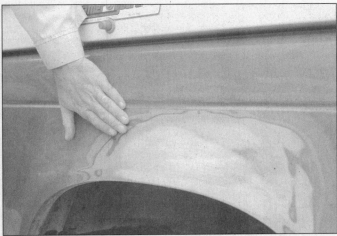

9 You shouldn't be able to feel any ridge at the transition from the filler to the bare metal or from the bare metal to the old paint. As soon as the repair is flat and uniform, remove the dust and mask off the adjacent panels or trim pieces

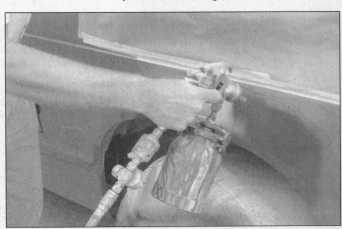

10 Apply several layers of primer to the area. Don't spray the primer on too heavy, so it sags or runs, and make sure each coat is dry before you spray on the next one. A professional-type spray gun is being used here, but aerosol spray primer is available inexpensively from auto parts stores

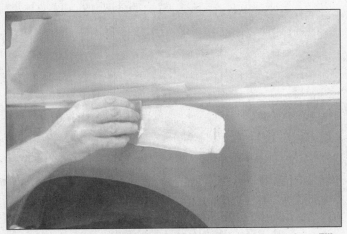

11 The primer will help reveal imperfections or scratches. Fill these with glazing compound. Follow the label instructions and sand it with 360 or 400-grit sandpaper until it's smooth. Repeat the glazing, sanding and respraying until the primer reveals a perfectly smooth surface

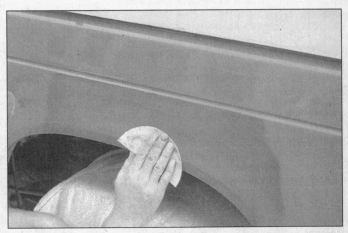

12 Finish sand the primer with very fine sandpaper (400 or 600-grit) to remove the primer overspray. Clean the area with water and allow it to dry. Use a tack rag to remove any dust, then apply the finish coat. Don't attempt to rub out or wax the repair area until the paint has dried completely (at least two weeks)

9.5 Remove the clips from the radiator support and insert them back into the the grille

4   Pull the grille forward to disengage it from the body and carefully lift out the grille.
5   Remove the grille retaining clips from the radiator support and reinstall the clips back into the grille **(see illustration)**.
6   To install the grille, position it in place and align the clips with the mating holes in the radiator support. Carefully press on the grille until the clips snap into place on the radiator support.

10   Hood - removal, installation and adjustment

Refer to illustrations 10.4, 10.10 and 10.11
Note: *The hood is heavy and somewhat awkward to remove and install - at least two people should perform this procedure.*

Removal and installation

1   Use blankets or pads to cover the cowl area of the body and the fenders. This will protect the body and paint as the hood is lifted off.
2   Scribe alignment marks around the

10.4 Remove the four retaining bolts and lift off the hood with the help of an assistant

hinge flanges to insure proper alignment during installation (paint or a permanent-type felt-tip marker also will work for this).
3   Disconnect the windshield washer hoses and any electrical connectors from the hood.
4   Have an assistant support the weight of the hood. Remove the hinge-to-hood bolts **(see illustration)**.
5   Lift off the hood.
6   Installation is the reverse of removal.

Adjustment

7   Fore-and-aft and side-to-side adjustment of the hood is done by moving the hood in relation to the hinge flanges after loosening the bolts.
8   Scribe or trace a line around the entire hinge plate so you can judge the amount of movement **(see illustration 10.4)**.
9   Loosen the bolts and move the hood into correct alignment. Move it only a little at a time. Tighten the hinge bolts and carefully lower the hood to check the alignment.
10   If necessary after installation, the entire hood latch assembly can be adjusted up-and-down as well as from side-to-side on the radiator support so the hood closes securely

and flush with the fenders. To make the adjustment, scribe a line around the hood latch mounting bolts to provide a reference point, then loosen them and reposition the latch assembly, as necessary **(see illustration)**. Following adjustment, retighten the mounting bolts.
11   Finally, adjust the hood bumpers on the radiator support so the hood, when closed, is flush with the fenders **(see illustration)**.
12   The hood latch assembly, as well as the hinges, should be periodically lubricated with white, lithium-base grease to prevent binding and wear.

11   Hood latch and release cable - removal and installation

Warning: *The models covered by this manual are equipped with Supplemental Restraint Systems (SRS), more commonly known as airbags. Always disable the airbag system before working in the vicinity of any airbag system components to avoid the possibility of accidental deployment of the airbags, which could cause personal injury (see Chapter 12).*

Latch

Refer to illustration 11.1
1   Remove the bolts and detach the latch assembly **(see illustration 10.10)**. Unhook the spring and pry the cable end from the latch **(see illustration)**.
2   Installation is the reverse of removal. Adjust the latch so the hood engages securely when closed and the hood bumpers are slightly compressed (see Section 10).

Release cable

Refer to illustrations 11.6 and 11.7
3   Disconnect the release cable from the hood latch assembly as described in Step 1.
4   Unclip the release cable from the engine wiring harness. Attach a length of wire to the cable to assist with the installation of the new cable.

10.10 Loosen the hood latch bolts (arrows), then move the latch as necessary to adjust the hood-closed position

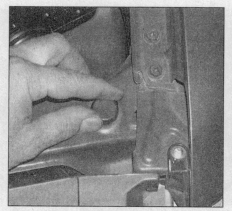

10.11 Twist the hood bumpers in-or-out to make fine adjustments to the hood closed height

11.1 To disconnect the cable from the hood latch mechanism, disengage the cable housing end (A) from the slot in the latch and twist the cable end (B) out of the latch arm

Chapter 11 Body

11.6 Hood release handle and cable mounting bolts - one piece design

11.7 To disconnect the hood release cable on two piece designs, disengage the cable housing end (A) from the slot in the bracket and twist the cable end (B) out of the handle

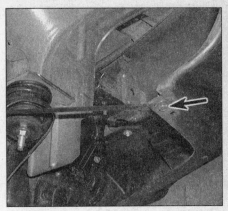

12.2 Remove the outer bolts (arrow) from the bumper support braces

5   Working in the passenger compartment, remove the driver's kick panel or the knee bolster (see Section 26). **Note:** *On some models the hood release handle is mounted to the side kick panel and on other models it's mounted to the dash. Remove the appropriate trim panel to allow access to the hood release handle.*

6   On some models the hood release handle and cable are a one piece design and must be replaced as a unit **(see illustration)**.

7   On other models the hood release cable is a two piece can be removed from the handle by using pliers to pull the cable housing end from the tab on the handle frame and rotating the cable end from the slot on the handle **(see illustration)**. If the handle assembly itself must be replaced, detach the cable as described above and simply remove the mounting bolts to remove the handle.

8   Trace the cable forward to the grommet where the cable goes through the firewall and pry the grommet out of the firewall. Pull the handle and cable rearward into the passenger compartment.

9   Disconnect the guide wire from the old cable and fasten it to the new cable.

10   With the new cable attached to the wire, pull the wire back through the firewall until the new cable reaches the latch assembly. Make sure that the grommet is properly seated on both sides of the hole in the firewall. Push on the grommet with your fingers from the passenger compartment side to seat the grommet in the firewall correctly.

11   The remainder of installation is the reverse of removal.

12 Bumpers - removal and installation

Warning: *The models covered by this manual are equipped with Supplemental Restraint Systems (SRS), more commonly known as airbags. Always disable the airbag system before working in the vicinity of any airbag system components to avoid the possibility of accidental deployment of the airbags, which could cause personal injury (see Chapter 12).*

Front bumper

Frontier and Xterra
Refer to illustrations 12.2, 12.3 and 12.5

1   Refer to Section 9 and remove the radiator grille.

2   Working in the front fenderwells, remove the bumper support brace bolts **(see illustration)**.

3   On 2000 and earlier Frontier models and all Xterra models, remove the bumper support bracket bolts **(see illustration)**. On later Frontier models, remove the bolts securing the bumper to the fender, which is similar to Pathfinder models **(see illustration 12.9b)**.

4   Disconnect any wiring harnesses or any other components that would interfere with removal of the bumper.

5   With an assistant supporting the bumper, remove the bolts/nuts retaining the bumper to the frame and detach the bumper **(see illustration)**.

6   Installation is the reverse of removal. Tighten the retaining bolts securely.

Pathfinder
Refer to illustrations 12.7, 12.9a, 12.9b, 12.9c and 12.11

7   Remove the inner fender shield from the wheelwell **(see illustration)**.

8   Refer to Section 9 and remove the radiator grille.

9   On 1999 and later models, remove the grille moulding from below the headlight and remove the bolts securing the bumper to the

12.3 Remove the bolts (arrow) securing the left and right support brackets

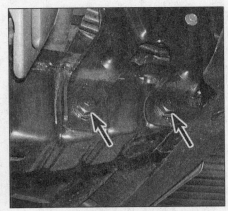

12.5 Remove the main bumper bracket bolts (arrows) on the side of each frame rail

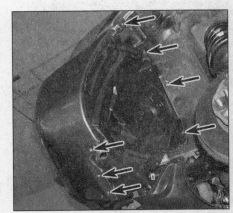

12.7 Front inner fender shield mounting screws (arrows) - Pathfinder models

11-8   Chapter 11 Body

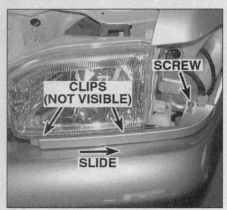

12.9a On 1999 and later Pathfinder models, remove screw and slide the grille moulding to the side of the vehicle to disengage it from the clips . . .

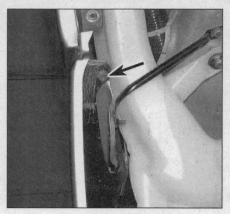

12.9b . . . then remove the front bumper-to-fender retaining bolts (arrow) . . .

12.9c . . . and the clips (arrows) securing the bumper to the radiator support

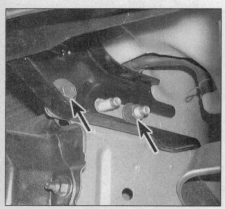

12.11 Bumper retaining bolts/nuts - Pathfinder models

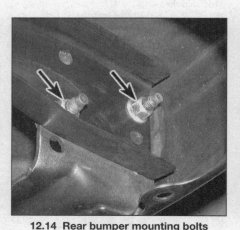

12.14 Rear bumper mounting bolts (arrows) - Frontier models

12.20 Mud guard mounting screws (arrows) - Pathfinder models

fender **(see illustrations)**. Also remove the clips securing the lower half of the bumper to the radiator support **(see illustration)**.

10 Referring to Chapter 12, remove the front turn signal lamps on 1998 and earlier models or the fog lights on 1999 and later models.

11 With an assistant supporting the bumper, work through openings at the front of the bumper and remove the bumper retaining bolts/nuts **(see illustration)**. Detach the bumper from the vehicle.

12 Installation is the reverse of removal. Tighten the retaining bolts securely.

Rear bumper
Frontier
Refer to illustration 12.14

13 Disconnect the electrical connectors from the back-up lights.
14 Remove the bolts securing the bumper to the bumper support bracket **(see illustration)**.
15 Installation is the reverse of removal. Tighten the retaining bolts securely.

Xterra

16 Open the rear liftgate.
17 Remove the screws securing the outer (plastic) bumper extensions. There are sev-

12.21 Rear inner fender shield mounting details - Pathfinder models

A Push pin fastener
B Screw

eral screws in the liftgate opening and four screws in each wheelwell opening.
18 With an assistant supporting the bumper, remove the nuts from the end of the steel bumper and detach the rear bumper from the vehicle.
19 Installation is the reverse of removal. Tighten the retaining bolts securely.

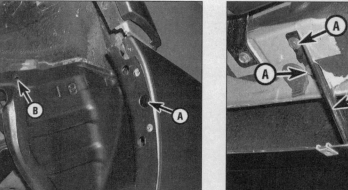

12.22 Remove the bolts (A) and the bumper support braces (B)

Pathfinder
Refer to illustrations 12.20, 12.21, 12.22, 12.24 and 12.25

20 Remove the mudguards **(see illustration)**.
21 Remove the inner fender shield **(see illustration)**.
22 Remove the bolts from the bumper support brace **(see illustration)**.
23 If equipped with a rear door tire carrier,

Chapter 11 Body

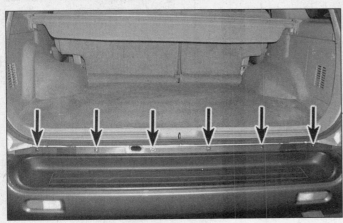

12.24 Rear bumper retaining clips (arrows) - Pathfinder models

12.25 Rear bumper mounting bolts (arrows) - Pathfinder models

remove the bolts securing the tire carrier striker, rubber bumper and the tire carrier guide.

24 Remove the clips securing the upper half of the bumper to the body **(see illustration)**.

25 With an assistant supporting the bumper, remove the nuts securing the bumper brackets to the frame and detach the rear bumper from the vehicle **(see illustration)**.

26 Installation is the reverse of removal. Tighten the retaining bolts securely.

13 Front fender - removal and installation

Refer to illustrations 13.5, 13.6a, 13.6b, 13.6c, 13.7 and 13.8

Warning: *The models covered by this manual are equipped with Supplemental Restraint Systems (SRS), more commonly known as airbags. Always disable the airbag system before working in the vicinity of any airbag system components to avoid the possibility of accidental deployment of the airbags, which could cause personal injury (see Chapter 12).*

1 Disconnect the negative cable from the battery. Loosen the wheel lug nuts, raise the vehicle, support it securely on jackstands and remove the front wheel.

2 Refer to Section 10 and remove the hood.

3 Refer to Section 12 and remove the front bumper. Also remove the fender flare and the mudguard from the outside of the wheelwell if equipped.

4 Refer to Section 14 and remove the cowl cover on the side from which you are removing the fender. **Note:** *On Pathfinder models, it will be necessary to remove the entire cowl, since it is a one piece design. It will also be necessary to remove the antenna when removing the right fender on earlier models.*

5 Pry out the plastic rivets, remove the screws and remove the inner fenderwell liner **(see illustration)**.

6 Remove the bolts securing the lower half of the fender **(see illustrations)**.

7 Open the door and remove the upper fender-to-door pillar bolt in the doorjamb **(see illustration)**.

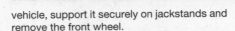

13.5 Remove the plastic pins and screws (arrows) and the fenderwell liner

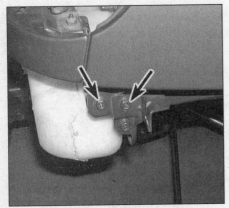

13.6a Remove the bolts at the front of the fender (arrows) ...

13.6b ... then remove the bolt (arrow) in the wheelwell securing the fender to the door pillar ...

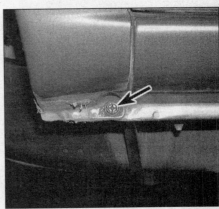

13.6c ... and the bolt (arrow) securing the bottom of the fender to the rocker panel

13.7 Remove the upper fender bolt in the door jamb

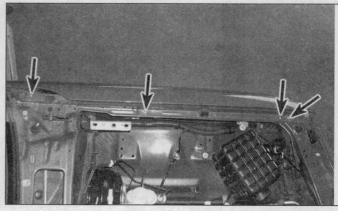

13.8 Unbolt the pencil brace bolts (arrows) at the front of the left fender

8   Remove the mounting bolts securing the top of the fender **(see illustration)**.
9   Detach the fender. It's a good idea to have an assistant support the fender while it's being moved away from the vehicle to prevent damage to the surrounding body panels.
10 Installation is the reverse of the removal procedure. Tighten all nuts, bolts and screws securely.

14 Cowl cover - removal and installation

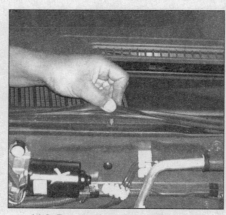

14.2 Pry out the clips and remove the hood seal

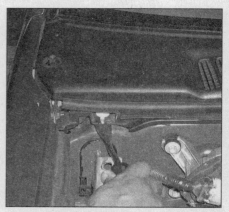

14.4 Pry upward on the clips at the front of the cowl cover to disengage it from the cowl

Refer to illustrations 14.2 and 14.4

1   Pry off the plastic trim cap on the windshield wiper arms, then detach the wiper arm retaining nuts and remove the wiper arms (see Chapter 12).
2   Carefully pry out the clips securing the hood seal to the cowl and remove the hood seal. The hood seal clips are approximately six to seven inches apart - pry directly underneath the clip being careful not to tear the hood seal **(see illustration)**.
3   On Frontier and Xterra models, remove the radio antenna from the cowl (see Chapter 12). On Pathfinder models, remove the retaining screws securing the rear of the cowl cover.
4   Pry up the clips securing the front of the cowl cover to the cowl and remove the cover from the vehicle **(see illustration)**. Once the cowl cover is raised, disconnect and plug the hose from the windshield washer reservoir attached to bottom of cowl cover.
5   Installation is the reverse of removal.

15 Door trim panels - removal and installation

Refer to illustrations 15.2a, 15.2b, 15.2c, 15.3a, 15.3b, 15.3c, 15.3d, 15.4, 15.5 and 15.6

1   On models with power door locks and/or power windows, disconnect the cable from the negative terminal of the battery (see Chapter 1).

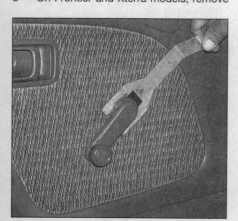

15.2a Use a hooked tool or a special window crank removal tool like this one to remove the retaining clip, then detach the window crank handle

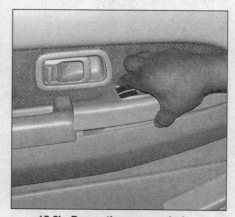

15.2b Pry up the power window switch assembly . . .

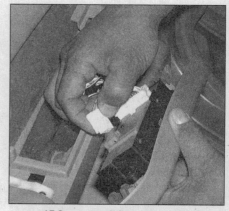

15.2c . . . and disconnect the electrical connectors

Chapter 11 Body

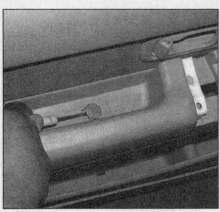

15.3a Pry off any screw trim caps . . .

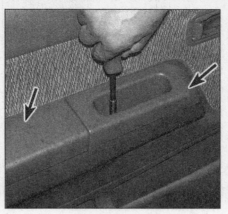

15.3b . . . remove the screws in the pull handle area

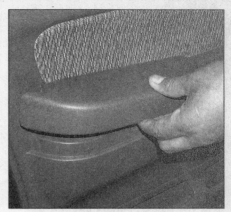

15.3c Detach the armrest pad . . .

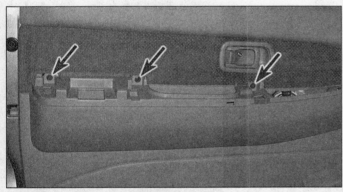

15.3d . . . and remove the screws at the center of the panel (Pathfinder shown, all others similar)

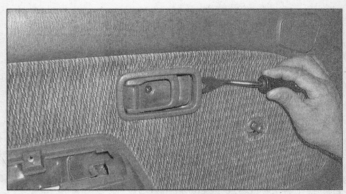

15.4 Pry off the inside handle bezel . . .

2   On manual window models, remove the window crank (see illustration). On power-window models, pry the power switch assembly out of the door panel with a flat-bladed trim tool and disconnect the electrical connectors from the switches (see illustrations).
3   Remove all door trim panel retaining screws and door pull/armrest assemblies (see illustrations).
4   Remove the bezel from the inside handle (see illustrations).
5   On Pathfinder and Xterra models, remove the clip securing the door trim panel near the outside mirror (see illustration).
6   Insert a wide putty knife, a thin screwdriver or a special trim panel removal tool between the trim panel and the door to disengage the retaining clips (see illustration). Work around the outside edge until the panel is free. Note: *Door trim panel retaining clips are approximately four to six inches apart. Pry at the clip location only. Prying between the clips will result in distorted or damaged trim panel(s).*
7   Once all of the clips are disengaged, raise the trim panel up and off the door. Disconnect any wiring harness connectors and remove the trim panel from the vehicle.
8   For access to the inner door, carefully peel back the plastic watershield.
9   Prior to installation of the door panel, be sure to reinstall any clips in the panel which

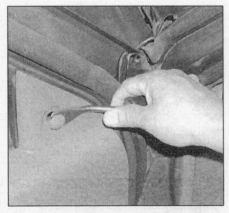

15.5 On Pathfinder and Xterra models, remove the clip adjacent to the outside mirror

may have come out during the removal procedure and remain in the door itself.
10   Connect the wiring harness connectors and place the panel in position on the door. Engage the trim panel with the top of the door first, then push downward on the panel until the clips on the trim panel align with all the holes in the door. Once the clips are aligned with the respective holes, push straight in on the panel until the clips are seated.
11   The remainder of the installation is the reverse of removal.

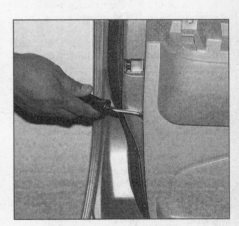

15.6 Use a trim panel removal tool to carefully pry out the clips around the outer edge

16 Door - removal, installation and adjustment

Note: *The door is heavy and somewhat awkward to remove and install - at least two people should perform this procedure.*

Removal and installation

Refer to illustrations 16.6 and 16.8

1   Lower the window completely in the door and then disconnect the negative cable

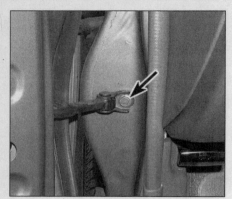

16.6 Remove the bolt (arrow) and detach the door stop strut from the body

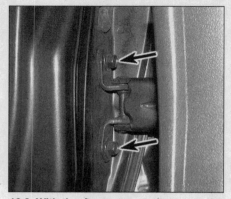

16.8 With the door supported, remove the hinge bolts (arrows indicate lower hinge bolts) and lift the door off

16.12 Adjust the door lock striker by loosening the mounting screws and gently tapping the striker in the desired direction

from the battery.

2 Open the door all the way and support it on jacks or blocks covered with rags to prevent damaging the paint.

3 Remove the door trim panel and water deflector as described in Section 15.

4 Disconnect all electrical connections, ground wires and harness retaining clips from the door. **Note:** *It is a good idea to label all connections to aid the reassembly process.*

5 From the door side, detach the rubber conduit between the body and the door. Then pull the wiring harness through conduit hole and remove from the door.

6 Remove the door stop strut at the body **(see illustration)**.

7 Mark around the door hinges with a pen or a scribe to facilitate realignment during reassembly.

8 With an assistant holding the door, remove the hinge to door bolts **(see illustration)** and lift the door off.

9 Installation is the reverse of removal.

Adjustment

Refer to illustration 16.12

10 Having proper door to body alignment is a critical part of a well functioning door assembly. First check the door hinge pins for excessive play. Fully open the door and lift up and down on the door without lifting the body. If a door has 1/16-inch (1.6 mm) or more excessive play, the hinges should be replaced.

11 Door-to-body adjustments are made by loosening the hinge-to-body bolts or hinge-to-door bolts and moving the door. Proper body alignment is achieved when the top of the doors are parallel with the roof section, the front door is flush with the fender, the rear door (SUV or cab models) is flush with the rear quarter panel and crew-cab and the bottom of the doors are aligned with the lower rocker panel. If these goals can't be reached by adjusting the hinge-to-body or hinge-to-door bolts, body alignment shims may have to be purchased and inserted behind the hinges to achieve correct alignment.

12 To adjust the door closed position, scribe a line or mark around the striker plate

17.2a Door latch retaining screws (Frontier model shown, other models similar)

to provide a reference point, then check that the door latch is contacting the center of the latch striker. If not adjust the up and down position first **(see illustration)**.

13 Finally adjust the latch striker sideways position, so that the door panel is flush with the center pillar or rear quarter panel or cab and provides positive engagement with the latch mechanism.

17 Door latch, lock cylinder and handles - removal and installation

1 Remove the door trim panel and the plastic watershield (see Section 15).

Door latch

Refer to illustrations 17.2a and 17.2b

2 Remove the latch retaining screws from the end of the door, then reach inside the door to release the latch from the control rods **(see illustration)**. **Note:** *Some models are equipped with a door stiffening brace that must be removed to allow access to the rear of the latch* **(see illustration)**.

3 On models with power door locks, disconnect the electrical connector.

4 Detach the door latch and (if equipped)

17.2b Remove the door stiffening brace (if equipped) to allow access to the rear of the latch

the door lock solenoid.

5 Installation is the reverse of removal.

Lock cylinder

Refer to illustration 17.7

6 Detach the electrical connector from the lock cylinder (if equipped).

7 Using a screwdriver or a pair of pliers, pry the lock cylinder retaining clip off the lock **(see illustration)**. Pull the lock outward to disengage it from the door and the lock cylinder control rod. **Note:** *On some models, it will be necessary to disengage the clip securing the control rod to the lock cylinder.*

17.7 Lock cylinder retaining clip (arrow)

Chapter 11 Body

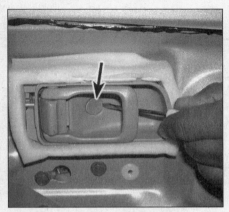

17.8a On front doors, the outside handle retaining nuts are accessed through the holes in the door frame

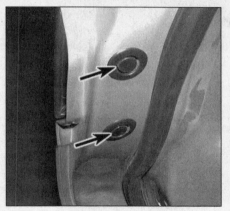

17.8b On vehicles with rear doors, pry out the trim caps (arrows) to access the outside handle retaining nuts

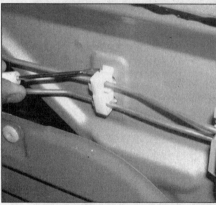

17.11 Release the control rod guide clip

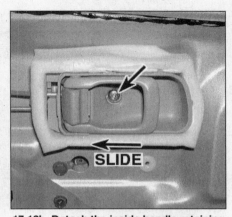

17.12a Pry out the retaining screw trim cover (if equipped)

17.12b Detach the inside handle retaining screw and slide the handle back to disengage it from the door

18.4 Remove the retaining bolts (arrows) and lift the glass out of the door

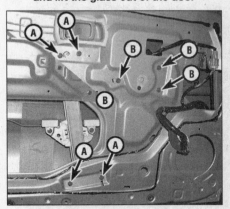

19.4 Door window regulator mounting details

A   Equalizer arm mounting bolts
B   Window regulator mounting bolts

Outside handle

Refer to illustrations 17.8a and 17.8b

8 Disconnect the control link rod from the outside handle, remove the outside handle retention nuts and pull the handle from the door **(see illustrations)**.
9 Installation is the reverse of removal.

Inside handle

Refer to illustrations 17.11, 17.12a and 17.12b

10 Remove the door trim panel.
11 Release the control rod guide clip **(see illustration)**.
12 Remove the retaining screws, pull the handle free from the door and disconnect the control rods from the rear of the inside handle **(see illustrations)**.
13 Installation is the reverse of removal.

18 Door window glass - removal and installation

Refer to illustration 18.4

1 Remove the door trim panel and watershield (see Section 15). Removal and installation of the window glass is essentially the same for front and rear doors (on models so equipped).
2 Lower the glass until the glass track bolts are visible through the holes in the door frame.
3 Place a rag inside the door panel to help prevent scratching the glass during removal.
4 Loosen the two bolts retaining the glass to the window regulator track and pull the glass from the door **(see illustration)**. **Note:** *When removing the rear door glass on SUV models, it will be necessary to remove the rear run channel between the stationary glass and the sliding glass to allow clearance for glass removal.*
5 To install, lower the glass into the door, slide it into position and tighten the nuts.
6 The remainder of installation is the reverse of removal.

19 Door window glass regulator - removal and installation

Refer to illustration 19.4

Note: *The procedure is the same for both front and rear window regulators.*

1 Remove the door trim panel and the plastic watershield (see Section 15). Also remove the door stiffener plate **(see illustration 17.2b)**.
2 Remove the window glass assembly (see Section 18).
3 On power operated windows, disconnect the electrical connector from the window regulator motor.
4 Remove the equalizer arm bracket and the regulator mounting bolts **(see illustration)**.

Chapter 11 Body

5 Pull the equalizer arm and regulator assemblies through the service hole in the door frame to remove it.
6 Installation is the reverse of removal.

20 Mirrors - removal and installation

Refer to illustrations 20.2, 20.4 and 20.6

Outside mirrors

1 Remove the door trim panel (see Section 15).
2 On Pathfinder models, remove the cover over the mirror mounting bolts **(see illustration)**.
3 If equipped with power mirrors, unplug the electrical connector.
4 Remove the bolts and detach the mirror from the door **(see illustration)**.
5 Installation is the reverse of removal.

Inside mirror

6 Insert a small screwdriver into the base of the mirror mounting bracket, and carefully pry rearward to disengage the mirror retaining spring from the mounting bracket **(see illustration)**. After the spring has been disengaged, slide the mirror up and off the mounting bracket.
7 Installation is the reverse of removal.
8 If the mounting bracket for the mirror has come off the windshield, it can be reattached with a special mirror adhesive kit available at auto parts stores. Clean the glass and support base thoroughly and follow the directions on the adhesive package, allowing the base to bond overnight before attaching the mirror.

21 Tailgate (pick-up models) – removal, installation and adjustment

Refer to illustrations 21.1 and 21.4

1 Open the tailgate and detach the retaining cables **(see illustration)**.
2 Lower the tailgate until the flat on the left

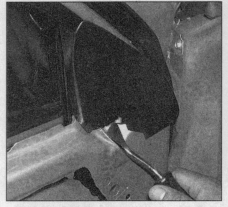

20.2 On Pathfinder models, pry off the mirror trim cover

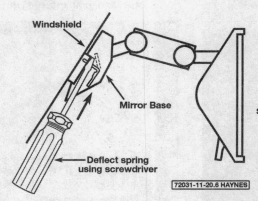

side hinge pin aligns with the slot in the hinge pocket. Lift the tailgate out of the pocket. With the help of an assistant to support the weight, withdraw the right hinge pin from the body and remove the tailgate from the vehicle.
3 Installation is the reverse of removal.
Note: *Apply some white grease to the mating parts of the tailgate hinge assembly before installation.*
4 If alignment is necessary, the latches (one on each side) can be loosened and moved (see Section 22), and/or the latch striker pins can be moved slightly for best tailgate fit **(see illustration)**.

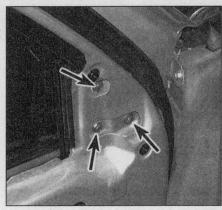

20.4 Remove the outside mirror mounting bolts (arrows)

20.6 Using a narrow screwdriver, pry the retaining spring away from the mounting bracket to free the mirror

22 Tailgate latch and handle (pick-up models) - removal and installation

Refer to illustrations 22.1, 22.2 and 22.4

1 Open the tailgate and remove the handle access cover **(see illustration)**.
2 Rotate the plastic retaining clips off the control rods and detach the rods from the handle **(see illustration)**.
3 To remove the handle, simply remove the mounting bolts and detach the handle from the tailgate.

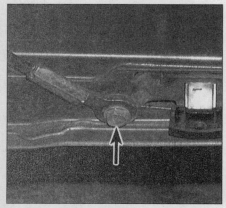

21.1 Remove the bolt (arrow) securing the cable to the tailgate (Frontier models)

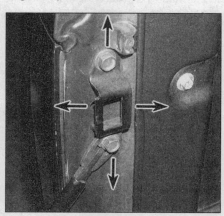

21.4 The tailgate latch strikers (left side shown) can be loosened and repositioned to obtain correct alignment

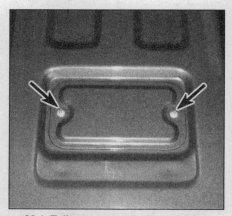

22.1 Tailgate access cover retaining screws (arrows)

Chapter 11 Body

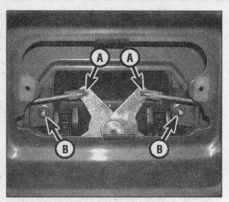

22.2 Tailgate control rods (A) and the handle mounting bolts (B)

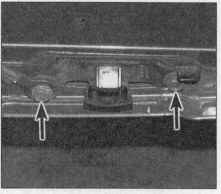

22.4 Remove the tailgate latch mounting bolts (arrows) and withdraw it with the control rod

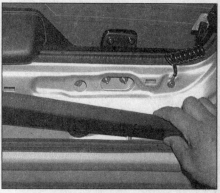

23.2 Pry off the upper trim panel if equipped

4   To remove the latch, simply remove the bolts and withdraw the latch from the side of the tailgate with the control rod attached **(see illustration)**. **Note:** *If your removing both latch assemblies it will be necessary to support the tailgate from below, since the tailgate support cables will be removed at the same time.*
5   Installation is the reverse of removal. Tighten all fasteners securely.

23 Liftgate and liftgate glass (SUV models) - removal and installation

Warning: *The liftgate is heavy and awkward to hold. At least two people should perform this procedure.*

Liftgate

Refer to illustrations 23.2, 23.3a, 23.3b and 23.6

1   Open the liftgate and support it fully in this position.
2   Remove the upper trim panel (if equipped) from the liftgate **(see illustration)**. Use a trim tool to pry out the fasteners on the rear of the trim panel.
3   Remove the liftgate pull handle, then pry out the plastic retaining clips and remove the lower trim panel from the liftgate **(see illustrations)**.
4   Disconnect the electrical connectors and remove the wiring harness from the liftgate.
5   Detach the support struts at the liftgate (see Section 24).
6   Remove the liftgate mounting bolts **(see illustration)**. Remove the bolts while at least one assistant, preferably two, helps you hold the liftgate.
7   Installation is the reverse of the removal procedure.

Liftgate glass

Refer to illustrations 23.10 and 23.11

8   Open the liftgate glass.
9   Disconnect the electrical connector for the rear window defogger.
10 Detach the glass support strut from the lower end of the glass **(see illustration)**.

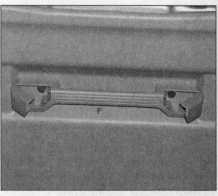

23.3a The pull handle retaining screws (arrows) on Pathfinder models are accessible after prying open the trim caps

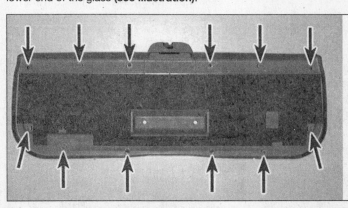

23.3b Liftgate lower trim panel fastener locations

23.6 Remove the liftgate mounting bolts (right side shown)

23.10 Disconnect the electrical connector (A) from the rear window defogger and detach the glass support strut (B) from the lower end of the glass

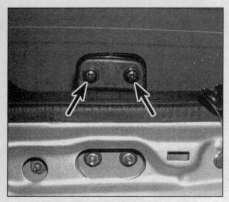

23.11 Liftgate glass mounting screws (arrows)

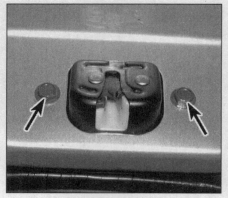

24.4a Liftgate door latch mounting bolts (arrows)

24.4b Liftgate glass latch mounting bolts (upper arrows) - lower arrow is the electrical connector found on electrically operated latch mechanisms

11 Remove the liftgate glass mounting screws **(see illustration)**. Remove the bolts while at least one assistant, preferably two, helps you hold the glass.
12 Installation is the reverse of the removal procedure. Do not overtighten the glass mounting screws.

24 Liftgate latch, lock cylinder, handle and support struts (SUV models) – removal, installation and adjustment

1 Remove the liftgate lower trim panel (see Section 23, Step 3) before proceeding with Steps 2 through 14 of this Section.

Latches

Refer to illustrations 24.4a and 24.4b

2 There are two latches, one mounted to the bottom edge of the liftgate, which is the latch for the liftgate door and one at the middle of the liftgate for the liftgate glass. Some models have electrically operated latch mechanisms and others have manually operated cable release mechanisms
3 To remove the latches, disconnect the control rods from the rear of the latch. Disconnect the electrical connector on electrically operated latch mechanisms or the release cable on manually operated mechanisms.
4 Remove the latch mounting bolts and detach the latch from the liftgate **(see illustrations)**.
5 Installation is the reverse of the removal procedure.

Lock cylinder

Refer to illustration 24.8

6 Detach the electrical connector from the lock cylinder (if equipped).
7 Disengage the clip securing the control rod to the lock cylinder.
8 Using a screwdriver or a pair of pliers, pry the lock cylinder retaining clip off the lock **(see illustration)**. Pull the lock outward to disengage it from the door.

Outside handle

Refer to illustration 24.9

9 Working through the access hole in the liftgate, remove the handle mounting nuts **(see illustration)**. Tilt the handle up and out to disengage it from the latch assembly.
10 Installation is the reverse of the removal procedure.

Support struts

Refer to illustrations 24.12 and 24.13

11 There are two support struts for the liftgate, and two separate struts to support the glass. Open the liftgate (or liftgate glass) and prop it securely in the full open position.
12 Release the small clip at the lower end of the strut, then pull the strut from the mounting ball **(see illustration)**.
13 Remove the bolt/screws securing the upper half of the support strut to the liftgate and detach the strut from the vehicle **(see illustration)**. Make sure the liftgate is supported properly before removing the struts or injury may occur.
14 Installation is the reverse of the removal procedure.

Adjustment

Refer to illustration 24.16

15 The closing position of the liftgate or the liftgate glass can be adjusted by moving either or both of the liftgate strikers.
16 Loosen the striker mounting bolts, move the striker slightly, retighten the bolts and check the closed position of the liftgate or the liftgate glass **(see illustration)**.

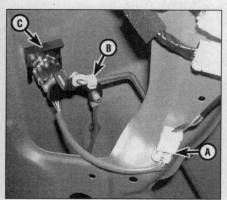

24.8 Liftgate lock cylinder mounting details

A Electrical connector
B Control rod retaining clip
C Lock cylinder retaining clip

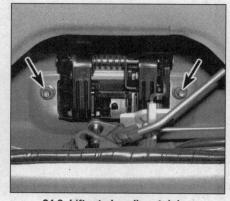

24.9 Liftgate handle retaining nuts (arrows)

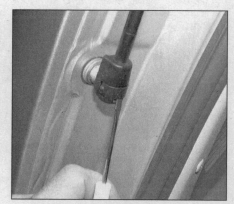

24.12 Detach the lower end of the liftgate support strut by pulling out the retainer clip with a small screwdriver, then pull the strut from the ballstud (liftgate strut shown, liftgate glass strut similar)

Chapter 11 Body

11-17

24.13 Upper mounting bolts (arrows) for the liftgate support strut

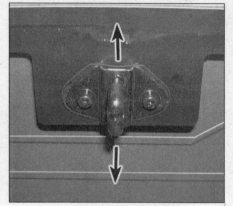

24.16 Striker mounting bolts (liftgate glass striker shown) - make a reference mark around the edges of the striker before repositioning it for adjustment

25.3a On Frontier models with column shift and Xterra models with manual transmissions, the front console consists of a DC power outlet and a passenger side airbag switch - detach the cup holder (A) if equipped and remove console retaining screws (arrow) from each side

25 Center console - removal and installation

Warning: *The models covered by this manual are equipped with Supplemental Restraint Systems (SRS), more commonly known as airbags. Always disable the airbag system before working in the vicinity of any airbag system components to avoid the possibility of accidental deployment of the airbags, which could cause personal injury (see Chapter 12).*

1 Disconnect the cable from the negative terminal of the battery and set the parking brake.
2 On vehicles equipped with manual transmissions, unscrew the knob from the shift lever.

Frontier and Xterra

Refer to illustrations 25.3a, 25.3b, 25.4 and 25.5

3 On V6 Frontier models and all Xterra models with floor shift, pry off the plastic trim panel surrounding the shift lever. The shift lever trim panel snaps into place with plastic clips. Locate the clips and pry up at the clip locations. Also remove the four screws (two on each side) of the front console and detach the front console from the vehicle. On column-shift Frontier models, remove the cup holder, then remove the small center console from the instrument panel **(see illustrations)**.
4 On Frontier and Xterra models with manual transmissions, remove the screws securing the shift lever console to the floor **(see illustration)**. Pull the shift lever console over the shift lever to remove it.
5 To remove the rear console/glove box, simply remove the four screws (two on each side) of the rear console and detach the rear console from the vehicle **(see illustration)**.

Pathfinder

Refer to illustrations 25.6a, 25.6b, 25.6c, 25.7a, 25.7b and 25.7c

6 Pull out the console ashtray and remove the screw located below the ashtray. Pry off the plastic trim panel surrounding the shift lever. The shift lever trim panel snaps into place with plastic clips. Locate the clips and pry up at the clip location **(see illustrations)**.

25.3b Pull the front console outward and disconnect the electrical connectors (arrow)

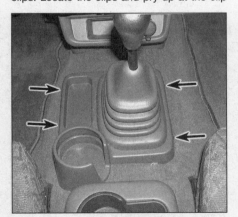

25.4 On Frontier and Xterra models equipped with manual transmissions, remove the four screws (arrows) on the side of the shift lever console

25.5 To remove the rear console/glove box on Frontier and Xterra models, simply remove the screws around the base of the console (arrow indicates one of four screws)

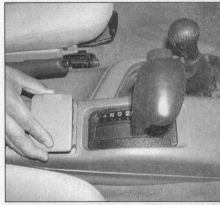

25.6a On Pathfinder models, it will be necessary to pull out the console ashtray and remove the screw in the opening . . .

11

Chapter 11 Body

25.6b ... then carefully pry up the shift lever trim panel ...

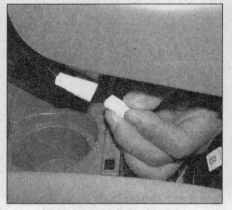

25.6c ... and disconnect the electrical connectors

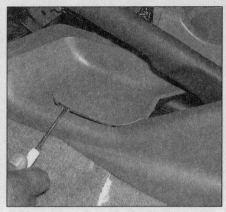

25.7a Pry out the trim cover below the parking brake handle ...

7   Pry out the trim cover below the parking brake handle and detach the screw located below the trim cover. Also remove the console mounting screws at the front and rear of the console **(see illustrations)**.
8   Unplug any electrical connections and remove the console assembly from the vehicle.
9   Installation is the reverse of the removal procedure.

26 Dashboard trim panels - removal and installation

25.7b ... then detach the screws securing the front of the console ...

25.7c ... and at the rear of the console on each side

Warning: *The models covered by this manual are equipped with Supplemental Restraint Systems (SRS), more commonly known as airbags. Always disable the airbag system before working in the vicinity of any airbag system components to avoid the possibility of accidental deployment of the airbags, which could cause personal injury (see Chapter 12).*

Instrument cluster bezel

Frontier and Xterra

Refer to illustration 26.4

1   With the wheels blocked, apply the parking brake. On Frontier models with column shift, lower the shift lever as far as possible.
2   On models equipped with tilt steering columns, tilt the steering wheel down as far as possible.
3   Remove the lower knee bolster (see Step 9).
4   Remove the bezel retaining screws and remove the bezel from the instrument panel **(see illustration)**.
5   Installation is the reverse of removal.

Pathfinder

Refer to illustration 26.7

6   Tilt the steering wheel down as far as possible.
7   Remove the bezel retaining screws and detach the bezel from the instrument panel **(see illustration)**.
8   Installation is the reverse of removal.

Knee bolster

Refer to illustrations 26.10 and 26.11

9   Remove the cover over the fuse/relay box at the left end of the dash.
10   Remove the screws securing the driver's knee bolster and pull it outward away from the dash **(see illustration)**. Disconnect any electrical connectors at the rear of the knee bolster and detach it from the instrument panel.
11   If the knee bolster reinforcement panel needs to be removed for any reason, remove the retaining bolts and take down the panel **(see illustration)**.
12   Installation is the reverse of removal.

Instrument panel cover

Frontier and Xterra

Refer to illustrations 26.13 and 26.14

13   Remove the screws at the center securing the cover to the instrument panel **(see illustration)**.

26.4 Instrument cluster bezel retaining screws (Frontier and Xterra models)

26.7 Instrument cluster bezel retaining screws (Pathfinder models)

Chapter 11 Body

26.10 The knee bolster is held in place by screws (arrows) at the bottom, and clips at the top (Frontier model shown, other models similar)

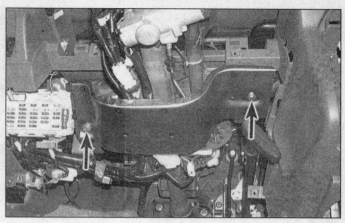

26.11 Knee bolster reinforcement plate mounting bolts (arrows)

26.13 Instrument panel cover mounting screws (arrows) - Frontier and Xterra models only

26.14 Pull the instrument panel cover out far enough to access the electrical connectors at the rear

26.16 Center trim panel retaining screws (arrows) - Frontier and Xterra models

14 Disengage the remaining clips and pull the cover outward slightly. Disconnect the electrical connectors and remove the cover from the vehicle (see illustration).
15 Installation is the reverse of removal.

Center trim panel

Frontier and Xterra

Refer to illustrations 26.16 and 26.17

16 Remove the screws from the lower edge of the trim panel (see illustration).
17 Pull outward to detach the retaining clips at the top of the panel and disconnect the electrical connectors from the rear of the trim panel (see illustration).
18 Installation is the reverse of removal.

Pathfinder

Refer to illustrations 26.20a, 26.20b and 26.20c

19 On 2000 and earlier models, remove the screws from the lower edge of the trim panel and pull outward to detach the retaining clips at the top of the panel. Disconnect the electrical connectors from the rear of the trim panel.
20 On 2001 models, pry out the center air duct grille, then remove the screws at the top of the center trim panel. Pull outward to detach the retaining clips at the bottom of the panel and disconnect the electrical connectors from the rear of the trim panel (see illustrations).
21 Installation is the reverse of removal.

26.17 Pull outward on the panel to detach the retaining clips and disconnect the electrical connectors (arrow)

26.20a On 2001 Pathfinder models, pry out the center air duct grille . . .

26.20b . . . and disconnect the electrical connector from the clock

11-20   Chapter 11 Body

26.20c Center trim panel retaining screws (arrows) - 2001 Pathfinder models

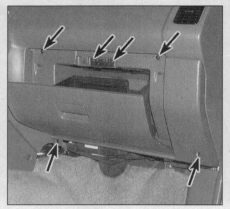

26.22 Glove box mounting screws (arrows) - Frontier model shown, other models similar

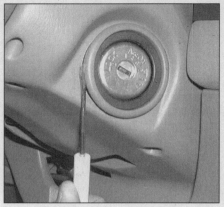

27.1 On Pathfinder models, pry off the key lock cylinder trim ring

Glove box

Refer to illustration 26.22

22 With the glove box open, remove the four screws at the top, two screws at the bottom, and pull the glove box housing from the instrument panel **(see illustration)**.
23 Installation is the reverse of the removal procedure.

27 Steering column covers - removal and installation

Refer to illustrations 27.1 and 27.2
Warning: *The models covered by this manual are equipped with Supplemental Restraint Systems (SRS), more commonly known as airbags. Always disable the airbag system before working in the vicinity of any airbag system components to avoid the possibility of accidental deployment of the airbags, which could cause personal injury (see Chapter 12).*

1 On Pathfinder models pry off the ignition lock cylinder trim ring **(see illustration)**.
2 On all models, remove the screws and detach the lower half of the steering column cover **(see illustration)**.
3 Pull the upper column cover straight up and off.
4 Installation is the reverse of removal.

28 Instrument panel - removal and installation

Refer to illustrations 28.7, 28.8, 28.9, 28.11, 28.12a, 28.12b, 28.12c and 28.12d
Warning: *The models covered by this manual are equipped with Supplemental Restraint Systems (SRS), more commonly known as airbags. Always disable the airbag system before working in the vicinity of any airbag system components to avoid the possibility of accidental deployment of the airbags, which could cause personal injury (see Chapter 12).*
Note: *This is a difficult procedure for the home mechanic, involving tedious disassembly and the disconnection/reconnection of numerous electrical connectors. If you do attempt this procedure, make sure you take good notes and mark all matching connectors (and their mounting points) to aid reassembly.*

1 Disconnect the cable from the negative terminal of the battery.
2 Remove the center console (see Section 25).
3 Remove the steering column covers (see Section 27).
4 Remove the dashboard trim panels (see Section 26).
5 Remove the instrument cluster (see Chapter 12).
6 Remove the radio and heater/air conditioner controls (see Chapter 3 and Chapter 12).
7 Remove the kick panels at each side **(see illustration)**. The panels simply pull out after removing the retaining screw at the bottom.
8 Remove the left and right trim strips along the interior of each windshield post **(see illustration)**.

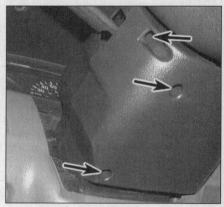

27.2 Steering column cover retaining screws (arrows)

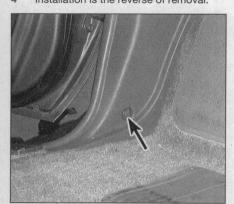

28.7 Remove the fastener (arrow) and pull out the kick panels at each side

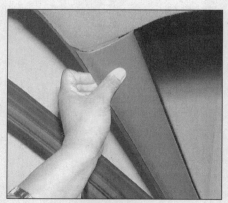

28.8 Pull back the rubber weatherstripping, then remove the windshield post trim (retained by clips)

28.9 Disconnect the passenger airbag wiring harness connector (A), then remove the two bolts (B)

Chapter 11 Body

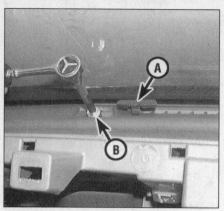

28.11 Some models will require prying up a trim cap to access the mounting bolts at the top of the instrument panel - other models will require prying out the entire defroster grille

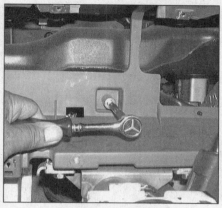

28.12a Remove the bolt securing the instrument panel in the instrument cluster opening (Frontier and Xterra instrument panel shown, Pathfinder models similar)

28.12b Remove bolts at the left . . .

9 The passenger airbag is attached to the back of the instrument panel. Disconnect the electrical connector and remove the bolts securing the bottom of the airbag to the instrument panel reinforcement beam (**see illustration**). **Note:** *The bolts are "tamper-proof" Torx type and require a special socket to remove or install.*
10 Refer to Chapter 10 and unbolt the steering column from the reinforcement beam and the firewall.
11 At the top of the instrument panel where it meets the windshield, pry up the defroster grilles to access the upper bolts of the instrument panel (**see illustration**).
12 Remove all remaining fasteners securing the instrument panel, and any electrical connectors still attached to the panel (**see illustrations**). Pull the panel back and out of the vehicle.
13 Installation is the reverse of removal.

29 Seats - removal and installation

Front bucket seats

Refer to illustrations 29.1a, 29.1b and 29.2
Warning: *The models covered by this manual are equipped with Supplemental Restraint Systems (SRS), more commonly known as airbags. Always disable the airbag system before working in the vicinity of any airbag system components to avoid the possibility of accidental deployment of the airbags, which could cause personal injury* (see Chapter 12).
1 Detach the trim caps (if equipped) and remove the seat track-to-floor bolts (**see illustrations**).
2 If equipped with power seats, tilt the seat back towards the rear of the vehicle and disconnect the electrical connectors (**see illustration**).
3 Remove the front seats from the vehicle.

28.12c . . . and the right lower edges of the instrument panel

28.12d Remove the bolts securing the center of the instrument panel (arrow indicates one of two bolts)

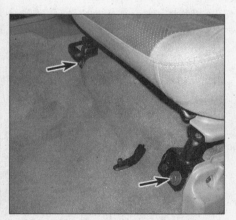

29.1a With the seat moved rearward, remove the trim caps (if equipped) and the front bolts (arrows)

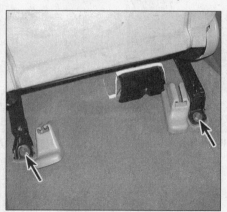

29.1b With the seat moved forward, detach the rear trim caps (if equipped) and the rear mounting bolts (arrows)

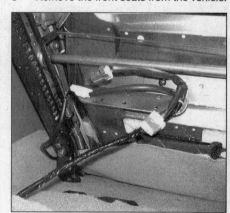

29.2 On vehicles equipped with power seats, disconnect the electrical connectors

Chapter 11 Body

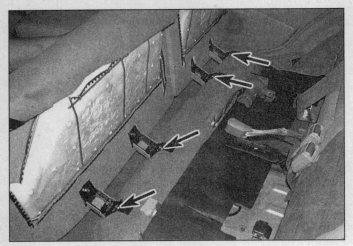

29.10 On Pathfinder models, fold the seat cushions forward, then remove the bracket mounting bolts (arrows)

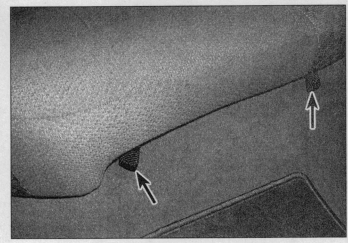

29.11 On Xterra models, lift up on the cushion straps to disengage the seat cushion

Bench seat (pick-up models)

4 The seat back and seat bottom on pick-up bench seats are part of an assembly that is removed or installed together.
5 Remove the seat track-to-floor bolts at the front of the seat.
6 Position the seat track in the most forward position, then fold the seat back down to access the rear mounting bolts.
7 Remove the seat track-to-floor bolts at the rear of the seat.
8 Pull the seat forward enough to clear the cab and slide the seat out of the vehicle. **Note:** *This is an awkward procedure, have an assistant help you.*
9 Installation is the reverse of the removal procedure.

Rear seat (SUV models)

Refer to illustrations 29.10, 29.11, 29.12a, 29.12b and 29.12c

10 On Pathfinder models, fold the seat cushions forward, then remove the mounting bolts **(see illustration)**.
11 On Xterra models, lift up on the cushion straps to disengage the seat cushion from the hook on the floor pan and remove the seat cushion from the vehicle **(see illustration)**.
12 Fold the seat back(s) forward to access the remaining mounting nuts **(see illustrations)**.
13 Remove the seat back from the vehicle
14 Installation is the reverse of removal.

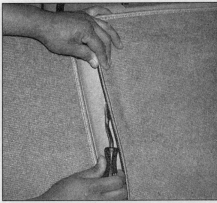

29.12a Pry off the seat back trim . . .

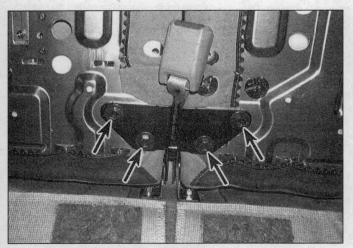

29.12b . . . to access the mounting bolts (arrows) at the center bracket . .

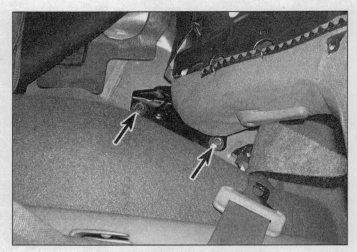

29.12c . . . and the end brackets

Chapter 12
Chassis electrical system

Contents

	Section		Section
Airbag system - general information	27	Horn - check and replacement	18
Antenna - removal and installation	12	Ignition switch and key lock cylinder - check and replacement	8
Bulb replacement	16	Instrument cluster - removal and installation	10
Circuit breakers - general information	4	Instrument panel gauges - check	9
Cruise control system - description, check and component replacement	21	Lighting system and windshield wiper/washer switches - check and replacement	7
Daytime Running Lights (DRL) - general information	19	Power door lock system - description and check	23
Electric side view mirrors - description and check	24	Power seats - description and check	25
Electric sunroof - description and check	26	Power window system - description and check	22
Electrical troubleshooting - general information	2	Radio and speakers - removal and installation	11
Fuses and fusible links - general information	3	Rear window defogger - check and repair	10
General information	1	Relays - general information and testing	5
Headlight bulb - replacement	13	Turn signal/hazard flashers - check and replacement	6
Headlight housing - replacement	15	Wiper motor and amplifier - check and replacement	17
Headlights - adjustment	14	Wiring diagrams - general information	28

1 General information

The electrical system is a 12-volt, negative ground type. Power for the lights and all electrical accessories is supplied by a lead/acid-type battery that is charged by the alternator.

This Chapter covers the various electrical components not associated with the engine. Information on the battery, alternator, distributor and starter motor can be found in Chapter 5.

It should be noted that when portions of the electrical system are serviced, the cable should be disconnected from the negative battery terminal to prevent electrical shorts and/or fires.
Warning: *The models covered by this manual are equipped with Supplemental Restraint Systems (SRS), more commonly known as airbags. Always disable the airbag system before working in the vicinity of any airbag system components to avoid the possibility of accidental deployment of the airbags, which could cause personal injury (see Section 27).*

2 Electrical troubleshooting - general information

Refer to illustrations 2.5a, 2.5b, 2.6, 2.9 and 2.15

A typical electrical circuit consists of an electrical component, any switches, relays, motors, fuses, fusible links or circuit breakers related to that component and the wiring and connectors that link the component to both the battery and the chassis. To help you pinpoint an electrical circuit problem, wiring diagrams are included at the end of this Chapter.

Before tackling any troublesome electrical circuit, first study the appropriate wiring diagrams to get a complete understanding of what makes up that individual circuit. Trouble spots, for instance, can often be narrowed down by noting if other components related to the circuit are operating properly. If several components or circuits fail at one time, chances are the problem is in a fuse or ground connection, because several circuits are often routed through the same fuse and ground connections.

Electrical problems usually stem from simple causes, such as loose or corroded connections, a blown fuse, a melted fusible link or a failed relay. Visually inspect the condition of all fuses, wires and connections in a problem circuit before troubleshooting the circuit.

If test equipment and instruments are going to be utilized, use the diagrams to plan

Chapter 12 Chassis electrical system

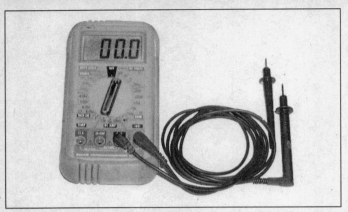

2.5a The most useful tool for electrical troubleshooting is a digital multimeter that can check volts, amps, and test continuity

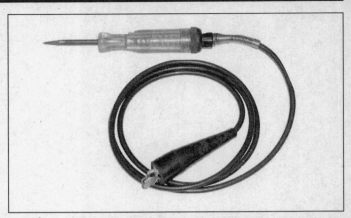

2.5b A simple test light is a very handy tool for checking voltage

ahead of time where you will make the necessary connections in order to accurately pinpoint the trouble spot.

The basic tools needed for electrical troubleshooting include a voltmeter, a test light (a 12-volt bulb with a set of test leads can also be used) or a continuity tester (which includes a bulb, battery and set of test leads) **(see illustrations)**. Also useful is a pair of jumper wires, preferably with a circuit breaker incorporated into one, which can be used to apply power and ground to electrical components. Before attempting to locate a problem with test instruments, use the wiring diagram(s) to decide where to make the connections.

Voltage checks

Voltage checks should be performed if a circuit is not functioning properly. Connect one lead of a test light to either the negative battery terminal or a known good ground. Connect the other lead to a connector in the circuit being tested, preferably nearest to the battery or fuse **(see illustration)**. If the bulb of the tester lights, voltage is present, which means that the part of the circuit between the connector and the battery is problem free. Continue checking the rest of the circuit in the same fashion. When you reach a point at which no voltage is present, the problem lies between that point and the last test point with voltage. Most of the time the problem can be traced to a loose connection. **Note:** *Keep in mind that some circuits receive voltage only when the ignition key is in the Accessory or Run position.*

Finding a short

A short-to-ground in a live circuit causes the fuse protecting the circuit to blow. When the fuse is replaced it will immediately blow again.

One method of finding the location of a short is to remove the fuse protecting the problem circuit and connect a test light in place of the fuse. Disconnect the load or ground from the circuit. Turn the ignition key on, if the circuit is shorted to ground there should be voltage present in the circuit and

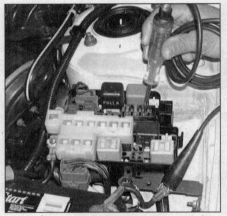

2.6 In use, a basic test light's lead is clipped to a known good ground, then the pointed probe can test connectors, wires or electrical sockets - if the bulb lights, the part being tested has battery voltage

the test light should light. Move the suspected wiring harness from side-to-side while watching the test light. If the bulb goes out, there is a short to ground somewhere in that area, probably where the insulation has rubbed through allowing the bare wire to contact the body. The same test can be performed on each component in the circuit, even a switch.

Ground check

Perform a ground test to check whether a component is properly grounded. Disconnect the battery and connect one lead of a continuity tester or multimeter (set to the ohms scale), to a known good ground. Connect the other lead to the wire or ground connection being tested. If the resistance is low (less than 5 ohms), the ground is good. If using a self-powered continuity tester, the bulb will light if the ground is good.

Continuity check

A continuity check is done to determine if there are any breaks in a circuit - if it is passing electricity properly. With the circuit

2.9 With a multimeter set to the ohms scale, resistance can be checked across two terminals - when checking for continuity, a low reading indicates continuity, a high reading indicates lack of continuity

off (no power in the circuit), a self-powered continuity tester or multimeter can be used to check the circuit. Connect the test leads to both ends of the circuit (or to the "power" end and a good ground), and if the self-powered test light comes on the circuit is passing current properly **(see illustration)**. If the resistance is low (less than 5 ohms), there is continuity; if the reading is 10,000 ohms or higher, there is a open somewhere in the circuit. The same procedure can be used to test a switch, by connecting the continuity tester to the switch terminals. With the switch turned On, the test light should come on (or low resistance should be indicated on a meter).

Finding an open circuit

When diagnosing for possible open circuits, it is often difficult to locate them by sight because the connectors hide oxidation or terminal misalignment. Merely wiggling a connector on a sensor or in the wiring harness may correct the open circuit condition. Remember this when an open circuit is indi-

Chapter 12 Chassis electrical system

2.15 To backprobe a connector, insert a small, sharp probe (such as a straight-pin) into the back of the connector alongside the desired wire until it contacts the metal terminal inside; connect your meter leads to the probes - this allows you to test a functioning circuit

3.1a The interior fuse box is located under the left (driver's) side of the instrument panel, under a cover

cated when troubleshooting a circuit. Intermittent problems may also be caused by oxidized or loose connections.

Electrical troubleshooting is simple if you keep in mind that all electrical circuits are basically electricity running from the battery, through the wires, switches, relays, fuses and fusible links to each electrical component (light bulb, motor, etc.) and to ground, from which it is passed back to the battery. Any electrical problem is an interruption in the flow of electricity to and from the battery.

Connectors

Most electrical connections on these vehicles are made with multi-wire plastic connectors. The mating halves of many connectors are secured with locking clips molded into the plastic connector shells. The mating halves of large connectors, such as some of those under the instrument panel, are held together by a bolt through the center of the connector.

To separate a connector with locking clips, use a small screwdriver to pry the clips apart carefully, then separate the connector halves. Pull only on the shell, never pull on the wiring harness as you may damage the individual wires and terminals inside the connectors. Look at the connector closely before trying to separate the halves. Often the locking clips are engaged in a way that is not immediately clear. Additionally, many connectors have more than one set of clips. One type of connector specific to this manufacturer is a slide-locking type connector. Actually, there are two variations of the slide-lock connector; weatherproof and non-weatherproof (generally, the non-weatherproof type is used in the interior of the vehicle, while the weatherproof type is used under the hood or other locations that are exposed to the elements. To disconnect a weatherproof slide-lock connector, push the slide-lock in towards the center of the connector, then pull the connector apart. To disconnect a non-weatherproof slide-lock connector, pull the slide-lock away from the center of the connector, then pull the connector apart. When connecting either type of slide lock connector, press the connector together until it "clicks" into place.

Each pair of connector terminals has a male half and a female half. When you look at the end view of a connector in a diagram, be sure to understand whether the view shows the harness side or the component side of the connector. Connector halves are mirror images of each other, and a terminal shown on the right side end-view of one half will be on the left side end view of the other half.

Backprobing a connector

It is often necessary to take circuit voltage measurements with a connector connected. Whenever possible, carefully insert a small straight pin (not your meter probe) into the rear of the connector shell to contact the metal terminal inside, then clip your meter lead to the pin. This kind of connection is called "backprobing" **(see illustration)**. When inserting a test probe into a male terminal, be careful not to distort the terminal opening. Doing so can lead to a poor connection and corrosion at that terminal later. Using the small straight pin instead of a meter probe results in less chance of deforming the terminal connector.

3 Fuses and fusible links - general information

Fuses

Refer to illustrations 3.1a, 3.1b, 3.1c and 3.3

The electrical circuits of the vehicle are protected by a combination of fuses, circuit breakers and fusible links. The fuse blocks are located under the instrument panel on the left side of the dashboard and in the engine compartment **(see illustrations)**.

Each of the fuses is designed to protect a specific circuit, and the various circuits are identified on the fuse panel itself.

Miniaturized fuses are employed in the fuse blocks. These compact fuses, with blade terminal design, allow fingertip removal and replacement. If an electrical component

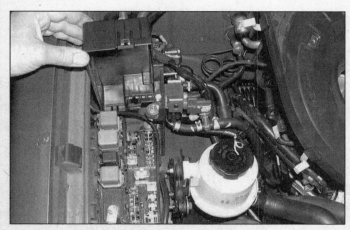

3.1b The engine compartment fuse and fusible link box is located behind the battery - this is a Frontier model . . .

3.1c . . . and this is a Pathfinder (note the fuse and relay location guide on the cover)

Chapter 12 Chassis electrical system

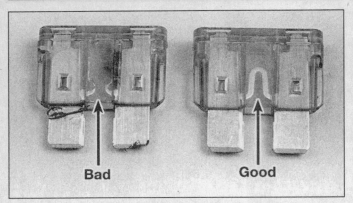

3.3 When a fuse blows, the element between the terminals melts - the fuse on the left is blown, the fuse on the right is good

5.1a The covers of the engine compartment relay boxes have a printed key to identify the relays

fails, always check the fuse first. The best way to check the fuses is with a test light. Check for power at the exposed terminal tips of each fuse. If power is present at one side of the fuse but not the other, the fuse is blown. A blown fuse can also be identified by visually inspecting it **(see illustration)**.

Be sure to replace blown fuses with the correct type. Fuses of different ratings are physically interchangeable, but only fuses of the proper rating should be used. Replacing a fuse with one of a higher or lower value than specified is not recommended. Each electrical circuit needs a specific amount of protection. The amperage value of each fuse is molded into the fuse body.

If the replacement fuse immediately fails, don't replace it again until the cause of the problem is isolated and corrected. In most cases, this will be a short circuit in the wiring caused by a broken or deteriorated wire.

Fusible links

Some circuits are protected by fusible links. The links are used in circuits that are not ordinarily fused, such as the ignition circuit.

The fusible links on these models are located in the engine compartment fuse block next to the battery **(see illustrations 3.1b and 3.1c)** and are similar to fuses, but larger.

To replace a fusible link, first disconnect the negative cable from the battery. Unplug the burned-out link and replace it with a new one (available from your dealer or auto parts store). Always determine the cause for the overload that melted the fusible link before installing a new one.

4 Circuit breakers - general information

Circuit breakers protect the power windows, power door locks, power seat and power sunroof. One or two circuit breakers are used, depending on options, they are located under the left end of the instrument panel above the fuse box.

Because the circuit breakers reset automatically, an electrical overload in a circuit breaker protected system will cause the circuit to fail momentarily, then come back on. If the circuit does not come back on, check it immediately.

5 Relays - general information and testing

General information

Refer to illustrations 5.1a and 5.1b

1 Several electrical accessories in the vehicle, such as the fuel injection system, horns, starter, and fog lamps use relays to transmit the electrical signal to the component. Relays use a low-current circuit (the control circuit) to open and close a high-current circuit (the power circuit). If the relay is defective, that component will not operate properly. Most relays are mounted in the engine compartment fuse/relay boxes, with some specialized relays located above the interior fuse box **(see illustrations)**. If a faulty relay is suspected, it can be removed and tested using the procedure below or by a dealer service department or a repair shop. Defective relays must be replaced as a unit.

Testing

Refer to illustrations 5.3a, 5.3b and 5.6

2 Refer to the wiring diagrams for the circuit to determine the proper connections for

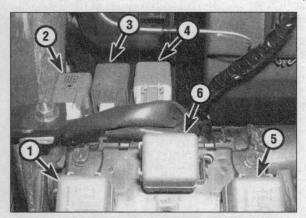

5.1b There are also several relays located above the interior fuse box

1	Ignition relay	4	Power window relay
2	Circuit breaker	5	Accessory relay
3	Circuit breaker	6	Blower relay

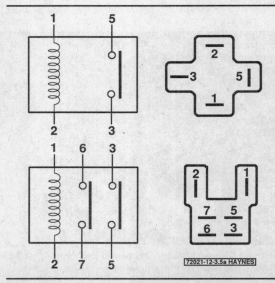

5.3a These two relays are typical normally open types; the one above completes a single circuit (terminal 5 to terminal 3) when energized - the lower relay type completes two circuits (6 and 7, and 3 and 5) when energized

Chapter 12 Chassis electrical system

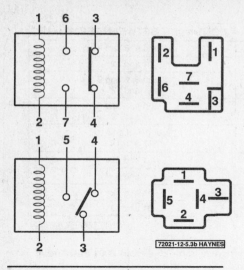

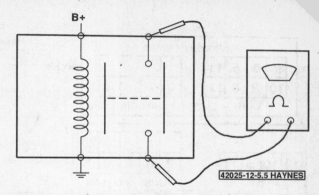

5.3b These relays are normally closed types, where current flows though one circuit until the relay is energized, which interrupts that circuit and completes the second circuit

5.6 To test a typical four-terminal normally open relay, connect an ohmmeter to the two terminals of the power circuit - the meter should indicate continuity with the relay energized and no continuity with the relay not energized

the relay you're testing. If you can't determine the correct connection from the wiring diagrams, however, you may be able to determine the test connections from the information that follows.

3 There are four basic types of relays used on these models (see illustrations). Some are normally open type and some normally closed, while others include a circuit of each type.

4 On most relays, two of the terminals are the relay control circuit (they connect to the relay coil which, when energized, closes the large contacts to complete the circuit). The other terminals are the power circuit (they are connected together within the relay when the control-circuit coil is energized).

5 Some relays may be marked as an aid to help you determine which terminals are the control circuit and which are the power circuit. If the relay is not marked, refer to the wiring diagrams at the end of this Chapter to determine the proper hook-ups for the relay you're testing.

6 To test a relay connect an ohmmeter across the two terminals of the power circuit, continuity should not be indicated (see illustration). Now connect a fused jumper wire between one of the two control circuit terminals and the positive battery terminal. Connect another jumper wire between the other control circuit terminal and ground. When the connections are made, the relay should click and continuity should be indicated on the meter. On some relays, polarity may be critical, so, if the relay doesn't click, try swapping the jumper wires on the control circuit terminals.

7 If the relay fails the above test, replace it.

6 Turn signal/hazard flashers - check and replacement

Refer to illustrations 6.4a and 6.4b
Warning: The models covered by this manual are equipped with Supplemental Restraint Systems (SRS), more commonly known as

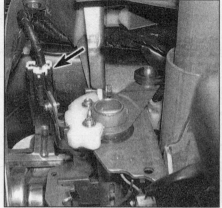

6.4a Turn signal/hazard flasher unit location (arrow) - Frontier

airbags. Always disable the airbag system before working in the vicinity of any airbag system components to avoid the possibility of accidental deployment of the airbags, which could cause personal injury (see Section 27).

1 The turn signal and hazard flasher is a single combination unit.

2 When the flasher unit is functioning properly, an audible click can be heard during its operation. If the turn signals fail on one side or the other and the flasher unit does not make its characteristic clicking sound, or if a bulb on one side of the vehicle flashes much faster than normal but the bulb at the other end of the vehicle (on the same side) doesn't light at all, a faulty turn signal bulb may be indicated.

3 If both turn signals fail to blink, the problem may be due to a blown fuse, a faulty flasher unit, a defective switch or a loose or open connection. If a quick check of the fuse box indicates that the turn signal fuse has blown, check the wiring for a short before installing a new fuse.

4 To replace the flasher, disconnect the electrical connector and remove the flasher unit from its mounting bracket located under the instrument panel to the right of the steer-

6.4b Location of the turn signal/hazard flasher (A); (B) is the smart entry control unit - Pathfinder

ing column (see illustrations).
5 Make sure that the replacement unit is identical to the original. Compare the old one to the new one before installing it.
6 Installation is the reverse of removal.

7 Lighting system and windshield wiper/washer switches - check and replacement

Warning: The models covered by this manual are equipped with Supplemental Restraint Systems (SRS), more commonly known as airbags. Always disable the airbag system before working in the vicinity of any airbag system components to avoid the possibility of accidental deployment of the airbags, which could cause personal injury (see Section 27).

Check

Refer to illustrations 7.3a, 7.3b, 7.3c, 7.3d, 7.3e and 7.3f

1 Disconnect the cable from the negative battery terminal.

2 Refer to the replacement procedure below and remove the switches for testing.

Chapter 12 Chassis electrical system

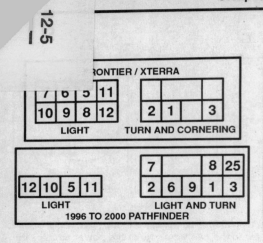

SWITCH POSITION	CONTINUITY BETWEEN TERMINALS
Headlight switch, Off, position C	5 and 6, 8 and 9
Park, positions A or B	11 and 12
Park, position C	5 and 6, 8 and 9 11 and 12
Head, position A or C	5 and 6, 8 and 9 11 and 12
Head, position B	5 and 7, 8 and 10 11 and 12
Turn signal, right turn	1 and 2
Turn signal, left turn	1 and 3
Cornering lamp, right turn	61 and 62
Cornering lamp, left turn	61 and 63

7.3a Terminal identification and continuity chart for the lighting portion of the combination switch - all except 2001 Pathfinder

SWITCH POSITION	CONTINUITY BETWEEN TERMINALS
Headlight switch, OFF, position A	5 and 6; 8,9 and 12
Headlight switch, OFF, position B	5 and 6, 8 and 9
Headlight switch, OFF, position C	5 and 7, 8 and 10
Headlight switch, PARK	8 and 42
Headligtht switch, AUTO	5 and 11
Headlight switch, ON	5 and 11, 8 and 12
Fog light switch, OFF	—
Fog light switch, ON	31 and 32
Turn signal, left turn	1 and 3
Turn signal, right turn	1 and 2

7.3b Terminal identification and continuity chart for the lighting portion of the combination switch - 2001 Pathfinder

SWITCH POSITION	CONTINUITY BETWEEN TERMINALS
Wiper switch, Off	13 and 14
Intermittent	13 and 14, 15 and 17
Low	14 and 17
High	16 and 17
Wash	17 and 18

7.3c Terminal identification and continuity chart for the front wiper portion of the combination switch

3 Using an ohmmeter, check for continuity between the indicated terminals of each switch with the switch in each of the indicated positions **(see illustrations)**.

4 If the continuity is not as specified, replace the defective switch.

Replacement

All except rear wiper washer switch on Xterras and 1998 and earlier Pathfinders

Refer to illustrations 7.7a and 7.7b

5 Disconnect the cable from the negative terminal of the battery.

6 Remove the knee bolster and steering column covers (see Chapter 11).

7 Remove the screws retaining the switch to the switch body and remove the switch **(see illustrations)**.

8 Installation is the reverse of removal.

Chapter 12 Chassis electrical system

4		3
5	2	1

SWITCH POSITION	CONTINUITY BETWEEN TERMINALS
Wiper switch, ON	1 and 3
Washer switch, depressed	2 and 3

72031-12-7.3d HAYNES

7.3d Terminal identification and continuity chart for the rear wiper switch - Xterra models

SWITCH POSITION	CONTINUITY BETWEEN TERMINALS
OFF	4 and 5
Intermittent	1 and 5, 8 and 9
ON	1 and 3, 1 and 5, 8 and 9
WASH	2 and 3, 4 and 5

72031-12-7.3e HAYNES

7.3e Terminal identification and continuity chart for the rear wiper switch - 1996 through 1998 Pathfinder models

21	22	23	24

SWITCH POSITION	CONTINUITY BETWEEN TERMINALS
Wiper off, WASH button depressed	23 and 24
Wiper on, WASH button depressed	22 and 23 and 24
Intermittent	21 and 24
ON	22 and 24

72031-12-7.3f HAYNES

7.3f Terminal identification and continuity chart for the rear wiper portion of the combination switch - 1999 and later Pathfinder models

Rear wiper/washer switch - Xterras and 1998 and earlier Pathfinders

9   Carefully pry the switch from the instrument panel.
10 Pull the switch out and unplug the electrical connector.
11 Installation is the reverse of removal.

8 Ignition switch and key lock cylinder - check and replacement

Warning: *The models covered by this manual are equipped with Supplemental Restraint Systems (SRS), more commonly known as airbags. Always disable the airbag system before working in the vicinity of any airbag system components to avoid the possibility of accidental deployment of the airbags, which could cause personal injury (see Section 27).*

Check

Refer to illustrations 8.4a and 8.4b
1 Disconnect the cable from the negative battery terminal.
2 Remove the knee bolster and steering column covers (see Chapter 11).
3 Follow the wiring harness down the column from the ignition switch and disconnect the ignition switch electrical connector.
4 Check for continuity between the indicated terminals of the connector with the key

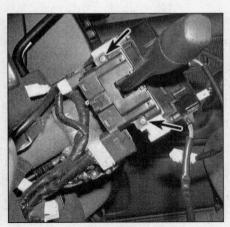

7.7a The turn signal/headlight switch is retained by these two screws (Frontier shown, others similar)

7.7b The wiper/washer switch is retained by these screws (Frontier shown, others similar)

	SWITCH POSITION	CONTINUITY BETWEEN TERMINALS
	OFF	none
	Acc	B and AC
B i1 ST	On	B, AC and i1
R AC i2	Start	B, i1, i2, ST and R

72031-12-8.a HAYNES

8.4a Ignition switch terminal identification and continuity chart - Frontier and Xterra models

Chapter 12 Chassis electrical system

e illustrations).
uity is not as specified,
k cylinder in each position
't worn or loose and that
the key position corresponds to the markings on the housing. If the lock cylinder is faulty, the entire steering column lock assembly will have to be replaced.

Replacement

Refer to illustration 8.8

7 Follow Steps 1 through 3 above to access the ignition switch.

8 The lock cylinder can't be replaced by itself - the whole housing must be replaced. Remove the shear-head bolts retaining the ignition switch/lock cylinder assembly and separate the bracket halves from the steering column. This can be accomplished by drilling a hole in the center of each bolt and unscrewing them with a screw extractor (see illustration).

9 Place the new switch assembly in position, install the new shear-head bolts and tighten them until the heads snap off. If the steering wheel was removed, be sure to center the airbag clockspring before installing the steering wheel (refer to Chapter 10, *Steering wheel - removal and installation*).

SWITCH POSITION	CONTINUITY BETWEEN TERMINALS
Off	none
Acc	1 and 2
On	1, 2 and 3
Start	1, 3, 4, and 6

Terminals:
```
3 5 1
4 2 6
```

72031-12-8.4b HAYNES

8.4b Ignition switch terminal identification and continuity chart - Pathfinder models

9 Instrument panel gauges - check

Note: *Refer to the Gauges and warning lights system wiring diagram at the end of this Chapter to familiarize yourself with the circuits before beginning this procedure.*

1 If the gauges and all the warning lights are inoperative, check the fuses. If the fuses are good, check the instrument panel power and ground circuits. Remove the instrument cluster and check for battery voltage and continuity to ground at the appropriate terminals of the instrument cluster connector. Refer to the wiring diagrams to determine the terminals for testing. For example; on Frontier and Xterra models, battery voltage should be present at the red/yellow wire terminal at all times and on the white/black wire terminal with the ignition key On. Continuity to a good chassis ground point should be indicated on the black and black/red wire terminals. If the power and ground circuits are good, the instrument cluster is probably defective.

2 If the temperature gauge does not respond or its accuracy is suspected, locate the temperature sending unit on the engine (see Chapter 3). Disconnect the connector from the sensor and connect the terminal in the harness connector to a good engine ground point with a jumper wire. Turn the ignition key On - the gauge pointer should move to the hot position. **Note:** *Turn the key Off right away; grounding the sending unit for too long could damage the gauge.* If the gauge responds correctly, the sending unit is probably defective. If the gauge does not respond, the wiring harness, gauge or instrument cluster is probably defective.

3 The same check can be performed on the fuel gauge. Locate the fuel gauge sending unit connector at the fuel tank (see Chapter 4), disconnect the connector and connect the fuel gauge terminal in the harness connector to a good engine ground point with a jumper wire. Turn the ignition key On, the gauge pointer should move to the Full position. **Note:** *Turn the key Off right away; grounding the sending unit for too long could damage the gauge.* If the gauge responds correctly, the sending unit is probably defective (refer to Chapter 4 for the fuel level sending unit check). If the gauge does not respond, the wiring harness, gauge or instrument cluster is probably defective.

4 If the engine oil pressure warning light comes on with the engine running, refer to Chapter 2D and check the engine oil pressure. If the oil pressure is good, disconnect the electrical connector from the oil pressure switch and connect the terminal in the harness connector to a good engine ground point. Turn the ignition key On - the warning light should illuminate. If it doesn't, the wiring harness or instrument cluster is defective or the bulb is burned out (refer to Section 16 for bulb replacement). If the bulb illuminates, check for continuity between the terminal on the oil pressure switch and the engine block; there should be continuity with the engine off and no continuity with the engine running. If continuity exists with the engine running, replace the oil pressure switch. Other warnings lights may be checked in a similar manner (refer to the wiring diagrams).

5 If the speedometer is inoperable, check the vehicle speed sensor (see Chapter 6).

10 Instrument cluster - removal and installation

Refer to illustrations 10.3a, 10.3b and 10.3c

Warning: *The models covered by this manual are equipped with Supplemental Restraint Systems (SRS), more commonly known as airbags. Always disable the airbag system before working in the vicinity of any airbag system components to avoid the possibility of accidental deployment of the airbags, which could cause personal injury* (see Section 27).

1 Disconnect the cable from the negative battery terminal.

2 Remove the instrument cluster bezel

8.8 To remove the ignition/lock cylinder assembly, drill a hole in the center of each bolt and unscrew them with a screw extractor

10.3a Instrument cluster mounting details - Frontier and Xterra models

Chapter 12 Chassis electrical system

10.3b Instrument cluster mounting details - Pathfinder models

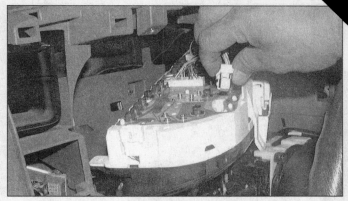

10.3c Pull the instrument cluster out and unplug the electrical connectors

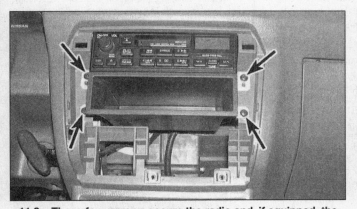

11.3a These four screws secure the radio and, if equipped, the CD player - Frontier model shown, Xterra similar

11.3b Radio/CD player mounting screws - Pathfinder

(see Chapter 11).
3   Remove the retaining screws and pull the cluster forward **(see illustrations)**.
4   Unplug the electrical connectors and remove the cluster from the vehicle.
5   Installation is the reverse of the removal procedure.

11   Radio and speakers - removal and installation

Warning: *The models covered by this manual are equipped with Supplemental Restraint Systems (SRS), more commonly known as airbags. Always disable the airbag system before working in the vicinity of any airbag system components to avoid the possibility of accidental deployment of the airbags, which could cause personal injury (see Section 27).*

Radio
Refer to illustrations 11.3a and 11.3b
1   Disconnect the cable from the negative battery terminal.
2   Remove the center bezel panel from the dash (see Chapter 11).
3   Remove the screws and pull the radio out of the dash **(see illustrations)**.
4   Disconnect the antenna lead and the electrical connectors and remove the radio.
5   Installation is the reverse of removal.

Speakers
Door speaker
Refer to illustration 11.7
6   Remove the door trim panel (see Chapter 11).
7   Remove the speaker retaining screws **(see illustration)**. Unplug the electrical connector and remove the speaker.
8   Installation is the reverse of removal.

Rear speaker (Xterra/Pathfinder)
9   Remove the rear quarter panel trim.
10 Remove the rear speaker mounting screws, unplug the electrical connector and remove the speaker.
11 Installation is the reverse of removal.

12   Antenna - removal and installation

Fixed antenna
Frontier/Xterra
Refer to illustrations 12.1 and 12.5
1   Remove the antenna mast and bezel **(see illustration)**.

11.7 Speaker mounting screws (front door)

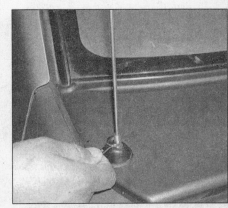

12.1 The fixed antenna mast can be unscrewed from its base with a wrench

Chapter 12 Chassis electrical system

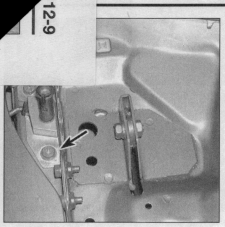

12.5 Remove the retaining screws (arrows), pull the antenna base out and disconnect it

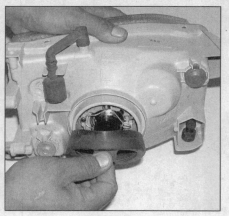

13.3 Remove the rubber boot from the back of the headlight housing (if equipped)

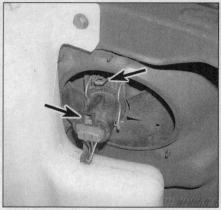

13.5a Unplug the electrical connector (A), detach the retaining clip (B) . . .

2 Remove the right-side windshield wiper arm and the right-side cowl grille.
3 Remove the right-side inner fender splash shield and detach the antenna cable from the fender.
4 Pull the radio out of the dash and disconnect the cable (see Section 11). Attach a pull wire to the end of the cable.
5 Remove the antenna base mounting screws **(see illustration)**. Remove the base and pull the cable through the body.
6 Detach the pull wire from the original cable and attach it to the new cable. Pull the cable through the body to the interior.
7 The remainder of installation is the reverse of removal.

Pathfinder

8 Remove the antenna mast and bezel.
9 Remove the right-side inner fender splash shield.
10 Pull the radio out of the dash and disconnect the cable (see Section 11). Attach a pull wire to the end of the cable.
11 Remove the antenna base mounting screws and pull the cable out.
12 Detach the pull wire from the original cable and attach it to the new cable. Pull the cable through the body to the interior.
13 The remainder of installation is the reverse of removal.

Power antenna

Antenna mast

Note: *Two people must perform this procedure.*

14 With the mast lowered, remove the antenna mast retaining nut and bezel.
15 Turn the ignition key and the radio to the On position. Withdraw the antenna mast and cable as it unwinds from the motor. Turn the radio Off.
16 Insert the antenna cable with the teeth facing the rear of the vehicle. Turn the radio On. Feed the antenna cable and mast into the tube as the motor winds the cable. When the mast is inside the tube, turn the radio Off.
17 Install the antenna mast retaining nut and bezel.

Antenna motor

18 Remove the antenna mast retaining nut and bezel.
19 Remove the right-side interior kick panel and disconnect the antenna motor electrical connector. Disconnect the cable from the junction block.
20 Remove the right-side inner fender splash shield and pull the motor harness and cable through the body grommet.
21 Remove the motor retaining screws and remove the antenna motor from the wheel well.
22 Installation is the reverse of removal.

13 Headlight bulb - replacement

Refer to illustrations 13.3, 13.5a and 13.5b
Warning: *These models are equipped with halogen gas-filled bulbs, which are under pressure and may shatter if the surface is scratched or the bulb is dropped. Wear eye protection, and handle the bulbs carefully, grasping only the base whenever possible. Do not touch the surface of the bulb with your fingers because the oil from your skin could cause it to overheat and fail prematurely. If you do touch the bulb surface, clean it with rubbing alcohol.*

1 If you're working on a Pathfinder model, remove the headlight housing (see Section 15).
2 If you're working on a Frontier or Xterra, unplug the electrical connector.
3 Remove the rubber boot (if equipped) **(see illustration)**.
4 On 1998 and earlier Pathfinder models, rotate the plastic retaining ring counterclockwise until loose, then withdraw the bulb from the housing.
5 On all other models detach the retaining clip and withdraw the bulb from the housing **(see illustrations)**.
6 Without touching the glass with your bare fingers, insert the new bulb assembly into the headlight housing and secure it with the retaining ring or clip.

13.5b . . . then pull out the headlight bulb (Frontier shown, Xterra similar)

7 If you're working on a Pathfinder, install the headlight housing.
8 Plug in the electrical connector and install the rubber boot (if equipped). Test headlight operation, then close the hood.

14 Headlights - adjustment

Refer to illustrations 14.1a, 14.1b and 14.2
Note: *It is important that the headlights are aimed correctly. If adjusted incorrectly they could blind the driver of an oncoming vehicle and cause a serious accident or seriously reduce your ability to see the road. The headlights should be checked for proper aim every 12 months and any time a new headlight is installed or front end body work is performed. It should be emphasized that the following procedure is only an interim step that will provide temporary adjustment until the headlights can be adjusted by a properly equipped shop.*

1 All models are equipped with two adjustment screws, one controlling left-and-right movement and one for up-and-down movement **(see illustrations)**.
2 There are several methods of adjusting the headlights. The simplest method requires

14.1a Headlight adjusting screw locations - Frontier (2000 and earlier) and Xterra models

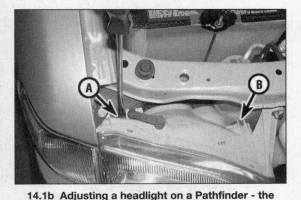

14.1b Adjusting a headlight on a Pathfinder - the adjuster closest to the fender (A) controls the up-and-down movement and the one closest to the radiator (B) controls side-to-side movement

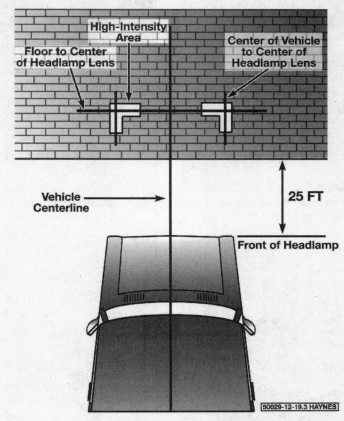

14.2 Headlight adjustment details

an open area with a blank wall and a level floor (see illustration).

3 Position masking tape vertically on the wall in reference to the vehicle centerline and the centerlines of both headlights.

4 Position a horizontal tape line in reference to the centerline of all the headlights. **Note:** *It may be easier to position the tape on the wall with the vehicle parked only a few inches away.*

5 Adjustment should be made with the vehicle parked 25 feet from the wall, sitting level, the gas tank half-full and no unusually heavy load in the vehicle.

6 Starting with the low beam adjustment, position the high intensity zone so it is two inches below the horizontal line and two inches to the side of the vertical headlight line away from oncoming traffic. Turn the adjustment screws until the desired level has been achieved.

7 With the high beams on, the high intensity zone should be vertically centered with the exact center just below the horizontal line. **Note:** *It may not be possible to position the headlight aim exactly for both high and low beams. If a compromise must be made, keep in mind that the low beams are the most used and have the greatest effect on driver safety.*

8 Have the headlights adjusted by a dealer service department at the earliest opportunity.

15 Headlight housing - replacement

Refer to illustration 15.4

1 Disconnect the cable from the negative battery terminal.

2 Remove the radiator grille/front fascia (see Chapter 11).

3 Remove the parking light housing (see Section 16).

4 Remove the retaining nuts and bolts, detach the headlight housing and withdraw it from the vehicle (see illustration).

5 Installation is the reverse of removal.

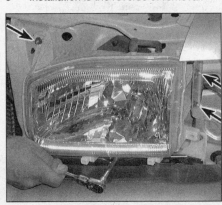

15.4 Headlight housing mounting details

16 Bulb replacement

Front side marker/turn signal lights

Refer to illustrations 16.1 and 16.2

1 Remove the screw and pull the turn signal and side marker light housing off the fender (see illustration).

2 To replace a turn signal bulb, twist the bulb socket to remove it from the housing, then push in on the bulb and turn it 1/8-turn counterclockwise to remove it (see illustra-

16.1 Remove this screw...

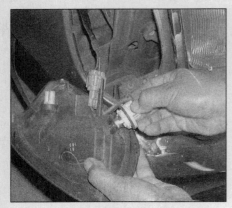

16.2 ... then rotate the light housing out for access to the bulbs

16.3 Taillight housing mounting bolts - Frontier models

16.4a Remove these two bolts ...

16.4b ... then pull the taillight housing straight back to disengage the ball socket retainers (Xterra/Pathfinder)

16.5 Twist the bulb holder and remove it from the housing; on later models the bulb pulls straight out of the holder

16.7 Carefully pry the license plate light out of the bumper (Xterra models)

16.8 License plate light mounting screws - Pathfinder

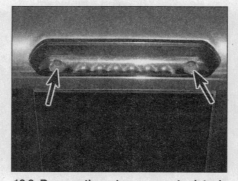

16.9 Remove these two screws to detach the high-mounted brake light lens for access to the bulbs (Frontier)

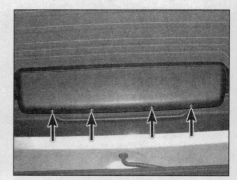

16.11a Remove these screws and detach the cover ...

tion). The marker light bulb holder is removed by twisting it in the housing, but once removed the bulb can be pulled straight out.

Tail and back-up lights

Refer to illustrations 16.3, 16.4a, 16.4b and 16.5

3   On Frontier models, open the tailgate. Remove the bolts and pull the taillight housing from the bed side (see illustration).

4   On Xterra and Pathfinder models, remove the bolts and pull the taillight housing from the body (see illustrations).

5   Remove the bulb socket from the housing and replace the bulb (see illustration).

License plate light

Refer to illustrations 16.7 and 16.8

6   On Frontier models, remove the screws and pull the housing from the bumper. Pull the bulb straight out to replace it.

7   On Xterra models, use a small screwdriver to unsnap the housing from the bumper (see illustration). Remove the bulb holder and replace the bulb.

8   On Pathfinder models, remove the screws and pull the housing from the liftgate (see illustration). Remove the bulb holder and replace the bulb.

High-mounted brake light

Refer to illustrations 16.9, 16.11a and 16.11b

9   On Frontier models, remove the screws and pull the lens from the roof (see illustration). Pull the bulb straight out of the socket to replace it.

10   On Xterra models, open the liftgate and use a small screwdriver to unsnap the cover from the trim panel. Remove the bulb

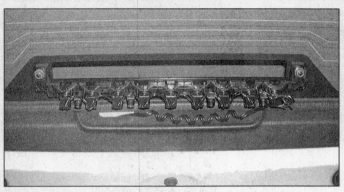

16.11b ... for access to the high-mounted brake light bulbs (Pathfinder)

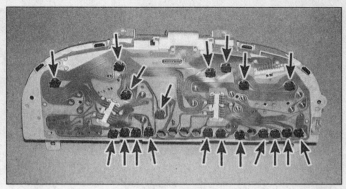

16.12 To replace a bulb on the back side of the instrument cluster, twist the bulb holder counterclockwise and remove it from the cluster; the bulb can then be pulled straight out of the holder

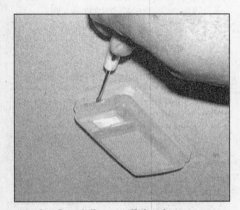

16.13a Carefully pry off the dome or map light lens using the notch provided

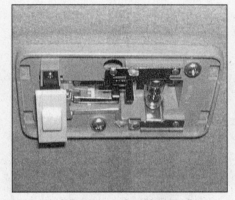

16.13b When removing "festoon" type bulbs, pry on their ends only (otherwise the glass may break)

16.14 On 2001 Pathfinders and Frontiers, the fog light housing is secured by two bolts

holder(s) and replace the bulb(s).
11 On Pathfinder models, remove the screws and pull the housing from the liftgate **(see illustrations)**. Remove the bulb holder(s) and replace the bulb(s).

Instrument cluster illumination
Refer to illustration 16.12
12 To gain access to the instrument cluster illumination lights, the instrument cluster will have to be removed (see Section 10). The bulbs can then be removed and replaced from the rear of the cluster **(see illustration)**.

Interior lights
Refer to illustrations 16.13a and 16.13b
13 Remove the lenses for the map lights or dome light by prying the cover off with a small screwdriver. Replace the bulb **(see illustrations)**.

Fog lights

Pathfinder and Frontier - 2001 models
Refer to illustrations 16.14 and 16.16
14 Remove the small grille from the lower part of the bumper, then remove the fog light mounting bolts **(see illustration)**.
15 Pull the light out and unplug the electrical connector.
16 Disengage the retaining clip, then remove the bulb from the fog light housing **(see illustration)**.
17 Installation is the reverse of removal.
Caution: *Do not touch the surface of the new bulb with your fingers because the oil from your skin could cause it to overheat and fail prematurely. If you do touch the bulb surface, clean it with rubbing alcohol.*

Pathfinder - 1999 and 2000 models
18 Carefully pry the fog light out of the front bumper fascia.
19 Remove the bulb holder from the light housing, then remove the bulb from the holder.
20 Installation is the reverse of removal.

Pathfinder - 1996 through 1998 models
21 Remove the mounting fastener, pull the light housing out and disconnect the electrical connector.
22 Remove the bulb holder from the light housing, then remove the bulb from the holder.
23 Installation is the reverse of removal.

All other models
Refer to illustration 16.27
24 Detach the clips from the sides of the fog light lens, then pull out the lens and disconnect the electrical connector.
25 Disengage the retaining spring and detach the bulb from the lens. The bulb can't be separated from the wire - it is replaced as an assembly.
26 Installation is the reverse of removal.
Caution: *Do not touch the surface of the new bulb with your fingers because the oil from your skin could cause it to overheat and fail prematurely. If you do touch the bulb surface, clean it with rubbing alcohol.*

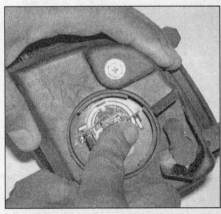

16.16 Release the clip and remove the fog light bulb from the housing

12-14   Chapter 12 Chassis electrical system

16.27 On Frontiers and Xterras, the fog light is attached to its bracket with one nut

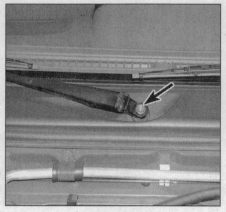

17.7 Use a small screwdriver to pry off the wiper arm nut cover, then remove the nut and pull the arm straight off its splined shaft

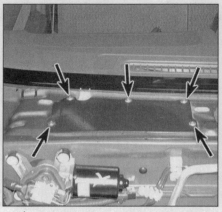

17.8 Remove these screws and take off the wiper linkage access plate

27   If it is necessary to replace the fog light assembly, remove the nut securing the fog light to its retaining bracket **(see illustration)**.

17 Wiper motor and amplifier - check and replacement

Check

Note: *Refer to the wiring diagrams for wire colors and locations in the following checks. When checking for voltage, probe a grounded 12-volt test light to each terminal at a connector until it lights; this verifies voltage (power) at the terminal. If the following checks fail to locate the problem, have the system diagnosed by a dealer service department or other properly equipped repair facility.*

1   If the wipers work slowly, make sure the battery is in good condition and has a strong charge (see Chapter 1). If the battery is in good condition, remove the wiper motor (see below) and operate the wiper arms by hand. Check for binding linkage and pivots. Lubricate or repair the linkage or pivots as necessary. Reinstall the wiper motor. If the wipers still operate slowly, check for loose or corroded connections, especially the ground connection. If all connections look OK, replace the motor.
2   If the wipers fail to operate when activated, check the fuse. If the fuse is OK, connect a jumper wire between the wiper motor and ground, then retest. If the motor works now, repair the ground connection. If the motor still doesn't work, turn the wiper switch to the HI position and check for voltage at the motor. **Note:** *The cowl cover will have to be removed (see Chapter 11). If there's voltage at the connector, remove the motor and check it off the vehicle with fused jumper wires from the battery. If the motor now works, check for binding linkage (see Step 1). If the motor still doesn't work, replace it. If there's no voltage to the motor, check the switch for continuity (see Section 7). If the switch is good, the wiper amplifier might be*

faulty (models with intermittent wipers only).
3   If the interval (delay) function is inoperative, check the continuity of all the wiring between the switch and wiper amplifier. If the wiring is OK, check the resistance of the delay control knob of the multi-function switch (see Section 7).
4   If the wipers stop at the position they're in when the switch is turned off (fail to park), check for a good ground on the connector side at the motor. With an ohmmeter connected between any of the black wire terminals and a known ground, resistance should be less than 5 ohms.
5   If the wipers won't shut off unless the ignition is OFF, disconnect the wiring from the wiper control switch. If the wipers stop, replace the switch. If the wipers keep running, there's a defective limit switch in the motor; replace the motor.
6   If the wipers won't retract below the hood line, check for mechanical obstructions in the wiper linkage or on the vehicle's body that would prevent the wipers from parking. If there are no obstructions, check the wiring between the switch and motor for continuity. If the wiring is OK, replace the wiper motor.

Wiper motor replacement
Front
Refer to illustrations 17.7, 17.8, 17.9 and 17.10

7   Disconnect the cable from the negative terminal of the battery, then remove the right side windshield wiper arm **(see illustration)**.
8   Remove the cowl cover (see Chapter 11) and the wiper linkage access plate **(see illustration)**.
9   Detach the wiper linkage from the wiper motor crank arm **(see illustration)**.
10   Unplug the electrical connector, remove the wiper motor retaining bolts and remove the motor **(see illustration)**.
11   Installation is the reverse of removal.

Rear (Xterra/Pathfinder)
Refer to illustrations 17.14, 17.16a and 17.16b

12   Disconnect the cable from the negative terminal of the battery.
13   Open the liftgate and remove the liftgate trim panel (see Chapter 11).
14   If you're working on a Pathfinder, remove the nut and detach the wiper linkage from the motor crank arm **(see illustration)**.
15   If you're working on an Xterra, remove

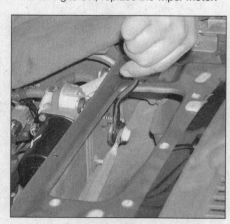

17.9 Pry the linkage rod socket off the motor crank arm

17.10 Unplug the electrical connector, unscrew the bolts and remove the windshield wiper motor

Chapter 12 Chassis electrical system

12-15

17.14 Unscrew this nut and detach the wiper linkage from the motor shaft (Pathfinder)

17.16a Remove the rear wiper motor mounting screws . . .

17.16b . . . and maneuver the motor out of the liftgate (Pathfinder)

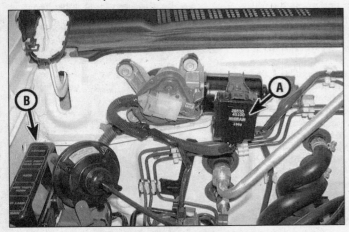

17.17 On Frontiers and Xterras the windshield wiper motor amplifier is mounted on the wiper motor (A); on 2000 and earlier Pathfinders it's located in the vicinity of arrow (B)

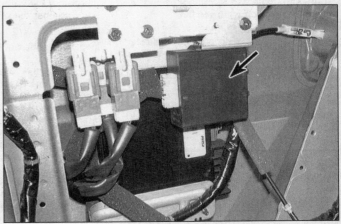

17.21 The rear wiper motor amplifier is located behind the right rear quarter panel trim

the wiper arm, nut cap, nut and washers.
16   Remove the motor mounting screws, detach the motor from the liftgate and unplug the electrical connector **(see illustrations)**.

Amplifier replacement
Front
Refer to illustration 17.17

Note: *On 2001 Pathfinder models, the windshield wiper amplifier is integral with the windshield wiper/washer switch.*
17   Locate the wiper amplifier. On Frontier and Xterra models it's mounted on the windshield wiper motor **(see illustration)**; on 2000 and earlier Pathfinders it's mounted to the right inner fender panel, in the right rear corner of the engine compartment.
18   Unplug the electrical connector, remove the mounting fastener and detach the amplifier.
19   Installation is the reverse of removal.

Rear (Pathfinder)
Refer to illustration 17.21
20   Remove the right rear quarter panel trim.
21   Unplug the electrical connector, unscrew the mounting bolt and remove the amplifier **(see illustration)**.
22   Installation is the reverse of removal.

18 Horn - check and replacement

Refer to illustration 18.3
Note: *Check the fuses before beginning electrical diagnosis.*
1   Unplug the electrical connector from the horn.
2   To test the horn, connect battery voltage to the terminal with a jumper wire. If the horn doesn't sound, replace it. If it does sound, the problem lies in the switch, relay or the wiring between the components.
3   To replace the horn, unplug the electrical connector and remove the bracket bolt **(see illustration)**.
4   Installation is the reverse of removal.

19 Daytime Running Lights (DRL) - general information

The Daytime Running Lights (DRL) sys-

tem used on Canadian models turns the headlights on whenever the engine is started. The only exception is when the engine is turned on when the parking brake is

18.3 Unplug the electrical connector and remove the bolt (arrow), then detach the horn

12

20.2 Using an ohmmeter, check the continuity of the rear window defogger switch; there should be continuity between these two terminals when the switch is turned On

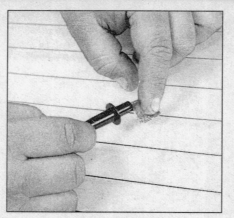

20.5 When measuring the voltage at the rear window defogger grid, wrap a piece of aluminum foil around the negative probe of the voltmeter and press the foil against the wire with your finger

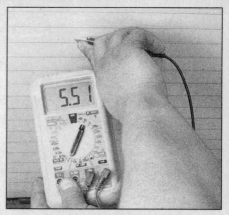

20.6 To determine if a heating element has broken, check the voltage at the center of each element - if the voltage is 6-volts, the element is unbroken

engaged. Once the parking brake is released, the lights will remain on as long as the ignition switch is on, even if the parking brake is later applied.

The DRL system supplies reduced power to the headlights so they won't be too bright for daytime use while prolonging headlight life.

20 Rear window defogger - check and repair

1 The rear window defogger consists of a number of horizontal heating elements baked onto the inside surface of the glass. Power is supplied through a large fuse from the power distribution box in the engine compartment. The heater is controlled by the instrument panel switch. Small breaks in the element can be repaired without removing the rear window.

Switch check

Refer to illustration 20.2

2 Remove the center bezel panel from the dash (see Chapter 11). Test the switch for continuity (see illustration).

Defogger grid check

Refer to illustrations 20.5 and 20.6

3 Turn the ignition switch and defogger switch to the ON position.
4 Using a voltmeter, place the positive probe against the defogger grid positive terminal and the negative probe against the ground terminal. If battery voltage is not indicated, check the fuse, defogger switch and related wiring. If voltage is indicated, but all or part of the defogger doesn't heat, proceed with the following tests.
5 When measuring voltage during the next two tests, wrap a piece of aluminum foil around the tip of the voltmeter positive probe and press the foil against the heating element with your finger (see illustration). Place the

negative probe on the defogger grid ground terminal.
6 Check the voltage at the center of each heating element (see illustration). If the voltage is 5 to 6 volts, the element is okay (there is no break). If the voltage is 0 volts, the element is broken between the center of the element and the positive end. If the voltage is 10 to 12 volts the element is broken between the center of the element and the ground side. Check each heating element.
7 If none of the elements are broken, connect the negative probe to a good chassis ground. The voltage reading should stay the same, if it doesn't the ground connection is bad.
8 To find the break, place the voltmeter negative probe against the defogger ground terminal. Place the voltmeter positive probe with the foil strip against the heating element at the positive side and slide it toward the negative side. The point at which the voltmeter deflects from several volts to zero is the point where the heating element is broken.

Repair

Refer to illustration 20.14

9 Repair the break in the element using a repair kit specifically for this purpose. The kit includes conductive plastic epoxy.
10 Before repairing a break, turn off the system and allow it to cool for a few minutes.
11 Lightly buff the element area with fine steel wool; then clean it thoroughly with rubbing alcohol.
12 Use masking tape to mask off the area being repaired.
13 Thoroughly mix the epoxy, following the kit instructions.
14 Apply the epoxy material to the slit in the masking tape, overlapping the undamaged area about 3/4-inch on either end (see illustration).
15 Allow the repair to cure for 24 hours before removing the tape and using the system.

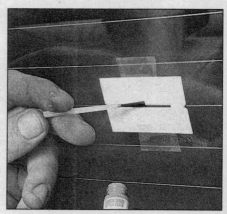

20.14 To use a defogger repair kit, apply masking tape to the inside of the window at the damaged area, then brush on the special conductive coating

21 Cruise control system - description, check and component replacement

Refer to illustration 21.5
Warning: *The models covered by this manual are equipped with Supplemental Restraint Systems (SRS), more commonly known as airbags. Always disable the airbag system before working in the vicinity of any airbag system components to avoid the possibility of accidental deployment of the airbags, which could cause personal injury (see Section 27).*
1 The cruise control system maintains vehicle speed with a vacuum-actuated servo motor, which is connected to the throttle linkage by a cable. The system consists of the servo motor, brake switch, vacuum pump, control switches, a relay and associated vacuum hoses. Some features of the system require special testers and diagnostic procedures that are beyond the scope of the home mechanic. Listed below are some general procedures that may be used to locate common problems.

Chapter 12 Chassis electrical system

21.5 The cruise control servo (A) is located on the right side of the engine compartment - make sure the cruise control cable (B) is not damaged and that it operates smoothly

2 Check the fuse (see Section 3).
3 The shift lock/cruise control brake switch (not the brake light switch) deactivates the cruise control system. Refer to Chapter 7B for the switch checking procedure. On manual transmission models, the clutch switch (at the top of the clutch pedal, not the clutch start switch) also deactivates the system. This switch must have continuity when the pedal is at rest, and no continuity as soon as the pedal is depressed.
4 Adjust or replace the switch, as necessary, if it isn't working properly.
5 Check the control cable between the cruise control servo/amplifier and the throttle linkage and adjust/replace as necessary **(see illustration)**. See Chapter 4 for the cable adjustment procedure, which is the same as for the accelerator cable.
6 The cruise control system uses a speed sensing device. The speed sensor is located in the transmission. To test the speed sensor, see Chapter 6.

Cruise control switches

Note: Make sure the horn works before testing the cruise control system, as the cruise control switch gets power from the horn relay.

Check

Refer to illustrations 21.8a and 21.8b

7 The cruise control switch pod is located on the right side of the steering wheel. On all except 2001 Pathfinders the cruise control main switch (on/off) is located on the instrument panel. Refer to the replacement procedure below to remove the cruise control switches for testing.
8 Using an ohmmeter, check for continuity between the indicated terminals with the various switches in each of the indicated positions **(see illustrations)**.
9 If the continuity is not as specified, replace the defective switch.

SWITCH POSITION	CONTINUITY BETWEEN TERMINALS
Off	none
Neutral	2 and 3, 2 and 4
On	1 and 3, 2 and 3, 2 and 4

21.8a Terminal identification and continuity chart for the instrument panel-mounted cruise control switch

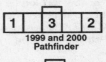

SWITCH POSITION	CONTINUITY BETWEEN TERMINALS
Resume / Accel	1 and 3
Set / Coast	1 and 2
Cancel	1 and 3*, 1 and 2* (with diodes)
Main switch, ON (2001 Pathfinder)	4 and 5

* Diodes are directional, test both directions

21.8b Terminal identification and continuity chart for the steering wheel-mounted cruise control switches

Replacement

Steering wheel-mounted switch

Refer to illustrations 21.11 and 21.12

10 Disconnect the cable from the negative terminal of the battery.
11 Remove the cruise control switch pod cover **(see illustration)**.
12 Remove the switch mounting screws, remove the switch and unplug the electrical connector **(see illustration)**.
13 Installation is the reverse of removal.

Instrument panel-mounted switch

14 Carefully pry the switch carrier out of the instrument panel.
15 Unplug the electrical connector from the switch, then squeeze the tabs on the side of the switch and push it out of the carrier.
16 Installation is the reverse of removal.

Automatic Speed Control Device (ASCD) control unit

Refer to illustration 21.18

17 Remove the left side under-dash panel

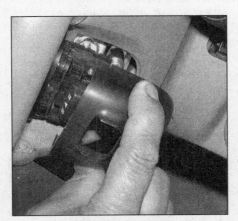

21.11 Grasp the switch cover and pull it off . . .

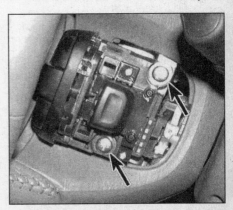

21.12 . . . then remove the switch screws, detach the switch and unplug the electrical connector

21.18 The control unit for the cruise control system is located under the left side of the instrument panel (Pathfinder shown)

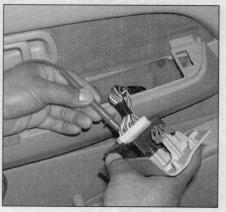

22.12a Using a voltmeter or a test light, check for power at the window switch by backprobing while operating the switch

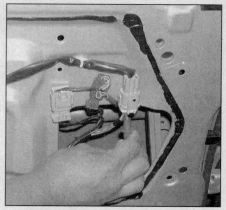

22.12b If there is power at the switch, check for power getting to the window motor by probing at the motor connector (arrow)

and, if you're working on a Pathfinder, the kick panel (see Chapter 11).
18 Unscrew the mounting fasteners, unplug the electrical connector and remove the control unit (see illustration).
19 Installation is the reverse of removal.

22 Power window system - description and check

Refer to illustrations 22.12a and 22.12b

1 The power window system operates the electric motors mounted in the doors which lower and raise the windows. The system consists of the control switches, the motors (regulators), glass mechanisms and associated wiring.
2 Power windows are wired so they can be lowered and raised from the master control switch by the driver or by remote switches located at the individual windows. Each window has a separate motor that is reversible. The position of the control switch determines the polarity and therefore the direction of operation. Some systems are equipped with relays that control current flow to the motors.
3 Some vehicles are equipped with a separate circuit breaker for each motor in addition to the fuse or circuit breaker protecting the whole circuit. This prevents one stuck window from disabling the whole system.
4 The power window system will only operate when the ignition switch is ON. In addition, many models have a window lockout switch at the master control switch which, when activated, disables the switches at the rear windows and, sometimes, the switch at the passenger's window also. Always check these items before troubleshooting a window problem.
5 These procedures are general in nature, so if you can't find the problem using them, take the vehicle to a dealer service department, automotive electrical specialist or other qualified repair shop.

6 If the power windows don't work at all, check the fuse or circuit breaker.
7 If only the rear windows are inoperative, or if the windows only operate from the master control switch, check the rear window lockout switch for continuity in the unlocked position. Replace it if it doesn't have continuity.
8 Check the wiring between the switches and fuse panel for continuity. Repair the wiring, if necessary.
9 If only one window is inoperative from the master control switch, try the other control switch at the window. Note: *This doesn't apply to the drivers door window.*
10 If the same window works from one switch, but not the other, check the switch for continuity.
11 If the switch tests OK, check for a short or open in the wiring between the affected switch and the window motor.
12 If one window is inoperative from both switches, check for voltage at the switch and at the motor while the switch is operated (see illustrations).
13 If voltage is reaching the motor, disconnect the glass from the regulator (see Chapter 11). Move the window up and down by hand while checking for binding and damage. Also check for binding and damage to the regulator. If the regulator is not damaged and the window moves up and down smoothly, replace the motor. If there's binding or damage, lubricate, repair or replace parts, as necessary.
14 If voltage isn't reaching the motor, check the wiring in the circuit for continuity between the switches and motors. You'll need to consult the wiring diagram for the vehicle. Some power window circuits are equipped with relays. If equipped, check that the relays are grounded properly and receiving voltage from the switches. Also check that each relay sends voltage to the motor when the switch is turned on. If it doesn't, replace the relay.
15 Test the windows after you are done to confirm proper repairs.

23 Power door lock system - description and check

Refer to illustration 23.5

1 The power door lock system operates the door lock actuators mounted in each door. The system consists of the door lock control unit, door switches, actuators and associated wiring (see illustration). Diagnosis can usually be limited to simple checks of the wiring connections, switches and actuators for minor faults which can be easily repaired.
2 The doors are locked by bi-directional solenoids (actuators) located in the doors. The lock switches have two operating positions: Lock and Unlock. These switches signal the door lock control unit which in turn connects voltage to the door lock actuators. Depending on which way the switch is activated, the control unit reverses polarity to the actuators, allowing the two sides of the circuit to be used alternately as the feed (positive) and ground side, locking or unlocking the doors.
3 Always check the circuit protection first. Some vehicles use a combination of circuit breakers and fuses.
4 Operate the door lock switches in both directions (Lock and Unlock) with the engine off. Listen for the sound of the actuators operating. If the system is completely inoperable (none of the actuators operate), the door lock control unit is defective or a serious wiring problem exists in the system wiring harness. Have the control unit diagnosed by a dealer service department or other qualified repair shop.
5 If one or more (but not all) the actuators are inoperable, remove the door trim panel from the inoperable door (see Chapter 11) and disconnect the electrical connector from the actuator. Using a voltmeter or test light, check for power at the connector while operating the door lock switch (see illustration). One of the terminals should have power in the Lock position; the other should have

23.5 Check for power at the door lock actuator connector (arrow) with the switch depressed - check the door lock actuator itself by disconnecting the connector and using jumper wires to temporarily apply battery voltage and ground directly to the actuator terminals

power in the unlock position. If power is present and the door lock actuator does not operate when connected, replace the door lock actuator.

6 If power is not present at the door lock actuator, test the switch for continuity. Replace the switch if there's no continuity in either switch position.

7 If the switch tests good but the actuator does not receive power, check the wiring for continuity between the control unit and the switch and actuator. Repair the wiring if there's no continuity.

8 If the switches, actuators and wiring are good, have the control unit diagnosed by a dealer service department or other qualified repair shop.

24 Electric side view mirrors - description and check

Refer to illustrations 24.6 and 24.7

1 Most electric side view mirrors use two motors to move the glass; one for up and down adjustments and one for left-right adjustments.

2 The control switch has a selector portion that sends voltage to the left or right side mirror. With the ignition ON but the engine OFF, roll down the windows and operate the mirror control switch through all functions (left-right and up-down) for both the left and right side mirrors.

3 Listen carefully for the sound of the electric motors running in the mirrors.

4 If the motors can be heard but the mirror glass doesn't move, there's probably a problem with the drive mechanism inside the mirror. Remove and disassemble the mirror to locate the problem.

5 If the mirrors don't operate and no sound comes from the mirrors, check the fuse.

6 If the fuse is OK, remove the mirror control switch from its mounting without disconnecting the wires attached to it **(see illustration)**. Turn the ignition ON and check for voltage at the switch. There should be voltage at one terminal. If there's no voltage at the switch, check for an open or short in the wiring between the fuse panel and the switch.

7 If there's voltage at the switch, disconnect it. Check the switch for continuity in all its operating positions **(see illustration)**. If the switch does not have continuity, replace it.

8 Re-connect the switch. Locate the wire going from the switch to ground. Leaving the switch connected, connect a jumper wire between this wire and ground. If the mirror works normally with this wire in place, repair the faulty ground connection.

9 If the mirror still doesn't work, remove the mirror and check the wires at the mirror for voltage. Check with ignition ON and the mirror selector switch on the appropriate side. Operate the mirror switch in all its positions. There should be voltage at one of the switch-to-mirror wires in each switch position (except the neutral "off" position).

10 If voltage isn't present in each switch position, check the wiring between the mirror and control switch for opens and shorts.

11 If there's voltage, remove the mirror and test it off the vehicle with jumper wires. Replace the mirror if it fails this test.

25 Power seats - description and check

1 The optional power seats on these models adjust forward and backward, up and down and tilt forward and backward.

2 The power seat system consists of a motor, a switch on the seat, a circuit breaker and the 40-amp F fuse in the engine compartment fuse/fusible link block.

3 Look under the seat for any objects which may be preventing the seat from moving.

4 If the seat won't work at all, check the fuse (see Section 3).

5 With the engine off to reduce the noise level, operate the seat controls in all directions and listen for sound coming from the seat motor(s).

6 If the motor runs or clicks but the seat doesn't move, the integral seat drive mechanism is damaged and the motor assembly must be replaced.

7 If the motor doesn't work or make noise, check for voltage at the motor while an assis-

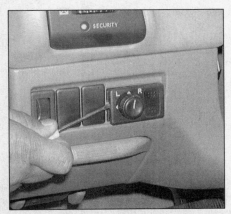

24.6 To remove the power mirror switch, carefully pry it out of the dash

2	3	9			1	8
5	7	4	12	11	6	10

Frontier / Xterra

2	3			1	9
5	7	4		6	8

1996 to 1999 Pathfinder

7	6			5	4
9	3	10	2	8	1

2000 and later Pathfinder

SWITCH POSITION	CONTINUITY BETWEEN TERMINALS
Mirror switch, OFF	1 and 2, 6 and 3
Left mirror, move RIGHT	2 and 3, 1 and 6
Left mirror, move LEFT	2 and 3, 1 and 7
Left mirror, move DOWN	1 and 2, 3 and 6
Right mirror, move RIGHT	1 and 2, 3 and 4
Right mirror, move LEFT	2 and 3, 1 and 4
Right mirror, move UP	2 and 3, 1 and 5
Right mirror, move DOWN	1 and 2, 3 and 5

24.7 Power mirror switch terminal identification and continuity chart

tant operates the switch.

8 If the motor is getting voltage but doesn't run, test it off the vehicle with jumper wires. If it still doesn't work, replace it.

9 If the motor isn't getting voltage, check for voltage at the switch. If there's no voltage at the switch, check the wiring between the fuse panel and the switch. If there's voltage at the switch, check the switch for continuity in all its operating positions. Replace the switch if there's no continuity.

10 If the switch is OK, check for a short or open in the wiring between the switch and motor.

11 Test the completed repairs.

26 Electric sunroof - description and check

1 The electric sunroof is powered by a single motor located in the roof behind the overhead console. The power circuit is protected by a circuit breaker. When sunlight isn't desired, an interior sliding panel can be closed.

2 The control switches (tilt and slide) send a ground signal to the sunroof motor when the switches are pressed. Power is supplied to the motor from the sunroof relay. With the ignition On but the engine Off, operate the sunroof control switch through the tilt and slide functions.

3 Listen carefully for the sound of the sunroof motor running in the roof.

4 If the motors can be heard but the sunroof glass doesn't move, there's probably a problem with the drive mechanism or drive cables.

5 If the sunroof does not operate and no sound comes from the motor, check the fuses (in the interior fuse panel *and* in the engine compartment fuse/fusible link box).

6 If the fuses are OK, pull down the overhead interior light/switch panel. Turn the ignition On and check for voltage to the motor. If there's voltage at the motor, check for power and ground at the switch. If power and ground exist at the motor and there's still no voltage at the switch replace the switch. If there's no voltage at the motor, check the power window relay or look for an open or short in the wiring.

7 If there's voltage at the switch, disconnect it. Check the switch for continuity in all its operating positions. If the switch does not have continuity, replace it.

8 If the switch has continuity re-connect the switch. Locate the wire going from the switch to ground. Leaving the switch connected, connect a jumper wire between this wire and ground. If the motor works normally with this wire in place, repair the faulty ground connection.

9 The sunroof can be closed manually by inserting a wrench into the motor shaft and rotating it clockwise. If your vehicle is equipped with a factory sunroof, the wrench comes in the factory toolbag in the trunk.

27 Airbag system - general information

1 These models are equipped with a Supplemental Restraint System (SRS), more commonly known as airbags, designed to protect the driver and front seat passenger from serious injury in the event of a head-on or frontal collision. Additionally, some later model Pathfinders are equipped with side-impact airbags, located in the outside rear corners of the driver and passenger front seats; these airbags are designed to operate primarily in side-impact collisions, though they may also activate in some other types of collisions.

2 All models have a diagnosis/sensor unit located inside the passenger compartment. 4WD Frontier and Xterra models also have a crash zone sensor mounted on the front of the upper radiator support.

Airbag modules

3 The airbag modules consist of a housing incorporating the cushion (airbag) and inflator unit. The inflator assembly is mounted on the back of the housing over a hole through which gas is expelled, inflating the bag almost instantaneously when an electrical signal is sent from the system. The specially wound wire on the driver's side that carries this signal to the module is called a "clockspring." The clockspring is a flat, ribbon-like electrically conductive tape that is wound many times so that it can transmit an electrical signal regardless of steering wheel position.

Diagnosis/sensor unit

Refer to illustration 27.4

4 The diagnosis/sensor unit contains an on-board microprocessor which monitors the operation of the system, and also contains a crash sensor. It checks this system every time the vehicle is started, causing the "AIRBAG" light to go on then off, if the system is operating properly. If there is a fault in the system, the light will go on and stay on and the unit will store fault codes indicating the nature of the fault. If the AIRBAG light goes on and stays on, the vehicle should be taken to your dealer immediately for service. The diagnosis/sensor unit is located under the center console **(see illustration)**. Models with side airbags also have a "satellite" crash sensor in each door "B" pillar.

Operation

5 For the airbags to deploy, the diagnosis/sensor unit must detect a deceleration force great enough to warrant deployment, up to a 30-degree angle from the centerline of the vehicle. On 4WD models, the crash zone sensor must also close its circuit. When this condition occurs, the circuit to the airbag inflator is closed and the airbag inflates. If the battery is destroyed by the impact, or is too low to power the inflator, a capacitor inside the diagnosis/sensor unit provides power.

27.4 The diagnosis/sensor unit is located under the rear of the floor console - if it is removed for any procedure, new factory bolts must be used to install it (they are specially coated; regular bolts must not be used)

27.11 To disable the passenger's side airbag, Remove the left lower trim panel/glove box and disconnect this electrical connector

Self-diagnosis system

6 A self-diagnosis circuit in the control unit displays a light on the instrument panel when the ignition switch is turned to the On position. If the system is operating normally, the light should go out after about seven seconds. If the light doesn't come on, or doesn't go out after seven seconds, or if it comes on while you're driving the vehicle, or if it blinks at any time, there's a malfunction in the SRS system. Have it inspected and repaired as soon as possible. Do not attempt to troubleshoot or service the SRS system yourself. Even a small mistake could cause the SRS system to malfunction when you need it. Servicing components near the SRS system

7 Nevertheless, there are times when you need to remove the steering wheel, radio or service other components on or near the instrument panel or front seats. At these times, you'll be working around components and wire harnesses for the SRS system. **ALWAYS DISABLE THE SRS SYSTEM BEFORE WORKING NEAR AIRBAG SYSTEM COMPONENTS OR RELATED WIRING.** **Warning:** *Never use electrical test equipment on any of the airbag system wiring or electrical connectors; it could cause the airbag(s) to deploy.*

Disabling the SRS system

Warning: *Any time you are working in the vicinity of airbag wiring or components, DISABLE THE SRS SYSTEM.*

8 Disconnect the battery negative and positive cables, then wait ten minutes before proceeding with any work.

Driver's side airbag

9 Remove the access panel in the steering wheel below the airbag and unplug the two-pin connector between the airbag and the clockspring (see Chapter 10, *Steering wheel - removal and installation*).

Passenger's side airbag

Refer to illustration 27.11

10 Remove the lower trim panel/glove box from the passenger's side of the instrument panel.

11 Disconnect the two-pin electrical connector between the passenger side airbag and the SRS main wiring harness **(see illustration)**.

Side-impact airbags

12 Using a trim tool, carefully pry around the perimeter of the finish panel on the back of each front seat. **Warning:** *Work only from the side of the seat opposite the airbag, and be carefully when prying near the airbag so you don't contact the airbag harness.*

13 With the seat back cover removed, disconnect the airbag electrical connector by sliding the locking tab then pulling the connector apart.

Enabling the system

14 After you've disabled the airbag and performed the necessary service, reconnect the two-pin airbag connector into the two-pin clockspring connector (driver's side), the SRS main harness (passenger's side) or the side-impact airbag. Reinstall the lid to the underside of the steering wheel or reinstall the glove box/trim panel or seat back panel.

15 Turn the ignition switch to the Off position.

16 Reattach the battery cables (see Chapter 1).

Removal and installation

Warning: *The bolts used throughout the airbag system to mount the airbag modules and diagnosis/sensor unit (and crash zone sensor on 4WD Frontier and Xterra models) are special tamper-proof Torx bolts, which have a special coating. These bolts are designed to be used once. Replace them with new factory bolts, and never use a substitute fastener.*

Driver's side airbag and clockspring

17 Refer to Chapter 10, *Steering wheel - removal and installation*, for removal and installation of the driver's side airbag and clockspring. **Warning:** *When installing the clockspring, be sure to follow the centering instructions carefully.*

Passenger side airbag

18 Disable the airbag system (beginning with Step 8).

19 Refer to Chapter 11 and remove the passenger-side lower dash panel and the glove box.

20 Disconnect the two-pin connector. Remove the special Torx tamper-proof bolts and remove unit from the top of the instrument panel. **Caution:** *The airbag assembly is heavier than it looks - use both hands when removing it from the dash.*

21 Installation is the reverse of the removal procedure. Tighten the bolts to 11 to 18 ft-lbs (15 to 25 Nm).

Side-impact airbags

22 Disable the airbag system (beginning with Step 8).

23 Refer to Steps 12 and 13 and disconnect the airbag electrical connector.

24 Remove the two special Torx nuts and remove the airbag.

25 Installation is the reverse of the removal procedure. Tighten the nuts to 54 to 68 in-lbs (6.1 to 7.7 Nm).

28 Wiring diagrams - general information

Since it isn't possible to include all wiring diagrams for every year and model covered by this manual, the following diagrams are those that are typical and most commonly needed.

Prior to troubleshooting any circuits, check the fuses and circuit breakers (if equipped) to make sure they are in good condition. Make sure the battery is properly charged and has clean, tight cable connections (see Chapter 1).

When checking the wiring system, make sure that all electrical connectors are clean, with no broken or loose pins. When unplugging an electrical connector, do not pull on the wires, only on the connector housings themselves.

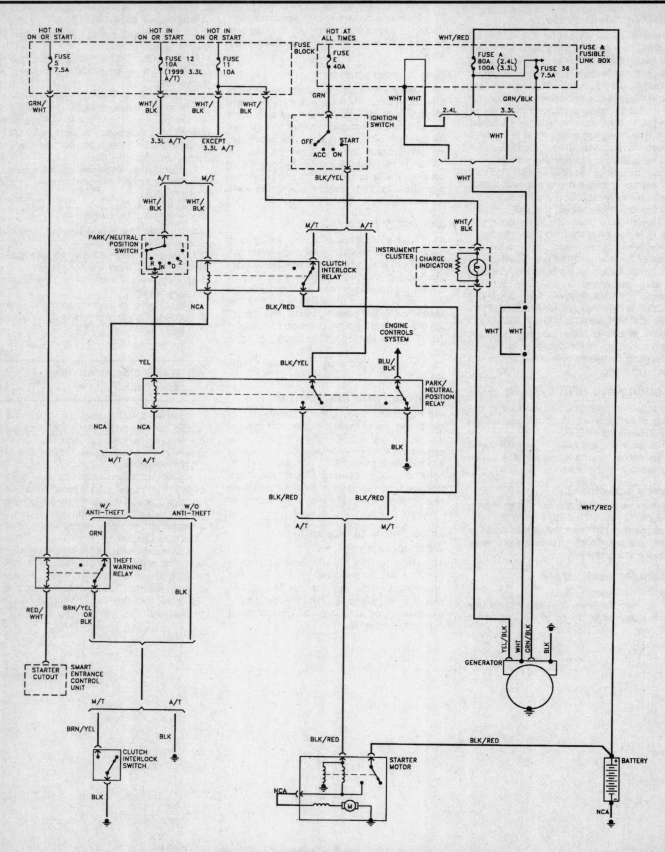

Starting and charging system - Frontier/Xterra

Chapter 12 Chassis electrical system

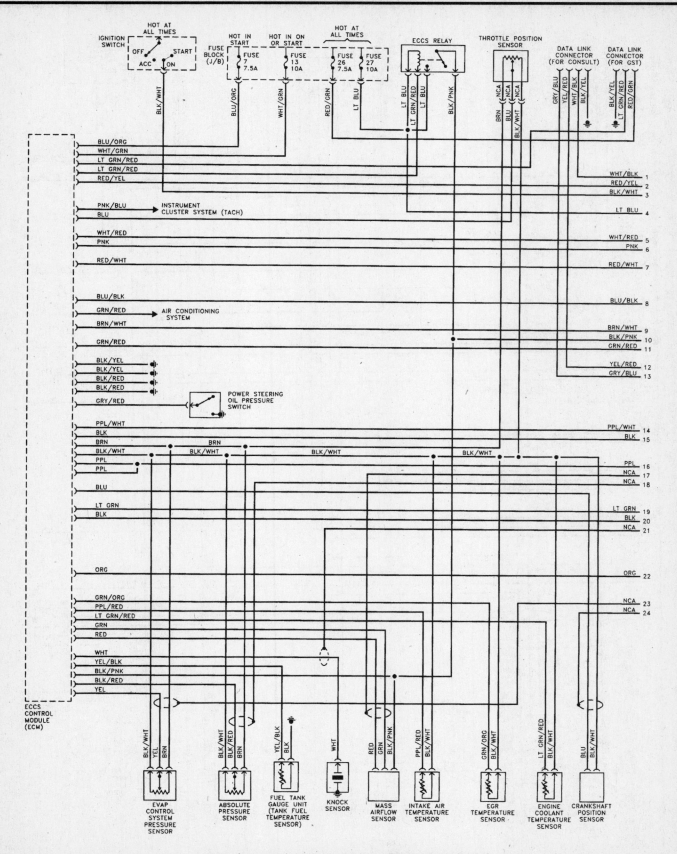

Engine control system - 2.4L four-cylinder engine (1 of 3)

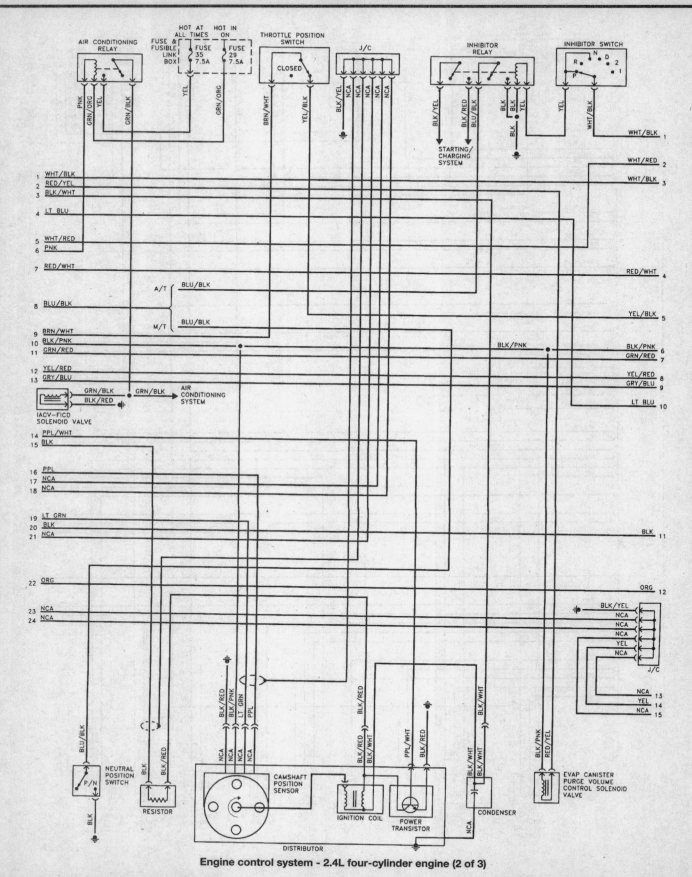

Engine control system - 2.4L four-cylinder engine (2 of 3)

Chapter 12 Chassis electrical system

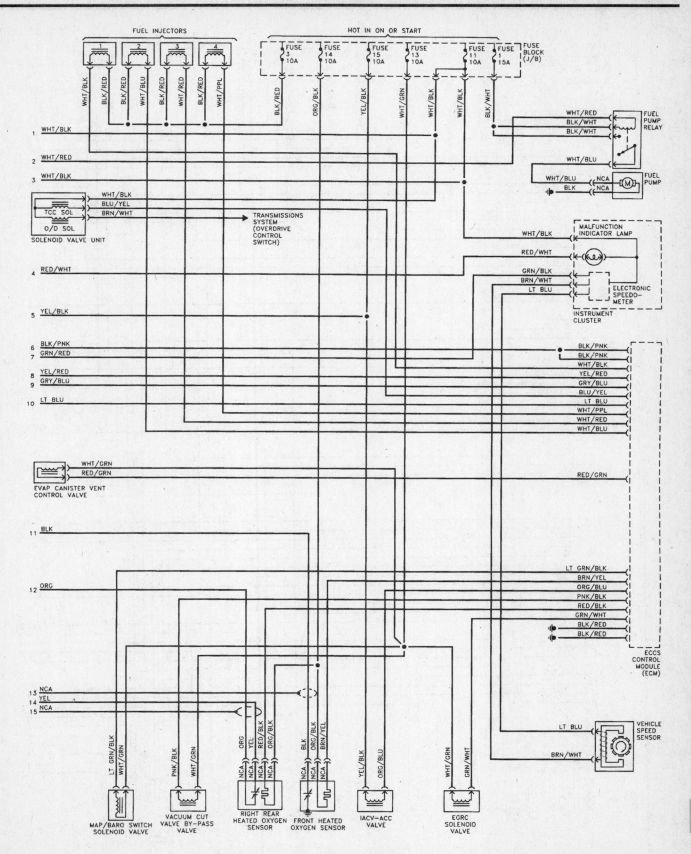

Engine control system - 2.4L four-cylinder engine (3 of 3)

12-26 Chapter 12 Chassis electrical system

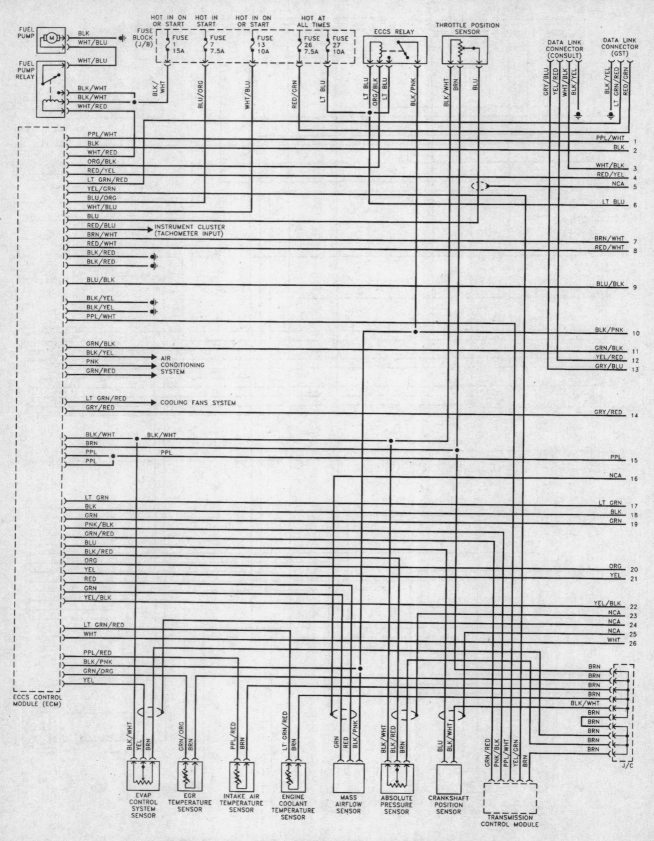

Engine control system - 3.3L V6 engine, Frontier/Xterra (1 of 3)

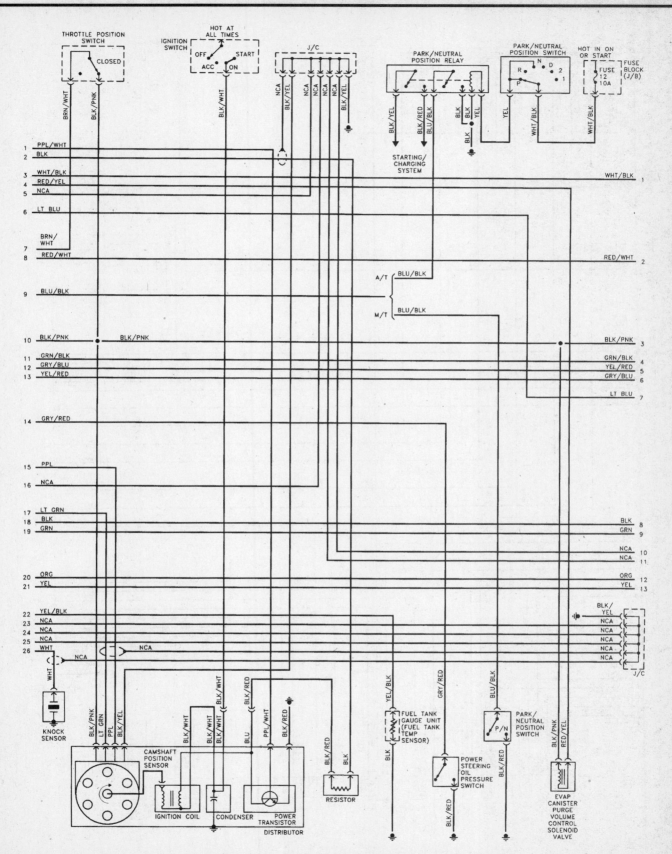

Engine control system - 3.3L V6 engine, Frontier/Xterra (2 of 3)

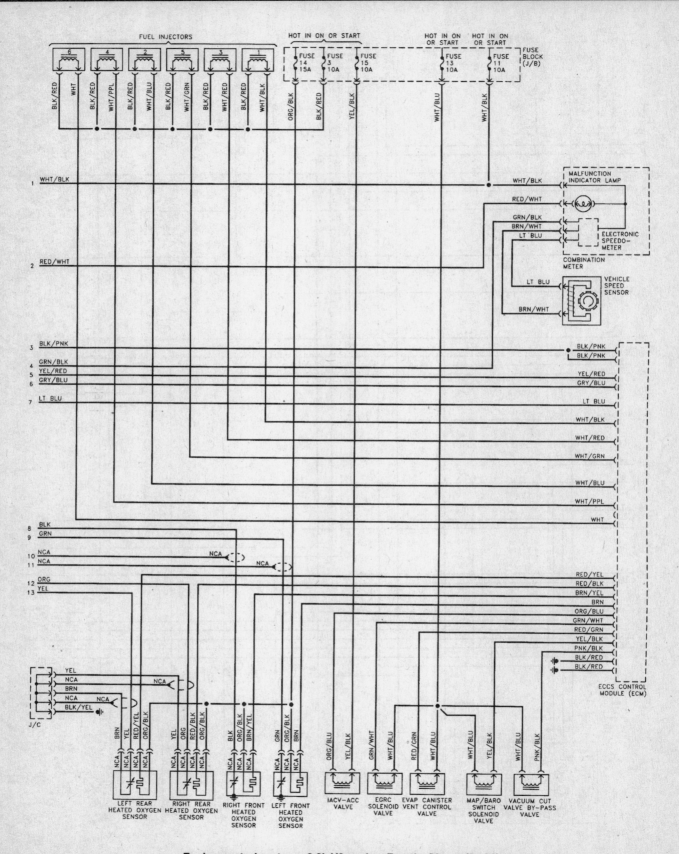

Engine control system - 3.3L V6 engine, Frontier/Xterra (3 of 3)

Chapter 12 Chassis electrical system

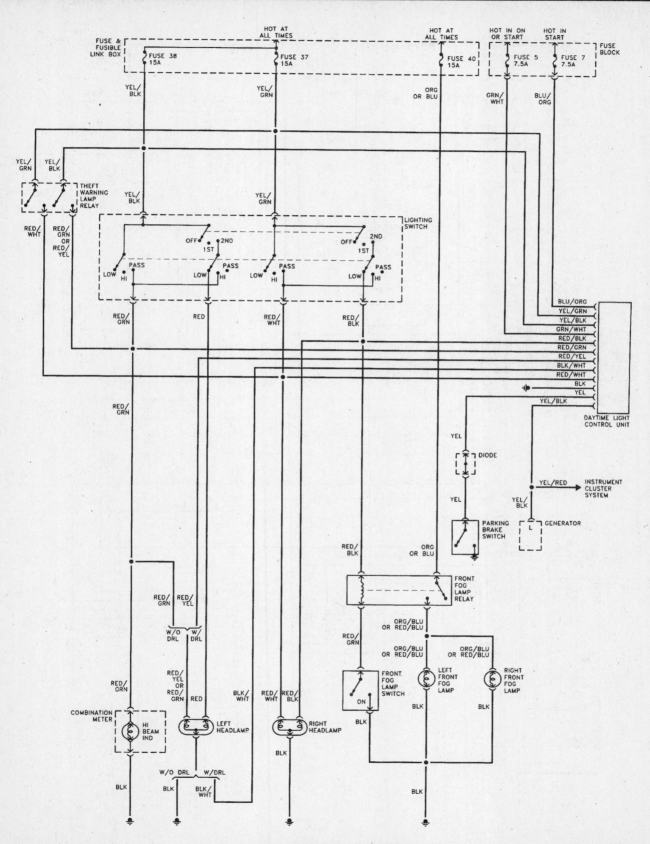

Headlight system - Frontier/Xterra

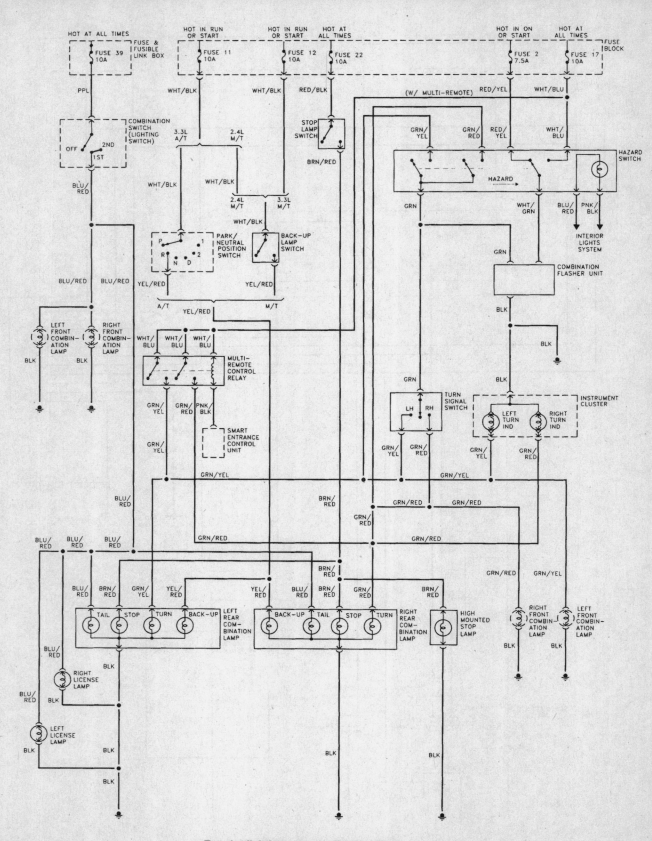

Exterior lighting system - Frontier/Xterra

Chapter 12 Chassis electrical system

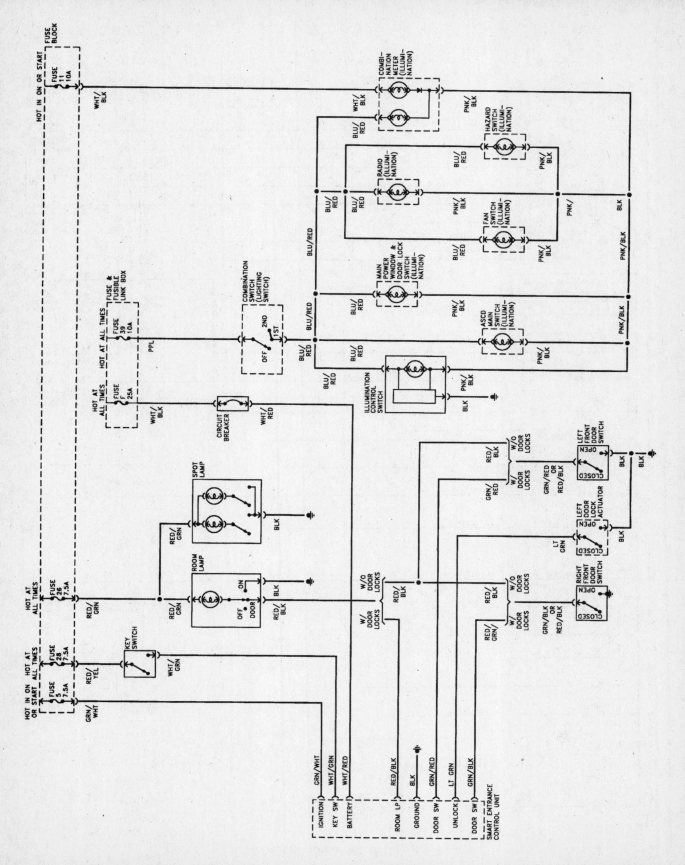

Interior lighting system - Frontier/Xterra

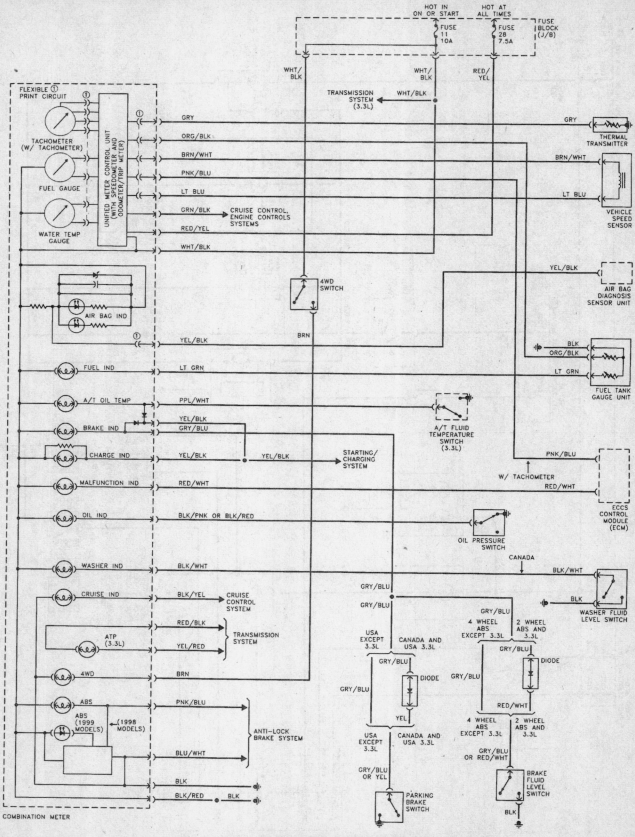

Warning lights and gauge system - Frontier/Xterra

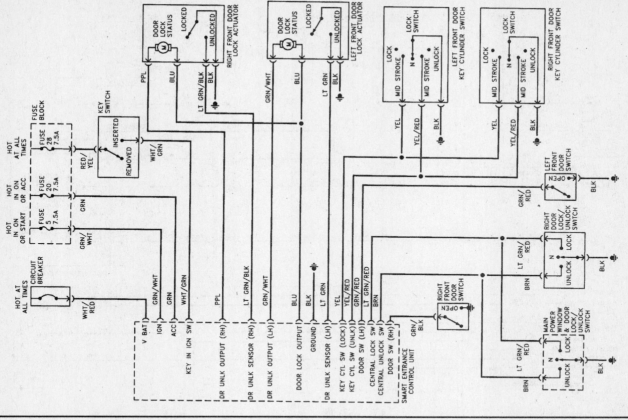

Power door lock system (front doors only) - Frontier/Xterra

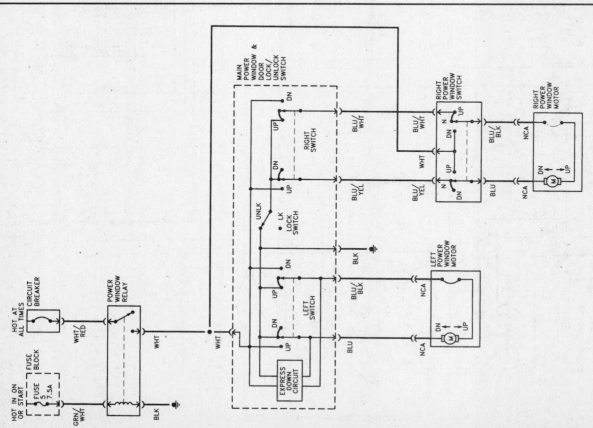

Power window system (front windows only) - Frontier/Xterra

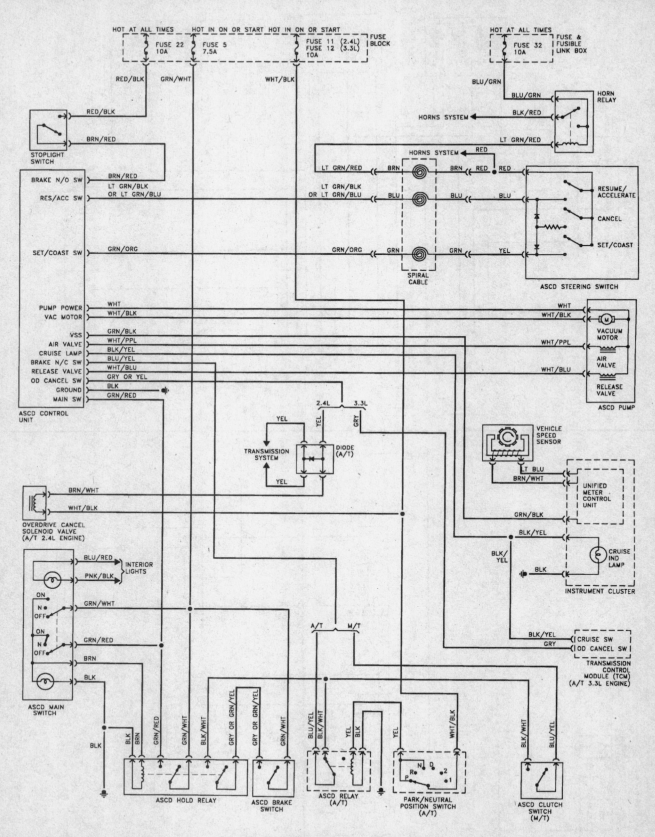

Cruise control system - Frontier/Xterra

Chapter 12 Chassis electrical system

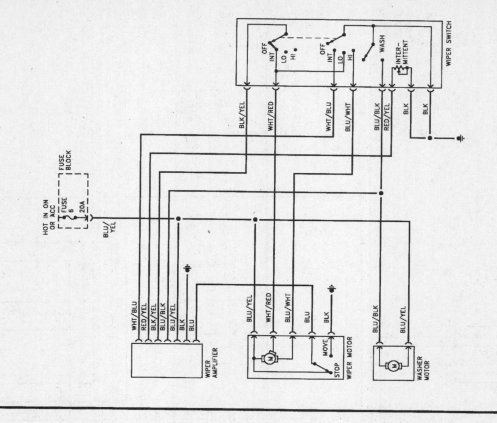

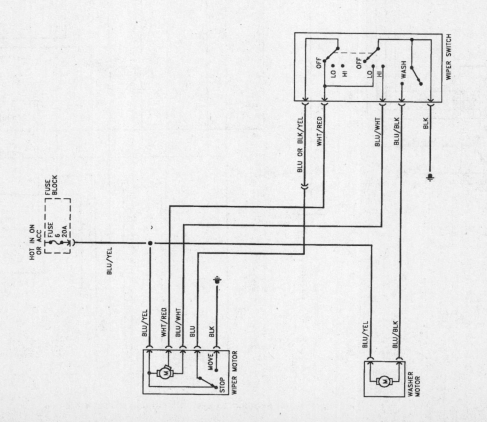

Windshield wiper/washer system - Frontier/Xterra

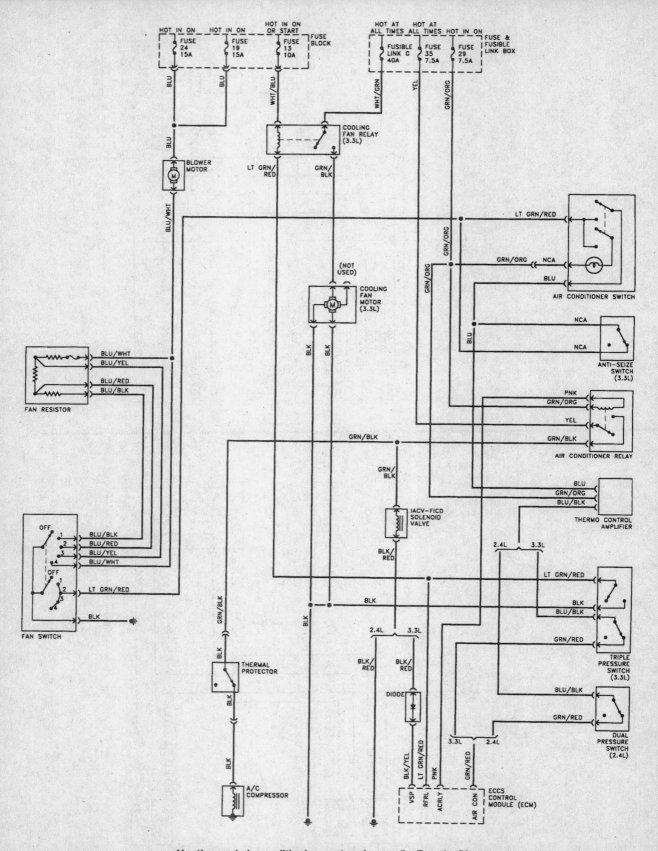

Heating and air conditioning system (manual) - Frontier/Xterra

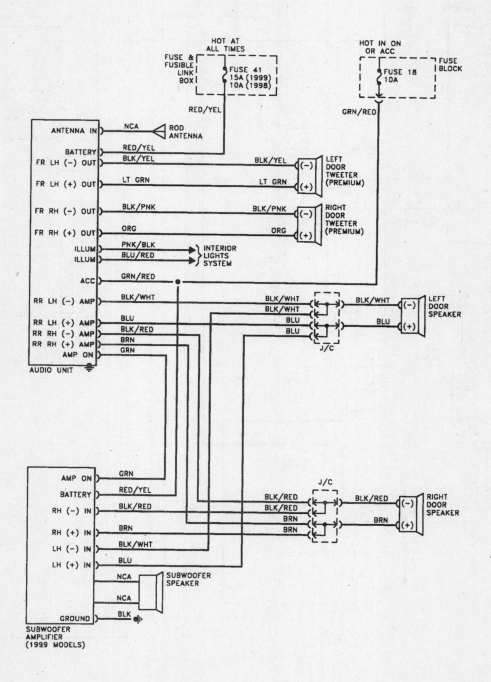

Audio system - two-door Frontier

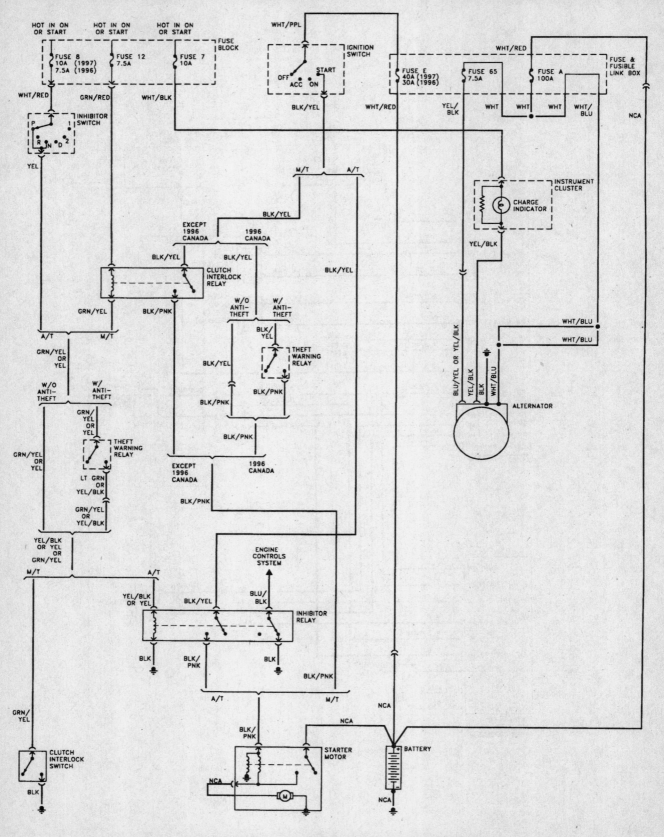

Starting and charging system - 1996 and 1997 Pathfinder

Chapter 12 Chassis electrical system

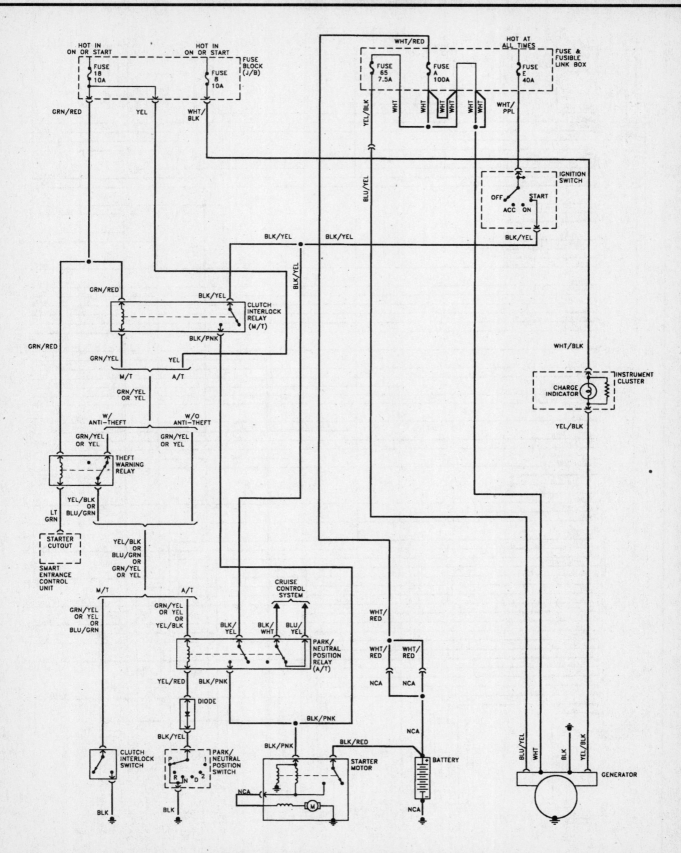

Starting and charging system - 1997 and later Pathfinder

12-40 Chapter 12 Chassis electrical system

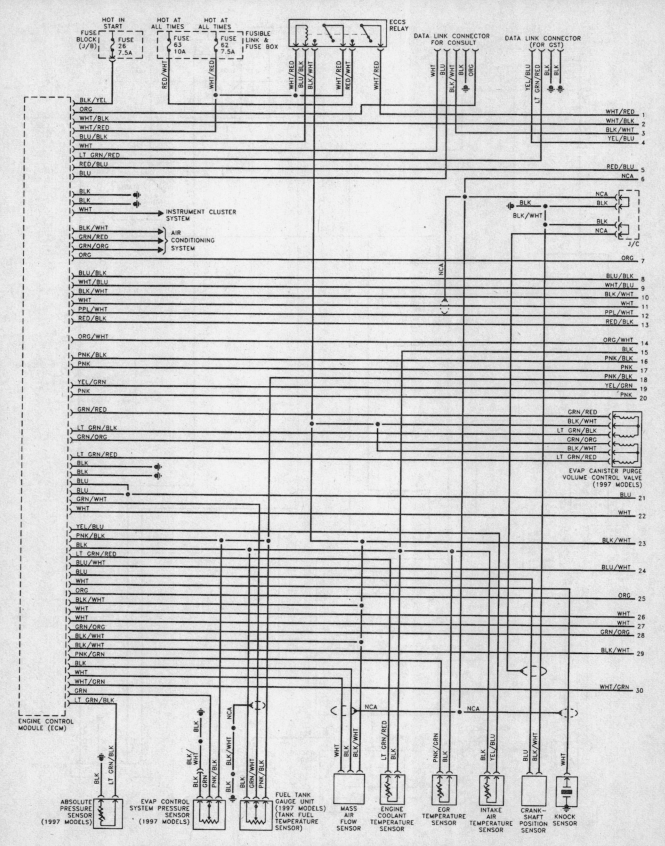

Engine control system - 3.3L V6 engine, 1996 and 1997 Pathfinder (1 of 3)

Chapter 12 Chassis electrical system

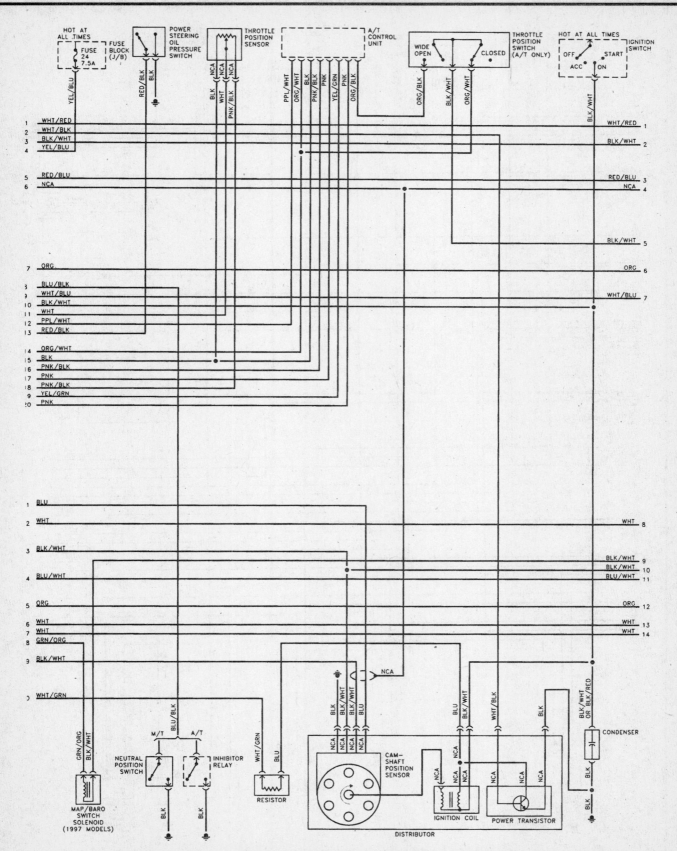

Engine control system - 3.3L V6 engine, 1996 and 1997 Pathfinder (2 of 3)

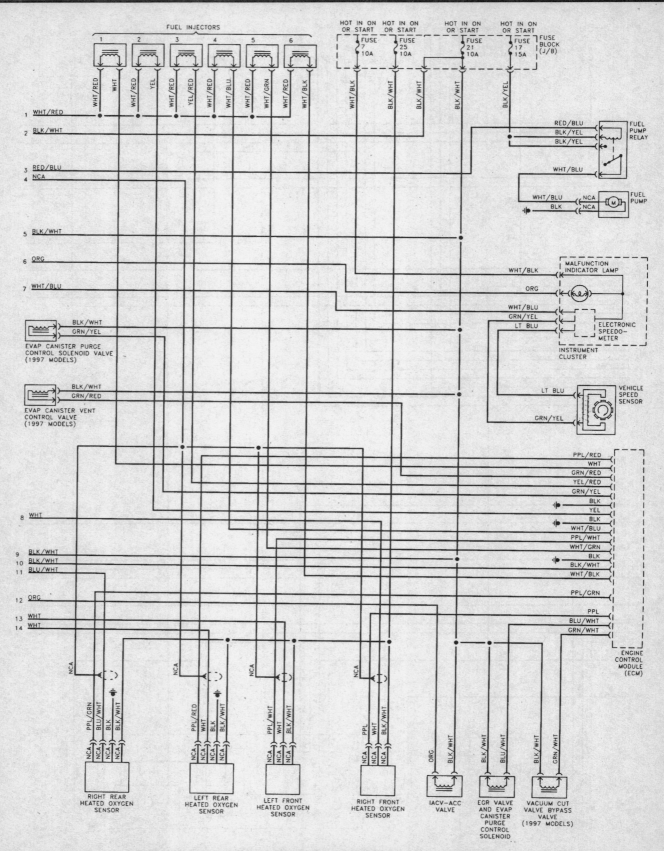

Engine control system - 3.3L V6 engine, 1996 and 1997 Pathfinder (3 of 3)

Chapter 12 Chassis electrical system

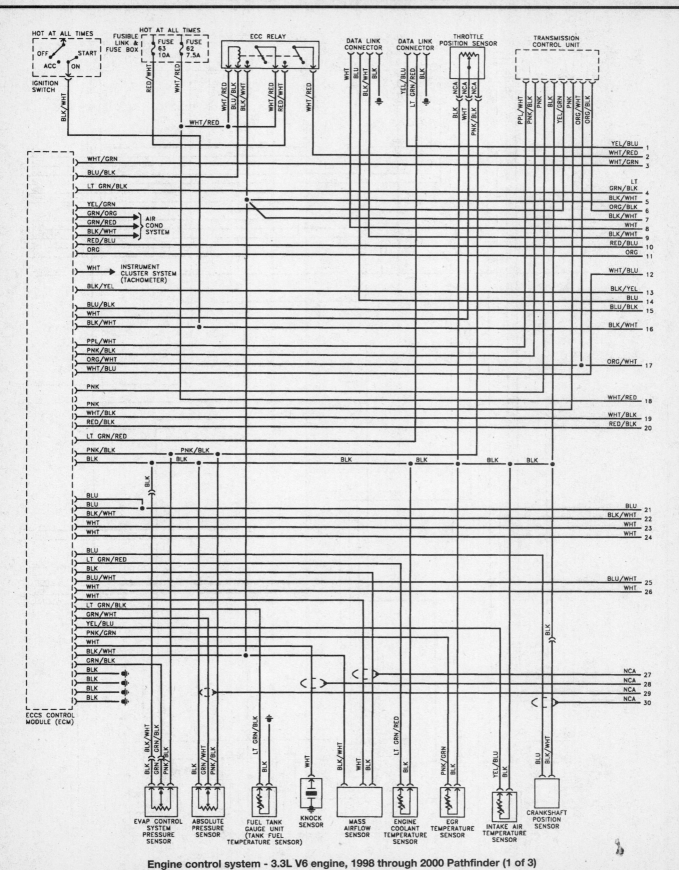

Engine control system - 3.3L V6 engine, 1998 through 2000 Pathfinder (1 of 3)

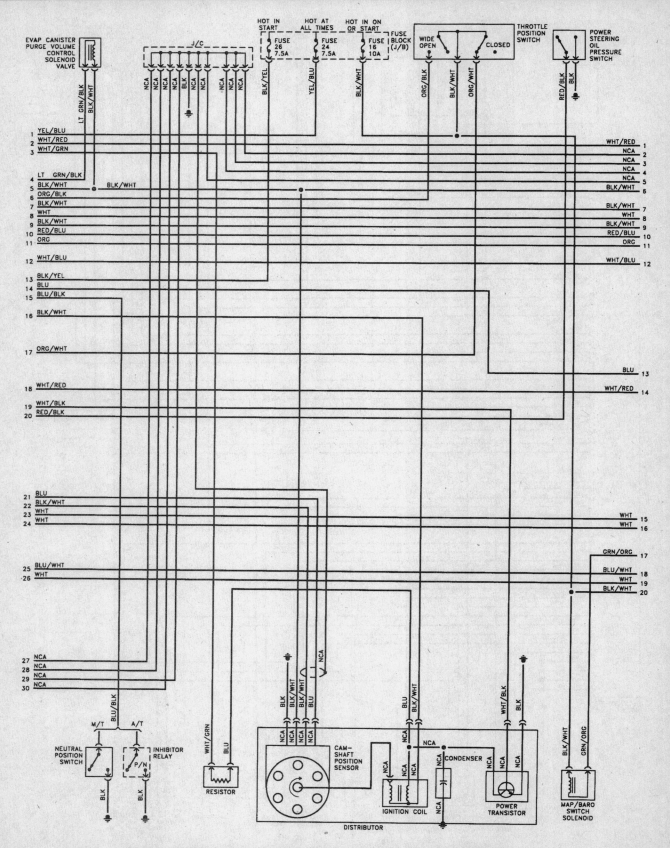

Engine control system - 3.3L V6 engine, 1998 through 2000 Pathfinder (2 of 3)

Chapter 12 Chassis electrical system

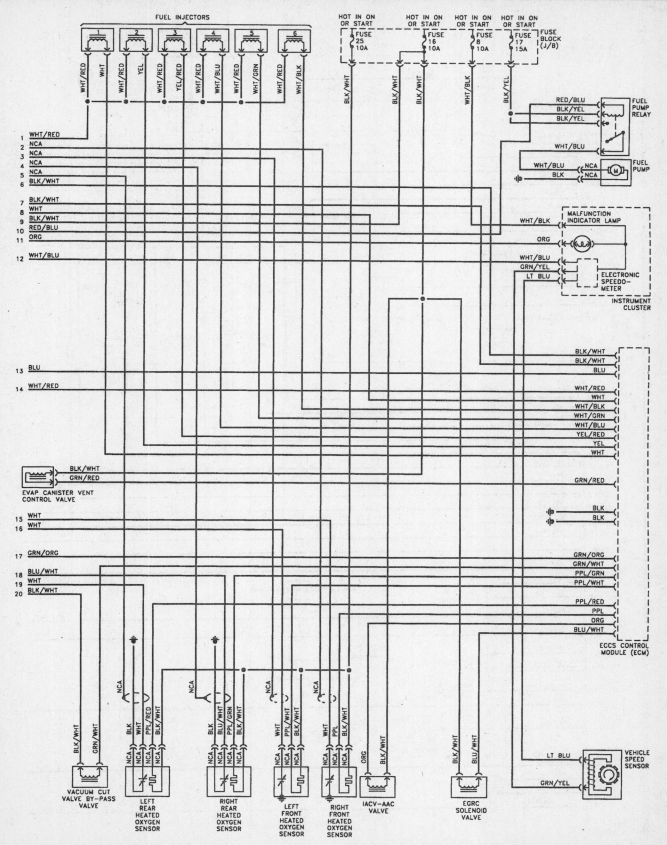

Engine control system - 3.3L V6 engine, 1998 through 2000 Pathfinder (3 of 3)

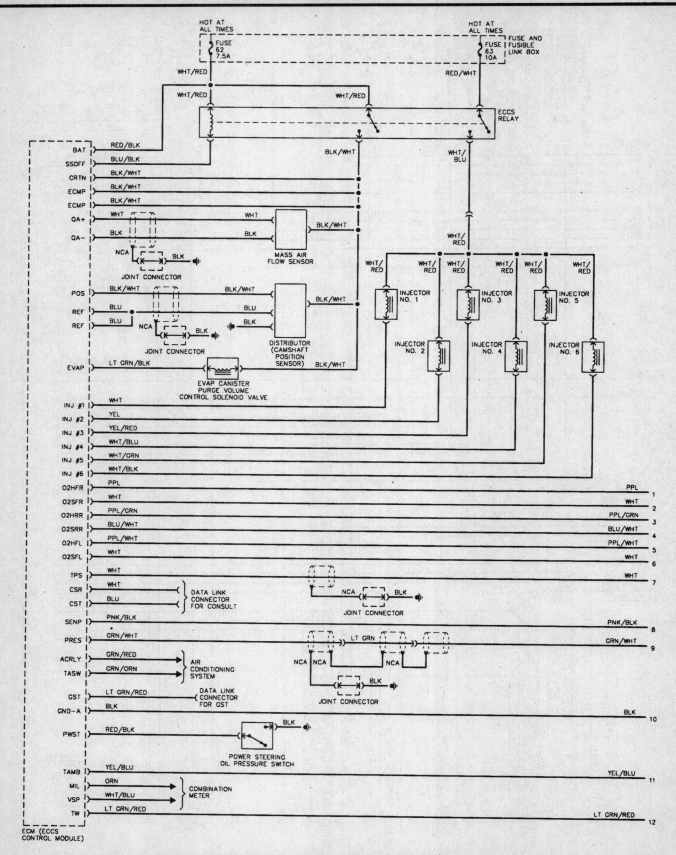

Engine control system - 3.5L V6 engine, Pathfinder (1 of 4)

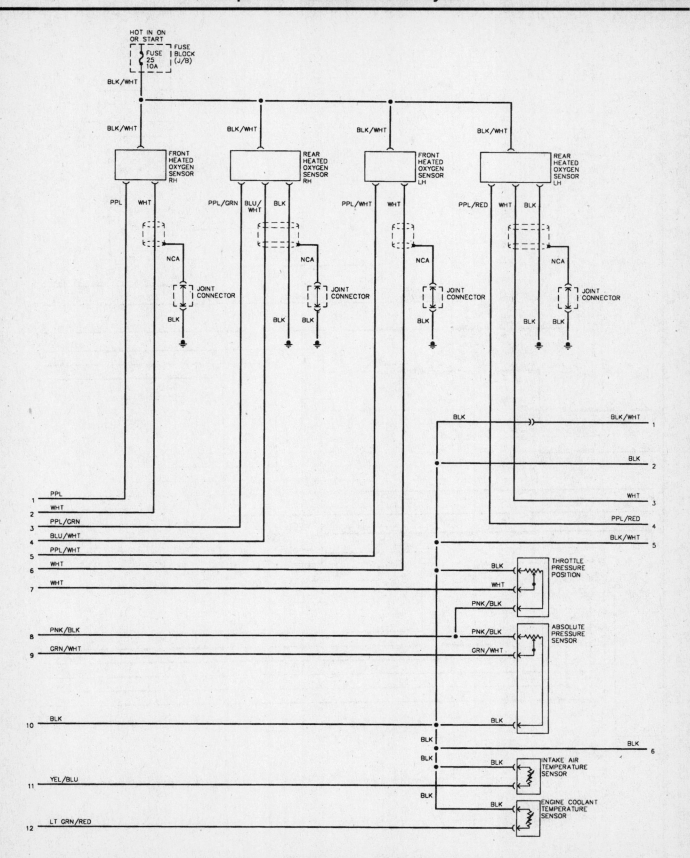

Engine control system - 3.5L V6 engine, Pathfinder (2 of 4)

Chapter 12 Chassis electrical system

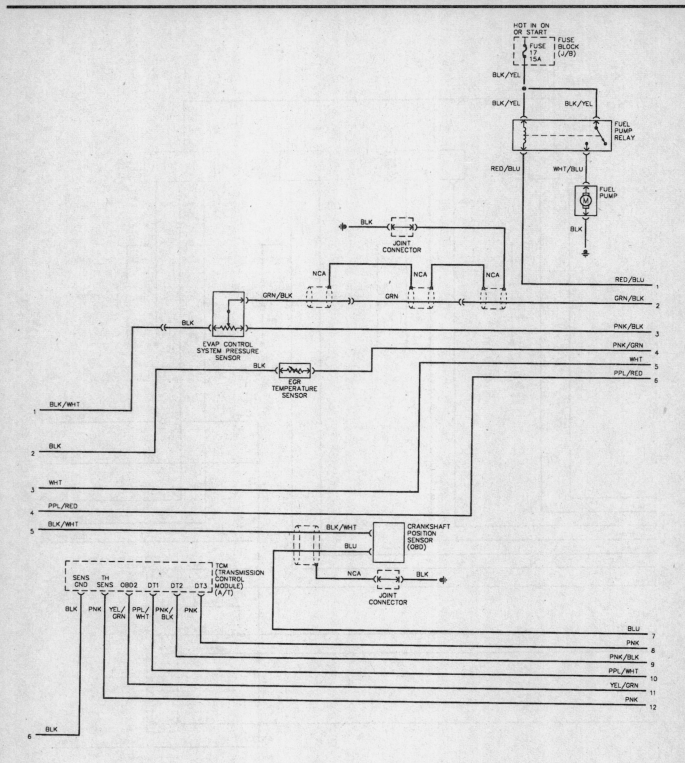

Engine control system - 3.5L V6 engine, Pathfinder (3 of 4)

Chapter 12 Chassis electrical system

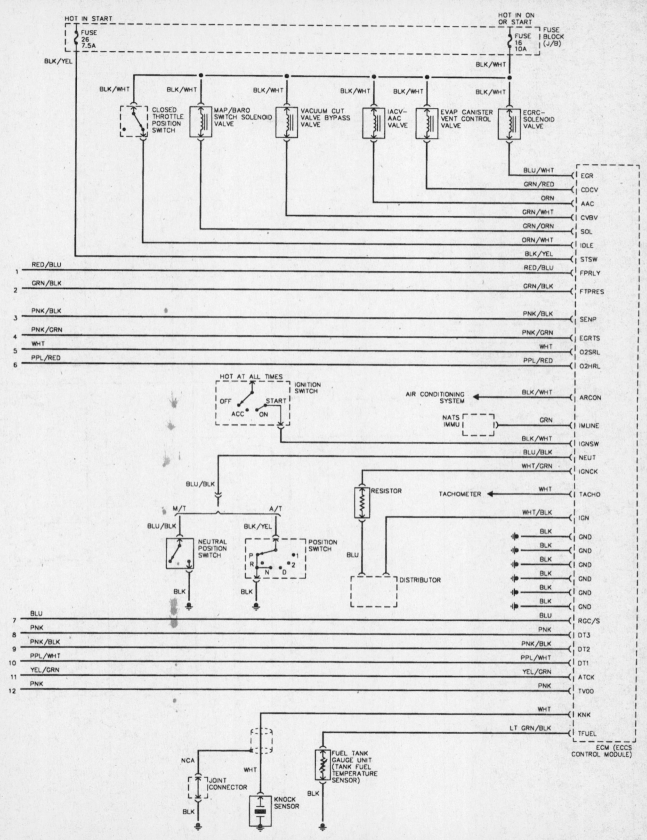

Engine control system - 3.5L V6 engine, Pathfinder (4 of 4)

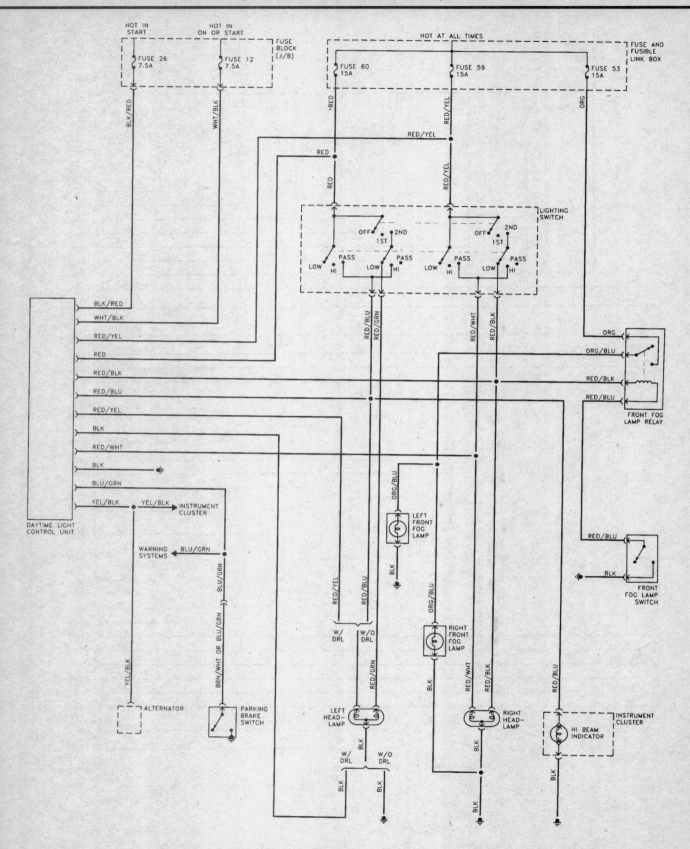

Headlight system - 1996 and 1997 Pathfinder

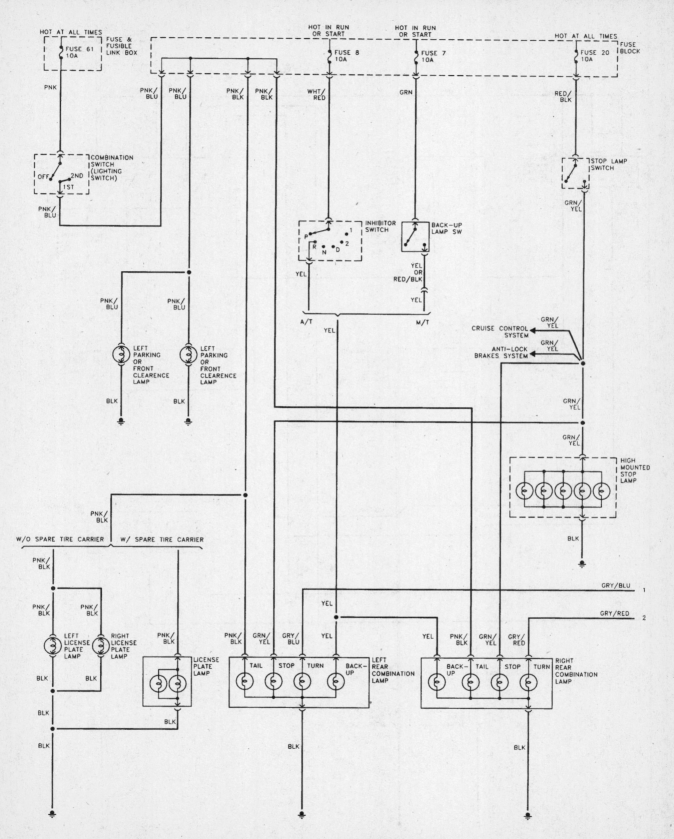

Exterior lighting system - 1996 and 1997 Pathfinder (1 of 2)

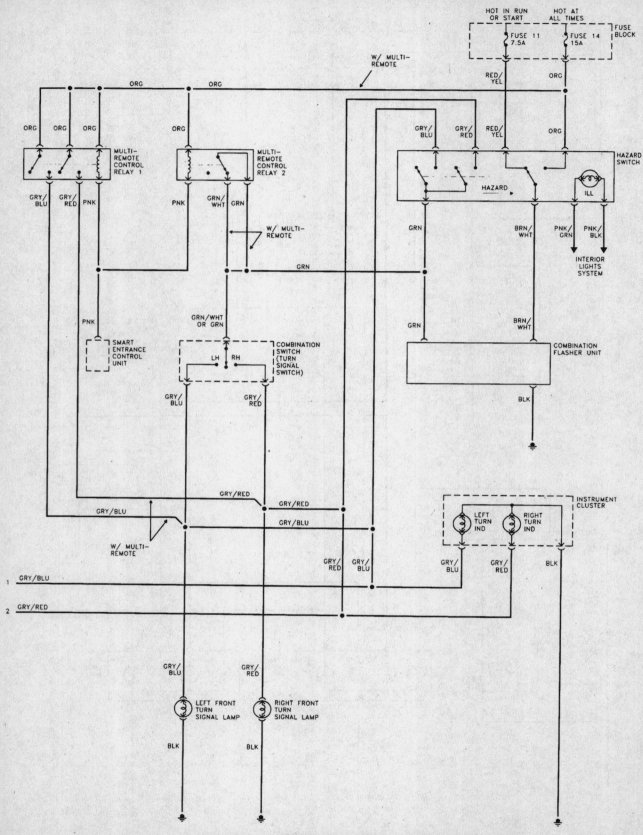

Exterior lighting system - 1996 and 1997 Pathfinder (2 of 2)

Chapter 12 Chassis electrical system

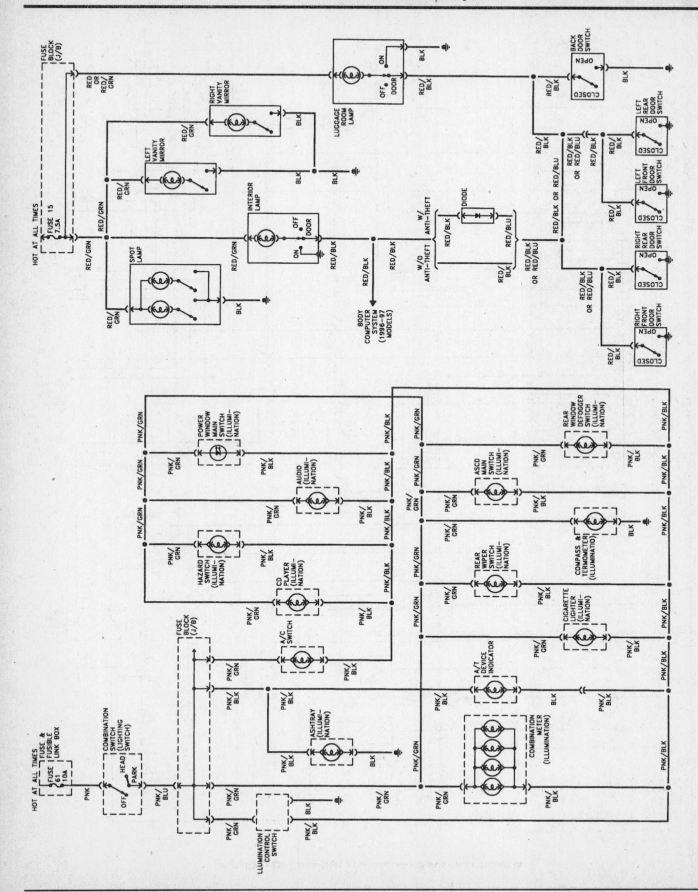

Interior lighting system - Pathfinder

12-54 Chapter 12 Chassis electrical system

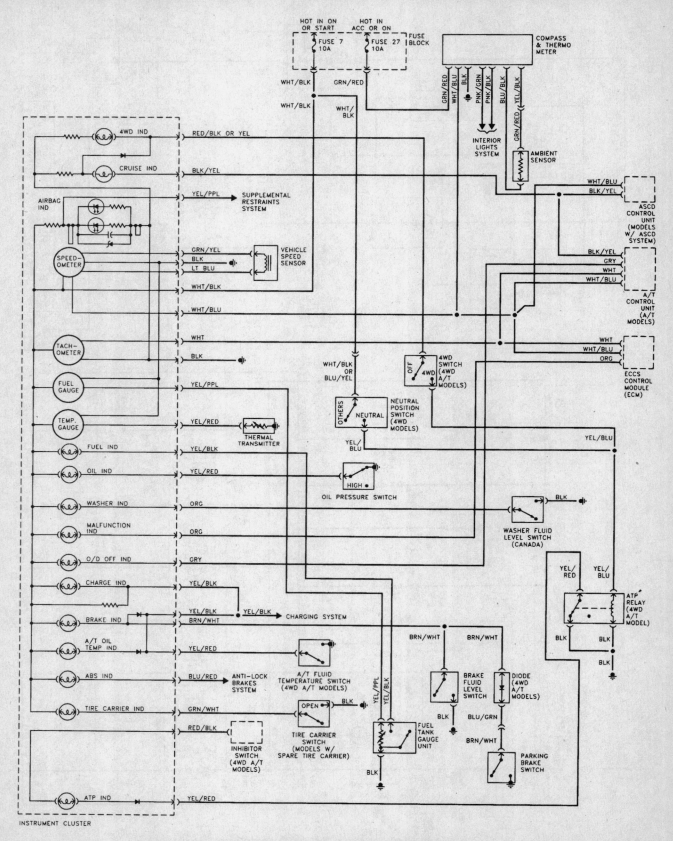

Warning lights and gauge system - 1996 and 1997 Pathfinder

Chapter 12 Chassis electrical system 12-55

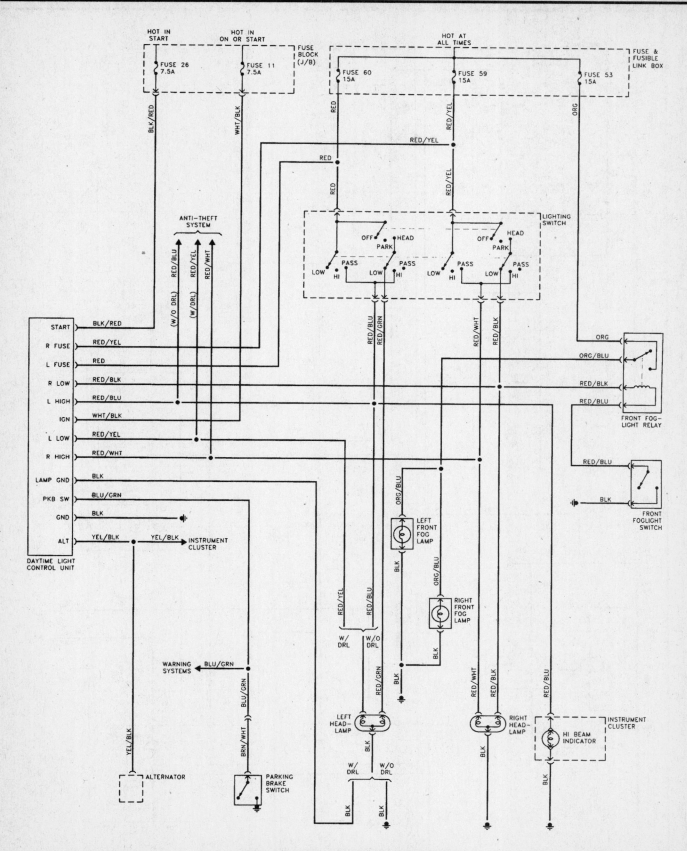

Headlight system - 1998 and later Pathfinder

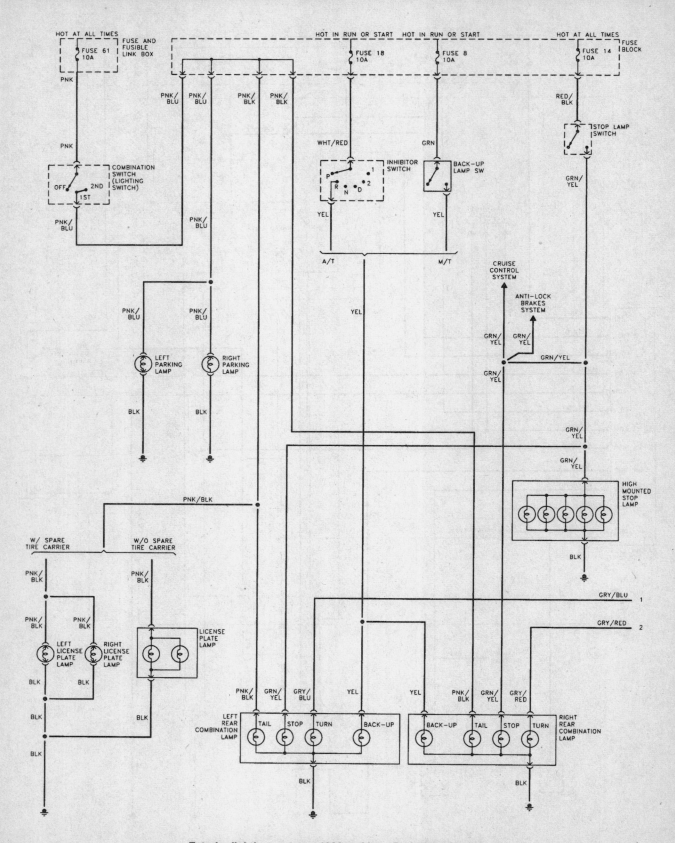

Exterior lighting system - 1998 and later Pathfinder (1 of 2)

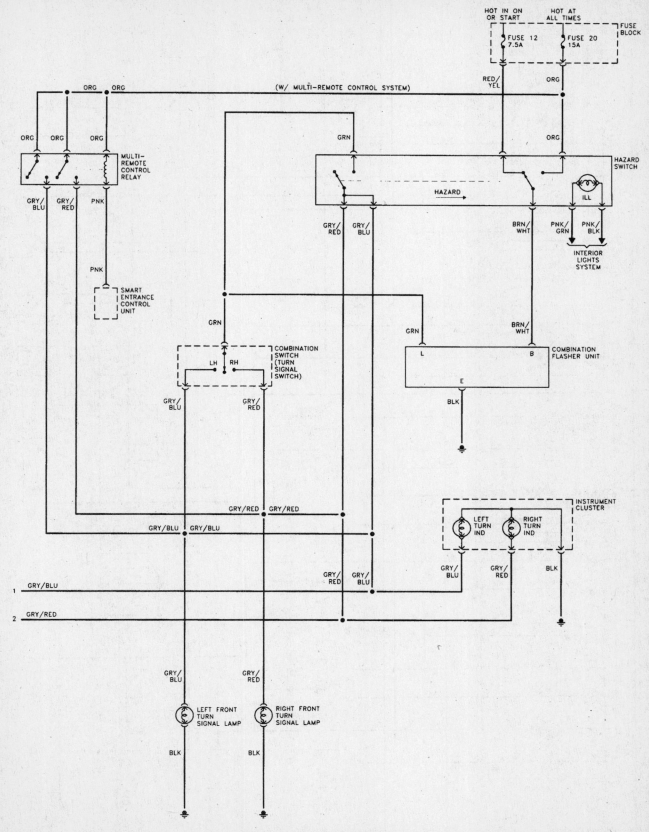

Exterior lighting system - 1998 and later Pathfinder (2 of 2)

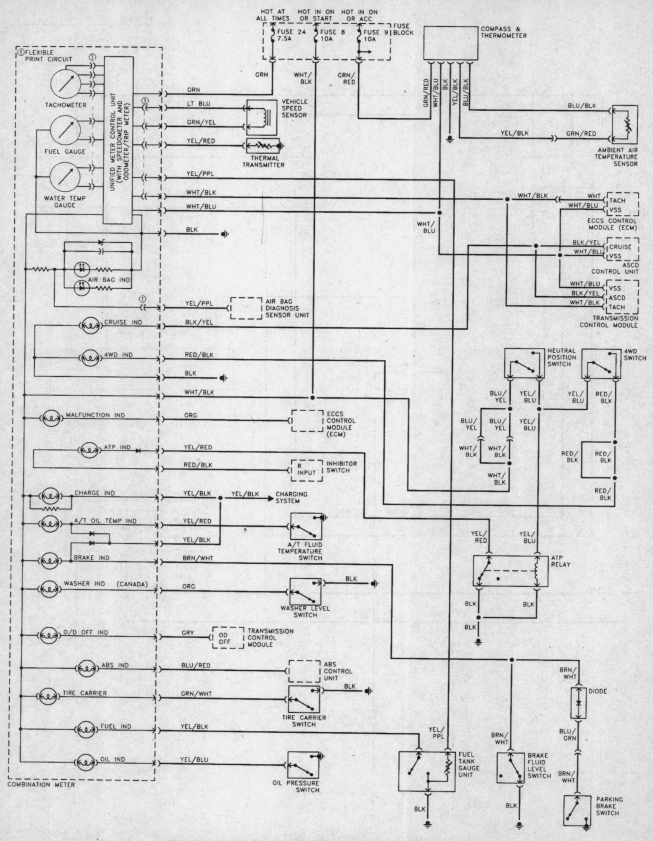

Warning lights and gauge system - 1998 and later Pathfinder

Chapter 12 Chassis electrical system

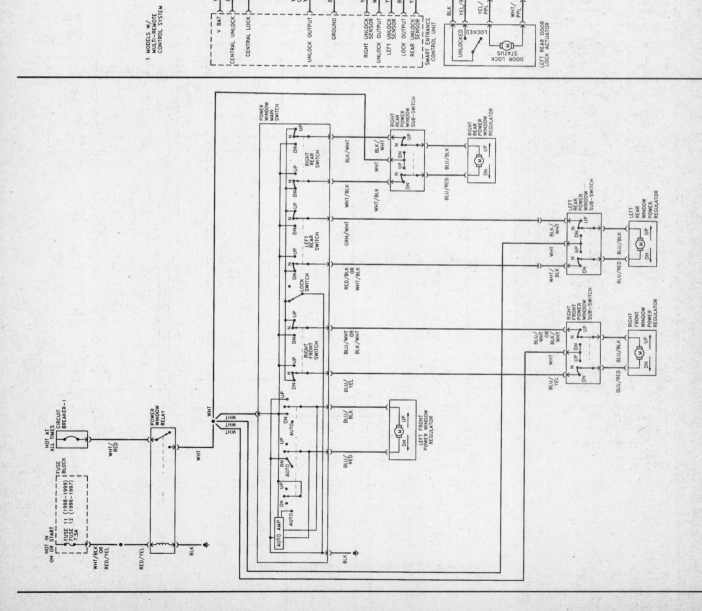

Power door lock system - 1996 and 1997 Pathfinder

Power window system - Pathfinder

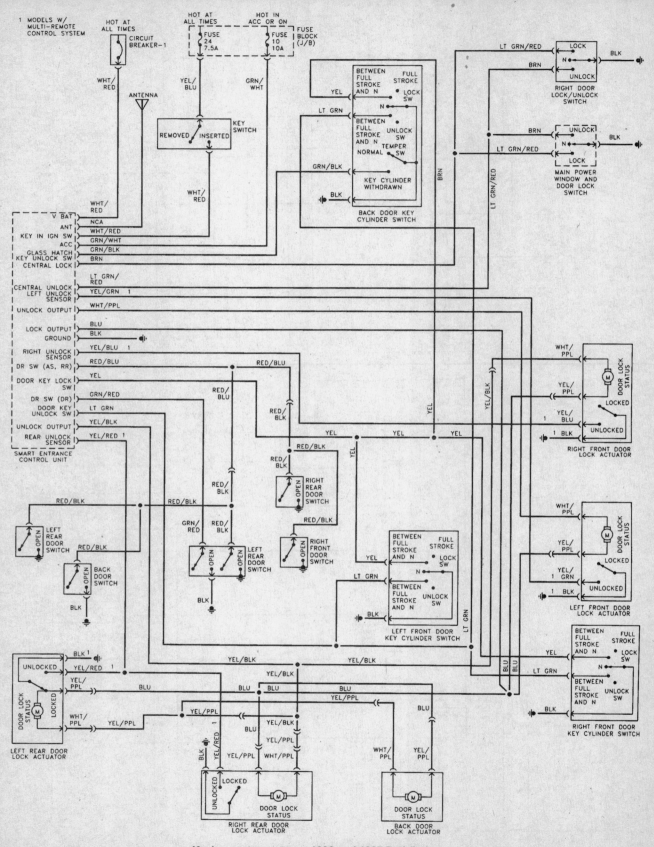

Keyless entry system - 1996 and 1997 Pathfinder

Chapter 12 Chassis electrical system

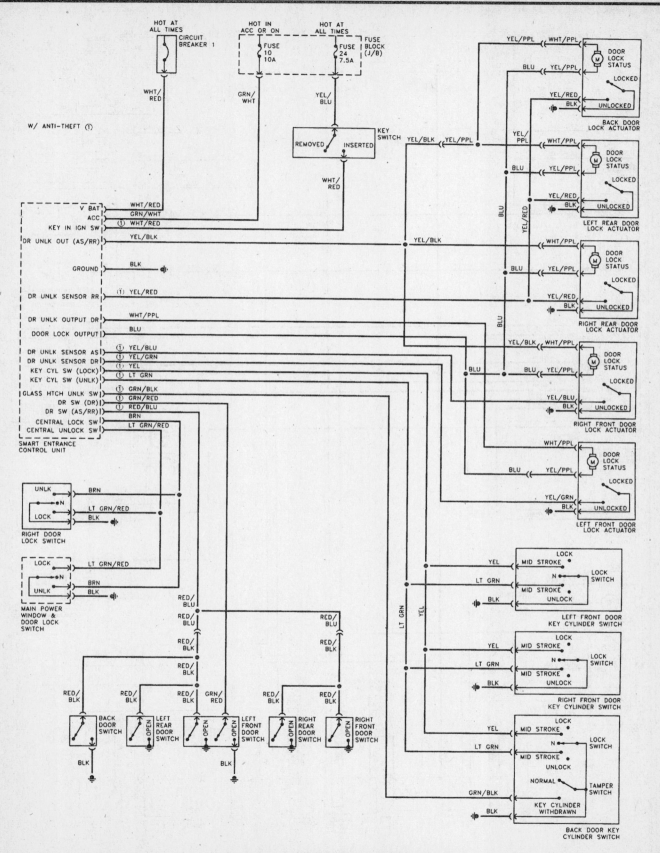

Power door lock system - 1998 and later Pathfinder

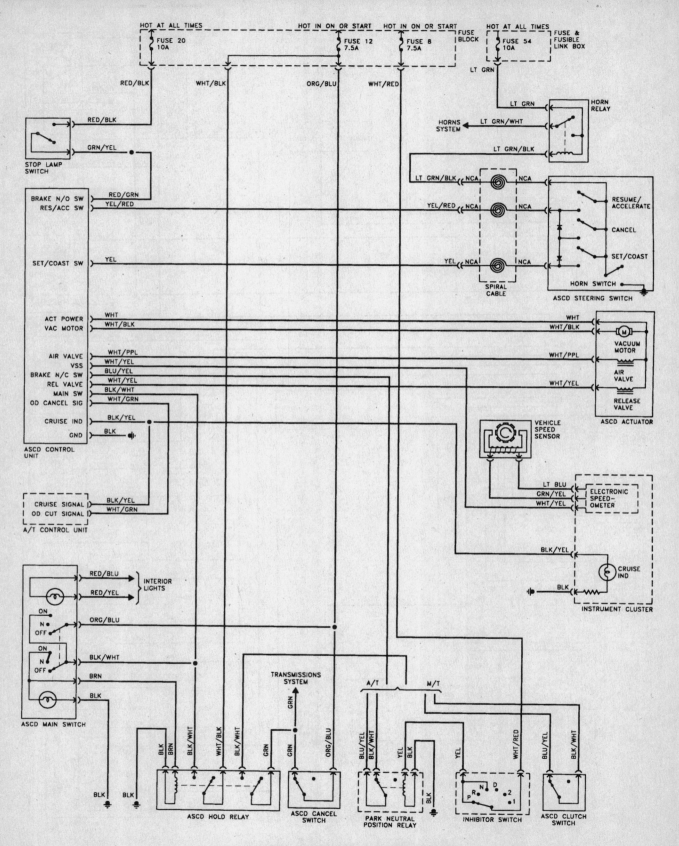

Cruise control system - 1996 and 1997 Pathfinder

Chapter 12 Chassis electrical system

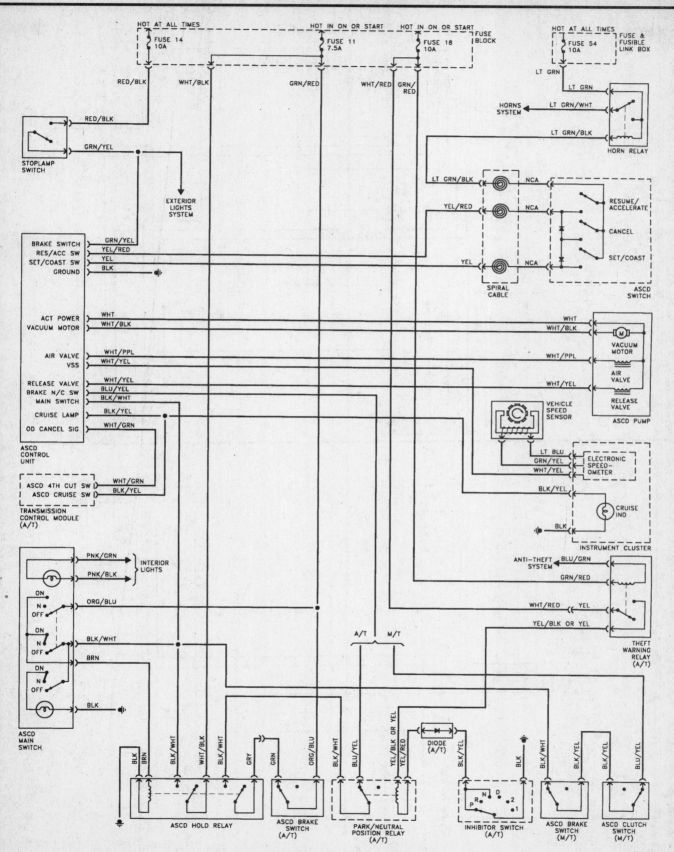

Cruise control system - 1998 and later Pathfinder

12-64　Chapter 12　Chassis electrical system

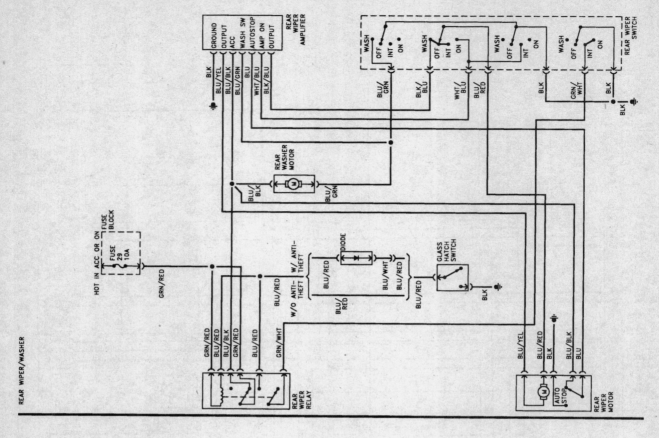

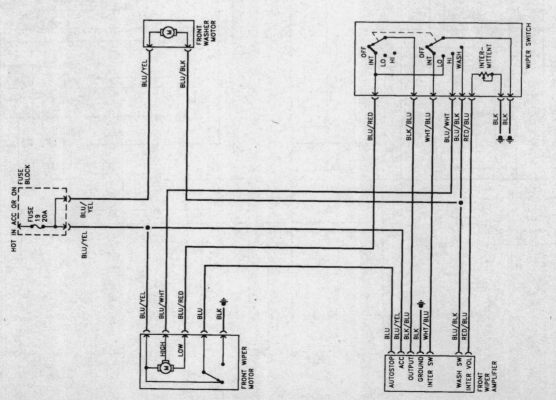

Windshield and rear window wiper/washer system - Pathfinder

Chapter 12 Chassis electrical system 12-65

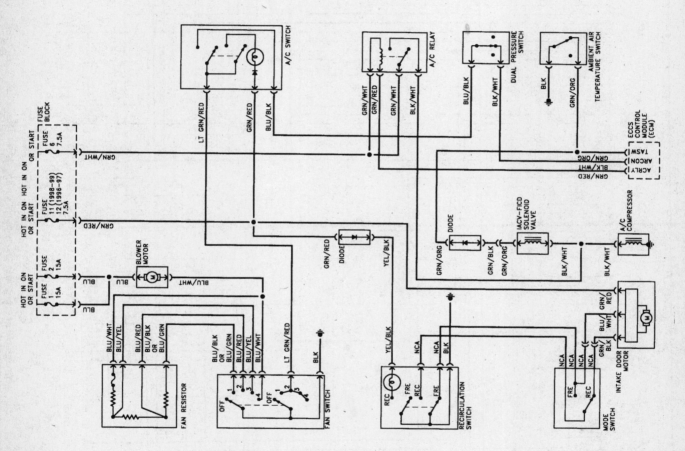

Heating and air conditioning system (manual) - Pathfinder

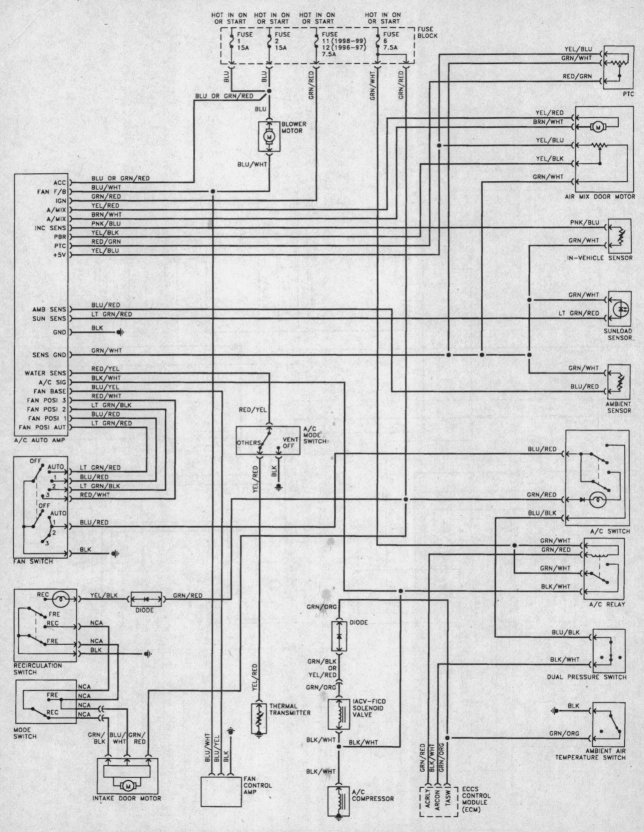

Heating and air conditioning system (automatic) - Pathfinder

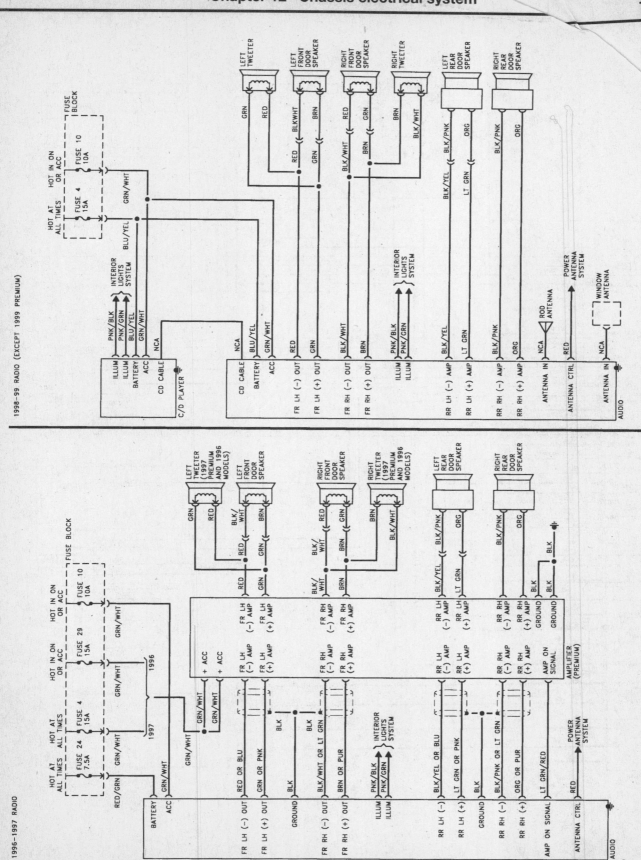

Audio system (except premium) - Pathfinder

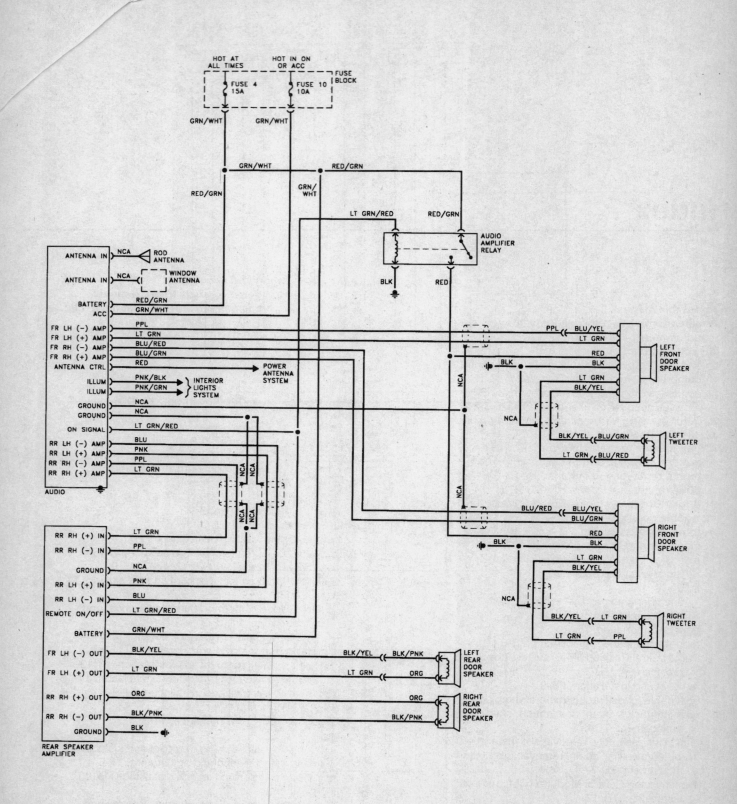

Audio system (premium) - Pathfinder

Index

A

About this manual, 0-5
Accelerator cable, removal, installation and adjustment, 4-8
Air conditioning
   and heating system, check and maintenance, 3-12
   compressor, removal and installation, 3-15
   condenser, removal and installation, 3-16
   receiver/drier, removal and installation, 3-15
Air filter housing, removal and installation, 4-8
Air filter replacement, 1-27
Airbag system, general information, 12-20
Alternator, removal and installation, 5-10
Antenna, removal and installation, 12-9
Antifreeze, general information, 3-2
Anti-lock Brake System (ABS), general information and trouble codes, 9-2
Automatic transmission, 7B-1 through 7B-12
   diagnosis and trouble codes, 7B-2
   extension housing oil seal (2WD models), replacement, 7B-8
   fluid
      and filter change, 1-31
      level check, 1-14
   general information, 7B-2
   mount, check and replacement, 7B-9
   Park/Neutral Position (PNP) switch, check, adjustment and replacement, 7B-8
   removal and installation, 7B-9
   shift cable, check, adjustment and replacement, 7B-3
   shift interlock system, description, check and component replacement, 7B-6
   shift lever assembly, removal and installation, 7B-5
   Throttle Valve (TV) cable, removal, installation and adjustment, 7B-7
   Transmission Control Module (TCM), removal and installation, 7B-12
Automotive chemicals and lubricants, 0-15
Axle assembly, removal and installation
   front, 8-19
   rear, 8-13
Axles, description and check, 8-11
Axleshaft, bearing and oil seals (rear), removal, bearing/seal replacement and installation, 8-11

B

Balljoints, check and replacement, 10-12
Battery
   cables, check and replacement, 5-5
   check and replacement, 5-3
   check, maintenance and charging, 1-16
   emergency jump starting, 0-13
Blower motor
   circuit, check, 3-8
   removal and installation, 3-9
Body, 11-1 through 11-22
   bumpers, removal and installation, 11-7
   center console, removal and installation, 11-17
   cowl cover, removal and installation, 11-10
   dashboard trim panels, removal and installation, 11-18
   door
      latch, lock cylinder and handles, removal and installation, 11-12
      removal, installation and adjustment, 11-11
      trim panels, removal and installation, 11-10
      window glass, removal and installation, 11-13
      window regulator, removal and installation, 11-13
   front fender, removal and installation, 11-9
   general information, 11-1
   hinges and locks, maintenance, 11-3
   hood
      latch and release cable, removal and installation, 11-6
      removal, installation and adjustment, 11-6
   instrument panel, removal and installation, 11-20
   liftgate
      and liftgate glass (SUV models), removal and installation, 11-15
      latch, lock cylinder, handle and support struts (SUV models), removal, installation and adjustment, 11-16

IND-2

maintenance... [obscured]
mirrors, removal and installation, 11-[obscured]
radiator...
repair damage, 11-3
minor damage, 11-2
seats, removal and installation, 11-21
steering column covers, removal and installation, 11-20
tailgate (pick-up models)
   removal, installation and adjustment, 11-14
   latch and handle, removal and installation, 11-14
upholstery and carpets, maintenance, 11-1
vinyl trim, maintenance, 11-1
windshield and fixed glass, replacement, 11-3

Booster battery (jump) starting, 0-13
Brakes, 9-1 through 9-26
  Anti-lock Brake System (ABS), general information and trouble codes, 9-2
  caliper, removal and installation, 9-8
  check, 1-24
  disc
    brake pads, replacement, 9-6
    inspection, removal and installation, 9-8
  drum brake shoes, replacement, 9-9
  fluid change, 1-26
  general information, 9-2
  hoses and lines, inspection and replacement, 9-20
  hydraulic system, bleeding, 9-22
  light switch, check and replacement, 9-26
  Load Sensing Valve (LSV), description and replacement, 9-21
  master cylinder, removal, installation and reservoir/seal replacement, 9-19
  parking brake
    adjustment, 9-23
    cables, replacement, 9-24
  power brake booster, check, replacement and adjustment, 9-23
  wheel cylinder, removal and installation, 9-19
Bulb replacement, 12-11
Bumpers, removal and installation, 11-7
Buying Parts, 0-7

C

Camshaft Position (CMP) sensor, check and replacement, 6-15
Camshaft(s) and lifters, removal and installation
  2.4L four-cylinder engine, 2A-9
  3.3L V6 engine, 2B-7
  3.5L V6 engine, 2C-10
Camshafts, lifters and bearings, inspection, 2D-21
Catalytic converter, 6-27
Center console, removal and installation, 11-17
Charging system
  check, 5-9
  general information and precautions, 5-9

Chassis electrical system, 12-1 through 12-68
  airbag system, general information, 12-20
  antenna, removal and installation, 12-9
  bulb replacement, 12-11
  circuit breakers, general information, 12-4
  cruise control system, description, check and component replacement, 12-16
  Daytime Running Lights (DRL), general information, 12-15
  electric
    side view mirrors, description and check, 12-19
    sunroof, description and check, 12-20
  electrical troubleshooting, general information, 12-1
  fuses and fusible links, general information, 12-3
  general information, 12-1
  headlight
    bulb, replacement, 12-10
    housing, replacement, 12-11
  headlights, adjustment, 12-10
  horn, check and replacement, 12-15
  ignition switch and key lock cylinder, check and replacement, 12-7
  instrument
    cluster, removal and installation, 12-8
    panel gauges, check, 12-8
  lighting system and windshield wiper/washer switches, check and replacement, 12-5
  power
    door lock system, description and check, 12-18
    seats, description and check, 12-19
    window system, description and check, 12-18
  radio and speakers, removal and installation, 12-9
  rear window defogger, check and repair, 12-16
  relays, general information and testing, 12-4
  turn signal/hazard flashers, check and replacement, 12-5
  wiper motor and amplifier, check and replacement, 12-14
  wiring diagrams, general information, 12-21
Chassis lubrication, 1-22
Circuit breakers, general information, 12-4
Clutch
  components, removal, inspection and installation, 8-4
  description and check, 8-3
  hydraulic system, bleeding, 8-4
  master cylinder, removal, installation and reservoir/seal replacement, 8-3
  pedal, adjustment, 8-7
  release bearing, removal, inspection and installation, 8-6
  release cylinder, removal and installation, 8-4
  start switch, check and replacement, 8-8
Clutch and driveline, 8-1 through 8-20
Coil spring, rear (Pathfinder models), removal and installation, 10-15
Conversion factors, 0-16
Coolant temperature gauge sending unit, check and replacement, 3-8
Cooling system check, 1-20
Cooling system servicing (draining, flushing and refilling), 1-29
Cooling, heating and air conditioning systems, 3-1 through 3-16
  air conditioning and heating system, check and maintenance, 3-12
  antifreeze, general information, 3-2

blower motor
  circuit, check, 3-8
  motor, removal and installation, 3-9
compressor, removal and installation, 3-15
condenser, removal and installation, 3-16
coolant temperature gauge sending unit, check and replacement, 3-8
engine cooling fan, check and replacement, 3-5
general information, 3-2
heater and air conditioning control assembly, removal and installation, 3-10
heater core, replacement, 3-10
radiator and coolant reservoir, removal and installation, 3-5
receiver/drier, removal and installation, 3-15
thermostat, check and replacement, 3-3
water pump, check and replacement, 3-6

Cowl cover, removal and installation, 11-10
Crankshaft
  inspection, 2D-19
  installation and main bearing oil clearance check, 2D-23
  removal, 2D-15
Crankshaft front oil seal, replacement
  2.4L four-cylinder engine, 2A-13
  3.3L V6 engine, 2B-11
  3.5L V6 engine, 2C-16
Crankshaft Position (CKP) sensor, check and replacement, 6-14
Crankshaft pulley, removal and installation
  2.4L four-cylinder engine, 2A-13
  3.3L V6 engine, 2B-11
  3.5L V6 engine, 2C-16
Cruise control system, description, check and component replacement, 12-16
Cylinder compression check, 2D-7
Cylinder head
  cleaning and inspection, 2D-12
  disassembly, 2D-12
  reassembly, 2D-14
  removal and installation
    2.4L four-cylinder engine, 2A-12
    3.3L V6 engine, 2B-10
    3.5L V6 engine, 2C-14
Cylinder honing, 2D-17

D

Dashboard trim panels, removal and installation, 11-18
Daytime Running Lights (DRL), general information, 12-15
Differential lubricant
  change, 1-36
  level check, 1-21
Disc brake pads, replacement, 9-6
Distributor (four-cylinder and 3.3L V6 models), removal and installation, 5-8
Door
  latch, lock cylinder and handles, removal and installation, 11-12
  removal, installation and adjustment, 11-11
  trim panels, removal and installation, 11-10

Door window glass
  regulator, removal and installation, 11-13
  removal and installation, 11-13
Driveaxle (4WD models)
  boot, replacement, 8-16
  general information and inspection, 8-14
  removal and installation, 8-15
Drivebelt check, adjustment and replacement, 1-19
Driveshaft
  and universal joints, general information and inspection, 8-8
  center support bearing, replacement, 8-9
  removal and installation, 8-9
Drum brake shoes, replacement, 9-9

E

Electric
  side view mirrors, description and check, 12-19
  sunroof, description and check, 12-20
Electrical troubleshooting, general information, 12-1
Emissions and engine control systems 6-1 through 6-28
  Camshaft Position (CMP) sensor, check and replacement, 6-15
  Crankshaft Position (CKP) sensor, check and replacement, 6-14
  Engine Coolant Temperature (ECT) sensor, check and replacement, 6-13
  Evaporative Emissions Control (EVAP) system, 6-26
  Exhaust Gas Recirculation (EGR) system, 6-24
  fuel temperature sensor, check and replacement, 6-18
  Idle Air Control (IAC) system, 6-20
  Intake Air Temperature (IAT) sensor, check and replacement, 6-12
  intake manifold runner control system, 6-21
  Knock Sensor (KS), check and replacement, 6-18
  Manifold Absolute Pressure (MAP) sensor and solenoid valve, check and replacement, 6-11
  Mass Airflow (MAF) sensor, check and replacement, 6-10
  On-Board Diagnosis (OBD) system and trouble codes, 6-2
  Oxygen (O2) sensor, check and replacement, 6-17
  Positive Crankcase Ventilation (PCV) system, 6-23
  Power steering pressure switch, check and replacement, 6-16
  Powertrain Control Module (PCM), removal and installation, 6-8
  Throttle Position Sensor (TPS), check, replacement and adjustment, 6-9
  variable valve timing control system, 6-22
  Vehicle Speed Sensor (VSS), check and replacement, 6-19
Engine cooling fan, check and replacement, 3-5
Engine electrical systems, 5-1 through 5-12
  alternator, removal and installation, 5-10
  battery
    cables, check and replacement, 5-5
    check and replacement, 5-3
    emergency jump starting, 0-13
  charging system
    check, 5-9
    general information and precautions, 5-9
  distributor (four-cylinder and 3.3L V6 models), removal and installation, 5-8

ignition coil(s) (3.5L V6 models), removal and installation, 5-8
ignition system
    check, 5-6
    general information, 5-5
ignition timing, check and adjustment, 5-9
starter motor
    and circuit, in-vehicle check, 5-11
    removal and installation, 5-12
starting system, general information and precautions, 5-11

Engine idle speed and fast idle cam (four-cylinder and 3.3L V6 models), check and adjustment, 4-12

Engine mounts, check and replacement
    2.4L four-cylinder engine, 2A-16
    3.3L V6 engine, 2B-14
    3.5L V6 engine, 2C-21

Engine oil and filter change, 1-14

Engine oil cooler and adapter (3.5L V6 engine), general information and replacement, 2C-20

Engines
    2.4L four-cylinder engine, 2A-1 through 2A-16
        camshaft and lifters, removal, inspection and installation, 2A-9
        crankshaft
            front oil seal, replacement, 2A-13
            pulley, removal and installation, 2A-13
        cylinder head, removal and installation, 2A-12
        exhaust manifold, removal and installation, 2A-11
        flywheel/driveplate, removal and installation, 2A-16
        general information, 2A-2
        intake manifold, removal and installation, 2A-10
        mounts, check and replacement, 2A-16
        oil
            pan, removal and installation, 2A-14
            pump and pick-up tube, removal, inspection and installation, 2A-14
        rear main oil seal, replacement, 2A-16
        repair operations possible with the engine in the vehicle, 2A-2
        timing chain and sprockets, removal, inspection and installation, 2A-6
        Top Dead Center (TDC) for number one piston, locating, 2A-2
        valve
            clearance, check and adjustment, 2A-4
            cover, removal and installation, 2A-3
            springs, retainers and seals, replacement, 2A-5
    3.3L V6 engine, 2B-1 through 2B-14
        camshafts, lifters and seals, removal and installation, 2B-7
        crankshaft front oil seal, replacement, 2B-11
        crankshaft pulley, removal and installation, 2B-11
        cylinder heads, removal and installation, 2B-10
        exhaust manifolds, removal and installation, 2B-10
        flywheel/driveplate, removal and installation, 2B-14
        intake manifold, removal and installation, 2B-9
        mounts, check and replacement, 2B-14
        oil
            pan, removal and installation, 2B-12
            pump, removal, inspection and installation, 2B-13
        rear main oil seal, replacement, 2B-14
        repair operations possible with the engine in the vehicle, 2B-2
        rocker arm assembly, removal, inspection and installation, 2B-3
        timing belt and sprockets, removal and installation, 2B-5
        Top Dead Center (TDC) for number one piston, locating, 2B-2
        valve
            covers, removal and installation, 2B-3
            springs, retainers and seals, replacement, 2B-4
    3.5L V6 engine, 2C-1 through 2C-22
        camshafts and lifters, removal and installation, 2C-10
        crankshaft front oil seal, replacement, 2C-16
        crankshaft pulley, removal and installation, 2C-16
        cylinder head, removal and installation, 2C-14
        exhaust manifold, removal and installation, 2C-13
        flywheel/driveplate, removal and installation, 2C-20
        intake manifold, removal and installation, 2C-12
        mounts, check and replacement, 2C-21
        oil
            cooler and adapter, general information and replacement, 2C-20
            pan, removal and installation, 2C-17
            pump, removal, inspection and installation, 2C-18
        rear main oil seal, replacement, 2C-21
        repair operations possible with the engine in the vehicle, 2C-2
        timing chain and sprockets, removal, inspection and installation, 2C-7
        Top Dead Center (TDC) for number one piston, locating, 2C-3
        valve
            clearance, check and adjustment, 2C-4
            covers, removal and installation, 2C-3
            springs, retainers and seals, replacement, 2C-6
    general engine overhaul procedures, 2D-1 through 2D-26
        camshafts, lifters and bearings, inspection, 2D-21
        crankshaft
            inspection, 2D-19
            installation and main bearing oil clearance check, 2D-23
            removal, 2D-15
        cylinder compression check, 2D-7
        cylinder head
            cleaning and inspection, 2D-12
            disassembly, 2D-12
            reassembly, 2D-14
        cylinder honing, 2D-17
        engine block
            cleaning, 2D-16
            inspection, 2D-17
        general information, engine overhaul, 2D-6
        initial start-up and break-in after overhaul, 2D-26
        main and connecting rod bearings, inspection and main bearing selection, 2D-20
        piston rings, installation, 2D-22
        pistons/connecting rods
            inspection, 2D-18
            installation and rod bearing oil clearance check, 2D-25
            removal, 2D-14
        rear main oil seal installation, 2D-25

Index

IND-5

vacuum gauge diagnostic checks, 2D-8
valves, servicing, 2D-14
Evaporative Emissions Control (EVAP) system, 1-28, 6-26
Exhaust Gas Recirculation (EGR) system, 1-36, 6-24
Exhaust manifold, removal and installation
   2.4L four-cylinder engine, 2A-11
   3.3L V6 engine, 2B-10
   3.5L V6 engine, 2C-13
Exhaust system
   check, 1-21
   servicing, general information, 4-13

F

Fluid level checks
   automatic transmission, 1-14
   battery electrolyte, 1-12
   brake and clutch fluid, 1-12
   differential, 1-21
   engine
      coolant, 1-11
      oil, 1-10
   manual transmission, 1-22
   transfer case, 1-22
   windshield washer fluid, 1-11
Flywheel/driveplate, removal and installation
   2.4L four-cylinder engine, 2A-16
   3.3L V6 engine, 2B-14
   3.5L V6 engine, 2C-20
Fraction/decimal/millimeter equivalents, 0-17
Free-running hubs (4WD models), removal and installation, 8-13
Front end alignment, general information, 10-23
Front fender, removal and installation, 11-9
Front hub and wheel bearing check, repack and adjustment, 1-32
Fuel and exhaust systems, 4-1 through 4-14
   accelerator cable, removal, installation and adjustment, 4-8
   air filter housing, removal and installation, 4-8
   engine idle speed and fast idle cam (four-cylinder and 3.3L V6 models), check and adjustment, 4-12
   exhaust system servicing, general information, 4-13
   fuel
      filter replacement, 1-27
      level sending unit, check and replacement, 4-6
      lines and fittings, replacement, 4-4
      pressure
         regulator, removal and installation, 4-10
         relief procedure, 4-3
      pump, removal and installation, 4-5
      pump/fuel pressure, check, 4-4
      rail and injectors, removal and installation, 4-11
      system check, 1-23
      tank
         cleaning and repair, 4-7
         removal and installation, 4-6
   general information, 4-3
   throttle body, removal and installation, 4-10
Fuel injection system
   check, 4-9
   general information, 4-9

Fuel temperature sensor, check and replacement, 6-18
Fuses and fusible links, general information, 12-3

G

General engine overhaul procedures, 2D-1 through 2D-26
   camshafts, lifters and bearings, inspection, 2D-21
   crankshaft
      inspection, 2D-19
      installation and main bearing oil clearance check, 2D-23
      removal, 2D-15
   cylinder compression check, 2D-7
   cylinder head
      cleaning and inspection, 2D-12
      disassembly, 2D-12
      reassembly, 2D-14
   cylinder honing, 2D-17
   disassembly sequence, 2D-11
   engine block
      cleaning, 2D-16
      inspection, 2D-17
   engine
      rebuilding alternatives, 2D-9
      removal and installation, 2D-10
      removal, methods and precautions, 2D-9
   general information, engine overhaul, 2D-6
   initial start-up and break-in after overhaul, 2D-26
   main and connecting rod bearings, inspection and main bearing selection, 2D-20
   oil pressure check, 2D-7
   piston rings, installation, 2D-22
   pistons/connecting rods
      inspection, 2D-18
      installation and rod bearing oil clearance check, 2D-25
      removal, 2D-14
   rear main oil seal installation, 2D-25
   reassembly sequence, 2D-22
   vacuum gauge diagnostic checks, 2D-8
   valves, servicing, 2D-14

H

Headlight
   adjustment, 12-10
   bulb, replacement, 12-10
   housing, replacement, 12-11
Heater
   and air conditioning control assembly, removal and installation, 3-10
   core, replacement, 3-10
Hinges and locks, maintenance, 11-3
Hood
   latch and release cable, removal and installation, 11-6
   removal, installation and adjustment, 11-6
Horn, check and replacement, 12-15

I

Idle Air Control (IAC) system, 6-20

Ignition
  coil(s) (3.5L V6 models), removal and installation, 5-8
  switch and key lock cylinder, check and replacement, 12-7
  system
    check, 5-6
    general information, 5-5
  timing, check and adjustment, 5-9
Initial start-up and break-in after overhaul, 2D-26
Instrument
  cluster, removal and installation, 12-8
  panel gauges, check, 12-8
  panel, removal and installation, 11-20
Intake Air Temperature (IAT) sensor, check and replacement, 6-12
Intake manifold runner control system, 6-21
Intake manifold, removal and installation
  2.4L four-cylinder engine, 2A-10
  3.3L V6 engine, 2B-9
  3.5L V6 engine, 2C-12
Intermediate shaft, removal and installation, 10-17
Introduction to the Nissan Frontier, Xterra and Pathfinder, 0-5

J

Jacking and towing, 0-14

K

Knock Sensor (KS), check and replacement, 6-18

L

Leaf spring (Frontier and Xterra models), removal and installation, 10-14
Liftgate (SUV models)
  and liftgate glass, removal and installation, 11-15
  latch, lock cylinder, handle and support struts, removal, installation and adjustment, 11-16
Lighting system and windshield wiper/washer switches, check and replacement, 12-5
Load Sensing Valve (LSV), description and replacement, 9-21
Lower control arm, removal and installation, 10-11

M

Main and connecting rod bearings, inspection and main bearing selection, 2D-20
Maintenance schedule, 1-4
Maintenance techniques, tools and working facilities, 0-7
Manifold Absolute Pressure (MAP) sensor and solenoid valve, check and replacement, 6-11
Manual transmission, 7A-1 through 7A-4
  back-up light switch, check and replacement, 7A-2
  general information, 7A-1
  lubricant
    change, 1-36
    level check, 1-22
  neutral position switch, check and replacement, 7A-2
  oil seals, replacement, 7A-3
  overhaul, general information, 7A-4
  removal and installation, 7A-3
  shift lever, removal and installation, 7A-2
Mass Airflow (MAF) sensor, check and replacement, 6-10
Master cylinder, removal, installation and reservoir/seal replacement, 9-19
Mirrors, removal and installation, 11-14

N

Neutral position switch, check and replacement, 7A-2

O

Oil pan, removal and installation
  2.4L four-cylinder engine, 2A-14
  3.3L V6 engine, 2B-12
  3.5L V6 engine, 2C-17
Oil pump, removal, inspection and installation
  2.4L four-cylinder engine, 2A-14
  3.3L V6 engine, 2B-13
  3.5L V6 engine, 2C-18
On-Board Diagnosis (OBD) system and trouble codes, 6-2
Oxygen (O2) sensor, check and replacement, 6-17

P

Park/Neutral Position (PNP) switch, check, adjustment and replacement, 7B-8
Parking brake
  adjustment, 9-23
  cables, replacement, 9-24
Pilot bushing, replacement, 8-7
Pinion oil seal, replacement, 8-12
Piston rings, installation, 2D-22
Pistons/connecting rods
  inspection, 2D-18
  installation and rod bearing oil clearance check, 2D-25
  removal, 2D-14
Positive Crankcase Ventilation (PCV) system, 6-23
Positive Crankcase Ventilation (PCV) valve and hose check and replacement, 1-34
Power brake booster, check, replacement and adjustment, 9-23
Power door lock system, description and check, 12-18
Power seats, description and check, 12-19
Power steering
  fluid level check, 1-14
  pressure switch, check and replacement, 6-16
  pump, removal and installation, 10-21
  system, bleeding, 10-22
Power window system, description and check, 12-18
Powertrain Control Module (PCM), removal and installation, 6-8

R

Radiator
- and coolant reservoir, removal and installation, 3-5
- grille, removal and installation, 11-3

Radio and speakers, removal and installation, 12-9

Rear main oil seal
- installation, 2D-25
- replacement
  - 2.4L four-cylinder engine, 2A-16
  - 3.3L V6 engine, 2B-14
  - 3.5L V6 engine, 2C-21

Rear window defogger, check and repair, 12-16

Relays, general information and testing, 12-4

Repair operations possible with the engine in the vehicle
- 2.4L four-cylinder engine, 2A-2
- 3.3L V6 engine, 2B-2
- 3.5L V6 engine, 2C-2

Rocker arm assembly (3.3L V6 engine), removal, inspection and installation, 2B-3

S

Safety first!, 0-18
Seat belt check, 1-16
Seats, removal and installation, 11-21
Shift
- cable, check, adjustment and replacement, 7B-3
- interlock system, description, check and component replacement, 7B-6
- lever assembly, removal and installation, 7B-5

Shock absorber, removal and installation
- front (Frontier and Xterra models), 10-6
- rear, 10-13

Spark plug
- replacement, 1-28
- wire, distributor cap and rotor check and replacement, 1-35

Stabilizer bar and bushings, removal and installation
- front, 10-6
- rear, 10-14

Starter motor
- and circuit, in-vehicle check, 5-11
- removal and installation, 5-12

Starting system, general information and precautions, 5-11
Steering column
- covers, removal and installation, 11-20
- removal and installation, 10-17
- transfer gear (Pathfinder models), removal and installation, 10-18

Steering gear
- boots (Pathfinder models), replacement, 10-19
- removal and installation, 10-20

Steering knuckle, removal and installation, 10-13
Steering linkage (Frontier and Xterra models), removal and installation, 10-19
Steering system
- intermediate shaft, removal and installation, 10-17
- power steering
  - pump, removal and installation, 10-21
  - system, bleeding, 10-22

steering column
- removal and installation, 10-17

steering gear
- boots (Pathfinder models), replacement, 10-19
- removal and installation, 10-20

steering linkage (Frontier and Xterra models), removal and installation, 10-19

steering wheel, removal and installation, 10-16

transfer gear (Pathfinder models), removal and installation, 10-18

tie-rod ends, removal and installation, 10-18

wheel studs, replacement, 10-22

Strut or coil spring (Pathfinder models), replacement, 10-9
Strut/coil spring assembly (Pathfinder models), removal, inspection and installation, 10-8
Suspension and steering systems, 10-1 through 10-24
Suspension system
- balljoints, check and replacement, 10-12
- coil spring, rear (Pathfinder models), removal and installation, 10-15
- leaf spring (Frontier and Xterra models), removal and installation, 10-14
- lower control arm, removal and installation, 10-11
- shock absorber, removal and installation
  - front (Frontier and Xterra models), 10-6
  - rear, 10-13
- stabilizer bar and bushings, removal and installation
  - front, 10-6
  - rear, 10-14
- steering knuckle, removal and installation, 10-13
- strut or coil spring (Pathfinder models), replacement, 10-9
- strut/coil spring assembly (Pathfinder models), removal, inspection and installation, 10-8
- suspension arms, rear (Pathfinder models), removal and installation, 10-15
- tension rod (2WD Frontier models), removal and installation, 10-12
- torsion bar (Frontier and Xterra models), removal and installation, 10-7
- upper control arm (Frontier and Xterra models), removal and installation, 10-10

Suspension, steering and driveaxle boot check, 1-23

T

Tailgate (pick-up models)
- removal, installation and adjustment, 11-14
- latch and handle, removal and installation, 11-14

Tension rod (2WD Frontier models), removal and installation, 10-12
Thermostat, check and replacement, 3-3
Throttle body, removal and installation, 4-10
Throttle Position Sensor (TPS), check, replacement and adjustment, 6-9
Throttle Valve (TV) cable, removal, installation and adjustment, 7B-7
Tie-rod ends, removal and installation, 10-18
Timing belt and sprockets (3.3L V6 engine), removal and installation, 2B-5

Timing chain and sprockets, removal, inspection and installation
  2.4L four-cylinder engine, 2A-6
  3.5L V6 engine, 2C-7
Tire and tire pressure checks, 1-12
Tire rotation, 1-21
Top Dead Center (TDC) for number one piston, locating
  2.4L four-cylinder engine, 2A-2
  3.3L V6 engine, 2B-2
  3.5L V6 engine, 2C-3
Torsion bar (Frontier and Xterra models), removal and installation, 10-7
Transfer case, 7C-1 through 7C-4
  general information, 7C-1
  lubricant
    change, 1-36
    level check, 1-22
  oil seals, replacement, 7C-2
  overhaul, general information, 7C-4
  position switches, check and replacement, 7C-2
  removal and installation, 7C-3
  shift lever, removal and installation, 7C-1
Transmission, automatic, 7B-1 through 7B-12
  diagnosis and trouble codes, 7B-2
  extension housing oil seal (2WD models), replacement, 7B-8
  fluid
    and filter change, 1-31
    level check, 1-14
  general information, 7B-2
  mount, check and replacement, 7B-9
  Park/Neutral Position (PNP) switch, check, adjustment and replacement, 7B-8
  removal and installation, 7B-9
  shift cable, check, adjustment and replacement, 7B-3
  shift interlock system, description, check and component replacement, 7B-6
  shift lever assembly, removal and installation, 7B-5
  Throttle Valve (TV) cable, removal, installation and adjustment, 7B-7
  Transmission Control Module (TCM), removal and installation, 7B-12
Transmission, manual, 7A-1 through 7A-4
  back-up light switch, check and replacement, 7A-2
  general information, 7A-1
  lubricant
    change, 1-36
    level check, 1-22
  neutral position switch, check and replacement, 7A-2
  oil seals, replacement, 7A-3
  removal and installation, 7A-3
  shift lever, removal and installation, 7A-2
Troubleshooting, 0-19
Tune-up and routine maintenance 1-1 through 1-36
Tune-up general information, 1-10
Turn signal/hazard flashers, check and replacement, 12-5

U

Underhood hose check and replacement, 1-20
Universal joints, replacement, 8-10
Upholstery and carpets, maintenance, 11-1
Upper control arm (Frontier and Xterra models), removal and installation, 10-10

V

Vacuum gauge diagnostic checks, 2D-8
Valve clearance, check and adjustment
  2.4L four-cylinder engine, 2A-4
  3.5L V6 engine, 2C-4
Valve cover, removal and installation
  2.4L four-cylinder engine, 2A-3
  3.3L V6 engine, 2B-3
  3.5L V6 engine, 2C-3
Valve springs, retainers and seals, replacement
  2.4L four-cylinder engine, 2A-5
  3.3L V6 engine, 2B-4
  3.5L V6 engine, 2C-6
Valves, servicing, 2D-14
Variable valve timing control system, 6-22
Vehicle identification numbers, 0-6
Vehicle Speed Sensor (VSS), check and replacement, 6-19
Vinyl trim, maintenance, 11-1

W

Water pump, check and replacement, 3-6
Wheel cylinder, removal and installation, 9-19
Wheel studs, replacement, 10-22
Wheels and tires, general information, 10-22
Windshield and fixed glass, replacement, 11-3
Wiper blade inspection and replacement, 1-16
Wiper motor and amplifier-check and replacement, 12-14
Wiring diagrams, general information, 12-21

Haynes Automotive Manuals

NOTE: New manuals are added to this list on a periodic basis. If you do not see a listing for your vehicle, consult your local Haynes dealer for the latest product information.

ACURA
- 12020 Integra '86 thru '89 & Legend '86 thru '90
- 12021 Integra '90 thru '93 & Legend '91 thru '95

AMC
- Jeep CJ - see JEEP (50020)
- 14020 Mid-size models '70 thru '83
- 14025 (Renault) Alliance & Encore '83 thru '87

AUDI
- 15020 4000 all models '80 thru '87
- 15025 5000 all models '77 thru '83
- 15026 5000 all models '84 thru '88

AUSTIN-HEALEY
- Sprite - see MG Midget (66015)

BMW
- *18020 3/5 Series not including diesel or all-wheel drive models '82 thru '92
- 18021 3-Series incl. Z3 models '92 thru '98
- 18025 320i all 4 cyl models '75 thru '83
- 18050 1500 thru 2002 except Turbo '59 thru '77

BUICK
- *19010 Buick Century '97 thru '02
- Century (front-wheel drive) - see GM (38005)
- *19020 Buick, Oldsmobile & Pontiac Full-size (Front-wheel drive) '85 thru '02
- Buick Electra, LeSabre and Park Avenue; Oldsmobile Delta 88 Royale, Ninety Eight and Regency; Pontiac Bonneville
- 19025 Buick Oldsmobile & Pontiac Full-size (Rear wheel drive)
- Buick Estate '70 thru '90, Electra '70 thru '84, LeSabre '70 thru '85, Limited '74 thru '79
- Oldsmobile Custom Cruiser '70 thru '90, Delta 88 '70 thru '85, Ninety-eight '70 thru '84
- Pontiac Bonneville '70 thru '81, Catalina '70 thru '81, Grandville '70 thru '75, Parisienne '83 thru '86
- 19030 Mid-size Regal & Century all rear-drive models with V6, V8 and Turbo '74 thru '87
- Regal - see GENERAL MOTORS (38010)
- Riviera - see GENERAL MOTORS (38030)
- Roadmaster - see CHEVROLET (24046)
- Skyhawk - see GENERAL MOTORS (38015)
- Skylark - see GM (38020, 38025)
- Somerset - see GENERAL MOTORS (38025)

CADILLAC
- 21030 Cadillac Rear Wheel Drive all gasoline models '70 thru '93
- Cimarron - see GENERAL MOTORS (38015)
- DeVille - see GM (38031 & 38032)
- Eldorado - see GM (38030 & 38031)
- Fleetwood - see GM (38031)
- Seville - see GM (38030, 38031 & 38032)

CHEVROLET
- *24010 Astro & GMC Safari Mini-vans '85 thru '02
- 24015 Camaro V8 all models '70 thru '81
- 24016 Camaro all models '82 thru '92
- 24017 Camaro & Firebird '93 thru '00
- Cavalier - see GENERAL MOTORS (38016)
- Celebrity - see GENERAL MOTORS (38005)
- 24020 Chevelle, Malibu & El Camino '69 thru '87
- 24024 Chevette & Pontiac T1000 '76 thru '87
- Citation - see GENERAL MOTORS (38020)
- 24032 Corsica/Beretta all models '87 thru '96
- 24040 Corvette all V8 models '68 thru '82
- 24041 Corvette all models '84 thru '96
- 10305 Chevrolet Engine Overhaul Manual
- 24045 Full-size Sedans Caprice, Impala, Biscayne, Bel Air & Wagons '69 thru '90
- 24046 Impala SS & Caprice and Buick Roadmaster '91 thru '96
- Impala - see LUMINA (24048)
- Lumina '90 thru '94 - see GM (38010)
- *24048 Lumina & Monte Carlo '95 thru '01
- Lumina APV - see GM (38035)
- 24050 Luv Pick-up all 2WD & 4WD '72 thru '82
- Malibu '97 thru '00 - see GM (38026)
- 24055 Monte Carlo all models '70 thru '88
- Monte Carlo '95 thru '01 - see LUMINA (24048)
- 24059 Nova all V8 models '69 thru '79
- 24060 Nova and Geo Prizm '85 thru '92
- 24064 Pick-ups '67 thru '87 - Chevrolet & GMC, all V8 & in-line 6 cyl, 2WD & 4WD '67 thru '87; Suburbans, Blazers & Jimmys '67 thru '91
- 24065 Pick-ups '88 thru '98 - Chevrolet & GMC, full-size pick-ups '88 thru '98, C/K Classic '99 & '00, Blazer & Jimmy '92 thru '94; Suburban '92 thru '99; Tahoe & Yukon '95 thru '99
- *24066 Pick-ups '99 thru '01 - Chevrolet Silverado & GMC Sierra full-size pick-ups '99 thru '01, Suburban/Tahoe/Yukon/Yukon XL '00 thru '01
- 24070 S-10 & S-15 Pick-ups '82 thru '93, Blazer & Jimmy '83 thru '94,
- *24071 S-10 & S-15 Pick-ups '94 thru '01, Blazer & Jimmy '95 thru '01, Hombre '96 thru '01
- 24075 Sprint '85 thru '88 & Geo Metro '89 thru '01
- 24080 Vans - Chevrolet & GMC '68 thru '96

CHRYSLER
- 25015 Chrysler Cirrus, Dodge Stratus, Plymouth Breeze '95 thru '00
- 10310 Chrysler Engine Overhaul Manual
- 25020 Full-size Front-Wheel Drive '88 thru '93
- K-Cars - see DODGE Aries (30008)
- Laser - see DODGE Daytona (30030)
- 25025 Chrysler LHS, Concorde, New Yorker, Dodge Intrepid, Eagle Vision, '93 thru '97
- *25026 Chrysler LHS, Concorde, 300M, Dodge Intrepid, '98 thru '03
- 25030 Chrysler & Plymouth Mid-size front wheel drive '82 thru '95
- Rear-wheel Drive - see Dodge (30050)
- *25035 PT Cruiser all models '01 thru '03
- *25040 Chrysler Sebring, Dodge Avenger '95 thru '02

DATSUN
- 28005 200SX all models '80 thru '83
- 28007 B-210 all models '73 thru '78
- 28009 210 all models '79 thru '82
- 28012 240Z, 260Z & 280Z Coupe '70 thru '78
- 28014 280ZX Coupe & 2+2 '79 thru '83
- 300ZX - see NISSAN (72010)
- 28016 310 all models '78 thru '82
- 28018 510 & PL521 Pick-up '68 thru '73
- 28020 510 all models '78 thru '81
- 28022 620 Series Pick-up all models '73 thru '79
- 720 Series Pick-up - see NISSAN (72030)
- 28025 810/Maxima all gasoline models, '77 thru '84

DODGE
- 400 & 600 - see CHRYSLER (25030)
- 30008 Aries & Plymouth Reliant '81 thru '89
- 30010 Caravan & Plymouth Voyager '84 thru '95
- *30011 Caravan & Plymouth Voyager '96 thru '02
- 30012 Challenger/Plymouth Saporro '78 thru '83
- 30016 Colt & Plymouth Champ '78 thru '87
- 30020 Dakota Pick-ups all models '87 thru '96
- *30021 Durango '98 & '99, Dakota '97 thru '99
- 30025 Dart, Demon, Plymouth Barracuda, Duster & Valiant 6 cyl models '67 thru '76
- 30030 Daytona & Chrysler Laser '84 thru '89
- Intrepid - see CHRYSLER (25025, 25026)
- *30034 Neon all models '95 thru '99
- 30035 Omni & Plymouth Horizon '78 thru '90
- 30040 Pick-ups all full-size models '74 thru '93
- *30041 Pick-ups all full-size models '94 thru '01
- 30045 Ram 50/D50 Pick-ups & Raider and Plymouth Arrow Pick-ups '79 thru '93
- 30050 Dodge/Plymouth/Chrysler RWD '71 thru '89
- 30055 Shadow & Plymouth Sundance '87 thru '94
- 30060 Spirit & Plymouth Acclaim '89 thru '95
- *30065 Vans - Dodge & Plymouth '71 thru '99

EAGLE
- Talon - see MITSUBISHI (68030, 68031)
- Vision - see CHRYSLER (25025)

FIAT
- 34010 124 Sport Coupe & Spider '68 thru '78
- 34025 X1/9 all models '74 thru '80

FORD
- 10355 Ford Automatic Transmission Overhaul
- 36004 Aerostar Mini-vans all models '86 thru '97
- 36006 Contour & Mercury Mystique '95 thru '00
- 36008 Courier Pick-up all models '72 thru '82
- *36012 Crown Victoria & Mercury Grand Marquis '88 thru '00
- 10320 Ford Engine Overhaul Manual
- 36016 Escort/Mercury Lynx all models '81 thru '90
- 36020 Escort/Mercury Tracer '91 thru '00
- 36024 Explorer & Mazda Navajo '91 thru '01
- 36028 Fairmont & Mercury Zephyr '78 thru '83
- 36030 Festiva & Aspire '88 thru '97
- 36032 Fiesta all models '77 thru '80
- *36034 Focus all models '00 and '01
- 36036 Ford & Mercury Full-size '75 thru '87
- 36040 Granada & Mercury Monarch '75 thru '80
- 36044 Ford & Mercury Mid-size '75 thru '86
- 36048 Mustang V8 all models '64-1/2 thru '73
- 36049 Mustang II 4 cyl, V6 & V8 models '74 thru '78
- 36050 Mustang & Mercury Capri all models Mustang, '79 thru '93; Capri, '79 thru '86
- *36051 Mustang all models '94 thru '00
- 36054 Pick-ups & Bronco '73 thru '79
- 36058 Pick-ups & Bronco '80 thru '96
- *36059 F-150 & Expedition '97 thru '02, F-250 '97 thru '99 & Lincoln Navigator '98 thru '02
- *36060 Super Duty Pick-ups, Excursion '97 thru '02
- 36062 Pinto & Mercury Bobcat '75 thru '80
- 36066 Probe all models '89 thru '92
- 36070 Ranger/Bronco II gasoline models '83 thru '92
- *36071 Ranger '93 thru '00 & Mazda Pick-ups '94 thru '00
- 36074 Taurus & Mercury Sable '86 thru '95
- *36075 Taurus & Mercury Sable '96 thru '01
- 36078 Tempo & Mercury Topaz '84 thru '94
- 36082 Thunderbird/Mercury Cougar '83 thru '88
- 36086 Thunderbird/Mercury Cougar '89 and '97
- 36090 Vans all V8 Econoline models '69 thru '91
- *36094 Vans full size '92 thru '01
- *36097 Windstar Mini-van '95 thru '01

GENERAL MOTORS
- 10360 GM Automatic Transmission Overhaul
- 38005 Buick Century, Chevrolet Celebrity, Oldsmobile Cutlass Ciera & Pontiac 6000 all models '82 thru '96
- *38010 Buick Regal, Chevrolet Lumina, Oldsmobile Cutlass Supreme & Pontiac Grand Prix (FWD) '88 thru '02
- 38015 Buick Skyhawk, Cadillac Cimarron, Chevrolet Cavalier, Oldsmobile Firenza & Pontiac J-2000 & Sunbird '82 thru '94
- *38016 Chevrolet Cavalier & Pontiac Sunfire '95 thru '00
- 38020 Buick Skylark, Chevrolet Citation, Olds Omega, Pontiac Phoenix '80 thru '85
- 38025 Buick Skylark & Somerset, Oldsmobile Achieva & Calais and Pontiac Grand Am all models '85 thru '98
- *38026 Chevrolet Malibu, Olds Alero & Cutlass, Pontiac Grand Am '97 thru '00
- 38030 Cadillac Eldorado '71 thru '85, Seville '80 thru '85, Oldsmobile Toronado '71 thru '85, Buick Riviera '79 thru '85
- *38031 Cadillac Eldorado & Seville '86 thru '91, DeVille '86 thru '93, Fleetwood & Olds Toronado '86 thru '92, Buick Riviera '86 thru '93
- 38032 Cadillac DeVille '94 thru '02 & Seville - '92 thru '02
- 38035 Chevrolet Lumina APV, Olds Silhouette & Pontiac Trans Sport all models '90 thru '95
- *38036 Chevrolet Venture, Olds Silhouette, Pontiac Trans Sport & Montana '97 thru '01
- General Motors Full-size Rear-wheel Drive - see BUICK (19025)

GEO
- Metro - see CHEVROLET Sprint (24075)
- Prizm - '85 thru '92 see CHEVY (24060), '93 thru '02 see TOYOTA Corolla (92036)

(Continued on other side)

---

* Listings shown with an asterisk (*) indicate model coverage as of this printing. These titles will be periodically updated to include later model years - consult your Haynes dealer for more information.

Haynes North America, Inc., 861 Lawrence Drive, Newbury Park, CA 91320-1514 • (805) 498-6703

Haynes Automotive Manuals (continued)

NOTE: New manuals are added to this list on a periodic basis. If you do not see a listing for your vehicle, consult your local Haynes dealer for the latest product information.

40030	Storm all models '90 thru '93	
	Tracker - see SUZUKI Samurai (90010)	

GMC
Vans & Pick-ups - see CHEVROLET

HONDA
- 42010 Accord CVCC all models '76 thru '83
- 42011 Accord all models '84 thru '89
- 42012 Accord all models '90 thru '93
- 42013 Accord all models '94 thru '97
- *42014 Accord all models '98 and '99
- 42020 Civic 1200 all models '73 thru '79
- 42021 Civic 1300 & 1500 CVCC '80 thru '83
- 42022 Civic 1500 CVCC all models '75 thru '79
- 42023 Civic all models '84 thru '91
- 42024 Civic & del Sol '92 thru '95
- *42025 Civic '96 thru '00, CR-V '97 thru '00, Acura Integra '94 thru '00
- 42040 Prelude CVCC all models '79 thru '89

HYUNDAI
- *43010 Elantra all models '96 thru '01
- 43015 Excel & Accent all models '86 thru '98

ISUZU
- Hombre - see CHEVROLET S-10 (24071)
- *47017 Rodeo '91 thru '02; Amigo '89 thru '94 and '98 thru '02; Honda Passport '95 thru '02
- 47020 Trooper & Pick-up '81 thru '93

JAGUAR
- 49010 XJ6 all 6 cyl models '68 thru '86
- 49011 XJ6 all models '88 thru '94
- 49015 XJ12 & XJS all 12 cyl models '72 thru '85

JEEP
- 50010 Cherokee, Comanche & Wagoneer Limited all models '84 thru '00
- 50020 CJ all models '49 thru '86
- *50025 Grand Cherokee all models '93 thru '00
- 50029 Grand Wagoneer & Pick-up '72 thru '91 Grand Wagoneer '84 thru '91, Cherokee & Wagoneer '72 thru '83, Pick-up '72 thru '88
- *50030 Wrangler all models '87 thru '00

LEXUS
ES 300 - see TOYOTA Camry (92007)

LINCOLN
- Navigator - see FORD Pick-up (36059)
- *59010 Rear-Wheel Drive all models '70 thru '01

MAZDA
- 61010 GLC Hatchback (rear-wheel drive) '77 thru '83
- 61011 GLC (front-wheel drive) '81 thru '85
- 61015 323 & Protegé '90 thru '00
- *61016 MX-5 Miata '90 thru '97
- 61020 MPV all models '89 thru '94
- Navajo - see Ford Explorer (36024)
- 61030 Pick-ups '72 thru '93
- Pick-ups '94 thru '00 - see Ford Ranger (36071)
- 61035 RX-7 all models '79 thru '85
- 61036 RX-7 all models '86 thru '91
- 61040 626 (rear-wheel drive) all models '79 thru '82
- 61041 626/MX-6 (front-wheel drive) '83 thru '91
- 61042 626 '93 thru '01, MX-6/Ford Probe '93 thru '01

MERCEDES-BENZ
- 63012 123 Series Diesel '76 thru '85
- 63015 190 Series four-cyl gas models, '84 thru '88
- 63020 230/250/280 6 cyl sohc models '68 thru '72
- 63025 280 123 Series gasoline models '77 thru '81
- 63030 350 & 450 all models '71 thru '80

MERCURY
- 64200 Villager & Nissan Quest '93 thru '01
- All other titles, see FORD Listing.

MG
- 66010 MGB Roadster & GT Coupe '62 thru '80
- 66015 MG Midget, Austin Healey Sprite '58 thru '80

MITSUBISHI
- 68020 Cordia, Tredia, Galant, Precis & Mirage '83 thru '93
- 68030 Eclipse, Eagle Talon & Ply. Laser '90 thru '94
- *68031 Eclipse '95 thru '01, Eagle Talon '95 thru '98
- 68040 Pick-up '83 thru '96 & Montero '83 thru '93

NISSAN
- 72010 300ZX all models including Turbo '84 thru '89
- 72015 Altima all models '93 thru '01
- 72020 Maxima all models '85 thru '92
- *72021 Maxima all models '93 thru '01
- 72030 Pick-ups '80 thru '97 Pathfinder '87 thru '95
- *72031 Frontier Pick-up '98 thru '01, Xterra '00 & '01, Pathfinder '96 thru '01
- 72040 Pulsar all models '83 thru '86
- Quest - see MERCURY Villager (64200)
- 72050 Sentra all models '82 thru '94
- 72051 Sentra & 200SX all models '95 thru '99
- 72060 Stanza all models '82 thru '90

OLDSMOBILE
- 73015 Cutlass V6 & V8 gas models '74 thru '88
- *For other OLDSMOBILE titles, see BUICK, CHEVROLET or GENERAL MOTORS listing.*

PLYMOUTH
For PLYMOUTH titles, see DODGE listing.

PONTIAC
- 79008 Fiero all models '84 thru '88
- 79018 Firebird V8 models except Turbo '70 thru '81
- 79019 Firebird all models '82 thru '92
- 79040 Mid-size Rear-wheel Drive '70 thru '87
- *For other PONTIAC titles, see BUICK, CHEVROLET or GENERAL MOTORS listing.*

PORSCHE
- 80020 911 except Turbo & Carrera 4 '65 thru '89
- 80025 914 all 4 cyl models '69 thru '76
- 80030 924 all models including Turbo '76 thru '82
- 80035 944 all models including Turbo '83 thru '89

RENAULT
Alliance & Encore - see AMC (14020)

SAAB
- *84010 900 all models including Turbo '79 thru '88

SATURN
- *87010 Saturn all models '91 thru '99

SUBARU
- 89002 1100, 1300, 1400 & 1600 '71 thru '79
- 89003 1600 & 1800 2WD & 4WD '80 thru '94

SUZUKI
- 90010 Samurai/Sidekick & Geo Tracker '86 thru '01

TOYOTA
- 92005 Camry all models '83 thru '91
- 92006 Camry all models '92 thru '96
- *92007 Camry, Avalon, Solara, Lexus ES 300 '97 thru '01
- 92015 Celica Rear Wheel Drive '71 thru '85
- 92020 Celica Front Wheel Drive '86 thru '99
- 92025 Celica Supra all models '79 thru '92
- 92030 Corolla all models '75 thru '79
- 92032 Corolla all rear wheel drive models '80 thru '87
- 92035 Corolla all front wheel drive models '84 thru '92
- 92036 Corolla & Geo Prizm '93 thru '02
- 92040 Corolla Tercel all models '80 thru '82
- 92045 Corona all models '74 thru '82
- 92050 Cressida all models '78 thru '82
- 92055 Land Cruiser FJ40, 43, 45, 55 '68 thru '82
- 92056 Land Cruiser FJ60, 62, 80, FZJ80 '80 thru '96
- 92065 MR2 all models '85 thru '87
- 92070 Pick-up all models '69 thru '78
- 92075 Pick-up all models '79 thru '95
- *92076 Tacoma '95 thru '00, 4Runner '96 thru '00, & T100 '93 thru '98
- *92078 Tundra '00 thru '02 & Sequoia '01 thru '02

- 92080 Previa all models '91 thru '95
- *92082 RAV4 all models '96 thru '02
- 92085 Tercel all models '87 thru '94

TRIUMPH
- 94007 Spitfire all models '62 thru '81
- 94010 TR7 all models '75 thru '81

VW
- 96008 Beetle & Karmann Ghia '54 thru '79
- *96009 New Beetle '98 thru '00
- 96016 Rabbit, Jetta, Scirocco & Pick-up gas models '74 thru '91 & Convertible '80 thru '92
- 96017 Golf, GTI & Jetta '93 thru '98 & Cabrio '95 thru '98
- *96018 Golf, GTI, Jetta & Cabrio '99 thru '02
- 96020 Rabbit, Jetta & Pick-up diesel '77 thru '84
- 96023 Passat '98 thru '01, Audi A4 '96 thru '01
- 96030 Transporter 1600 all models '68 thru '79
- 96035 Transporter 1700, 1800 & 2000 '72 thru '79
- 96040 Type 3 1500 & 1600 all models '63 thru '73
- 96045 Vanagon all air-cooled models '80 thru '83

VOLVO
- 97010 120, 130 Series & 1800 Sports '61 thru '73
- 97015 140 Series all models '66 thru '74
- 97020 240 Series all models '76 thru '93
- 97040 740 & 760 Series all models '82 thru '88
- 97050 850 Series all models '93 thru '97

TECHBOOK MANUALS
- 10205 Automotive Computer Codes
- 10210 Automotive Emissions Control Manual
- 10215 Fuel Injection Manual, 1978 thru 1985
- 10220 Fuel Injection Manual, 1986 thru 1999
- 10225 Holley Carburetor Manual
- 10230 Rochester Carburetor Manual
- 10240 Weber/Zenith/Stromberg/SU Carburetors
- 10305 Chevrolet Engine Overhaul Manual
- 10310 Chrysler Engine Overhaul Manual
- 10320 Ford Engine Overhaul Manual
- 10330 GM and Ford Diesel Engine Repair Manual
- 10340 Small Engine Repair Manual, 5 HP & Less
- 10341 Small Engine Repair Manual, 5.5 - 20 HP
- 10345 Suspension, Steering & Driveline Manual
- 10355 Ford Automatic Transmission Overhaul
- 10360 GM Automatic Transmission Overhaul
- 10405 Automotive Body Repair & Painting
- 10410 Automotive Brake Manual
- 10411 Automotive Anti-lock Brake (ABS) Systems
- 10415 Automotive Detailing Manual
- 10420 Automotive Eelectrical Manual
- 10425 Automotive Heating & Air Conditioning
- 10430 Automotive Reference Manual & Dictionary
- 10435 Automotive Tools Manual
- 10440 Used Car Buying Guide
- 10445 Welding Manual
- 10450 ATV Basics

SPANISH MANUALS
- 98903 Reparación de Carrocería & Pintura
- 98905 Códigos Automotrices de la Computadora
- 98910 Frenos Automotriz
- 98915 Inyección de Combustible 1986 al 1999
- 99040 Chevrolet & GMC Camionetas '67 al '87 Incluye Suburban, Blazer & Jimmy '67 al '91
- 99041 Chevrolet & GMC Camionetas '88 al '98 Incluye Suburban '92 al '98, Blazer & Jimmy '92 al '94, Tahoe y Yukon '95 al '98
- 99042 Chevrolet & GMC Camionetas Cerradas '68 al '95
- 99055 Dodge Caravan & Plymouth Voyager '84 al '95
- 99075 Ford Camionetas y Bronco '80 al '94
- 99077 Ford Camionetas Cerradas '69 al '91
- 99083 Ford Modelos de Tamaño Grande '75 al '87
- 99088 Ford Modelos de Tamaño Mediano '75 al '86
- 99091 Ford Taurus & Mercury Sable '86 al '95
- 99095 GM Modelos de Tamaño Grande '70 al '90
- 99100 GM Modelos de Tamaño Mediano '70 al '88
- 99110 Nissan Camioneta '80 al '96, Pathfinder '87 al '95
- 99118 Nissan Sentra '82 al '94
- 99125 Toyota Camionetas y 4Runner '79 al '95

Over 100 Haynes motorcycle manuals also available

3-03

* Listings shown with an asterisk (*) indicate model coverage as of this printing. These titles will be periodically updated to include later model years - consult your Haynes dealer for more information.

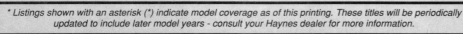

Haynes North America, Inc., 861 Lawrence Drive, Newbury Park, CA 91320-1514 • (805) 498-6703